Progress in Molecular and Subcellular Biology

Volume 59

This series gives an insight into the most current, cutting edge topics in molecular biology, including applications in biotechnology and molecular medicine. In the recent years, the progress of research in the frontier area of molecular and cell biology has resulted in an overwhelming amount of data on the structural components and molecular machineries of the cell and its organelles and the complexity of intra- and intercellular communication. The molecular basis of hereditary and acquired diseases is beginning to be unravelled, and profound new insights into development and evolutionary biology, as well as the genetically driven formation of 3D biological architectures, have been gained from molecular approaches. Topical volumes, written and edited by acknowledged experts in the field, present the most recent findings and their implications for future research. This series is indexed in PubMed.

More information about this series at http://www.springer.com/series/388

Luis B. Agellon • Marek Michalak
Editors

Cellular Biology of the Endoplasmic Reticulum

Springer

Editors
Luis B. Agellon
School of Human Nutrition
McGill University
Ste. Anne de Bellevue, QC, Canada

Marek Michalak
Department of Biochemistry
University of Alberta
Edmonton, AB, Canada

ISSN 0079-6484 ISSN 2197-8484 (electronic)
Progress in Molecular and Subcellular Biology
ISBN 978-3-030-67698-8 ISBN 978-3-030-67696-4 (eBook)
https://doi.org/10.1007/978-3-030-67696-4

This Springer imprint is published by the registered company Springer Nature Switzerland AG.
The registered company address is: Gewerbestrasse 11, 6330 Cham, Switzerland

Preface

The idea for this book project originated at the 13th International Calreticulin Workshop organized by Luis B. Agellon and held in Montreal, Quebec, Canada, in 2019, nearly 45 years after the discovery of calreticulin, which was known then as the high affinity calcium binding protein.

The "calreticulin workshop," as it became known, grew out of an informal scientific meeting that was held in 1994 in Banff, Alberta, Canada, attended by a group of researchers working on diverse biological questions and model systems but have encountered proteins during the course of their investigations that exhibited remarkable similarities in characteristics to calreticulin. At that time, these proteins were referred to by different names, but once the molecular cloning of their cDNAs and genes was achieved and their primary structures deduced, it became clear that the researchers were all working with the same protein. The initial 1994 Banff meeting served as the inspiration for a recurring calreticulin workshop. Since then, the International Calreticulin Workshop has been held in Canada, Chile, Denmark, Italy, Switzerland, the UK, and the USA. In 2019, the calreticulin workshop held in Montreal celebrated a quarter of a century of enduring collaborations among scientists with a shared interest in calreticulin. Through the years, the calreticulin workshop served as a regular forum for information exchange between the world's leading specialists from a wide range of scientific disciplines working on diverse research topics but with a common interest in calreticulin. The discussions at the calreticulin workshops have helped to develop a molecular understanding of endoplasmic reticulum structure and function from the point of view of calreticulin.

The International Calreticulin Workshop has witnessed remarkable progress toward the understanding of the biology of intracellular membranes, calcium signaling, and cellular stress coping strategies. Some of the important contributions in these areas are highlighted in this book.

Ste. Anne de Bellevue, QC, Canada — Luis B. Agellon
Edmonton, AB, Canada — Marek Michalak

Contents

Chapter 1
A View of the Endoplasmic Reticulum Through the Calreticulin Lens

Luis B. Agellon and Marek Michalak

Abstract Calreticulin is well known as an ER-resident protein that serves as the major endoplasmic reticulum (ER) Ca^{2+} binding protein. This protein has been the major topic of discussion in an international workshop that has been meeting for a quarter of a century. In sharing information about this protein, the field also witnessed remarkable insights into the importance of the ER as an organelle and the role of ER Ca^{2+} in coordinating ER and cellular functions. Recent technological advances have helped to uncover the contributions of calreticulin in maintaining Ca^{2+} homeostasis in the ER and to unravel its involvement in a multitude of cellular processes as highlighted in this collection of articles. The continuing revelations of unexpected involvement of calreticulin and Ca^{2+} in many critical aspects of cellular function promises to further improve insights into the significance of this protein in the promotion of physiology as well as prevention of pathology.

Keywords Endoplasmic reticulum · Calcium homeostasis · Calreticulin · ER stress

Abbreviations

ER	Endoplasmic reticulum
UPR	Unfolded protein responses
PDI	Protein disulfide isomerase
BiP	Immunoglobulin binding protein
GRP	Glucose regulated protein
$InsP_3R$	Inositol trisphosphate receptor
RyR	Ryanodine receptor/Ca^{2+} channel

L. B. Agellon (✉)
School of Human Nutrition, McGill University, Ste. Anne de Bellevue, QC, Canada
e-mail: luis.agellon@mcgill.ca

M. Michalak
Department of Biochemistry, University of Alberta, Edmonton, AB, Canada
e-mail: marek.michalak@ualberta.ca

L. B. Agellon, M. Michalak (eds.), *Cellular Biology of the Endoplasmic Reticulum*, Progress in Molecular and Subcellular Biology 59,
https://doi.org/10.1007/978-3-030-67696-4_1

SERCA	Sarcoplasmic/endoplasmic reticulum Ca^{2+}-ATPase
SOCE	Store-operated Ca^{2+} entry
STIM1	Stromal interaction molecule 1
ORAI1	Ca^{2+} release-activated Ca^{2+} channel protein 1
ERAD	Endoplasmic reticulum associated degradation

1.1 Introduction

The idea for this book project originated at the 13th International Calreticulin Workshop organized by L. Agellon and held in Montreal, Quebec, Canada in 2019, nearly 45 years after the discovery of calreticulin, which was known then as the Ca^{2+} binding protein (Ostwald and MacLennan 1974). The International Calreticulin Workshop has witnessed remarkable progress towards the understanding of the cell biology of intracellular membranes, Ca^{2+} signaling as well as protein folding, trafficking and secretion. Some of the important contributions focusing on structural and functional analysis of calreticulin and calreticulin mutants, protein quality control, calnexin function, ER-phagy, Ca^{2+} signaling, ER stress and the unfolded protein response (UPR), are highlighted in this book.

The International Calreticulin Workshop grew out of an informal scientific meeting that was held in 1994 in Banff, Alberta, Canada, attended by a group of researchers working on diverse biological questions and model systems but have encountered proteins during the course of their investigations that exhibited remarkable similarities in characteristics to calreticulin. At that time, these proteins were referred to by different names (Michalak et al. 1992) but once the molecular cloning of their cDNAs and genes was achieved and their primary structures deduced it became clear that the researchers were all working with the same protein (Waisman et al. 1985; Damiani et al. 1988; Macer and Koch 1988; Fliegel et al. 1989a; Smith and Koch 1989; Van et al. 1989). The initial 1994 Banff meeting served as the inspiration for a regular calreticulin workshop. Since then, the International Calreticulin Workshop has been held in Canada, Chile, Denmark, Italy, Switzerland, UK, USA. In 2019, the CRT2019 Workshop held in Montreal celebrated a quarter of a century of enduring collaborations among scientists with a shared interest in calreticulin (Eggleton and Michalak 2013; Gold et al. 2015; Charonis et al. 2017). Through the years, the workshop served as a regular forum for information exchange between world leading scientists working on the ER, ER chaperones, ER stress, and protein folding diseases, cellular signaling and Ca^{2+} homeostasis. These workshops have continued to bring together specialists in protein structure, human genetics, molecular biology, biochemistry, immunology, biophysics and physiology, and the discussions helped to develop a molecular understanding of ER structure and function from the point of view of calreticulin.

1.2 The ER and ER Ca^{2+}

Since Porter's description of the ER 75 years ago (Porter et al. 1945), the importance of this cell organelle in many vital cellular processes have been appreciated, including Ca^{2+} storage and release; lipid synthesis as well as the synthesis of membrane-bound and secreted proteins, their folding, and post-translational modifications. The ER is the first compartment in the protein secretory pathway. It is a multifunctional and dynamic organelle able to detect and integrate incoming signals, modulate its own luminal environment and generate output signals in response to a variety of physical, chemical, and nutritional challenges (Corbett and Michalak 2000; Yoshida 2007; Rutkowski and Kaufman 2007; Schroder 2008; Ron and Hubbard 2008; Wang et al. 2019). To drive these diverse cellular functions, the ER contains proteins localized within the lumen and dispersed as integral membrane proteins that participate in an array of cellular processes (Sweeney et al. 2017; Michalak and Agellon 2018; Beese et al. 2019; Hetz et al. 2020; Shergalis et al. 2020). In particular, the ER is intimately involved in cellular Ca^{2+} signaling and communication with the plasma membrane Ca^{2+} channels and nuclear transcriptional machinery (Ye et al. 2000; Putney 2007; Hetz et al. 2020). The ER is well equipped with Ca^{2+}transport and storage systems comprised of Ca^{2+} channels, Ca^{2+}-ATPase pumps, Ca^{2+} sensors, Ca^{2+} buffers and Ca^{2+} storage proteins (Krebs et al. 2015). There is a variety of multifunctional ER resident proteins that bind and buffer large amounts of Ca^{2+} in the lumen of ER but also serve as chaperones, such as BiP, calreticulin, PDI and GRP94 (Prins and Michalak 2011; Wang et al. 2019). However, a portion of ER luminal Ca^{2+} is free and is available for many Ca^{2+}-dependent processes in the lumen of the ER such as protein and lipid synthesis and quality control, including the regulation of the UPR (Corbett and Michalak 2000; Groenendyk et al. 2004; Krebs et al. 2015; Jung et al. 2017a; Wang et al. 2017). Large reductions in luminal ER Ca^{2+} concentration lead to the accumulation of misfolded proteins and trigger increased production of ER chaperones, ER to nucleus/ER to plasma membrane signaling to inhibit ER protein synthesis and facilitate protein degradation to reestablish ER homeostasis. Thus, Ca^{2+} buffering within the ER lumen is an essential component of cellular Ca^{2+} homeostasis (Corbett and Michalak 2000; Krebs et al. 2015; Wang et al. 2019).

1.3 Calreticulin, a Key Ca^{2+}-Binding ER-Resident Protein Required for Optimal ER Function

Calreticulin was first identified in skeletal muscle sarcoplasmic reticulum (Ostwald and MacLennan 1974). It took over 15 years to realize that calreticulin is only a minor component of muscle sarcoplasmic reticulum but rather the major Ca^{2+} binding protein of non-muscle ER (Fliegel et al. 1989b), as over 50% of Ca^{2+} in the lumen of the ER is bound to calreticulin (Mesaeli et al. 1999; Nakamura et al.

2001b). Another ER-associated chaperone whose structure looks very much like calreticulin is calnexin but, unlike calreticulin, calnexin is a type I integral membrane protein with an extended C-terminal cytoplasmic tail. Calreticulin and calnexin are encoded by two distinct genes (Mesaeli et al. 1999; Kraus et al. 2010). Calreticulin and calnexin are both composed of distinct structural and functional domains: a globular N-domain, an extended P-domain forming a flexible loop are responsible for substrate binding and protein folding and a C-domain. The P-domain of both proteins forms heterodimeric complexes with PDIA3 (ERp57) (Hebert and Molinari 2007) and other chaperones and this is regulated by changes in Ca^{2+} concentration in the lumen of the ER (Corbett et al. 2000; Corbett and Michalak 2000). The C-domain of calreticulin is highly acidic, binds Ca^{2+} with high capacity and interacts with the P-domain of the protein in a Ca^{2+}-dependent manner (Boelt et al. 2016). In contrast, calnexin has a single transmembrane helix followed by an extended cytoplasmic C-domain that undergoes several posttranslational modifications (phosphorylation, sumoylation, myristoylation), and interacts with several cytoplasmic proteins (Roderick et al. 2000; Myhill et al. 2008; Cameron et al. 2009; Lynes et al. 2012, 2013; Lakkaraju and van der Goot 2013; Lee et al. 2015; Dudek et al. 2015; Li et al. 2011; Jung et al. 2017b). Both calreticulin and calnexin function as lectin-like chaperones with similar substrate specificity (Hebert and Molinari 2007; Michalak et al. 2009; Halperin et al. 2014), and both proteins participate in the so-called calreticulin/calnexin cycle that ensures the correct processing of glycoproteins in the ER. Inactivation of calreticulin and calnexin genes manifest very different phenotypes (Mesaeli et al. 1999; Rauch et al. 2000; Kraus et al. 2010; Denzel et al. 2002; Jung et al. 2018), indicating that despite sharing many structural and biochemical features, each of these proteins have distinct roles in the ER and in ER Ca^{2+} homeostasis.

Calreticulin interacts with other ER resident and integral membrane proteins to regulate their functions (Michalak et al. 2009). For example, interaction of calreticulin with PDIA1 (Baksh et al. 1995) affects luminal ER redox conditions upon Ca^{2+} depletion (Avezov et al. 2015), and its interaction with PDIA3 supports ER protein folding (Hebert and Molinari 2007). Calreticulin works together with other ER resident chaperones BiP, GRP94 and PDI to maintain the steady state free Ca^{2+} concentration in the lumen of the ER. Importantly, disruption of Ca^{2+} binding by calreticulin has profound effects on ER Ca^{2+} homeostasis, intracellular communication and cell growth resulting in unfavorable consequences at the cellular and organismal levels. As observed for other ER luminal proteins, the expression of calreticulin is increased by Ca^{2+} depletion, cellular stress, and nutrient starvation (Waser et al. 1997; Michalak et al. 2009). The forced over-expression of calreticulin results in expanded ER Ca^{2+} storage capacity, increased Ca^{2+} release via the $InsP_3R$, and delayed store-operated Ca^{2+} entry (SOCE) consistent with the key role of this protein in ER Ca^{2+} homeostasis (Nakamura et al. 2001a; Arnaudeau et al. 2002). Calreticulin effects on SOCE activity (Arnaudeau et al. 2002) likely involve complex formation between calreticulin, PDIA3 and STIM1 (Prins et al. 2011; Di Buduo et al. 2020).

Ca^{2+} in ER stores also serves as an important regulator of several ER enzymes and proteins, including feedback regulation of the $InsP_3R$, RyR, and SERCA. This is readily demonstrated by the acute but varied phenotypes associated with targeted inactivation of Ca^{2+} binding proteins as seen in mice (Mesaeli et al. 1999; Knollmann et al. 2006; Paolini et al. 2007; Schwaller 2012). Whole body calreticulin deficiency in mice is embryonic lethal due to failure of the Ca^{2+}-dependent phosphatase calcineurin to activate the MEF2C transcription factor (Guo et al. 2002; Mesaeli et al. 1999; Lynch et al. 2005), transgene expression of a constitutively active calcineurin rescues calreticulin-deficient mice from embryonic lethality (Guo et al. 2002) although surviving mice exhibit severe metabolic defects (Wang et al. 2017). These studies demonstrate the intimate interdependence of calreticulin and ER Ca^{2+} in modulating ER function and Ca^{2+} dependent signalling. Calreticulin deficiency in humans has not been observed and naturally occurring variants of calreticulin are rare (Nangalia et al. 2013; Klampfl et al. 2013). Notably, two sets of calreticulin variants are associated with altered cell calcium homeostasis along with the pathogenesis of myeloproliferative neoplasms (Di Buduo et al. 2020).

1.4 Calreticulin, Looking Outside from the ER and the Cell

Although calreticulin is abundant in the ER, it is now evident that this protein also participates in various biological functions outside the ER, such as cell adhesion/migration, cell proliferation, and immunogenic cell death (Gold et al. 2010). In 1994, calreticulin was detected on the surface of neutrophils (Eggleton et al. 1994) and subsequently on the surfaces of melanoma cells (White et al. 1995) and fibroblasts (Gray et al. 1995). Calreticulin was identified as a potential antithrombotic agent by virtue of its ability to bind vitamin K-dependent coagulation factors (Kuwabara et al. 1995). Indeed, intracoronary infusion of calreticulin prevents occlusion of the coronary artery as extracellular calreticulin interacts with the endothelium to stimulate release of nitric oxide and inhibit platelet-dependent clot formation (Dai et al. 1997). Conditioned media containing the N-terminal fragment of calreticulin (amino acids 1–180; N + P-domain) released by cultured immortalized B cell-derived cells kills tumor cells and inhibits angiogenesis (Pike et al. 1998). Extracellular calreticulin has been seen to participate in cell signaling leading to the assembly and disassembly of focal adhesions (Goicoechea et al. 2000, 2002). Cell surface-localized calreticulin may also serve as a marker for clearance of apoptotic cells by both macrophages and dendritic cells (Gardai et al. 2005; Obeid et al. 2007a, b; Panaretakis et al. 2008). Perhaps one of the most unexpected finding is the apparent ability of extracellular calreticulin to promote repair of cutaneous wounds (Nanney et al. 2008; Greives et al. 2012; Gold et al. 2010). This is accomplished via activation of TGFβ signaling pathway, induction of collagen synthesis, stimulation of fibroblast and keratinocyte migration and increased phagocytosis by macrophages at the site of the wound (Nanney et al. 2008; Greives et al. 2012; Gold et al. 2010; Suto et al. 2020). Although calreticulin was initially described as an ER resident high-affinity Ca^{2+} binding protein, the growing list of examples argue that calreticulin can also be found in the extracellular space where it functions to regulate a variety of physiological and pathological processes.

1.5 Calreticulin, the View to the Future

The last quarter century of studying the features of calreticulin and its functions witnessed remarkable discoveries illustrating the importance of the ER and ER Ca^{2+} in the life of the cell and cellular function. Many new links between calreticulin and cellular signalling pathways involved in metabolism and cellular homeostasis have been established. The expression of the calreticulin gene itself is regulated by the status of ER Ca^{2+} stores (Waser et al. 1997), underscoring the intimate relationship between Ca^{2+} and calreticulin. In addition, many insights about calreticulin function have been gleaned from studies done in non-mammalian systems such as parasites (*Trypanosoma cruzi*, *Entamoeba histolyticazi*, *Trichinella spiralis*, *Leishmania mansoni*, *Schistosoma manosni*) and worms (*C. elegans*, *Taenia solium*, *Necatus americanus*). Noteworthy is the observation that the enhanced lipid synthesis phenotype exhibited by calreticulin-deficient mice rescued by expression of activated calcineurin is reproduced in calreticulin-deficient worms (Wang et al. 2017), indicating that the calreticulin serves an ancient function. Tardigrades are emerging as exciting animal models for investigating novel stress coping strategies because of their remarkable ability to survive environmental extremes. Recent proteome analysis of the tardigrade *Milnesium tardigardum* indicates the presence of ER quality control components (calreticulin, calnexin and PDIA3) as well as regulators of the UPR signalling pathway (Schokraie et al. 2012) in this organism. It will be interesting to learn about the unique strategies that tardigrades have adopted, and whether these involve yet unknown roles for calreticulin, Ca^{2+} signalling and the ER, to mitigate cellular stress and overcome stressful conditions. Sunny ways ahead for calreticulin!

References

Arnaudeau S, Frieden M, Nakamura K, Castelbou C, Michalak M, Demaurex N (2002) Calreticulin differentially modulates calcium uptake and release in the endoplasmic reticulum and mitochondria. J Biol Chem 277(48):46696–46705. https://doi.org/10.1074/jbc.M202395200

Avezov E, Konno T, Zyryanova A, Chen W, Laine R, Crespillo-Casado A, Melo EP et al (2015) Retarded PDI diffusion and a reductive shift in poise of the calcium depleted endoplasmic reticulum. BMC Biol 13:2. https://doi.org/10.1186/s12915-014-0112-2

Baksh S, Burns K, Andrin C, Michalak M (1995) Interaction of calreticulin with protein disulfide isomerase. J Biol Chem 270:31338–31344

Beese CJ, Brynjolfsdottir SH, Frankel LB (2019) Selective autophagy of the protein homeostasis machinery: ribophagy, proteaphagy and ER-phagy. Front Cell Dev Biol 7:373. https://doi.org/10.3389/fcell.2019.00373

Boelt SG, Norn C, Rasmussen MI, Andre I, Ciplys E, Slibinskas R, Houen G, Hojrup P (2016) Mapping the Ca^{2+} induced structural change in calreticulin. J Proteome 142:138–148. https://doi.org/10.1016/j.jprot.2016.05.015

Cameron PH, Chevet E, Pluquet O, Thomas DY, Bergeron JJ (2009) Calnexin phosphorylation attenuates the release of partially misfolded alpha1-antitrypsin to the secretory pathway. J Biol Chem 284(50):34570–34579. https://doi.org/10.1074/jbc.M109.053165

Charonis AS, Michalak M, Groenendyk J, Agellon LB (2017) Endoplasmic reticulum in health and disease: the 12th International Calreticulin Workshop, Delphi, Greece. J Cell Mol Med 21 (12):3141–3149. https://doi.org/10.1111/jcmm.13413

Corbett EF, Michalak M (2000) Calcium, a signaling molecule in the endoplasmic reticulum? Trends Biochem Sci 25(7):307–311

Corbett EF, Michalak KM, Oikawa K, Johnson S, Campbell ID, Eggleton P, Kay C, Michalak M (2000) The conformation of calreticulin is influenced by the endoplasmic reticulum lumenal environment. J Biol Chem 275:27177–27185

Dai E, Stewart M, Ritchie B, Mesaeli N, Raha S, Kolodziejczyk D, Hobman ML et al (1997) Calreticulin, a potential vascular regulatory protein, reduces intimal hyperplasia after arterial injury. Arterioscler Thromb Vasc Biol 17:2359–2368

Damiani E, Spamer C, Heilmann C, Salvatori S, Margreth A (1988) Endoplasmic reticulum of rat liver contains two proteins closely related to skeletal sarcoplasmic reticulum Ca-ATPase and calsequestrin. J Biol Chem 263:340–343

Denzel A, Molinari M, Trigueros C, Martin JE, Velmurgan S, Brown S, Stamp G, Owen MJ (2002) Early postnatal death and motor disorders in mice congenitally deficient in calnexin expression. Mol Cell Biol 22(21):7398–7404

Di Buduo CA, Abbonante V, Marty C, Moccia F, Rumi E, Pietra D, Soprano PM et al (2020) Defective interaction of mutant calreticulin and SOCE in megakaryocytes from patients with myeloproliferative neoplasms. Blood 135(2):133–144. https://doi.org/10.1182/blood.2019001103

Dudek E, Millott R, Liu WX, Beauchamp E, Berthiaume LG, Michalak M (2015) N-myristoyltransferase 1 interacts with calnexin at the endoplasmic reticulum. Biochem Biophys Res Commun 468(4):889–893. https://doi.org/10.1016/j.bbrc.2015.11.052

Eggleton P, Michalak M (2013) Calreticulin for better or for worse, in sickness and in health, until death do us part. Cell Calcium 54(2):126–131. https://doi.org/10.1016/j.ceca.2013.05.006

Eggleton P, Lieu TS, Zappi EG, Sastry K, Coburn J, Zaner KS, Sontheimer RD, Capra JD, Ghebrehiwet B, Tauber AI (1994) Calreticulin is released from activated neutrophils and binds to C1q and mannan-binding protein. Clin Immunol Immunopathol 72:405–409

Fliegel L, Burns K, MacLennan DH, Reithmeier RA, Michalak M (1989a) Molecular cloning of the high affinity calcium-binding protein (calreticulin) of skeletal muscle sarcoplasmic reticulum. J Biol Chem 264(36):21522–21528

Fliegel L, Burns K, Opas M, Michalak M (1989b) The high-affinity calcium binding protein of sarcoplasmic reticulum. Tissue distribution, and homology with calregulin. Biochim Biophys Acta 982:1–8

Gardai SJ, McPhillips KA, Frasch SC, Janssen WJ, Starefeldt A, Murphy-Ullrich JE, Bratton DL, Oldenborg PA, Michalak M, Henson PM (2005) Cell-surface calreticulin initiates clearance of viable or apoptotic cells through trans-activation of LRP on the phagocyte. Cell 123 (2):321–334. https://doi.org/10.1016/j.cell.2005.08.032

Goicoechea S, Orr AW, Pallero MA, Eggleton P, Murphy-Ullrich JE (2000) Thrombospondin mediates focal adhesion disassembly through interactions with cell surface calreticulin. J Biol Chem 275(46):36358–36368. https://doi.org/10.1074/jbc.M005951200

Goicoechea S, Pallero MA, Eggleton P, Michalak M, Murphy-Ullrich JE (2002) The anti-adhesive activity of thrombospondin is mediated by the N-terminal domain of cell surface calreticulin. J Biol Chem 277(40):37219–37228. https://doi.org/10.1074/jbc.M202200200

Gold LI, Eggleton P, Sweetwyne MT, Van Duyn LB, Greives MR, Naylor SM, Michalak M, Murphy-Ullrich JE (2010) Calreticulin: non-endoplasmic reticulum functions in physiology and disease. FASEB J 24(3):665–683. https://doi.org/10.1096/fj.09-145482

Gold L, Williams D, Groenendyk J, Michalak M, Eggleton P (2015) Unfolding the complexities of ER chaperones in health and disease: report on the 11th international calreticulin workshop. Cell Stress Chaperones 20(6):875–883. https://doi.org/10.1007/s12192-015-0638-4

Gray AJ, Park PW, Broekelmann TJ, Laurent GJ, Reeves JT, Stenmark KR, Mecham RP (1995) The mitogenic effects of the B beta chain of fibrinogen are mediated through cell surface calreticulin. J Biol Chem 270:26602–26606

Greives MR, Samra F, Pavlides SC, Blechman KM, Naylor SM, Woodrell CD, Cadacio C et al (2012) Exogenous calreticulin improves diabetic wound healing. Wound Repair Regen 20 (5):715–730. https://doi.org/10.1111/j.1524-475X.2012.00822.x

Groenendyk J, Lynch J, Michalak M (2004) Calreticulin, calcium, and calcineurin - signaling from the endoplasmic reticulum. Mol Cells 17(3):383–389

Guo L, Nakamura K, Lynch J, Opas M, Olson EN, Agellon LB, Michalak M (2002) Cardiac-specific expression of calcineurin reverses embryonic lethality in calreticulin-deficient mouse. J Biol Chem 277(52):50776–50779

Halperin L, Jung J, Michalak M (2014) The many functions of the endoplasmic reticulum chaperones and folding enzymes. IUBMB Life 66:318–326. https://doi.org/10.1002/iub.1272

Hebert DN, Molinari M (2007) In and out of the ER: protein folding, quality control, degradation, and related human diseases. Physiol Rev 87(4):1377–1408

Hetz C, Zhang K, Kaufman RJ (2020) Mechanisms, regulation and functions of the unfolded protein response. Nat Rev Mol Cell Biol 21(8):421–438. https://doi.org/10.1038/s41580-020-0250-z

Jung J, Michalak M, Agellon LB (2017a) endoplasmic reticulum malfunction in the nervous system. Front Neurosci 11:220. https://doi.org/10.3389/fnins.2017.00220

Jung J, Wang J, Groenendyk J, Lee D, Michalak M, Agellon LB (2017b) Fatty acid binding protein (Fabp) 5 interacts with the calnexin cytoplasmic domain at the endoplasmic reticulum. Biochem Biophys Res Commun 493(1):202–206. https://doi.org/10.1016/j.bbrc.2017.09.046

Jung J, Eggleton P, Robinson A, Wang J, Gutowski N, Holley J, Newcombe J et al (2018) Calnexin is necessary for T cell transmigration into the central nervous system. JCI Insight 3(5). https://doi.org/10.1172/jci.insight.98410

Klampfl T, Gisslinger H, Harutyunyan AS, Nivarthi H, Rumi E, Milosevic JD, Them NC et al (2013) Somatic mutations of calreticulin in myeloproliferative neoplasms. N Engl J Med 369 (25):2379–2390. https://doi.org/10.1056/NEJMoa1311347

Knollmann BC, Chopra N, Hlaing T, Akin B, Yang T, Ettensohn K, Knollmann BE et al (2006) Casq2 deletion causes sarcoplasmic reticulum volume increase, premature Ca^{2+} release, and catecholaminergic polymorphic ventricular tachycardia. J Clin Invest 116(9):2510–2520. https://doi.org/10.1172/JCI29128

Kraus A, Groenendyk J, Bedard K, Baldwin TA, Krause KH, Dubois-Dauphin M, Dyck J et al (2010) Calnexin deficiency leads to dysmyelination. J Biol Chem 285(24):18928–18938. https://doi.org/10.1074/jbc.M110.107201

Krebs J, Agellon LB, Michalak M (2015) Ca^{2+} homeostasis and endoplasmic reticulum (ER) stress: an integrated view of calcium signaling. Biochem Biophys Res Commun 460(1):114–121. https://doi.org/10.1016/j.bbrc.2015.02.004

Kuwabara K, Pinsky DJ, Schmidt AM, Benedict C, Brett J, Ogawa S, Broekman MJ et al (1995) Calreticulin, an antithrombotic agent which binds to vitamin K- dependent coagulation factors, stimulates endothelial nitric oxide production, and limits thrombosis in canine coronary arteries. J Biol Chem 270:8179–8187

Lakkaraju AK, van der Goot FG (2013) Calnexin controls the STAT3-mediated transcriptional response to EGF. Mol Cell 51(3):386–396. https://doi.org/10.1016/j.molcel.2013.07.009

Lee D, Kraus A, Prins D, Groenendyk J, Aubry I, Liu WX, Li HD et al (2015) UBC9-dependent association between calnexin and protein tyrosine phosphatase 1B (PTP1B) at the endoplasmic reticulum. J Biol Chem 290(9):5725–5738. https://doi.org/10.1074/jbc.M114.635474

Li HD, Liu WX, Michalak M (2011) Enhanced clathrin-dependent endocytosis in the absence of calnexin. PLoS One 6(7):e21678. https://doi.org/10.1371/journal.pone.0021678

Lynch J, Guo L, Gelebart P, Chilibeck K, Xu J, Molkentin JD, Agellon LB, Michalak M (2005) Calreticulin signals upstream of calcineurin and MEF2C in a critical Ca^{2+}-dependent signaling cascade. J Cell Biol 170(1):37–47

Lynes EM, Bui M, Yap MC, Benson MD, Schneider B, Ellgaard L, Berthiaume LG, Simmen T (2012) Palmitoylated TMX and calnexin target to the mitochondria-associated membrane. EMBO J 31(2):457–470. https://doi.org/10.1038/emboj.2011.384

Lynes EM, Raturi A, Shenkman M, Ortiz Sandoval C, Yap MC, Wu J, Janowicz A et al (2013) Palmitoylation is the switch that assigns calnexin to quality control or ER calcium signaling. J Cell Sci 126:3893–3903. https://doi.org/10.1242/jcs.125856

Macer DRJ, Koch GEL (1988) Identification of a set of calcium-binding proteins in reticuloplasm, the luminal content of the endoplasmic reticulum. J Cell Sci 91:61–70

Mesaeli N, Nakamura K, Zvaritch E, Dickie P, Dziak E, Krause KH, Opas M, MacLennan DH, Michalak M (1999) Calreticulin is essential for cardiac development. J Cell Biol 144 (5):857–868

Michalak M, Agellon LB (2018) Stress coping strategies in the heart: an integrated view. Front Cardiovasc Med 5:168. https://doi.org/10.3389/fcvm.2018.00168

Michalak M, Milner RE, Burns K, Opas M (1992) Calreticulin. Biochem J 285:681–692

Michalak M, Groenendyk J, Szabo E, Gold LI, Opas M (2009) Calreticulin, a multi-process calcium-buffering chaperone of the endoplasmic reticulum. Biochem J 417(3):651–666. https://doi.org/10.1042/BJ20081847

Myhill N, Lynes EM, Nanji JA, Blagoveshchenskaya AD, Fei H, Carmine Simmen K, Cooper TJ, Thomas G, Simmen T (2008) The subcellular distribution of calnexin is mediated by PACS-2. Mol Biol Cell 19(7):2777–2788. https://doi.org/10.1091/mbc.E07-10-0995

Nakamura K, Robertson M, Liu G, Dickie P, Guo JQ, Duff HJ, Opas M, Kavanagh K, Michalak M (2001a) Complete heart block and sudden death in mouse over-expressing calreticulin. J Clin Invest 107(10):1245–1253

Nakamura K, Zuppini A, Arnaudeau S, Lynch J, Ahsan I, Krause R, Papp S et al (2001b) Functional specialization of calreticulin domains. J Cell Biol 154(5):961–972. https://doi.org/10.1083/jcb.200102073

Nangalia J, Massie CE, Baxter EJ, Nice FL, Gundem G, Wedge DC, Avezov E et al (2013) Somatic CALR mutations in myeloproliferative neoplasms with nonmutated JAK2. N Engl J Med 369 (25):2391–2405. https://doi.org/10.1056/NEJMoa1312542

Nanney LB, Woodrell CD, Greives MR, Cardwell NL, Pollins AC, Bancroft TA, Chesser A et al (2008) Calreticulin enhances porcine wound repair by diverse biological effects. Am J Pathol 173(3):610–630. https://doi.org/10.2353/ajpath.2008.071027

Obeid M, Tesniere A, Ghiringhelli F, Fimia GM, Apetoh L, Perfettini JL, Castedo M et al (2007a) Calreticulin exposure dictates the immunogenicity of cancer cell death. Nat Med 13(1):54–61. https://doi.org/10.1038/nm1523

Obeid M, Tesniere A, Panaretakis T, Tufi R, Joza N, van Endert P, Ghiringhelli F et al (2007b) Ecto-calreticulin in immunogenic chemotherapy. Immunol Rev 220(1):22–34

Ostwald TJ, MacLennan DH (1974) Isolation of a high affinity calcium-binding protein from sarcoplasmic reticulum. J Biol Chem 249:974–979

Panaretakis T, Joza N, Modjtahedi N, Tesniere A, Vitale I, Durchschlag M, Fimia GM et al (2008) The co-translocation of ERp57 and calreticulin determines the immunogenicity of cell death. Cell Death Differ 15(9):1499–1509. https://doi.org/10.1038/cdd.2008.67

Paolini C, Quarta M, Nori A, Boncompagni S, Canato M, Volpe P, Allen PD, Reggiani C, Protasi F (2007) Reorganized stores and impaired calcium handling in skeletal muscle of mice lacking calsequestrin-1. J Physiol 583(Pt 2):767–784. https://doi.org/10.1113/jphysiol.2007.138024

Pike SE, Yao L, Jones KD, Cherney B, Appella E, Sakaguchi K, Nakhasi H et al (1998) Vasostatin, a calreticulin fragment, inhibits angiogenesis and suppresses tumor growth. J Exp Med 188:2349–2356

Porter KR, Claude A, Fullam EF (1945) A study of tissue culture cells by electron microscopy: Methods and Preliminary Observations. J Exp Med 81(3):233–246. https://doi.org/10.1084/jem.81.3.233

Prins D, Michalak M (2011) Organellar calcium buffers. Cold Spring Harb Perspect Biol 3(3): a004069. https://doi.org/10.1101/cshperspect.a004069

Prins D, Groenendyk J, Touret N, Michalak M (2011) Modulation of STIM1 and capacitative Ca^{2+} entry by the endoplasmic reticulum luminal oxidoreductase ERp57. EMBO Rep 12 (11):1182–1188. https://doi.org/10.1038/embor.2011.173

Putney JW Jr (2007) Recent breakthroughs in the molecular mechanism of capacitative calcium entry (with thoughts on how we got here). Cell Calcium 42(2):103–110

Rauch F, Prud'homme J, Arabian A, Dedhar S, St-Arnaud R (2000) Heart, brain, and body wall defects in mice lacking calreticulin. Exp Cell Res 256:105–111

Roderick HL, Lechleiter JD, Camacho P (2000) Cytosolic phosphorylation of calnexin controls intracellular Ca^{2+} oscillations via an interaction with SERCA2b. J Cell Biol 149(6):1235–1248

Ron D, Hubbard SR (2008) How IRE1 reacts to ER stress. Cell 132(1):24–26. https://doi.org/10.1016/j.cell.2007.12.017

Rutkowski DT, Kaufman RJ (2007) That which does not kill me makes me stronger: adapting to chronic ER stress. Trends Biochem Sci 32(10):469–476

Schokraie E, Warnken U, Hotz-Wagenblatt A, Grohme MA, Hengherr S, Forster F, Schill RO, Frohme M, Dandekar T, Schnolzer M (2012) Comparative proteome analysis of Milnesium tardigradum in early embryonic state versus adults in active and anhydrobiotic state. PLoS One 7(9):e45682. https://doi.org/10.1371/journal.pone.0045682

Schroder M (2008) Endoplasmic reticulum stress responses. Cell Mol Life Sci 65(6):862–894

Schwaller B (2012) The use of transgenic mouse models to reveal the functions of Ca^{2+} buffer proteins in excitable cells. Biochim Biophys Acta 1820(8):1294–1303. https://doi.org/10.1016/j.bbagen.2011.11.008

Shergalis AG, Hu S, Bankhead A 3rd, Neamati N (2020) Role of the ERO1-PDI interaction in oxidative protein folding and disease. Pharmacol Ther 210:107525. https://doi.org/10.1016/j.pharmthera.2020.107525

Smith MJ, Koch GL (1989) Multiple zones in the sequence of calreticulin (CRP55, calregulin, HACBP), a major calcium binding ER/SR protein. EMBO J 8(12):3581–3586

Suto MJ, Gupta V, Mathew B, Zhang W, Pallero MA, Murphy-Ullrich JE (2020) Identification of inhibitors of thrombospondin 1 activation of TGF-beta. ACS Med Chem Lett 11(6):1130–1136. https://doi.org/10.1021/acsmedchemlett.9b00540

Sweeney P, Park H, Baumann M, Dunlop J, Frydman J, Kopito R, McCampbell A et al (2017) Protein misfolding in neurodegenerative diseases: implications and strategies. Transl Neurodegener 6:6. https://doi.org/10.1186/s40035-017-0077-5

Van PN, Peter F, Soling H-D (1989) Four intracisternal calcium-binding glycoproteins from rat liver microsomes with high affinity for calcium. No indication for calsequestrin-like proteins in inositol 1,4,5-trisphosphate-sensitive calcium sequestering rat liver vesicles. J Biol Chem 264:17494–17501

Waisman DM, Salimath BP, Anderson MJ (1985) Isolation and characterization of CAB-63, a novel calcium-binding protein. J Biol Chem 260:1652–1660

Wang WA, Liu WX, Durnaoglu S, Lee SK, Lian J, Lehner R, Ahnn J, Agellon LB, Michalak M (2017) Loss of calreticulin uncovers a critical role for calcium in regulating cellular lipid homeostasis. Sci Rep 7(1):5941. https://doi.org/10.1038/s41598-017-05734-x

Wang WA, Agellon LB, Michalak M (2019) Organellar calcium handling in the cellular reticular network. Cold Spring Harb Perspect Biol 11(12):a038265. https://doi.org/10.1101/cshperspect.a038265

Waser M, Mesaeli N, Spencer C, Michalak M (1997) Regulation of calreticulin gene expression by calcium. J Cell Biol 138:547–557
White TK, Zhu Q, Tanzer ML (1995) Cell surface calreticulin is a putative mannoside lectin which triggers mouse melanoma cell spreading. J Biol Chem 270:15926–15929
Ye J, Rawson RB, Komuro R, Chen X, Dave UP, Prywes R, Brown MS, Goldstein JL (2000) ER stress induces cleavage of membrane-bound ATF6 by the same proteases that process SREBPs. Mol Cell 6(6):1355–13564
Yoshida H (2007) ER stress and diseases. FEBS J 274(3):630–658

Chapter 2
Structural Analysis of Calreticulin, an Endoplasmic Reticulum-Resident Molecular Chaperone

Gunnar Houen, Peter Højrup, Evaldas Ciplys, Christine Gaboriaud, and Rimantas Slibinskas

Abstract Calreticulin (Calr) is an endoplasmic reticulum (ER) chaperone involved in protein quality control, Ca^{2+} regulation and other cellular processes. The structure of Calr is unusual, reflecting different functions of the protein: a proline-rich β-hairpin arm and an acidic C-terminal tail protrude from a globular core, composed of a β-sheet sandwich and an α-helix. The arm and tail interact in the presence of Ca^{2+} and cover the upper β-sheet, where a carbohydrate-binding site gives the chaperone glycoprotein affinity. At the edge of the carbohydrate-binding site is a conserved, strained disulphide bridge, formed between C^{106} and C^{137} of human Calr, which lies in a polypeptide-binding site. The lower β-sheet has several conserved residues, comprised of a characteristic triad, D^{166}-H^{170}-D^{187}, Tyr^{172} and the free C^{163}. In addition to its role in the ER, Calr translocates to the cell surface upon stress and functions as an immune surveillance marker. In some myeloproliferative neoplasms, the acidic Ca^{2+}-binding C-terminal tail is transformed into a polybasic sequence.

G. Houen (✉)
Department of Neurology, Glostrup Research Institute, Glostrup, Denmark

Institute of Biochemistry and Molecular Biology, University of Southern Denmark, Odense, Denmark
e-mail: gunnar.houen@regionh.dk

P. Højrup
Institute of Biochemistry and Molecular Biology, University of Southern Denmark, Odense, Denmark
e-mail: php@bmb.sdu.dk

E. Ciplys · R. Slibinskas
Department of Eucaryote Gene Engineering, Life Sciences Center, Vilnius University, Vilnius, Lithuania
e-mail: evaldas.ciplys@bti.vu.lt; rimantas.slibinskas@bti.vu.lt

C. Gaboriaud
Institut de Biologie Structurale Jean-Pierre Ebel, CEA, Grenoble, France
e-mail: christine.gaboriaud@ibs.fr

L. B. Agellon, M. Michalak (eds.), *Cellular Biology of the Endoplasmic Reticulum*, Progress in Molecular and Subcellular Biology 59,
https://doi.org/10.1007/978-3-030-67696-4_2

Keywords Calreticulin · Calnexin · Chaperone · Lectin · Protein synthesis · Protein stability · Protein structure

Abbreviations

Calr	Calreticulin
Calr3	Calsperin
Canx	Calnexin
CDS	Circular dichroism spectroscopy
Clgn	Calmegin
DTT	Dithiothreitol
ECM	Electron cryo-microscopy
ER	Endoplasmic reticulum
INDELS	Insertions and deletions
JAK	Janus kinase
MHC	Major histocompatibility complex
MM	Molecular modelling
MPN	Myeloproliferative neoplasms
NMRS	Nuclear magnetic resonance spectroscopy
SAXS	Small angle X-ray scattering
XL-MS	Cross-linking/mass spectrometry
XRC	X-ray crystallography

2.1 Introduction

In eukaryotic cells, ER is the site for synthesis, maturation and quality control of secreted, membrane-bound and ER-resident proteins (Halperin et al. 2014; Kepp and Galluzzi 2020). This process involves translation of mRNAs on ER membrane-bound ribosomes, translocation of the nascent polypeptide across the ER membrane and co-translational folding to the native three-dimensional structure. To assist in these processes, the ER contains elaborate systems involved in post-translational modification and quality control of newly synthesized proteins (Vincenz-Donnelly L and Hipp MS 2017; Halperin et al. 2014). An important protein quality control system in the ER is constituted by the chaperones calreticulin (Calr) and calnexin (Canx), which act in concert with protein disulfide isomerases (e.g. Pdia3/ERp57) to facilitate the synthesis and maturation of proteins, and to prevent accumulation of misfolded proteins (Bergeron et al. 1994; Michalak et al. 1998; Coe and Michalak 2010; Halperin et al. 2014; Matsusaki et al. 2020). Canx is a transmembrane protein, with its major part in the ER lumen, while Calr is an ER luminal protein (Bergeron et al. 1994; Wada et al. 1995; Michalak et al. 1998; Danilczyk et al. 2000). In humans, these proteins are encoded by separate genes with 9 (*Calr*) or 20 (*Canx*) exons (McCauliffe et al. 1992; Tjoelker et al. 1994). In mammals, Calr and Canx furthermore play important roles in the loading of peptides on MHC class I (Blees et al. 2017). This chapter reviews current knowledge on the structure of human Calr (hCalr).

2.2 Primary Structures of Calr and Related Proteins

Analyses of Canx and Calr amino acid sequences have revealed that they are evolutionarily conserved proteins encoded by single genes. In the testicles a Canx homologue, calmegin (Clgn) and a Calr homologue, calsperin (Calr3) are also present (Ohsako et al. 1994; Watanabe et al. 1994; Ikawa et al. 2011). In plants, several Calr isoforms are encoded by separate genes (Jia et al. 2009; Del Bem 2011; Wasąg et al. 2019). In yeasts, a Canx homologue is present (de Virgilio et al. 1993; Parlati et al. 1995).

Figure 2.1 shows an alignment of amino acid sequences of selected Calr/Canx family members. Common to both Canx and Calr is an N-terminal domain of approximately 200 amino acid residues, followed by a proline-rich domain of approximately 100 (Calr and Calr3) or 145 (Canx and Clgn) amino acid residues, which is followed by a C-terminal domain. In Calr (and Calr3) the C-terminal (C) domain is highly acidic and consists of approximately 100 amino acids ending in an ER retrieval sequence of four residues (KDEL in hCalr). In Canx (and Clgn) the C-terminal domain has a transmembrane stretch of 20 amino acids and a cytoplasmic "tail" of approximately 90 amino acids.

Relatively few residues are invariant among all Calr and related proteins (Fig. 2.1) and these mainly reside in the N-terminal (N) domain and the middle proline-rich (P) domain. Conserved stretches/clusters of amino acids in the N-domain are found in the vicinity of the conserved disulfide bridge (C^{106}-C^{137} in Calr) and in the P-domain in the neighborhood of conserved W residues, being part of characteristic repeat sequences. A third interesting residue (C^{163} in hCalr) is conserved in all known mammalian and some other Calrs.

2.3 Higher Order Structures of Calr and Related Proteins

Table 2.1 shows a list of Calr and Canx proteins from different species with solved three dimensional (3D) structures. Figure 2.2 shows two partial structures for human Calr, plus one view of its full-length elongated structure in the context of the MHC-I peptide-loading complex. The 3D structures of Canx and Calr show close similarity but also some important differences. Both proteins have a globular core consisting of a β-sheet sandwich, where the strands are arranged in a "jelly roll" fold, and an α-helix. The sandwich is composed of two β-sheets, where the N-terminal 200 amino acids contribute 6 strands to the first, "upper" β-sheet and 6 strands to the second "lower" β-sheet. Intertwined in the two sheets are two β-strands derived from the

```
A ---------- ---------- ---------- ---------- ---------- ---------M FTLFLLIALS SAKVYFHETF
B ---------- ---------- ---------- ---------- ---------M RLLLCLIFLV FVFNFALSTV HFKDTFDNDW
C ---------- ---------- ---------- ---------- ---------M LLSVPLLLGL LGLAAADPAI YFKEQFLDGD
D ---------- ---------- ---------- ---------- ---------M LLSVPLLLGL LGLAAADPAI YFKEQFLDGD
E ---------- ---------- ---------- ---------- ---------M LLSVPLLLGL LGLAVAEPAV YFKEQFLDGD
F ---------- ---------- ---------- ---------- ---------- ---MKSLCLL AIVAVVSAEV YFKEEFNDAS
G ---------- ---------- ---------- ---------- ---------M ARALVQLWAI CMLRVALATV YFQEEFLDGE
H ---------- ---------- ---------- ---------- ---------- -MRAAIFFCA LLSLATLSAV HGTVYFHEEF
I MEGKWLLCML LVLGTTIVQA HEGHDDDMID IEDDLDDVIE EVEDSKSKPD TSAPTSPKVT YKAPVPTGEV YFADSFDRGT
J MEGKWLLCML LVLGTAIVEA HDGHDDDVID IEDDLDDVIE EVEDSKPD-T TAPPSSPKVT YKAPVPTGEV YFADSFDRGT
K ---MHFQAFW LCLGLLFISI NAEFMDDDVE TEDFEENSEE IDVN------ -ESELSSEIK YKTPQPIGEV YFAETFDSGR
                                                                                   *   |

A ENRDKWIDST SSGKALGPFK IV-----SGK WYGD-ANNKG LQTSEDNKFY IAAAKLDEEF SNKDKNLIVQ YNLKFEQGID
B ESRWVVSDWH KEDGKSGKLV HT-----AGK WFGD-ENQKG IQTSEDARFY AVSAKFP-SF SNKGKDLVLQ YTVKNEQKVD
C AWTNRWVESK HKSD-FGKFV LS-----SGK FYGDLEKDKG LQTSQDARFY ALSAKFEP-F SNKGQTLVVQ FTVKHEQNID
D AWTNRWVESK HKSD-FGKFV LS-----SGK FYGDQEKDKG LQTSQDARFY ALSARFEP-F SNKGQTLVVQ FTVKHEQNID
E GWTSRWIESK HKSD-FGKFV LS-----SGK FYGDEEKDKG LQTSQDARFY ALSASFEP-F SNKGQTLVVQ FTVKHEQNID
F -WEKRWVQSK HKDD-FGAFK LS-----AGK FFDVESRDQG IQTSQDAKFY SRAAKFDKDF SNKGKTLVIQ YTVKHEQGID
G HWRNRWLQST NDSR-FGHFR LS-----SGK FYGHKEKDKG LQTTQNGRFY AISARFKP-F SNKGKTLVIQ YTVKHEQKMD
H KSMEHWTTSK HRDD-FGKVE IS-----AGK FYADAEKSKG LRLTEDARFY ALSTAFPTPI TNEKKSLVVS FSVKHEQDLK
I LSGWILSKAK KDDTDDEIAK YDGKWEVDEM KETKLPGDKG LVLMSRAKHH AISAKLNKPF LFDTKPLIVQ YEVNFQNGIE
J LSGWILSKAK KDDTDDEIAK YDGKWEVEEM KESKLPGDKG LVLMSRAKHH AISAKLNKPF LFDTKPLIVQ YEVNFQNGIE
K LAGWVLSKAK KDDMDEEISI YDGRWEIEEL KENQVPGDRG LVLKSRAKHH AISAVLAKPF IFADKPLIVQ YEVNFQDGID
                 .                                 .:* :   .  :.:   :: :    :     : *::. : :: :: :

A CGGGYIKLLP KKSIESEEKF TPESEYNIMF GPDVCGG-SK RTHVIMNYKG KNNLIRKEI- --------KC ESDDISHLYT
B CGGSYIKLLP SKLDQS--AF DGESEYSIMF GPDVCGA-SK RVHVILNYKG KNHLIKKEIN --------KV ETDQLTHQYT
C CGGGYVKLFP SGLDQK--DM HGDSEYNIMF GPDICGPGTK KVHVIFNYKG KNVLINKDI- --------RC KDDEFTHLYT
D CGGGYVKLFP GGLDQK--DM HGDSEYNIMF GPDICGPGTK KVHVIFNYKG KNVLINKDI- --------RC KDDEFTHLYT
E CGGGYVKLFP NSLDQT--DM HGDSEYNIMF GPDICGPGTK KVHVIFNYKG KNVLINKDI- --------RC KDDEFTHLYT
F CGGGYVKVMR ADADLG--DF HGETPYNVMF GPDICGP-TR RVHVILNYKG ENKLIKKEI- --------TC KSDELTHLYT
G CGGGYIKVFP ADIDQK--NL NGKSQYYIMF GPDICGFDIK KVHVILHFKN KYHENKKLI- --------RC KVDGFTHLYT
H CGGGYIKLLP SMDPEK---F HGETKYWLMF GPDRCGS-QN RVHIILHYNG ENREWSKRI- --------RF PEDKLTHVYT
I CGGAYVKLLS KTPELNLDQF HDKTPYTIMF GPDKCGEDYK LHFIFRHKNP KTGVYEEKHA KRPDADLKTY FTDKKTHLYT
J CGGAYVKLLS KTPELNLDQF HDKTPYTIMF GPDKCGEDYK LHFIFRHKNP KTGIYEEKHA KRPDADLKTY FTDKKTHLYT
K CGGAYIKLLA DTDDLILENF YDKTSYIIMF GPDKCGEDYK LHFIFRHKHP KTGVFEEKHA KPPDVDLKKF FTDRKTHLYT
  ***.*:*::          :    .: * :** *** **     .  .:: : :  :      :                 *   :* **

A LIIRPNNTYV VKIDGVEKQE GKFDEDWD-- MLAPKEIDDP NVSKPADWVD EKEIDDPNDK KPEGWDDIP- ----------
B LVISPDNTYK VLVDNKEIQA GNLADDWE-- LLPSKQIKDP KQSKPVDWVD VKEIDDPEDV KPAGHDDIP- ----------
C LIVRPDNTYE VKIDNSQVES GSLEDDWD-- FLPPKKIKDP DAAKPEDWDE RAKIDDPTDS KPEDWDKP-- ----------
D LIVRPDNTYE VKIDNSQVES GSLEDDWD-- FLPPKKIKDP DAAKPEDWDE RAKIDDPTDS KPEDWDKP-- ----------
E LIVRPDNTYE VKIDNSQVES GSLEDDWD-- FLPPKKIKDP DASKPEDWDE RAKIDDPTDS KPEDWDKP-- ----------
F LILNSDNTYE VKIDGESAQT GSLEEDWD-- LLPAKKIKDP DAKKPEDWDE REYIDDAEDA KPEDWEKP-- ----------
G LILRPDLSYD VKIDGQSIES GSIEYDWN-- LTSLKKETSP AESKD----- ---WEQTKDN KAQDWEK--- ----------
H LHIAADNSYE FFLDGESKAK GQLDGDWT-- LLPPREIVDE TDKKPEDWVD EETMDDPEDK KPEDWDNEP- ----------
I LILNPDNSFE ILVDQSIVNS GNLLNDMTPP VNPSREIEDP EDQKPEDWDE RPKIPDPDAV KPDDWNEDAP AKIPDEEATK
J LILNPDNSFE ILVDQSVVNS GNLLNDMTPP VNPSREIEDP EDRKPEDWDE RPKIPDPEAV KPDDWDEDAP AKIPDEEATK
K LVMNPDDTFE VLVDQTVVNK GSLLEDVVPP IKPPKEIEDP NDKKPEEWDE RAKIPDPSAV KPEDWDESEP AQIEDSSVVK
  * : .: ::   . :*      *.:  *          . . ::  .       *          :.     *. . :.         |
```

Fig. 2.1 Alignment of Calrs and related proteins from various species. (**a**) Calr *E. histolytica* XP_655241; (**b**) Calr *D. discoideum* Q23858; (**c**) Calr *M. musculus* P14211; (**d**) Calr *R. norvegicus* P18418; (**e**) Calr *H. sapiens* P27797; (**f**) Calr *C. elegans* P27798; (**g**) Calr3 *H. sapiens* Q96L12; (**h**) Calr *T. cruzi* XP_812571; (**i**) Canx *C. lupus* P24643; (**j**) Canx *H. sapiens* P27824; (**k**) Clgn *H. sapiens* O14967. The alignment was carried out using Clustal 2.1 and default parameters. Coloring: core region—blue, P-domain—yellow. C-domain—green

C-terminal domain of the molecules, the first constituting the Calr β19 strand of the upper β-sheet and the second constituting the Calr β20a/b strand of the lower β-sheet (Fig. 2.2a). From the core extend the proline-rich stretches as β-hairpins (P-domain, Fig. 2.2c), stabilised by hydrogen bonds between four (Canx) or three (Calr) sets of repeat sequences. The α-helix of the globular core is longer in Calr, where it continues into the acidic tail (Fig. 2.2c), while in Canx the α-helix is followed by a transmembrane sequence and the cytoplasmic domain.

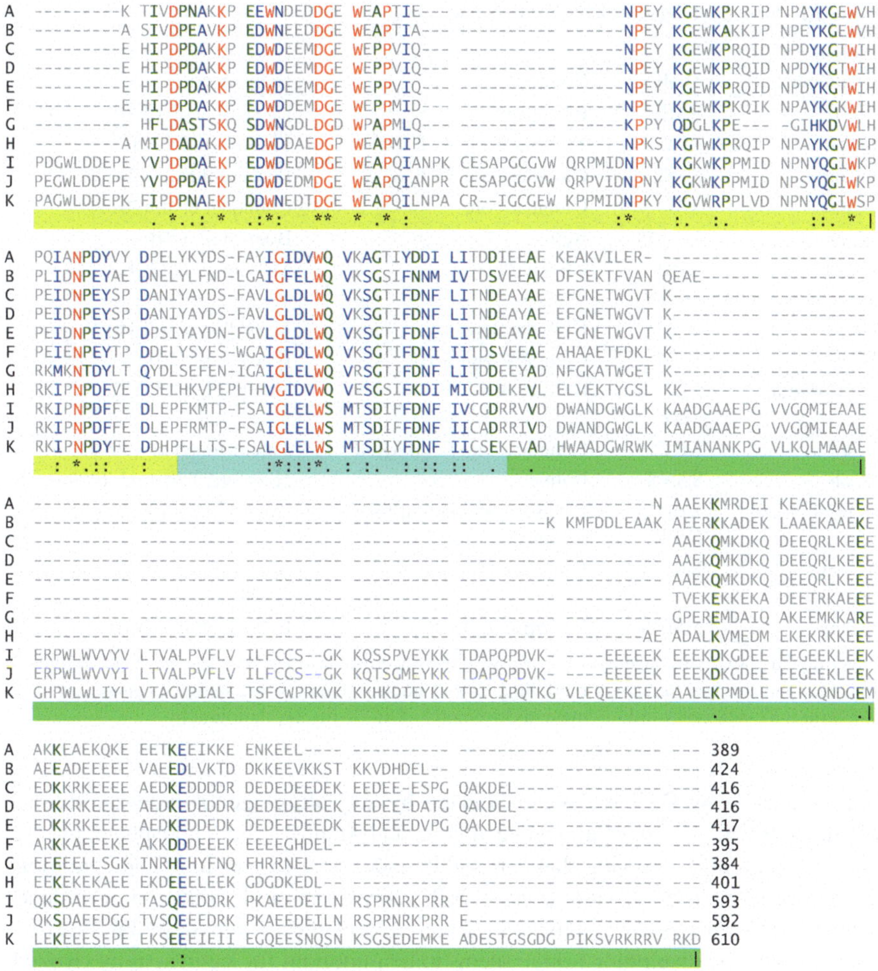

Fig. 2.1 (continued)

There are several interesting characteristic features of the conserved structure of Calr. A carbohydrate-binding site on the upper concave β-sheet capable of accommodating a tetrasaccharide and having affinity for terminal glucose residues. This site contains several conserved residues (cluster 1) of importance for carbohydrate binding (Kozlov et al. 2010a, b; Chouquet et al. 2011; Moreau et al. 2016).

An exposed disulphide bridge (C^{106}–C^{137} in hCalr) is located on the edge of the carbohydrate-binding site. This bridge is unusual due to its surface location and its geometry, which reveals an almost parallel, strained conformation with a sulphur atom distance of 2.8–3 Å (in contrast to normal unstrained disulphide bonds, which have a torsion angle of approximately 90° and a sulphur atom distance of 2 Å (Thornton 1981). The disulphide bridge is important for thermodynamic stability

Table 2.1 Structural data available for Calr and Canx

Protein (fragment)	Method	Comments	References
rCalr	NMRS	P hairpin	Ellgaard et al. (2001a, b, 2002)
mCalr	XRC	18-206, 301-368, C163S	Kozlov et al. (2010a)
mCalr	XRC	Globular core, C163S	Pocanschi et al. (2011)
hCalr	XRC	Globular core	Chouquet et al. (2011)
hCalr	XL-MS	Native	Boelt et al. (2016)
tcCalr	XRC, SAXS	Globular core	Moreau et al. (2016)
ehCalr	XRC	Globular core	Moreau et al. (2016)
hCalr	ECM	Complex with MHC	Blees et al. (2017)
Crt C-domain	CDS, MM	C-"tail", Ca^{2+} sensor	Villamil Giraldo et al. (2010)
hCalr	XRC	Complex with GABARAP	Thielmann et al. (2009)
hCalr	SAXS	Native	Nørgaard Toft et al. (2008)
m/hCalr, Canx, Clgn	XRC, NMR	P hairpin cyclophilin/ERp29 complex	Kozlov et al. 2010b, (2017)
dCanx	XRC	Fragment (47-468)	Schrag et al. (2001), Hahn et al. (1998)
dCanx	NMRS	P hairpin tip	Pollock et al. (2004)

CDS circular dichroism spectroscopy, *d* dog, *ECM* electron cryo-microscopy, *eh E. histolytica, h* human, *m* mouse, *MM* molecular modelling, *NMRS* nuclear magnetic resonance spectroscopy, *r* rat, *tc T. cruci, XL-MS* cross-linking/mass spectrometry, *XRC* X-ray crystallography

and reduceable by DTT but is not affected by alkylating agents under conditions, where the free Cys^{163} in hCalr can be alkylated (100 mM sodium phosphate, pH 7.2, room temperature), showing that it is stable under near-to-physiological conditions (Houen and Koch 1994; Højrup et al. 2001; Jørgensen et al. 2005), in accordance with its surface-exposed localisation at the edge of the carbohydrate-binding site. Interestingly, the disulphide bridge seems to be important for lectin activity (Kozlov et al. 2010a).

The lower β-sheet has a conserved structural Ca^{2+} ion-binding site (Fig. 2.2a) and also contains several residues conserved in published structures of Calr and Caln (cluster 2) and in primary sequences of Calr and related proteins (Fig. 2.1), including a triad composed of D^{166}, H^{170} and D^{187} (DHD triad). Finally, the C-terminal α-helix contains two conserved residues including W^{347}, close to the DHD triad (Fig. 2.2d) (Kozlov et al. 2010a, b; Chouquet et al. 2011; Moreau et al. 2016).

Chemical cross-linking in solution in combination with mass spectrometry (MS) of human Calr has confirmed the overall structure of the globular domain and has added some structural information on the parts of Calr absent in the crystallised "amputated" Calr (Fig. 2.2b). In the presence of Ca^{2+}, the P-domain hairpin folds back on itself and associates with the C-domain, which itself forms an α-helix extending "upwards" to meet the P-domain. Together, they cover the lectin site, however, in the absence of Ca^{2+}, they are more flexible and do not sterically

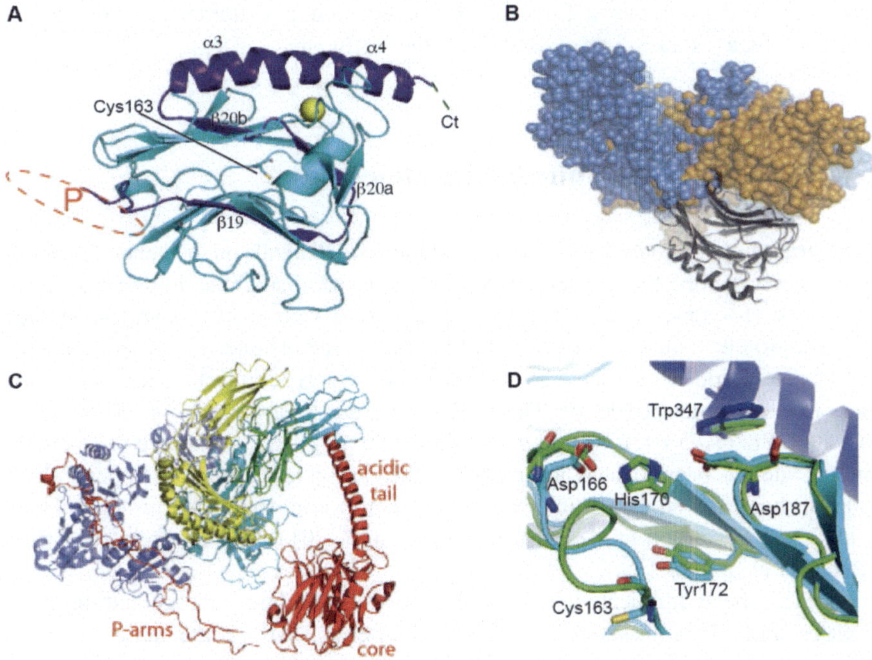

Fig. 2.2 *H. Sapiens* (Hs) Calr structures. (**a**) XRC core structure (Chouquet et al. 2011). The structural Ca^{2+} ion is shown in yellow. (**b**) XL-MS structure (Boelt et al. 2016). (**c**) HsCalr (red) in complex with MHC-I (yellow), Pdia3 (blue), tapasin (cyan) and ß2-microglobulin (green) (Blees et al. 2017). (**d**) Zoom on the cluster2 residues common to Calr (cyan) and Canx (green) proteins

limit access to the lectin site to the same extent. X-ray studies of *Trypanosoma cruzi* and *Entamoeba histolytica* Calrs have also shown for the first time a possible hinge motion at the basis of the Calr P-domain, as a switch between an open and a more compact conformation. This part of the Calr molecule differs in Canx (Moreau et al. 2016).

Structural studies of full-length Calr proteins have been hindered by its flexibility. However, its overall ordered structure could be trapped by cryo-electron microscopy in the context of the MHC-I peptide-loading complex (Blees et al. 2017). In this complex, Calr still shows substantial flexibility, as compared to its structurally stable binding partners. This flexibility remains despite the fact that it interacts with an MHC-class I glycan through its carbohydrate-binding site, with Pdia3 through the tip of its P-arm and with tapasin through its acidic tail.

The overall structure of Calr and Canx indicated by the structural data for various amputated forms of the molecules has been confirmed by several studies with other methods including electron cryo-microscopy (ECM), small angle X-ray scattering (SAXS) and nuclear magnetic resonance (NMR) spectroscopy (Table 2.1). Both molecules have a stable globular beta sandwich core with an α helix at the bottom.

From the core protrudes a β-hairpin P-domain and a C-domain, which in the presence of Ca^{2+} make contacts and cover the lectin site.

2.4 Calr Structure-Function Relationships

The thermodynamic stability of Calr is moderate and depends on the general packing of the core region, the conserved, surface-exposed disulphide bridge and Ca^{2+} association (Jørgensen et al. 2005; Duus et al. 2013). Calr and Canx are chaperones for glycoprotein synthesis and several of the conserved residues in cluster 1, including the disulphide bridge, have been shown to be involved in lectin activity of Calr/Canx (Kozlov et al. 2010a, b). These residues are also strikingly conserved in the parasite Calrs (*T. cruzi, E. histolytica*) (Moreau et al. 2016). A putative polypeptide binding site has been located at the edge of the carbohydrate-binding upper β-sheet and Calr has been shown to interact with several non-glycosylated proteins (Duus et al. 2009; Pocanschi et al. 2011; Chouquet et al. 2011; Møllegaard et al. 2011; Moreau et al. 2016).

Calr has been implicated in several diseases including cancers, autoimmune diseases, and others (Tesniere et al. 2008; Gold et al. 2010; Zamanian et al. 2013; Wiersma et al. 2015; Eggleton et al. 2016; Schcolnik-Cabrera et al. 2019; Houen 2019). Most evident in relation to structure are somatic alterations seen in myeloproliferative neoplasms (MPNs), including polycythemia vera, myelofibrosis and essential thrombocythemia (Klampfl et al. 2013; Nangalia et al. 2013; Pietra et al. 2016; Imai et al. 2017). These alterations are mainly insertions and deletions (INDELS) in exon 9 leading to a common polybasic stretch of amino acids in the C-terminus, in close conjunction to the α-helix in the C-domain. Since these INDELS have only been detected in MPNs, it can be speculated that the frameshifted Calr forms allow the affected myeloid precursor cells to remain viable, to grow and differentiate to some degree and to exhibit neoplastic properties. The structural polybasic alterations of the C-terminus can be speculated to mimic the binding of Ca^{2+} to the acidic C-terminus of wild type Calr, allowing the protein to remain stable and viable, but imparting some new properties on the molecule (i.e. an oncogenic alteration). In agreement with this, the frameshifted Calr has been shown to interact with the thrombopoietin receptor, which itself can cause MPN upon mutation and which signals through a pathway involving the Janus kinase (JAK), mutations of which is the most prominent cause of MPN (Vainchenker et al. 2019).

In relation to cancer in general, Calr has been shown to translocate to the cell surface in response to various forms of cellular stress, where it may serve as a signal for immune recognition (Tesniere et al. 2008; Wiersma et al. 2015; Eggleton et al. 2016; Schcolnik-Cabrera et al. 2019). However, it is presently unclear, whether the various involvements of Calr in cell transformation and immune surveillance are related to structural alterations or to changes in concentration and subcellular localization.

2.5 Summary Points

- Calr and Canx are conserved Ca^{2+}-binding molecular chaperones in eucaryotes.
- Both Calr and Canx have a globular β-sandwich/α-helix core with a structural Ca^{2+} ion. From the core extends a characteristic β-hairpin P-domain and a C-domain. In Calr, the C-domain is acidic and has a large capacity for Ca^{2+} binding. In Canx, the C-domain has a transmembrane stretch and a cytoplasmic domain.
- The upper β-sheet of the globular core sandwich has a carbohydrate-binding site and a disulphide bridge at the edge, which both are involved in carbohydrate interactions. A polypeptide binding site is located at the edge of the carbohydrate-binding site but has not been precisely mapped yet.
- The lower β-sheet of the globular sandwich has a conserved D^{166}-H^{170}-D^{187} triad, a free C^{163} and several other conserved residues, including Y^{172} (hCalr numbering) (Fig. 2.2d).
- The C-terminal α-helix at the bottom of the α-helix core has a conserved W^{347} residue facing the D^{166}-H^{170}-D^{187} triad (hCalr numbering) (Fig. 2.2d).
- The structure of Calr/Canx proteins reflects their multiple functions and interaction partners, including nascent (glycosylated) proteins, cyclophilin, Pdia3, and others, as illustrated by the role of Calr and Canx in the peptide loading complex.
- Characteristic mutations in the exon coding for the extreme C-terminus of hCalr changes the polyacidic Ca^{2+}-binding sequence to a polybasic sequence.

2.6 Future Issues

- Detailed structural analysis of complete Calr/Canx proteins in complex with substrates and binding partners.
- Detailed analysis of the role of individual amino acids in the thermodynamic stability and functions of Calr and Canx.
- Detailed understanding of the role of the polybasic sequence in frameshifted Calr variants involved in myeloproliferative neoplasms.
- Detailed understanding of cell surface functions of Calr.

Conflicts-of-Interest Disclosure Statement GH, PHP, EC, CG, RS declare no conflicts-of-interest.

References

Bergeron JJ, Brenner MB, Thomas DY, Williams DB (1994) Calnexin: a membrane-bound chaperone of the endoplasmic reticulum. Trends Biochem Sci 19:124–128. https://doi.org/10.1016/0968-0004(94)90205-4

Blees A, Januliene D, Hofmann T, Koller N, Schmidt C, Trowitzsch S, Moeller A, Tampé R (2017) Structure of the human MHC-I peptide-loading complex. Nature 551:525–528. https://doi.org/10.1038/nature24627

Boelt SG, Norn C, Rasmussen MI, André I, Čiplys E, Slibinskas R, Houen G, Højrup P (2016) Mapping the Ca^{2+} induced structural change in calreticulin. J Proteome 142:138–148. https://doi.org/10.1016/j.jprot.2016.05.015

Chouquet A, Païdassi H, Ling WL, Frachet P, Houen G, Arlaud GJ, Gaboriaud C (2011) X-ray structure of the human calreticulin globular domain reveals a peptide-binding area and suggests a multi-molecular mechanism. PLoS One 6:e17886. https://doi.org/10.1371/journal.pone.0017886

Coe H, Michalak M (2010) ERp57, a multifunctional endoplasmic reticulum resident oxidoreductase. Int J Biochem Cell Biol 42:796–799. https://doi.org/10.1016/j.biocel.2010.01.009

Danilczyk UG, Cohen-Doyle MF, Williams DB (2000) Functional relationship between calreticulin, calnexin, and the endoplasmic reticulum luminal domain of calnexin. J Biol Chem 275:13089–13097. https://doi.org/10.1074/jbc.275.17.13089

de Virgilio C, Bürckert N, Neuhaus JM, Boller T, Wiemken A (1993) CNE1, a Saccharomyces cerevisiae homologue of the genes encoding mammalian calnexin and calreticulin. Yeast 9:185–188. https://doi.org/10.1002/yea.320090209

Del Bem LE (2011) The evolutionary history of calreticulin and calnexin genes in green plants. Genetica 139:255–259. https://doi.org/10.1007/s10709-010-9544-y

Duus K, Sandhu N, Jørgensen CS, Hansen PR, Steinø A, Thaysen-Andersen M, Højrup P, Houen G (2009) Interaction of the chaperone calreticulin with proteins and peptides of different structural classes. Protein Pept Lett 16:1414–1423. https://doi.org/10.2174/092986609789353772

Duus K, Larsen N, Tran TA, Güven E, Skov LK, Jespersgaard C, Gajhede M, Houen G (2013) Chemical and thermal unfolding of calreticulin. Protein Pept Lett 20:562–568. https://doi.org/10.2174/0929866511320050009

Eggleton P, Bremer E, Dudek E, Michalak M (2016) Calreticulin, a therapeutic target? Expert Opin Ther Targets 20:1137–1147. https://doi.org/10.1517/14728222.2016.1164695

Ellgaard L, Riek R, Herrmann T, Güntert P, Braun D, Helenius A, Wüthrich K (2001a) NMR structure of the calreticulin P-domain. Proc Natl Acad Sci U S A 98:3133–3138. https://doi.org/10.1073/pnas.051630098

Ellgaard L, Riek R, Braun D, Herrmann T, Helenius A, Wüthrich K (2001b) Three-dimensional structure topology of the calreticulin P-domain based on NMR assignment. FEBS Lett 488:69–73. https://doi.org/10.1016/s0014-5793(00)02382-6

Ellgaard L, Bettendorff P, Braun D, Herrmann T, Fiorito F, Jelesarov I, Güntert P, Helenius A, Wüthrich K (2002) NMR structures of 36 and 73-residue fragments of the calreticulin P-domain. J Mol Biol 322:773–784. https://doi.org/10.1016/s0022-2836(02)00812-4

Gold LI, Eggleton P, Sweetwyne MT, Van Duyn LB, Greives MR, Naylor SM, Michalak M, Murphy-Ullrich JE (2010) Calreticulin: non-endoplasmic reticulum functions in physiology and disease. FASEB J 24:665–683. https://doi.org/10.1096/fj.09-145482

Hahn M, Borisova S, Schrag JD, Tessier DC, Zapun A, Tom R, Kamen AA, Bergeron JJ, Thomas DY, Cygler M (1998) Identification and crystallization of a protease-resistant core of calnexin that retains biological activity. J Struct Biol 123:260–264. https://doi.org/10.1006/jsbi.1998.4032

Halperin L, Jung J, Michalak M (2014) The many functions of the endoplasmic reticulum chaperones and folding enzymes. IUBMB Life 66:318–326. https://doi.org/10.1002/iub.1272

Højrup P, Roepstorff P, Houen G (2001) Human placental calreticulin characterization of domain structure and post-translational modifications. Eur J Biochem 268:2558–2565. https://doi.org/10.1046/j.1432-1327.2001.02138.x

Houen G (2019) COMMENTARY: calreticulin - oncogene, anti-oncogene, or both? Curr Protein Pept Sci 20:111–112. https://doi.org/10.2174/138920372001181031111149

Houen G, Koch C (1994) Human placental calreticulin: purification, characterization and association with other proteins. Acta Chem Scand 48:905–911. https://doi.org/10.3891/acta.chem.scand.48-0905

Ikawa M, Tokuhiro K, Yamaguchi R, Benham AM, Tamura T, Wada I, Satouh Y, Inoue N, Okabe M (2011) Calsperin is a testis-specific chaperone required for sperm fertility. J Biol Chem 286:5639–5646. https://doi.org/10.1074/jbc.M110.140152

Imai M, Araki M, Komatsu N (2017) Somatic mutations of calreticulin in myeloproliferative neoplasms. Int J Hematol 105:743–747. https://doi.org/10.1007/s12185-017-2246-9

Jia XY, He LH, Jing RL, Li RZ (2009) Calreticulin: conserved protein and diverse functions in plants. Physiol Plant 136:127–138. https://doi.org/10.1111/j.1399-3054.2009.1223.x

Jørgensen CS, Trandum C, Larsen N, Ryder LR, Gajhede M, Skov LK, Højrup P, Barkholt V, Houen G (2005) Conformational stability of calreticulin. Protein Pept Lett 12:687–693. https://doi.org/10.2174/0929866054696082

Kepp O, Galluzzi L (eds) (2020) Biology of the endoplasmic reticulum. Int Rev Cell Mol Biol 350:1–325. https://doi.org/10.1016/S1937-6448(20)30020-4

Klampfl T, Gisslinger H, Harutyunyan AS, Nivarthi H, Rumi E, Milosevic JD, Them NC, Berg T, Gisslinger B, Pietra D, Chen D, Vladimer GI, Bagienski K, Milanesi C, Casetti IC, Sant'Antonio E, Ferretti V, Elena C, Schischlik F, Cleary C, Six M, Schalling M, Schönegger A, Bock C, Malcovati L, Pascutto C, Superti-Furga G, Cazzola M, Kralovics R (2013) Somatic mutations of calreticulin in myeloproliferative neoplasms. N Engl J Med 369:2379–2390. https://doi.org/10.1056/NEJMoa1311347

Kozlov G, Pocanschi CL, Rosenauer A, Bastos-Aristizabal S, Gorelik A, Williams DB, Gehring K (2010a) Structural basis of carbohydrate recognition by calreticulin. J Biol Chem 285:38612–38620. https://doi.org/10.1074/jbc.M110.168294

Kozlov G, Bastos-Aristizabal S, Määttänen P, Rosenauer A, Zheng F, Killikelly A, Trempe JF, Thomas DY, Gehring K (2010b) Structural basis of cyclophilin B binding by the calnexin/calreticulin P-domain. J Biol Chem 285:35551–35557. https://doi.org/10.1074/jbc.M110.160101

Kozlov G, Muñoz-Escobar J, Castro K, Gehring K (2017) Mapping the ER interactome: the P domains of calnexin and calreticulin as plurivalent adapters for foldases and chaperones. Structure 25:1415–1422. https://doi.org/10.1016/j.str.2017.07.010

Matsusaki M, Kanemura S, Kinoshita M, Lee YH, Inaba K, Okumura M (2020) The protein disulfide isomerase family: from proteostasis to pathogenesis. Biochim Biophys Acta 1864:129338. https://doi.org/10.1016/j.bbagen.2019.04.003

McCauliffe DP, Yang YS, Wilson J, Sontheimer RD, Capra JD (1992) The 5'-flanking region of the human calreticulin gene shares homology with the human GRP78, GRP94, and protein disulfide isomerase promoters. J Biol Chem 267:2557–2562

Michalak M, Mariani P, Opas M (1998) Calreticulin, a multifunctional Ca^{2+} binding chaperone of the endoplasmic reticulum. Biochem Cell Biol 76:779–785. https://doi.org/10.1139/bcb-76-5-779

Møllegaard KM, Duus K, Træholt SD, Thaysen-Andersen M, Liu Y, Palma AS, Feizi T, Hansen PR, Højrup P, Houen G (2011) The interactions of calreticulin with immunoglobulin G and immunoglobulin Y. Biochim Biophys Acta 1814:889–899. https://doi.org/10.1016/j.bbapap.2011.03.015

Moreau C, Cioci G, Iannello M, Laffly E, Chouquet A, Ferreira A, Thielens NM, Gaboriaud C (2016) Structures of parasite calreticulins provide insights into their flexibility and dual carbohydrate/peptide-binding properties. IUCrJ 3:408–419. https://doi.org/10.1107/S2052252516012847

Nangalia J, Massie CE, Baxter EJ, Nice FL, Gundem G, Wedge DC, Avezov E, Li J, Kollmann K, Kent DG, Aziz A, Godfrey AL, Hinton J, Martincorena I, Van Loo P, Jones AV, Guglielmelli P, Tarpey P, Harding HP, Fitzpatrick JD, Goudie CT, Ortmann CA, Loughran SJ, Raine K, Jones DR, Butler AP, Teague JW, O'Meara S, McLaren S, Bianchi M, Silber Y, Dimitropoulou D, Bloxham D, Mudie L, Maddison M, Robinson B, Keohane C, Maclean C, Hill K, Orchard K,

Tauro S, Du MQ, Greaves M, Bowen D, BJP H, Harrison CN, NCP C, Ron D, Vannucchi AM, Papaemmanuil E, Campbell PJ, Green AR (2013) Somatic CALR mutations in myeloproliferative neoplasms with nonmutated JAK2. N Engl J Med 369:2391–2405. https://doi.org/10.1056/NEJMoa1312542

Nørgaard Toft K, Larsen N, Steen Jørgensen F, Højrup P, Houen G, Vestergaard B (2008) Small angle X-ray scattering study of calreticulin reveals conformational plasticity. Biochim Biophys Acta 1784:1265–1270. https://doi.org/10.1016/j.bbapap.2008.05.005

Ohsako S, Hayashi Y, Bunick D (1994) Molecular cloning and sequencing of calnexin-t. An abundant male germ cell-specific calcium-binding protein of the endoplasmic reticulum. J Biol Chem 269:14140–14148

Parlati F, Dignard D, Bergeron JJ, Thomas DY (1995) The calnexin homologue cnx1+ in Schizosaccharomyces pombe, is an essential gene which can be complemented by its soluble ER domain. EMBO J 14:3064–3072

Pietra D, Rumi E, Ferretti VV, Di Buduo CA, Milanesi C, Cavalloni C, Sant'Antonio E, Abbonante V, Moccia F, Casetti IC, Bellini M, Renna MC, Roncoroni E, Fugazza E, Astori C, Boveri E, Rosti V, Barosi G, Balduini A, Cazzola M (2016) Differential clinical effects of different mutation subtypes in CALR-mutant myeloproliferative neoplasms. Leukemia 30:431–438. https://doi.org/10.1038/leu.2015.277

Pocanschi CL, Kozlov G, Brockmeier U, Brockmeier A, Williams DB, Gehring K (2011) Structural and functional relationships between the lectin and arm domains of calreticulin. J Biol Chem 286:27266–27277. https://doi.org/10.1074/jbc.M111.258467

Pollock S, Kozlov G, Pelletier MF, Trempe JF, Jansen G, Sitnikov D, Bergeron JJ, Gehring K, Ekiel I, Thomas DY (2004) Specific interaction of ERp57 and calnexin determined by NMR spectroscopy and an ER two-hybrid system. EMBO J 23:1020–1029. https://doi.org/10.1038/sj.emboj.7600119

Schcolnik-Cabrera A, Oldak B, Juárez M, Cruz-Rivera M, Flisser A, Mendlovic F (2019) Calreticulin in phagocytosis and cancer: opposite roles in immune response outcomes. Apoptosis 24:245–255. https://doi.org/10.1007/s10495-019-01532-0

Schrag JD, Bergeron JJ, Li Y, Borisova S, Hahn M, Thomas DY, Cygler M (2001) The structure of calnexin, an ER chaperone involved in quality control of protein folding. Mol Cell 8:633–446. https://doi.org/10.1016/s1097-2765(01)00318-5

Tesniere A, Panaretakis T, Kepp O, Apetoh L, Ghiringhelli F, Zitvogel L, Kroemer G (2008) Molecular characteristics of immunogenic cancer cell death. Cell Death Differ 15:3–12. https://doi.org/10.1038/sj.cdd.4402269

Thielmann Y, Weiergräber OH, Mohrlüder J, Willbold D (2009) Structural framework of the GABARAP-calreticulin interface--implications for substrate binding to endoplasmic reticulum chaperones. FEBS J 276:1140–1152. https://doi.org/10.1111/j.1742-4658.2008.06857.x

Thornton JM (1981) Disulphide bridges in globular proteins. J Mol Biol 151:261–287

Tjoelker LW, Seyfried CE, Eddy RL Jr, Byers MG, Shows TB, Calderon J, Schreiber RB, Gray PW (1994) Human, mouse, and rat calnexin cDNA cloning: identification of potential calcium binding motifs and gene localization to human chromosome 5. Biochemistry 33:3229–3236. https://doi.org/10.1021/bi00177a013

Vainchenker W, Plo I, Marty C, Varghese LN, Constantinescu SN (2019) The role of the thrombopoietin receptor MPL in myeloproliferative neoplasms: recent findings and potential therapeutic applications. Expert Rev Hematol 12:437–448. https://doi.org/10.1080/17474086.2019.1617129

Villamil Giraldo AM, Lopez Medus M, Gonzalez Lebrero M, Pagano RS, Labriola CA, Landolfo L, Delfino JM, Parodi AJ, Caramelo JJ (2010) The structure of calreticulin C-terminal domain is modulated by physiological variations of calcium concentration. J Biol Chem 285:4544–4553. https://doi.org/10.1074/jbc.M109.034512

Vincenz-Donnelly L, Hipp MS (2017) The endoplasmic reticulum: a hub of protein quality control in health and disease. Free Radic Biol Med 108:383–393. https://doi.org/10.1016/j.freeradbiomed.2017.03.031

Wada I, Imai S, Kai M, Sakane F, Kanoh H (1995) Chaperone function of calreticulin when expressed in the endoplasmic reticulum as the membrane-anchored and soluble forms. J Biol Chem 270:20298–20304. https://doi.org/10.1074/jbc.270.35.20298

Wasąg P, Grajkowski T, Suwińska A, Lenartowska M, Lenartowski R (2019) Phylogenetic analysis of plant calreticulin homologs. Mol Phylogenet Evol 134:99–110. https://doi.org/10.1016/j.ympev.2019.01.014

Watanabe D, Yamada K, Nishina Y, Tajima Y, Koshimizu U, Nagata A, Nishimune Y (1994) Molecular cloning of a novel Ca^{2+}-binding protein (calmegin) specifically expressed during male meiotic germ cell development. J Biol Chem 269:7744–7749

Wiersma VR, Michalak M, Abdullah TM, Bremer E, Eggleton P (2015) Mechanisms of translocation of ER chaperones to the cell surface and immunomodulatory roles in cancer and autoimmunity. Front Oncol 5:7. https://doi.org/10.3389/fonc.2015.00007

Zamanian M, Veerakumarasivam A, Abdullah S, Rosli R (2013) Calreticulin and cancer. Pathol Oncol Res 19:149–154. https://doi.org/10.1007/s12253-012-9600-2

Chapter 3
The Role of Endoplasmic Reticulum Chaperones in Protein Folding and Quality Control

Benjamin M. Adams, Nathan P. Canniff, Kevin P. Guay, and Daniel N. Hebert

Abstract Molecular chaperones assist the folding of nascent chains in the cell. Chaperones also aid in quality control decisions as persistent chaperone binding can help to sort terminal misfolded proteins for degradation. There are two major molecular chaperone families in the endoplasmic reticulum (ER) that assist proteins in reaching their native structure and evaluating the fidelity of the maturation process. The ER Hsp70 chaperone, BiP, supports adenine nucleotide-regulated binding to non-native proteins that possess exposed hydrophobic regions. In contrast, the carbohydrate-dependent chaperone system involving the membrane protein calnexin and its soluble paralogue calreticulin recognize a specific glycoform of an exposed hydrophilic protein modification for which the composition is controlled by a series of glycosidases and transferases. Here, we compare and contrast the properties, mechanisms of action and functions of these different chaperones systems that work in parallel, as well as together, to assist a large variety of substrates that traverse the eukaryotic secretory pathway.

Keywords Endoplasmic reticulum · Molecular chaperones · Quality control

Abbreviations

AMP	Adenosine 5′-monophosphate
ATP	Adenosine triphosphate
BiP	Binding-immunoglobulin protein

Benjamin M. Adams and Nathan P. Canniff contributed equally.

B. M. Adams · N. P. Canniff · K. P. Guay · D. N. Hebert (✉)
Department of Biochemistry and Molecular Biology, University of Massachusetts, Amherst, MA, USA

Program in Molecular and Cellular Biology, University of Massachusetts, Amherst, Amherst, MA, USA
e-mail: dhebert@biochem.umass.edu

L. B. Agellon, M. Michalak (eds.), *Cellular Biology of the Endoplasmic Reticulum*, Progress in Molecular and Subcellular Biology 59,
https://doi.org/10.1007/978-3-030-67696-4_3

CNX	Calnexin
CRT	Calreticulin
CypB	Cyclophilin B
ER	Endoplasmic reticulum
ERdj	ER-localized DNAJ
NBD	Nucleotide binding domain
PDIA3	Protein disulfide-isomerase A3
PDIA9	Protein disulfide-isomerase A9
SBD	Substrate binding domain
Sep15	Selenoprotein F
TPR	Tetratrico peptide repeat
UGGT1(2)	UDP-glucose: glycoprotein glucosyltransferase I (II)

3.1 Introduction

Anfinsen's seminal studies demonstrated that the primary sequence of the protein determines the three-dimensional structure of the folded protein and that small proteins can fold in isolation in the test tube (Anfinsen 1973). The folding of more complex proteins; however, frequently requires assistance from cellular folding factors. By transiently interacting with the maturing nascent chains molecular chaperones increase the folding efficiency in the suboptimal folding environment of the cell. Prolonged interactions with chaperones can also play a central role in quality control decisions that supports the targeting of irreparable misfolded proteins for degradation.

Hsp70 is a well conserved classical family of chaperones that is ubiquitously expressed throughout the cell and performs diverse roles (Balchin et al. 2020). In the endoplasmic reticulum (ER) where proteins targeted to the eukaryotic secretory pathway mature the Hsp70 family member is BiP (Binding-immunoglobulin Protein or Grp78) (Pobre et al. 2019). In addition to assisting in protein folding, BiP aids in protein translocation, as well as the preparation and targeting of terminal misfolded for degradation.

Proteins that traverse the secretory pathway are frequently modified by N-linked glycans, and a chaperone system has evolved to utilize these glycans. The carbohydrate-dependent chaperone system binds to the N-glycan modification based on their dynamic composition to aid in folding and quality control of the cargo they are attached to (Hebert et al. 2014). Altogether, the Hsp70 and carbohydrate-dependent chaperone networks of the ER assist a broad range of clients that pass through the secretory pathway to help maintain cellular homeostasis and these chaperone networks are the focus of this chapter.

3.2 BiP Binding Cycle

BiP is a member of the classical Hsp70 molecular chaperone family. BiP substrate binding is regulated by adenine nucleotides and its associated co-factors. Because of its diverse roles in maintaining ER homeostasis, it has been referred to as a master regulator of the ER (Pobre et al. 2019). As a luminal chaperone of the ER, it assists in the translocation, folding and quality control of nascent chains as they emerge into the ER. Beyond its roles that involve interacting directly with secretory pathway client proteins, BiP also plays important roles in controlling the activity of ER resident signaling molecules that transcriptionally regulate the composition of the ER depending upon protein homeostasis demands (Hetz et al. 2020). Altogether, BiP is a critical protein homeostasis factor of the ER.

3.2.1 BiP Organization and Structure

BiP is composed of two discrete functional domains that communicate with each other in part through a flexible linker. The N-terminus of BiP forms a nucleotide binding domain (NBD; Fig. 3.1, green domain) followed by a C-terminal substrate binding domain (SBD; Fig. 3.1, purple domain) (Chang et al. 2008). The NBD is characterized by its ability to bind adenine nucleotides, its ATPase activity and its capacity to control the conformation or state of the SBD.

The NBD of BiP is a large globular domain organized into two large lobes with a deep cleft containing the nucleotide binding pocket between them (Pobre et al. 2019). The nucleotide binding pocket contains the adenine nucleotide sensing residue T^{37}, which interacts with the γ-phosphate of the nucleotide, distinguishing ADP and ATP (Wei et al. 1995). Hsp70 proteins occupy two main conformations, an open or docked state (ATP bound), or a closed or undocked state (ADP bound). Transition between the two states is regulated allosterically by the nucleotide bound state of the NBD. The ATPase activity of BiP can be stimulated via interaction with J-domain containing proteins through their defining conserved H-P-D or J-motif (Pobre et al. 2019). The invariant D of the J-motif interacts with the NBD and leads to a destabilization of the domain-domain interface of the NBD, resulting in ATP hydrolysis.

The SBD is comprised of both the substrate binding pocket, located between two loops, as well as the extended alpha-helical lid region to cap or trap the substrate. The substrate binding pocket is located between the two regions of the SBD: SBDα and SBDβ (Yang et al. 2015). The SBDα region is largely alpha-helical and comprises the lid of the Hsp70, while the SBDβ consists of eight beta strands (Fig. 3.1, N-terminal purple domain) (Yang et al. 2015; Chang et al. 2008). In the ATP bound state, the extended alpha-helical SBDα domain remains open, which allows for more promiscuous interactions with substrates, facilitating interactions with a high on- and off-rate. Hydrolysis of the bound ATP leaves the NBD in the

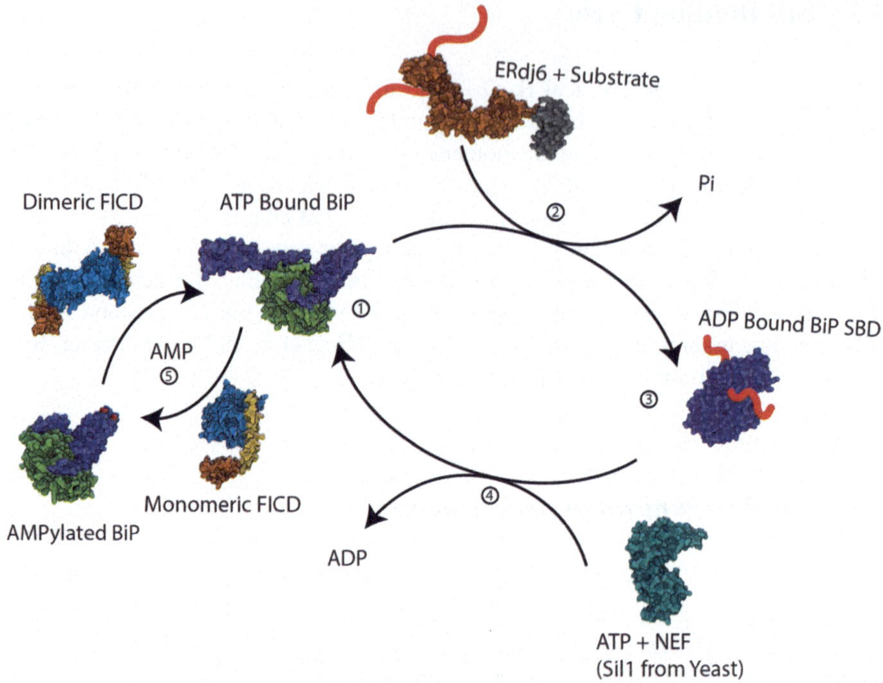

Fig. 3.1 *An overview of the BiP substrate binding cycle.* In the ATP-bound state, BiP is in the open conformation (1), with both the SBD (purple) and NBD (green) domains docked together. Substrate may be bound directly or transferred to BiP via a co-factor. Here, ERdj6 brings substrate (red) bound to the TPR domain (orange) of ERdj6 to pass to BiP (2), then stimulates the ATPase activity of BiP via the J-domain (grey) (PDB: 2Y4K). Upon stimulation, BiP hydrolyzes ATP to ADP and the SBD clamps onto the substrate (3), undocking from the NBD for more dynamic movement (PDB: 5E85). Nucleotide exchange is facilitated by a NEF (Sil1) (4), exchanging ADP for ATP, which causes BiP to unclamp from the substrate and occupy the ATP- bound open conformation once more (1) (PDB: 3QML). BiP sequestering can be achieved through modification, specifically AMPylation via FICD. FICD AMPylates BiP as a monomer and de-AMPylates BiP as a dimer (5) (AMP shown on the BiP structure in red) (PDB: 6I7K, SO4P, 6I7G, respectively)

ADP-bound state, which in turn causes an allosteric conformational shift in the SBD; the SBDα closes over the SBDβ as it undocks from the NBD. The two domains now only remain loosely connected by the extended flexible linker region, allowing for the dynamic independent movement of both domains in contrast to the docked ATP bound state (Yang et al. 2015). The substrate binding pocket resides between the SBDβ and the helical SBDα and the conformational shift in both domains leads to the trapping of the substrate ensuring prolonged interaction with the Hsp70 (Yang et al. 2015; Chang et al. 2008).

3.2.2 BiP Binding Specificity

Chaperones from the classical heat shock families bind exposed hydrophobic domains on substrates to drive proper folding and retard aggregation (Balchin et al. 2020). Using an affinity panning assay, BiP was found to preferentially bind hydrophobic domains containing alternating aromatic and hydrophobic amino acids (Blond-Elguindi et al. 1993; Rüdiger et al. 1997). These stretches of 7–9 amino acids tend to be buried in the mature protein fold and are indicative of an immature, unfolded protein or exposed non-native domains. BiP binding sites have been statistically predicted to be located every ~36 amino acids (Rüdiger et al. 1997).

BiP binds additional sites dispersed throughout the protein rather than just rare sites with higher aggregation potential, indicating a strong pro-folding function (Behnke et al. 2016). These binding motifs within nascent polypeptides often times are seven amino acids long, however BiP can also bind larger epitopes, sometimes comprising tertiary structural components (Behnke et al. 2016). These larger structures can interact directly with the lid domain of the SBD of BiP suggestive of a larger pool of substrates.

3.2.3 Substrate Delivery to BiP

BiP relies upon its cofactors to facilitate its ATPase cycle to transition it to its high affinity substrate binding state. The ER-localized DNAJ cofactors (ERdjs) facilitate the hydrolysis of ATP through the stimulation of BiP via an interaction with their J-domains; as such, all ERdj family members are marked by both ER localization and a luminal J-domain (Pobre et al. 2019). ERdj proteins can be broadly understood to interact with BiP primarily via their J-domain, but the majority harbor additional roles. The ERdjs can be split into two general families: (1) translocation regulators, or (2) substrate binders along with their J-domain functionality. The ERdj proteins which work to facilitate translocation are ERdj1 and ERdj2, while the other ERdjs (with the exception of ERdj7 for which exists no known function) bind directly to substrates for quality control. Both subfamilies of ERdj proteins facilitate the ATPase and substrate binding cycle, allowing for BiP to efficiently function.

The only mammalian homologue of *E. coli* DNAJ localized to the ER is ERdj3 as it contains all the domains found in bacterial DNAJ. It is also highly conserved with other soluble DNAJ homologues such as Ydj1 from *Saccharomyces cerevisiae*. ERdj3 has been found to form tetramers instead of the more common dimeric assemblies associated with the other Hsp40 co-chaperones (Chen et al. 2017). These tetramers appear to be created via the dimerization of dimers. By binding directly to an unfolded protein substrate, ERdj3 retards aggregation of the substrate until it is transferred to BiP (Shen and Hendershot 2005). The stimulation of the ATPase activity of BiP is necessary for ERdj3 to release its bound substrate (Jin et al.

2008). ERdj3 interacts with the lid of the SBD to facilitate passage of the substrate to BiP (Marcinowski et al. 2011).

ERdj4 is a soluble ERdj that contains an N-terminal J-domain followed by a G-F rich flexible linker region and a C-terminal SBD. The expression level of ERdj4 is low under normal conditions, but is upregulated upon ER stress (Shen and Hendershot 2005). It is hypothesized that ERdj4 functions primarily in a pro-ERAD process with emphasis on binding substrates to facilitate their selection, preparation or entrance to the degradation process (Pobre et al. 2019). ERdj4 also appears to help recruit BiP to IRE1 thus functioning to quench the activated dimeric state of IRE1 to return it to an inactive monomeric state (Amin-Wetzel et al. 2017). Thus, ERdj4 may be an integral part of the unfolded protein response (UPR) transduction signaling pathway, a function further supported by the relative lack of expression under normal growth conditions.

ERdj5 and ERdj6 also directly bind substrate. ERdj5 is a unique ER localized J-protein as it is also part of the protein disulfide isomerase (PDI) family of co-chaperones. It contains an ER luminal J-domain as well as six thioredoxin domains (Cunnea et al. 2003). The crystal structure of ERdj5 indicates that the J-domain and the four C-terminal thioredoxin domains are oriented in the same plane (Hagiwara et al. 2011). These domains may work in concert by allowing ERdj5 to act on substrates bound by BiP by acting as a reductant to reduce non-native substrates to prepare them for dislocation from the ER for proteasomal clearance in the cytosol (Ushioda et al. 2008). ERdj6, also known as DNAJC3/P58^{IPK}, is a soluble ERdj protein that contains an N-terminal ER targeting sequence followed by nine tetratricopeptide repeat domains (TPRs) and a C-terminal J-domain (Fig. 3.1 orange, grey domains respectively) (Pobre et al. 2019; Rutkowski et al. 2007). Under normal growth conditions, ERdj6 resides in the ER and binds to both unfolded protein substrates and BiP (Rutkowski et al. 2007; Petrova et al. 2008). In contrast, under stress, ERdj6 translocation into the ER is less efficient and consequently ERdj6 can accumulate in the cytosol (Rutkowski et al. 2007). The TPR domains in ERdj6 are organized into three subdomains of three TPRs each, where subdomain 2 binds to PKR and subdomain 1 is the primary substrate binding region (Melville et al. 1999; Svärd et al. 2011). The concerted effort between the TPR subdomain 1 and J-domain facilitates the exchange of bound unfolded substrate to BiP (Svärd et al. 2011). ERdj proteins play a critical role in the regulation of BiP, contributing to both the delivery of new substrate as well as facilitating the ATPase cycle of BiP.

3.2.4 Substrate Release from BiP

Release from the BiP binding cycle relies on the exchange of ADP for ATP. This supports the raising of the SBD alpha-helical lid and freeing of the trapped substrate. The co-chaperones that assist this process are called nucleotide exchange factors (NEFs). There are two ER resident NEFs that act as cofactors for BiP: BiP-associated protein (Bap), and glucose-regulated protein 170 (Grp170). Both

BiP cofactors were originally found in yeast. Bap is the human homologue to *Saccharomyces cerevisiae* Sil1 (Behnke et al. 2015). Similarly, luminal Hsp 70 (Lhs1) is the homologue of human GRP170 (Craven et al. 1996).

Bap shares tissue specific expression profiles and strongly co-localizes with BiP (Chung et al. 2002). Consequently, Bap was found to preferentially interact with the ADP-bound state of BiP and facilitate the exchange of ADP for ATP (Rosam et al. 2018). The interaction of the Bap C-terminal domain is sufficient for its weak NEF activity, however full length Bap is necessary for its complete nucleotide exchange function (Rosam et al. 2018). The C-terminus of Sil1 interacts with BiP, resulting in the opening its nucleotide binding pocket to facilitate nucleotide exchange (Yan et al. 2011). Disease-associated variants of Bap include missense mutations in the N- and C-termini impairing the interaction with BiP underscoring its role in ER homeostasis (Yan et al. 2011).

Grp170 stimulates the exchange of ADP for ATP on BiP. It shares a similar structure to that of the Hsp70 family as it consists of an N-terminal NBD and a C-terminal SBD (Andréasson et al. 2010). The NBD domain of Grp170 interacts directly with the NBD of BiP, opening the nucleotide binding pocket to facilitate nucleotide exchange (Andréasson et al. 2010). The SBD of Grp170 has a C-terminal extension as well as a small unstructured loop in the β-rich region (Andréasson et al. 2010). Both of these additional domains modulate the binding of Grp170, though the exact purpose of substrate binding by Grp170 remains elusive (Andréasson et al. 2010; Behnke and Hendershot 2014). Recent work has proposed that Grp170 acts as an aggregation deterrent, binding to less common, more aggregation prone epitopes rich in aromatic residues as compared to that of BiP (Behnke et al. 2016).

Terminal release from the BiP binding cycle occurs when the protein has either achieved its native mature fold and no longer presents interaction motifs, or the protein is deemed terminally misfolded. For irreparable substrates, exit from the BiP binding cycle is facilitated by the delivery of the misfolded protein to downstream ER-associated degradation (ERAD) machinery (Ushioda et al. 2013; Pobre et al. 2019). When unstressed, the ER has two pathways for ERAD entrance, one through persistent BiP association of non-glycosylated proteins, and the other through the calnexin/calreticulin pathway for glycosylated proteins (Lamriben et al. 2016). However, the two distinct pathways become interchangeable when the cell is under stress, leading to BiP facilitated binding and subsequent degradation of glycosylated proteins as well (Ushioda et al. 2013). ERAD machinery retro-translocates the BiP-associated cargo out of the ER where it is degraded by the ubiquitin-proteasome system. This release can be facilitated by J-domain proteins including ERdj5 contributing to their diverse roles in regulating the functions of BiP in the ER lumen.

3.2.5 *BiP Modification*

Under non-stressed growth conditions, BiP is post-translationally modified to sequester an inactive store of the protein for immediate deployment and activation upon signaling. The AMPylation of BiP by FICD (also known as HYPE) maintains an inactivated pool of BiP under normal conditions (Preissler et al. 2017b; Perera et al. 2019). AMPylation on T^{518} inactivates BiP by biasing it towards a conformation that mimics the ATP bound state (Preissler et al. 2017b). When client population is low, BiP can also be modulated by forming a multimeric state, which further contributes to creating an inactive pool of BiP. If the client load drastically increases, FICD can also facilitate the de-AMPylation of BiP and rapidly revert the chaperone to its functional state on a timescale faster than transcriptional regulation (Preissler et al. 2015, 2017a; Perera et al. 2019). An oligomeric switch for FICD has been implicated as the important regulatory mechanism for promotion of FICD AMPylation or de-AMPylation activity, as dimeric FICD functions to de-AMPylate BiP, whereas monomeric FICD acts to AMPylate BiP (Perera et al. 2019). FICD-mediated AMPylation is not limited to BiP as a global proteomics analysis identified 25 novel AMPylated substrates, suggesting that this regulatory method may be employed more widely (Broncel et al. 2016).

BiP is the canonical Hsp70 chaperone of the ER that directly contacts non-native substrates through protein-protein interactions. These substrate protein sequences are often aggregation-prone as BiP acts to bind exposed hydrophobic sequences. This can allow the rest of the protein to fold correctly or if terminally misfolded, the substrate may be delivered to downstream ERAD machinery for eventual turnover.

3.3 Calnexin and Calreticulin Binding Cycle

The majority of ER-targeted proteins are modified with at least one N-linked glycan, a carbohydrate group which plays multiple roles in ER protein maturation, quality control and trafficking. N-linked glycans are bulky and hydrophilic, significantly altering the interaction of a modified substrate with the surrounding environment, promoting solubility, productive folding, altering protein function, and acting as a reporter of glycoprotein folded status (Hebert et al. 2014). The role of the N-linked glycan in acting as a reporter of glycoprotein folding and age is key to glycoprotein quality control and proteostasis in the ER. N-linked glycans are dynamically modified during the folding and maturation pathway of the glycoprotein, and the glycoform promotes interaction with carbohydrate-binding proteins in the ER, altering the folding, trafficking, or degradation pathway of a glycoprotein (Adams et al. 2019b). At the center of glycoprotein quality control are the lectin chaperones calnexin and calreticulin and their associated proteins that act to promote proper glycoprotein folding through pro-folding interactions and the retention of non-native glycoproteins for repair or degradation targeting.

3.3.1 Calnexin/Calreticulin Organization and Structure

Calnexin and calreticulin share about 39% identity and play similar, though not identical, roles in glycoprotein quality control. Both are comprised of a signal sequence for ER targeting, an N-terminal lectin domain, a P-rich domain (P-domain), and a C-terminal domain. Calnexin possesses a transmembrane domain while calreticulin is soluble ER luminal protein.

The N-terminal cleavable signal sequence of both calnexin and calreticulin directs delivery to the ER and the type-I topology for calnexin (Ou et al. 1995). The lectin domains of both proteins adopt a globular fold composed of a beta-sandwich with two anti-parallel beta sheets, which comprise a single carbohydrate binding site. Crystal structures of both lectin domains demonstrate a binding site for calcium, potentially promoting stability and lectin activity (Schrag et al. 2001; Kozlov et al. 2010). The luminal domains of calnexin and calreticulin also possess an ATP binding site which, upon phosphorylation, may lead to destabilization and decreased lectin activity (Ou et al. 1995; Wijeyesakere et al. 2015).

Extending from the lectin domain is the P-domain. The P-domain is comprised of sequential Pro-rich motifs, with calnexin containing four repeats of motif 1 (17 amino acids) followed by four repeats of motif 2 (14 amino acids), and calreticulin containing three repeats each of motif 1 and 2. Between the two sets of motifs is a hairpin turn, leading to the P-domain adopting an extended, hook-like fold (Ellgaard et al. 2001; Schrag et al. 2001). Multiple folding factors interact with calnexin and calreticulin in the region of the hairpin turn, which will be discussed in more detail later.

The C-terminal region is the most dissimilar region between calnexin and calreticulin. Calnexin contains a C-terminal region composed of a transmembrane domain and a cytoplasmic tail, which contains sites for phosphorylation and palmitoylation, as well as a di-R sequence that promotes ER retention (Rajagopalan et al. 1994). Calreticulin has a C-terminal tail containing acidic residues, which provide low affinity calcium binding sites, therefore supporting the role of calreticulin as a more robust ER calcium buffer as compared to calnexin (Lamriben et al. 2016). ER retention of the soluble protein calreticulin is supported by its N-terminal K-D-E-L motif.

3.3.2 Calnexin/Calreticulin Binding Specificity

The lectin domains of calnexin and calreticulin support binding to glycoproteins that possess monoglucosylated glycans presented on substrates (Hebert et al. 2014) (Fig. 3.2). Crystallographic evidence has demonstrated that the lectin domain of calreticulin interacts with the A-branch tetrasaccharide Glcα1-3Manα1-2Manα1-2Man (Kozlov et al. 2010), while isothermal calorimetry (ITC) studies have shown that calreticulin has low micromolar affinity for the A-branch tetrasaccharide

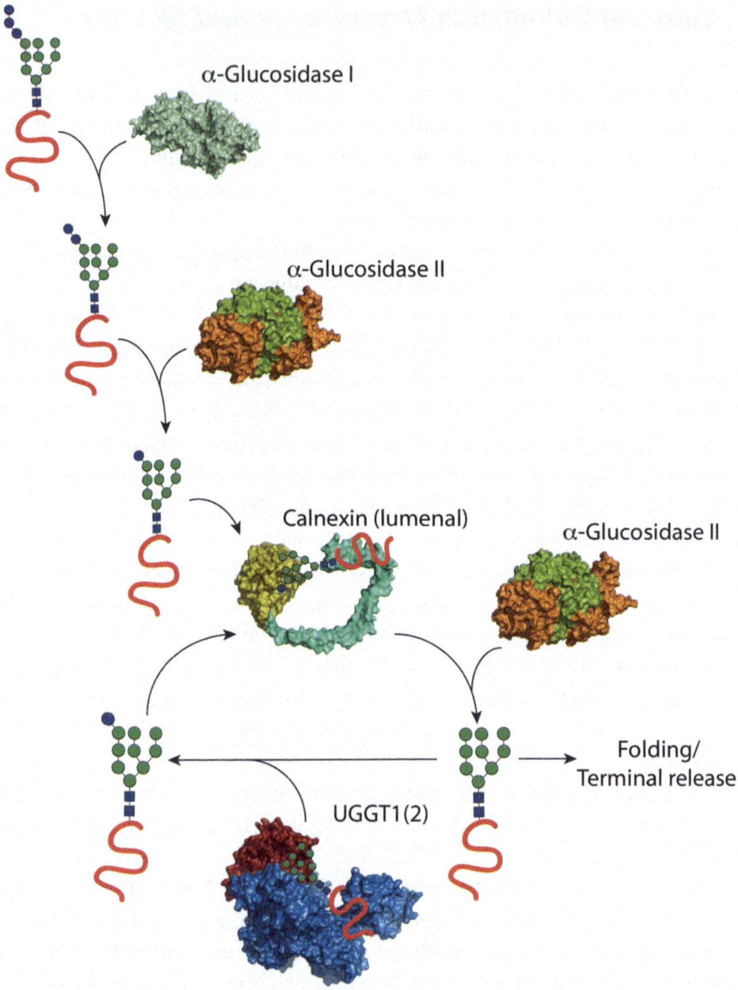

Fig. 3.2 *The calnexin and calreticulin cycle.* Nascent glycoproteins possess N-linked glycans with a $Glc_3Man_9GlcNAc_2$ structure. The terminal glucoses are then sequentially processed by α-glucosidase I (PDB 4J5T) and α-glucosidase II (PDB 5F0E; Catalytic GRH31 domain, green). In the monoglucosylated state, the substrate is then engaged by calnexin or calreticulin (for clarity, only the luminal domain of calnexin is shown) (PDB 1JHN; Lectin domain, yellow; P-domain, teal). Upon release from calnexin or calreticulin, the terminal glucose may then be trimmed by α-glucosidase II. Folding to a native state promotes release from the cycle, while non-native substrates are reglucosylated by UGGT1/2 (PDB 5MZO; Folding sensor domain, blue; Catalytic domain, red). Reglucosylation promotes re-engagement by calnexin or calreticulin

Glc_1Man_3 with decreasing affinity for Glc_1Man_2 and Glc_1Man_1 (Kapoor et al. 2003). The B- and C-branch mannoses do not appear to be required for interaction with calnexin and calreticulin, as removal of these residues does not abrogate interaction with calnexin. However, the B/C branch point mannose was found to be required for glycan interaction with calnexin and calreticulin, despite this mannose not being present in ITC experiments demonstrating calnexin and calreticulin lectin activity (Spiro et al. 1996; Vassilakos et al. 1998). More work is therefore required to determine the precise mannose composition required for interaction with calnexin and calreticulin, though it is clear a monoglucosylated glycan is necessary.

In addition to lectin-based interactions, calnexin and calreticulin display protein-protein based interactions as well, though to a lesser degree. Cell-based experiments have found that calnexin can interact with some substrates lacking monoglucosylated glycans (Danilcyk and Williams 2001). Disruption of protein-protein interaction sites mapped to the lectin domain led to a small decrease in chaperone function, while impairment of lectin activity produced a severe impact on this chaperone function demonstrating that the carbohydrate binding site is the predominant site responsible for substrate binding (Lum et al. 2016).

3.3.3 Delivery to the Calnexin/Calreticulin Cycle

Calnexin and calreticulin can engage substrates both co-translationally as well as post-translationally in mammalian cells. Upon translocation into the ER through the Sec61 channel, glycoproteins are modified with preassembled N-linked glycans as $Glc_3Man_9GlcNAc_2$ transferred *en bloc* from a dolichol phosphate in the ER membrane to the Asn site in the acceptor sequence N-X-S/T/C (where X is not a P) via the oligosaccharyltransferase (OST) complex. In mammals, there are two forms of the multimeric OST complex: one that contains the catalytic domains STT3A that transfers glycans co-translationally assisted by its proximity to the Sec61 translocon, and a second that contains STT3B that transfers N-glycans post-translationally (Cherepanova et al. 2019). Interestingly, *Saccharomyces cerevisia*e only possess the STT3B containing post-translational modification complex (Shrimal et al. 2019). The mode of glycan addition significantly impacts the temporal nature of glycan addition and initial lectin chaperone binding.

After glycosylation, the terminal glucose is rapidly trimmed off the glycan by glucosidase I, generating a glycan with two glucoses. Glucosidase I is a type I membrane protein requiring this initial trimming event to take place proximal to the membrane. The next step in the calnexin/calreticulin cycle is the trimming of the second glucose by glucosidase II, generating a monoglucosylated glycan. This glycoform can then be recognized by calnexin or calreticulin. Glucosidase II is a heterodimeric complex comprised of an α-domain that contains the catalytic domain and is highly expressed, and a β-domain that contains the ER retention sequence. Glucosidase IIβ also contributes to substrate recognition as it contains a mannose 6-phosphate receptor homologous (MRH) domain that supports binding to terminal mannose residues. Therefore, it helps to recruit the glucosidase II dimer to glycan rich regions for rapid trimming and generation of the monoglucosylated substrate (Deprez et al. 2005).

As glycosylation can occur co-translationally in mammals, co-translational interaction with the calnexin/calreticulin cycle is therefore possible (Daniels et al. 2003). This however is dependent on the location of the glycan on the substrate, as C-terminal glycosylation would likely not support co-translational interaction with calnexin and calreticulin because translation would be terminated shortly after glycosylation occurs. As calnexin and calreticulin has been demonstrated to interact co-translationally with substrates, any delay in glycoprotein entry into this cycle due to the requirement of glucose trimming by glucosidases I/II is brief. Since glucosidases I/II do not appear to actively select substrates based on folding status, this initial entry step into the calnexin/calreticulin cycle appears to be promiscuous.

3.3.4 Calnexin/Calreticulin Substrate Release and Rebinding

As calnexin and calreticulin require a monoglucosylated glycan for substrate recognition, glucose trimming of the final glucose by glucosidase II represents a release of the substrate from the calnexin/calreticulin cycle. It is notable that unlike BiP that has different cofactors that support delivery and release, glucosidase II acts to both deliver and release substrates from the calnexin/calreticulin cycle (Table 3.1).

After trimming of the final glucose, the glycoprotein has two general fates, depending on the folded status of the protein. If the protein has reached its native fold and oligomeric state, it can exit the ER to traffic to its proper destination (Hebert et al. 2014). Alternatively, non-native glycoproteins can be recognized by the folding sensor uridine diphosphate-glucose: glycoprotein glucosyltransferase I

Table 3.1 Comparison of BiP and carbohydrate-binding chaperone networks

	BiP	Calnexin/calreticulin
Binding specificity	Extended hydrophobic domains	$Glc_1Man_xGlcNAc_2$
Binding regulation		
Chaperone	ATP/ADP	
Substrate		Carbohydrate composition
Delivery		
Chaperone	ERdj1-7	
Substrate	ERdj3-6	Glucosidase II
Rebinding		UGGT1/2
Release		
Chaperone	Grp170 Bap/Sil1	
Substrate		Glucosidase II
Additional associated factors	EDEM1 (ERdj5)	PDIA3/PDIA9/cyclophilin B Sep15 (UGGT1/2)
Modifications	AMPylation	Phosphorylation (calnexin) Palmitoylation (calnexin)

(UGGT1) and be reglucosylated, placing the glycan back in a monoglucosylated state. Reglucosylation supports glycoprotein rebinding to calnexin and calreticulin, promoting further rounds of intervention and retention of non-native glycoproteins.

UGGT1 is suggested to preferentially recognize near-native substrates over terminally misfolded substrates, though how this fine-tuned selectivity is achieved is poorly understood (Caramelo et al. 2004; Adams et al. 2019a). UGGT1 is the central gatekeeper of the calnexin/calreticulin cycle, recognizing non-native glycoproteins that should be retained in the ER as these may be non-functional or aggregation prone, and therefore potentially deleterious to cellular health. Therefore, the initial delivery (glucosidases I and II) and re-delivery (UGGT1) to the calnexin/calreticulin cycle are roles played by separate factors, with glycoprotein folding status playing the determinant role in persistent lectin chaperone binding.

UGGT1 has a paralogue, UGGT2, with 55% sequence identity, which has been shown to be active *in vitro*, though no cellular substrates have been identified (Takeda et al. 2014). Both UGGT1 and UGGT2 are soluble, ER-resident proteins with similar domain organization, consisting of an N-terminal folding sensor domain and a C-terminal glucosyltransferase domain. Recent crystal structures of *Thermomyces dupontii* and *Chaetomium thermophilum* UGGT1 have demonstrated that the N-terminal folding sensor is composed of four thioredoxin domains that lack catalytically active C-X-X-C motifs (Satoh et al. 2017; Roversi et al. 2017). These thioredoxin domains likely play a role in substrate recognition. The thioredoxin domains are orientated into a curved and flexible structure with a large central cavity possessing hydrophobic patches, suggesting a binding region for non-native substrates. The folding-sensor domain is more dissimilar between UGGT1 and UGGT2, displaying 49% identity, as compared to 84% identity between the C-terminal glucosyltransferase domains. This suggests that substrates may not completely overlap, but UGGT1 is expected to carry the bulk of the quality control load as it is expressed at a 25-fold higher level in HeLa cells than UGGT2 (Itzhak et al. 2016). Further studies are needed to delineate substrate specificities of UGGT1 and 2, and their differential roles in glycoprotein folding and quality control.

3.3.5 Terminal Release of Substrates from Calnexin/Calreticulin Binding Cycle

Persistent calnexin and calreticulin interaction may occur as long as non-native substrates are recognized and reglucosylated by UGGT. However, to clear the ER of terminally misfolded glycoproteins, there are a number of possible mechanisms in place for extracting irreparable glycoproteins from the carbohydrate-dependent chaperone cycle. These mechanisms use two general concepts: (1) the recognition or pass-off to downstream machinery for ERAD targeting; and (2) the alteration of the substrate N-glycan so it is no longer a suitable structure for reglucosylation.

These mechanisms are not mutually exclusive as they have a number of overlapping components.

One early model for extraction is based on observations that misfolded substrates can interact with ER degradation-enhancing alpha-mannosidase-like protein 1 (EDEM1) (Oda et al. 2003; Molinari et al. 2003). This model is supported by the observation that a misfolded disease associated variant of alpha-1-antitrypsin was passed directly from calnexin to EDEM1, possibly assisted by the association of the transmembrane domains of calnexin and EDEM1 (Oda et al. 2003). Studies using a misfolded variant of BACE (BACE457) found that EDEM1 expression accelerated its deglucosylation and turnover (Molinari et al. 2003). These studies demonstrated a role for EDEM1 in extracting aberrant glycoproteins from the lectin chaperone binding cycle for subsequent targeting for dislocation to the cytosol and eventual proteasomal degradation.

The second general mechanism for removal of substrates from the calnexin/calreticulin binding cycle involves rendering the substrate unsuitable for reglucosylation. Human cells contain seven possible exo-mannosidases (MAN1a1/1a2/1b1/1c1 and EDEM1-3) and a single endomannosidase (MANEA) that might be responsible for trimming substrate glycans to make them poor substrates for UGGT. Removal of the terminal mannose on the A-branch could diminish reglucosylation, and ultimately reentry into the cycle, by creating a poorer substrate for UGGT (Avezov et al. 2008). Trimming of mannoses from the B- and C-branches also appears to diminish recognition by the UGGTs (Sousa et al. 1992). Alternatively, the trimming of α1,2-linked mannose residues exposes α1,6-linked mannose residues on the C-branch, which have been shown to be recognized by downstream ERAD receptors Os-9 and XTP3B to support accelerated degradation of aberrant clients (Roth and Zuber 2017). Future studies will be needed to delineate the roles of the mannosidases in extraction of terminally misfolded proteins from the calnexin/calreticulin cycle. As some of these mannosidases are localized to ER subcompartments, vesicles or the Golgi, the recycling or trafficking of substrate or mannosidases might play an important role in substrate targeting for degradation.

3.3.6 Calnexin/Calreticulin-Associated Factors

Calnexin and calreticulin associate with several co-chaperones in order to promote productive substrate folding. Interaction with co-chaperones is mediated via the P-domain (Kozlov et al. 2017). The P-domains of calnexin and calreticulin are structurally similar, though calnexin possesses four Pro-rich motifs while calreticulin possesses three, shortening its reach. The P-domain of calnexin and calreticulin functions as a binding site for three known co-chaperones: the protein disulfide isomerase (PDI) family members PDIA3 (ERp57) and PDIA9 (ERp29), and the peptidyl cis trans protein isomerase (PPI) family member cyclophilin B (CypB). Therefore, these lectin chaperones act as a platform to recruit foldases or protein folding catalysis to non-native nascent chains (Kozlov and Gehring 2020).

PDIA3 is an ER resident member of the protein disulfide isomerase family, which promotes native disulfide bond formation of substrates. It consists of four thioredoxin domains, a, b, a′, and b′, with a and a′ possessing catalytically active C-X-X-C motifs (Frickel et al. 2004). The PDIA3 binding site on calnexin is located on the hairpin turn of the P-domain (Frickel et al. 2002), interacting with a positively charged patch on the b′ domain of PDIA3 (Kozlov et al. 2006). A similar binding orientation has been shown by cryo-EM in the context of the peptide loading complex, a multi-protein complex including calreticulin and PDIA3 in which MHC class I is loaded with peptides for cell surface antigen presentation (Blees et al. 2017). Notably, PDIA3 has been shown to display increased oxidoreductase activity when associated with calnexin or calreticulin (Zapun et al. 1998). This increased activity can likely be attributed to the increased local concentration of PDIA3 and substrates when both are engaged by calnexin or calreticulin, highlighting the role of calnexin and calreticulin as both chaperones and scaffolds wherein substrates can be placed in close proximity to multiple chaperone co-factors.

PDIA9, though a member of the PDI family, does not possess a full catalytically active thioredoxin site and functions as a dimer, with the thioredoxin-like domain mediating dimerization (Liepinsh et al. 2001). Its active site contains a single C residue that is required for its ability to assist trafficking of substrates, such as insulin, but is not required for client interactions or disulfide formation or isomerization (Viviano et al. 2020). Calnexin and calreticulin interact with PDIA9 at the tip of the P-domain, in a similar region to the PDIA3 interaction domain, via a hydrophobic pocket in the C-terminal D-domain of PDIA9 (Kozlov et al. 2017).

CypB is a predominantly ER resident member of the PPI family, which has been suggested to have a pro-folding function especially important for collagen folding (Terajima et al. 2017). CypB also binds to the tip of the P-domain of calnexin and calreticulin (Kozlov et al. 2017). It is therefore notable that all known associated factors of calnexin and calreticulin bind to a similar region of the P-domain, which acts as a mechanism to allow a glycoprotein substrate to be engaged by multiple folding factors in an efficient manner.

In addition to the co-chaperones of the calnexin/calreticulin cycle, UGGT1/2 engages the co-factor Selenoprotein F (Sep15) within the ER. Sep15 is a 15-kDa protein containing a selenocysteine (Sec, U) encoded by a UGA codon (Kasaikina et al. 2011). The mRNA for all known selenoproteins contain a unique stem-loop structure, known as a Sec insertion sequence (SECIS), that prevents the termination of translation and instead inserts a Sec at the UGA position (Kryukov et al. 2003). Immediately C-terminal to the cleaved signal sequence is a C-rich domain with six well conserved C. Furthermore, there is a highly conserved C-G-U-K motif that is the predicted redox-active site for Sep15 (Korotkov et al. 2001). Interestingly, Sep15 does not contain an ER retention signal at its C-terminus, suggesting it is retained by some other means. Immunoprecipitation of overexpressed mouse Sep15, in the presence of overexpressed UGGT1/2, has shown that Sep15 co-precipitates and forms a 1:1 complex with both UGGT1 and UGGT2. Mutagenesis of the N-terminal C-rich domain revealed that all six C are essential for its interaction with UGGT1/2. Notably, mutating U to C does not ablate the interaction between Sep15 and UGGT1

(Labunskyy et al. 2005). Of note is the entirety of Sep15 is found in complex with UGGT1/2 and does not rely on intermolecular disulfide bonds to maintain the interaction (Korotkov et al. 2001). It is therefore likely that Sep15 is retained within the ER through its interaction with UGGT1 and UGGT2. Sep15 has been shown to increase the ability of UGGT1 and UGGT2 to reglucosylate a synthetic target in vitro (Takeda et al. 2014). Further investigations are needed to understand how Sep15 contributes to the calnexin/calreticulin cycle in a cellular context.

3.3.7 *Calnexin/Calreticulin Modification*

Three major modifications of calnexin have been described that target its cytosolically localized C-terminal tail: palmitoylation, phosphorylation, and SUMOylation. The sites of palmitoylation have been identified as two C residues in the C-terminal tail of calnexin. Its palmitoylation has been suggested to either promote interactions between calnexin and the ER translocon complex or to enrich calnexin at ER mitochondrial-associated-membrane sites (Lynes et al. 2012; Lakkaraju et al. 2012). Phosphorylation of calnexin has been shown to occur at three sites in the C-terminal tail, mediated primarily by two cytosolic kinases, CK2 and ERK1 MAPK (Chevet et al. 2010). Phosphorylation at S^{563} recruits calnexin to ribosome engaged translocons as well as promotes prolonged interaction with the model secretory pathway protein alpha 1-antitrypsin under misfolding conditions (Cameron et al. 2009). Phosphorylation of $S^{534/544}$ appears to promote the redistribution of peripheral ER calnexin to a perinuclear localization, potentially through disrupted interaction with the cytosolic protein PACS-2 (Myhill et al. 2008). UBC9-dependent SUMOylation of calnexin at L^{506} in the cytosolic C-terminal tail has been demonstrated to support interaction with protein tyrosine phosphatase 1B (PTP1B) (Lee et al. 2015). While PTP1B is primarily targeted to the ER via a C-terminal transmembrane region, the SUMO-dependent interaction between calnexin and PTP1B promotes the localization of PTP1B to the cytosolic face of the ER membrane while potentially altering the pool of substrates which PTP1B can engage. Modifications of the calnexin C-terminal tail therefore act to alter the localization of calnexin and binding partners and consequently may modify the functional role of calnexin, but the mechanism of these processes requires further characterization.

3.4 Concluding Remarks

The BiP and calnexin/calreticulin chaperone pathways aid in the folding and quality control of a wide range of cargo that traverse the eukaryotic secretory pathway using diverse mechanisms for the selection and the regulation of binding to cargo. The Hsp70 ER family member BiP binds to exposed hydrophobic motifs that are normally sequestered in the core of a properly folded protein. BiP binding is regulated by its adenine nucleotide binding state. In contrast, calnexin and calreticulin binding is dictated by covalent changes to the substrate itself or by controlling the N-glycan composition of cargo protein. These distinct pathways provide a broad range of assistance for the thousands of different client proteins of the secretory pathway.

These two chaperone systems use opposing methods for substrate recognition and binding. BiP, like other Hsp70 family members, binds to short stretches of exposed hydrophobic residues that should be hidden in the core of the folded proteins. In converse, the lectin chaperones bind to large hydrophilic modifications on the client that generally remain exposed to the aqueous environment on the natively folded protein. How can two such dissimilar mechanisms for binding be used to select proteins that require assistance for folding?

The exposure of hydrophobic residues is a ubiquitous hallmark of a misfolded protein and it is directly utilized by the Hsp70 system regardless of the location in the cell to mediate substrate recognition and binding. While binding to calnexin and calreticulin is through a monoglucosylated protein, there are two different mechanisms to support binding. Binding commences after the recently transferred N-glycan is rapidly trimmed sequentially by glucosidases I and II to generate the monoglucosylated side chain. This initial binding event, therefore, is controlled by the timing of glycan addition by the OST and its subsequent trimming shortly thereafter. As glycans can be added co-translationally in mammals, this allows for early intervention of the lectin chaperones in the folding process. If the trimming of the second and third glucose that are both released by glucosidase II is rapid, it is likely that lectin chaperone binding to this glycan would be bypassed. Rebinding or additional rounds of lectin chaperone binding are mediated by UGGT1, and likely UGGT2. These proteins use their folding sensor domain comprised of four thioredoxin domains to recognize non-native proteins likely through exposed hydrophobic segments. Therefore, while the mechanisms of binding differ greatly for both chaperone systems, both systems share that persistent chaperone binding is at least in part based on exposed hydrophobic motifs. BiP scans for these motifs directly to mediate binding, while UGGT does the scanning for the lectin chaperones and then modifies the glycan signal accordingly to direct chaperone binding.

The ER is a large organelle that can spread throughout the cytosol of many cells and perform diverse functions. The seven J-proteins of the ER help to direct BiP to various location in the ER, as well as to a variety of functions. For example, ERdj2 (yeast Sec63) directs BiP to the translocon that resides in the ER membrane to assist in protein translocation and position it for early engagement and assistance for nascent chains as they emerge into the ER lumen (Pobre et al. 2019). ERdj5 helps with preparation and potential targeting of misfolded proteins for ERAD. In contrast, the lectin chaperones diversify their localization by having two paralogues, one membrane associated and the other soluble bypassing the need to recruit a chaperone to the membrane. Post-translational modification of the cytoplasmic C-terminal tail of calnexin appears to be involved in its localization. While BiP is recruited and controlled by its J-domain proteins to dictate its function, the lectin chaperones in some respects act as a platform to recruit diverse folding factors to the non-native substrate for modification of C residues (PDIA3 and PDIA9) or P orientation (CypB) (Kozlov and Gehring 2020).

Both chaperone systems appear to act as holdases or folding factors that protect and delay the folding of the region of the protein they are bound to thereby controlling the folding trajectory of the nascent chain (Daniels et al. 2003; Wang et al. 2008; Mayer and Gierasch 2019). This results in a higher yield of properly folded protein or more efficient folding by which the rate of folding can either be accelerated or delayed. In addition, persistent binding of both systems to non-native clients supports ER retention and subsequent targeting for degradation. The binding mechanisms of BiP and calnexin/calreticulin vary; however, their functions in protein folding and quality control have a number of important overlapping roles.

While the lectin chaperone system works largely on glycosylated client proteins, there has been some evidence for it interacting with non-glycosylated targets though its biological significance is poorly understood. Glycoprotein folding and quality control can be helped by both BiP and calnexin/calreticulin. The timing for initial chaperone binding is dependent upon the location of the glycans on the protein as calnexin/calreticulin binding appears to be dominant over BiP binding. It was found that prior binding to BiP occurred if there was no N-glycan in the first 50 amino acids for co-translational translocating nascent chains (Molinari and Helenius 2000). However, once a glycosylation site emerged into the ER lumen, if it was modified and rapidly trimmed to support lectin chaperone binding that this binding would dominate over BiP binding. The analysis of the primary sequence of a protein using algorithms that search for BiP binding sites and N-glycosylation sites can provide important insights into the timing of chaperones interactions with the nascent chain; however, empirical demonstration is required as substrate features do not always successfully predict which quality control systems will be utilized (Adams et al. 2019a).

The ER serves as the site for the maturation of a large array of proteins. While small proteins likely fold efficiently unassisted by chaperones, larger proteins that fold and frequently assemble into multiprotein complexes in the ER are expected to be accommodated by region-specific engagement by BiP and calnexin/calreticulin along with their associated factors. These complex proteins that are reliant upon chaperones for their proper maturation and sorting are commonly the source of diseases rooted in misfolding underscoring the importance in understanding how chaperones direct the folding and trafficking of proteins in the early secretory pathway.

3.5 Summary Points

- The binding of the ER Hsp70 family member BiP to non-native proteins is regulated by adenine nucleotides and its associated cofactors (J-proteins for delivery and NEFs for release).

- The lectin chaperones bind to proteins possessing monoglucosylated N-linked glycans, therefore binding is regulated by glucosidases and transferases that dictate the glycan composition.
- Both chaperone systems bind to non-native substrates; however, BiP binds to exposed hydrophobic motifs that should be sequestered by a properly folded protein and the lectin chaperones bind to exposed bulky hydrophilic modifications.
- The classical and lectin chaperones prevent aggregation thereby increasing the properly folded protein yield.
- Persistent chaperone binding supports ER retention and subsequent delivery of terminally misfolded proteins for degradation.

3.6 Future Issues

- The identification of BiP and calnexin/calreticulin co-translational and post-translational binding sites using proteomic and ribosomal profiling approaches.
- The determination of cellular UGGT1 and UGGT2 substrates and modification sites.
- To understand the effects of the modifications of the C-terminal tail of calnexin on its cellular function.
- To investigate the mechanism of extraction from the calnexin/calreticulin cycle by the variety of exo- and endo-mannosidases.
- Are substrates engaged by the BiP and lectin chaperone cycles simultaneously or in a dedicated manner?
- What diseases states or therapeutic approaches can be addressed through modifications of the ER chaperone cycles?

Acknowledgements This work was supported by the National Institutes of Health ((GM086874 to D.N.H.) and a Chemistry-Biology Interface program training grant (T32GM008515 to B.M.A. and N.P.C.)).

Conflict-of-Interest Disclosure Statement The authors declare that they have no conflicts of interest.

References

Adams BM, Ke H, Gierash LM, Gershenson A, Hebert DN (2019a) Proper secretion of the serpin antithrombin relies strictly on thiol-dependent quality control. J Biol Chem 294 (50):18992–18911. https://doi.org/10.1074/jbc.RA119.010450

Adams BM, Oster ME, Hebert DN (2019b) Protein quality control in the endoplasmic reticulum. Protein J 38(3):317–329. https://doi.org/10.1007/s10930-019-09831-w

Amin-Wetzel N, Saunders RA, Kamphuis MJ, Rato C, Preissler S, Harding HP, Ron D (2017) A J-protein co-chaperone recruits BiP to monomerize IRE1 and repress the unfolded protein response. Cell 171(7):1625–1637.e13. https://doi.org/10.1016/j.cell.2017.10.040

Andréasson C, Rampelt H, Fiaux J, Druffel-Augustin S, Bukau B (2010) The endoplasmic reticulum Grp170 acts as a nucleotide exchange factor of Hsp70 via a mechanism similar to that of the cytosolic Hsp110. J Biol Chem 285(16):12445–12453. https://doi.org/10.1074/jbc.M109.096735

Anfinsen CB (1973) Principles that govern the folding of protein chains. Science 167 (3919):886–887. https://doi.org/10.1126/science.181.4096.223

Avezov E, Frenkel Z, Ehrlich M, Herscovics A, Lederkremer GZ (2008) Endoplasmic reticulum (ER) mannosidase i is compartmentalized and required for N-glycan trimming to Man5–6-GlcNAc2 in glycoprotein ER-associated degradation. Mol Biol Cell 19(1):216–225. https://doi.org/10.1091/mbc.E07-05-0505

Balchin D, Hayer-Hartl M, Ulrich Hartl F (2020) Recent advances in understanding catalysis of protein folding by molecular chaperones. FEBS Lett 594(17):2770–2781. https://doi.org/10.1002/1873-3468.13844

Behnke J, Hendershot LM (2014) The large Hsp70 Grp170 binds to unfolded protein substrates in vivo with a regulation distinct from conventional Hsp70s. J Biol Chem 289(5):2899–2907. https://doi.org/10.1074/jbc.M113.507491

Behnke J, Feige MJ, Hendershot LM (2015) BiP and its nucleotide exchange factors Grp170 and Sil1: mechanisms of action and biological functions. J Mol Biol 427(7):1589–1608. https://doi.org/10.1016/j.jmb.2015.02.011

Behnke J, Mann MJ, Scruggs F-L, Feige MJ, Hendershot LM (2016) Members of the Hsp70 family recognize distinct types of sequences to execute ER quality control. Mol Cell 63(5):739–752. https://doi.org/10.1016/j.molcel.2016.07.012

Blees A, Januliene D, Hofmann T, Koller N, Schmidt C, Trowitzsch S, Moeller A, Tampé R (2017) Structure of the human MHC-I peptide-loading complex. Nature 551(7681):525–528. https://doi.org/10.1038/nature24627

Blond-Elguindi S, Cwirla SE, Dower WJ, Lipshutz RJ, Sprang SR, Sambrook JF, Gething M-J (1993) Affinity panning of a library of peptides displayed on bacteriophages reveals the binding specificity of BiP. Cell 75(4):717–728. https://doi.org/10.1016/0092-8674(93)90492-9

Broncel M, Serwa RA, Bunney TD, Katan M, Tate EW (2016) Global profiling of huntingtin-associated protein E (HYPE)-mediated AMPylation through a chemical proteomic approach. Mol Cell Proteomics 15(2):715–725. https://doi.org/10.1074/mcp.O115.054429

Cameron PH, Chevet E, Pluquet O, Thomas DY, Bergeron JJM (2009) Calnexin phosphorylation attenuates the release of partially misfolded 1-antitrypsin to the secretory pathway. J Biol Chem 284(50):34570–34579. https://doi.org/10.1074/jbc.M109.053165

Caramelo JJ, Castro OA, de Prat-Gay G, Parodi AJ (2004) The endoplasmic reticulum glucosyltransferase recognizes nearly native glycoprotein folding intermediates. J Biol Chem 279(44):46280–46285. https://doi.org/10.1074/jbc.M408404200

Chang Y-W, Sun Y-J, Chung W, Hsiao C-D (2008) Crystal structures of the 70-KDa heat shock proteins in domain disjoining conformation. J Biol Chem 283(22):15502–15511. https://doi.org/10.1074/jbc.M708992200

Chen K-C, Qu S, Chowdhury S, Noxon IC, Schonhoft JD, Plate L, Powers ET, Kelly JW, Lander GC, Wiseman RL (2017) The endoplasmic reticulum HSP40 co-chaperone ERdj3/DNAJB11 assembles and functions as a tetramer. EMBO J 36(15):2296–2309. https://doi.org/10.15252/embj.201695616

Cherepanova NA, Venev SV, Leszyk JD, Shaffer SA, Gilmore R (2019) Quantitative proteomics reveals new classes of STT3A- and STT3B-dependent N-glycosylation sites. J Cell Biol 218 (8):2782–2796. https://doi.org/10.1083/jcb.201904004

Chevet E, Smirle J, Cameron PH, Thomas DY, Bergeron JJM (2010) Calnexin phosphorylation: linking cytoplasmic signalling to endoplasmic reticulum lumenal functions. Semin Cell Dev Biol 21(5):486–490. https://doi.org/10.1016/j.semcdb.2009.12.005

Chung KT, Shen Y, Hendershot LM (2002) BAP, a mammalian BiP-associated protein, is a nucleotide exchange factor that regulates the ATPase activity of BiP. J Biol Chem 277 (49):47557–47563. https://doi.org/10.1074/jbc.M208377200

Craven RA, Egerton M, Stirling CJ (1996) A novel Hsp70 of the yeast ER lumen is required for the efficient translocation of a number of protein precursors. EMBO J 15(11):2640–2650. https://doi.org/10.1002/j.1460-2075.1996.tb00624.x

Cunnea PM, Miranda-Vizuete A, Bertoli G, Simmen T, Damdimopoulos AE, Hermann S, Leinonen S et al (2003) ERdj5, an endoplasmic reticulum (ER)-resident protein containing DnaJ and thioredoxin domains, is expressed in secretory cells or following ER stress. J Biol Chem 278 (2):1059–1066. https://doi.org/10.1074/jbc.M206995200

Daniels R, Kurowski B, Johnson AE, Hebert DN (2003) N-linked glycans direct the cotranslational folding pathway of influenza hemagglutinin. Mol Cell 11(1):79–90. https://doi.org/10.1016/s1097-2765(02)00821-3

Danilcyk UG, Williams DB (2001) The lectin chaperone calnexin utilizes polypeptide-based interactions to associate with many of its substrates in vivo. J Biol Chem 276 (27):25532–25540. https://doi.org/10.1074/jbc.M100270200

Deprez P, Gautschi M, Helenius A (2005) More than one glycan is needed for ER glucosidase II to allow entry of glycoproteins into the calnexin/calreticulin cycle. Mol Cell 19(2):183–195. https://doi.org/10.1016/j.molcel.2005.05.029

Ellgaard L, Riek R, Herrmann T, Güntert P, Braun D, Helenius A, Wüthrich K (2001) NMR structure of the calreticulin P-domain. Proc Natl Acad Sci U S A 98(6):3133–3138. https://doi.org/10.1073/pnas.051630098

Frickel E-M, Riek R, Jelesarov I, Helenius A, Wüthrich K, Ellgaard L (2002) TROSY-NMR reveals interaction between ERp57 and the tip of the calreticulin Pdomain. Proc Natl Acad Sci U S A 99 (4):1954–1959. https://doi.org/10.1073/pnas.042699099

Frickel E-M, Frei P, Bouvier M, Stafford WF, Helenius A, Ellgaard L (2004) ERp57 is a multifunctional thiol-disulfide oxidoreductase. J Biol Chem 279(18):18277–18287. https://doi.org/10.1074/jbc.M314089200

Hagiwara M, Maegawa K-i, Suzuki M, Ushioda R, Araki K, Matsumoto Y, Hoseki J, Nagata K, Inaba K (2011) Structural basis of an ERAD pathway mediated by the ER-resident protein disulfide reductase ERdj5. Mol Cell 41(4):432–444. https://doi.org/10.1016/j.molcel.2011.01.021

Hebert DN, Lamriben L, Powers ET, Kelly JW (2014) The intrinsic and extrinsic effects of N-linked glycans on glycoproteostasis. Nat Chem Biol 10(11):902–910. https://doi.org/10.1038/nchembio.1651

Hetz C, Zhang K, Kaufman RJ (2020) Mechanisms, regulation and functions of the unfolded protein response. Nat Rev Mol Cell Biol 21(8):421–438. https://doi.org/10.1038/s41580-020-0250-z

Itzhak DN, Tyanova S, Cox J, Borner GHH (2016) Global, quantitative and dynamic mapping of protein subcellular localization. elife 5:e16950. https://doi.org/10.7554/eLife.16950

Jin Y, Awad W, Petrova K, Hendershot LM (2008) Regulated release of ERdj3 from unfolded proteins by BiP. EMBO J 27(21):2873–2882. https://doi.org/10.1038/emboj.2008.207

Kapoor M, Srinivas H, Kandiah E, Gemma E, Ellgaard L, Oscarson S, Helenius A, Surolia A (2003) Interactions of substrate with calreticulin, an endoplasmic reticulum chaperone. J Biol Chem 278(8):6194–6200. https://doi.org/10.1074/jbc.M209132200

Kasaikina MV, Fomenko DE, Labunskyy VM, Lachke SA, Qiu W, Moncaster JA, Zhang J et al (2011) Roles of the 15-KDa selenoprotein (Sep15) in redox homeostasis and cataract development revealed by the analysis of Sep 15 knockout mice. J Biol Chem 286(38):33203–33212. https://doi.org/10.1074/jbc.M111.259218

Korotkov KV, Kumaraswamy E, Zhou Y, Hatfield DL, Gladyshev VN (2001) Association between the 15-KDa selenoprotein and UDP-glucose: glycoprotein glucosyltransferase in the endoplasmic reticulum of mammalian cells. J Biol Chem 276(18):15330–15336. https://doi.org/10.1074/jbc.M009861200

Kozlov G, Gehring K (2020) Calnexin cycle - structural features of the ER chaperone system. FEBS J 287(20):4322–4340. https://doi.org/10.1111/febs.15330

Kozlov G, Määttänen P, Schrag JD, Pollock S, Cygler M, Nagar B, Thomas DY, Gehring K (2006) Crystal structure of the Bb' domains of the protein disulfide isomerase ERp57. Structure 14 (8):1331–1339. https://doi.org/10.1016/j.str.2006.06.019

Kozlov G, Pocanschi CL, Rosenauer A, Bastos-Aristizabal S, Gorelik A, Williams DB, Gehring K (2010) Structural basis of carbohydrate recognition by calreticulin. J Biol Chem 285 (49):38612–38620. https://doi.org/10.1074/jbc.M110.168294

Kozlov G, Muñoz-Escobar J, Castro K, Gehring K (2017) Mapping the ER interactome: the P domains of calnexin and calreticulin as plurivalent adapters for foldases and chaperones. Structure 25(9):1415–1422. https://doi.org/10.1016/j.str.2017.07.010

Kryukov GV, Castellano S, Novoselov SV, Lobanov AV, Zhetab O, Guigó R, Gladyshev VN (2003) Characterization of mammalian selenoproteomes. Science 300(5624):1439–1443. https://doi.org/10.1126/science.1083516

Labunskyy VM, Ferguson AD, Fomenko DE, Chelliah Y, Hatfield DL, Gladyshev VN (2005) A novel cysteine-rich domain of Sep15 mediates the interaction with UDP-glucose: glycoprotein glucosyltransferase. J Biol Chem 280(45):37839–37845. https://doi.org/10.1074/jbc.M508685200

Lakkaraju AKK, Abrami L, Lemmin T, Blaskovic S, Kunz B, Kihara A, Peraro MD, van der Goot F (2012) Palmitoylated calnexin is a key component of the ribosome–translocon complex. Eur Mol Biol Organ J 31(7):1823–1835. https://doi.org/10.1038/emboj.2012.15

Lamriben L, Graham JB, Adams BM, Hebert DN (2016) N-glycan-based ER molecular chaperone and protein quality control system: the calnexin binding cycle. Traffic (Copenhagen, Denmark) 17(4):308–326. https://doi.org/10.1111/tra.12358

Lee D, Kraus A, Prins D, Groenendyk J, Aubry I, Liu W-X, Li H-D et al (2015) UBC9-dependent association between calnexin and protein tyrosine phosphatase 1B (PTP1B) at the endoplasmic reticulum. J Biol Chem 290(9):5725–5738. https://doi.org/10.1074/jbc.M114.635474

Liepinsh E, Baryshev M, Sharipo A, Ingelman-Sundberg M, Otting G, Mkrtchian S (2001) Thioredoxin fold as homodimerization module in the putative chaperone ERp29: NMR structures of the domains and experimental model of the 51 KDa dimer. Structure 9(6):457–571. https://doi.org/10.1016/s0969-2126(01)00607-4

Lum R, Ahmad S, Hong SJ, Chapman DC, Kozlov G, Williams DB (2016) Contributions of the lectin and polypeptide binding sites of calreticulin to its chaperone functions in vitro and in cells. J Biol Chem 291(37):19631–19641. https://doi.org/10.1074/jbc.M116.746321

Lynes EM, Bui M, Yap MC, Benson MD, Schneider B, Ellgaard L, Berthiaume LG, Simmen T (2012) Palmitoylated TMX and calnexin target to the mitochondria-associated membrane. Eur Mol Biol Organ J 31(2):457–470. https://doi.org/10.1038/emboj.2011.384

Marcinowski M, Höller M, Feige MJ, Baerend D, Lamb DC, Buchner J (2011) Substrate discrimination of the chaperone BiP by autonomous and cochaperone-regulated conformational transitions. Nat Struct Mol Biol 18(2):150–158. https://doi.org/10.1038/nsmb.1970

Mayer MP, Gierasch LM (2019) Recent advances in the structural and mechanistic aspects of Hsp70 molecular chaperones. J Biol Chem 294(6):2085–2097. https://doi.org/10.1074/jbc.REV118.002810

Melville MW, Tan S-L, Wambach M, Song J, Morimoto RI, Katze MG (1999) The cellular inhibitor of the PKR protein kinase, P58IPK, is an influenza virus-activated co-chaperone that modulates heat shock protein 70 activity. J Biol Chem 274(6):3797–3803. https://doi.org/10.1074/jbc.274.6.3797

Molinari M, Helenius A (2000) Chaperone selection during glycoprotein translocation into the endoplasmic reticulum. Science 288(5464):331–333. https://doi.org/10.1126/science.288.5464.331

Molinari M, Calanca V, Galli C, Lucca P, Paganetti P (2003) Role of EDEM in the release of misfolded glycoproteins from the calnexin cycle. Science 299(5611):1397–1400. https://doi.org/10.1126/science.1079474

Myhill N, Lynes EM, Nanji JA, Blagoveshchenskaya AD, Fei H, Carmine Simmen K, Cooper TJ, Thomas G, Simmen T (2008) The subcellular distribution of calnexin is mediated by PACS-2. Mol Biol Cell 19(7):2777–2788. https://doi.org/10.1091/mbc.e07-10-0995

Oda Y, Hosokawa N, Wada I, Nagata K (2003) EDEM as an acceptor of terminally misfolded glycoproteins released from calnexin. Science 299(5611):1394–1397. https://doi.org/10.1126/science.1079181

Ou W-J, Bergeron JJM, Li Y, Kang YC, Thomas DY (1995) Conformational changes induced in the endoplasmic reticulum luminal domain of calnexin by Mg-ATP and Ca2+. J Biol Chem 270 (30):18051–18059. https://doi.org/10.1074/jbc.270.30.18051

Perera LA, Rato C, Yan Y, Neidhardt L, McLaughlin SH, Read RJ, Preissler S, Ron D (2019) An oligomeric state-dependent switch in the ER enzyme FICD regulates AMPylation and DeAMPylation of BiP. EMBO J 38(21):e102177–e102177. https://doi.org/10.15252/embj.2019102177

Petrova K, Oyadomari S, Hendershot LM, Ron D (2008) Regulated association of misfolded endoplasmic reticulum lumenal proteins with P58/DNAJc3. EMBO J 27(21):2862–2872. https://doi.org/10.1038/emboj.2008.199

Pobre KFR, Poet GJ, Hendershot LM (2019) The endoplasmic reticulum (ER) chaperone BiP is a master regulator of ER functions: getting by with a little help from ERdj friends. J Biol Chem 294(6):2098–2108. https://doi.org/10.1074/jbc.REV118.002804

Preissler S, Chambers JE, Crespillo-Casado A, Avezov E, Miranda E, Perez J, Hendershot LM, Harding HP, Ron D (2015) Physiological modulation of BiP activity by trans-protomer engagement of the interdomain linker. elife 4(Oct):e08961. https://doi.org/10.7554/eLife.08961

Preissler S, Rato C, Perera L, Saudek V, Ron D (2017a) FICD acts bifunctionally to AMPylate and De-AMPylate the endoplasmic reticulum chaperone BiP. Nat Struct Mol Biol 24(1):23–29. https://doi.org/10.1038/nsmb.3337

Preissler S, Rohland L, Yan Y, Chen R, Read RJ, Ron D (2017b) AMPylation targets the rate-limiting step of BiP's ATPase cycle for its functional inactivation. elife 6(Oct):e29428. https://doi.org/10.7554/eLife.29428

Rajagopalan S, Xu Y, Brenner MB (1994) Retention of unassembled components of integral membrane proteins by calnexin. Science 263(5145):387–390. https://doi.org/10.1126/science.8278814

Rosam M, Krader D, Nickels C, Hochmair J, Back KC, Agam G, Barth A et al (2018) bap (sil1) regulates the molecular chaperone BiP by coupling release of nucleotide and substrate. Nat Struct Mol Biol 25(1):90–100. https://doi.org/10.1038/s41594-017-0012-6

Roth J, Zuber C (2017) Quality control of glycoprotein folding and ERAD: the role of N-glycan handling, EDEM1 and OS-9. Histochem Cell Biol 147:269–284. https://doi.org/10.1007/s00418-016-1513-9

Roversi P, Marti L, Caputo AT, Alonzi DS, Hill JC, Dent KC, Kumar A et al (2017) Interdomain conformational flexibility underpins the activity of UGGT, the eukaryotic glycoprotein secretion checkpoint. Proc Natl Acad Sci U S A 114(32):8544–8549. https://doi.org/10.1073/pnas.1703682114

Rüdiger S, Germeroth L, Schneider-Mergener J, Bukau B (1997) Substrate specificity of the DnaK chaperone determined by screening cellulose-bound peptide libraries. EMBO J 16 (7):1501–1507. https://doi.org/10.1093/emboj/16.7.1501

Rutkowski DT, Kang S-W, Goodman AG, Garrison JL, Taunton J, Katze MG, Kaufman RJ, Hegde RS (2007) The role of P58IPK in protecting the stressed endoplasmic reticulum. Mol Biol Cell 18(9):3681–3691. https://doi.org/10.1091/mbc.e07-03-0272

Satoh T, Song C, Zhu T, Toshimori T, Murata K, Hayashi Y, Kamikubo H, Uchihashi T, Kato K (2017) Visualization of a flexible modular structure of the ER folding-sensor enzyme UGGT. Sci Rep 7(1):12142. https://doi.org/10.1038/s41598-017-12283-w

Schrag JD, Bergeron JJM, Li Y, Borisova S, Hahn M, Thomas DY, Cygler M (2001) The structure of calnexin, an ER chaperone involved in quality control of protein folding. Mol Cell 8 (3):633–644. https://doi.org/10.1016/s1097-2765(01)00318-5

Shen Y, Hendershot LM (2005) ERdj3, a stress-inducible endoplasmic reticulum DnaJ homologue, serves as a cofactor for BiP's interactions with unfolded substrates. Mol Biol Cell 16(1):40–50. https://doi.org/10.1091/mbc.e04-05-0434

Shrimal S, Cherepanova NA, Mandon ES, Venev SV, Gilmore R (2019) Asparagine-linked glycosylation is not directly coupled to protein translocation across the endoplasmic reticulum in Saccharomyces Cerevisiae. Mol Biol Cell 30(21):2626–2638. https://doi.org/10.1091/mbc.E19-06-0330

Sousa MC, Ferrero-Garcia MA, Parodi AJ (1992) Recognition of the oligosaccharide and protein moieties of glycoproteins by the UDP-Glc:glycoprotein glucosyltransferase. Biochemistry 31 (1):97–105. https://doi.org/10.1021/bi00116a015

Spiro RG, Zhu Q, Bhoyroo V, Söling H-D (1996) Definition of the lectin-like properties of the molecular chaperone, calreticulin, and demonstration of its copurification with endomannosidase from rat liver golgi. J Biol Chem 271(19):11588–11594. https://doi.org/10.1074/jbc.271.19.11588

Svärd M, Biterova EI, Bourhis J-M, Guy JE (2011) The crystal structure of the human co-chaperone P58(IPK). PLoS One 6(7):e22337–e22337. https://doi.org/10.1371/journal.pone.0022337

Takeda Y, Seko A, Hachisu M, Daikoku S, Izumi M, Koizumi A, Fujikawa K, Kajihara Y, Ito Y (2014) Both isoforms of human UDP-glucose:glycoprotein glucosyltransferase are enzymatically active. Glycobiology 24(4):344–350. https://doi.org/10.1093/glycob/cwt163

Terajima M, Taga Y, Cabral WA, Nagasawa M, Sumida N, Hattor S, Marini JC, Yamauchi M (2017) Cyclophilin B deficiency causes abnormal dentin collagen matrix. J Proteome Res 16 (8):2914–2293. https://doi.org/10.1021/acs.jproteome.7b00190

Ushioda R, Hoseki J, Araki K, Jansen G, Thomas DY, Nagata K (2008) ERdj5 is required as a disulfide reductase for degradation of misfolded proteins in the ER. Science 321(5888):569. https://doi.org/10.1126/science.1159293

Ushioda R, Hoseki J, Nagata K (2013) Glycosylation-independent ERAD pathway serves as a backup system under ER stress. Mol Biol Cell 24(20):3155–3163. https://doi.org/10.1091/mbc.E13-03-0138

Vassilakos A, Michalak M, Lehrman MA, Williams DB (1998) Oligosaccharide binding characteristics of the molecular chaperones calnexin and calreticulin. Biochemistry 37 (10):3480–3490. https://doi.org/10.1021/bi972465g

Viviano J, Brecker M, Ferrara-Cook C, Suaud L, Rubenstein RC (2020) ERp29 as a regulator of insulin biosynthesis. PLoS One 15(5):e0233502. https://doi.org/10.1371/journal.pone.0233502

Wang N, Glidden EJ, Murphy SR, Pearse BR, Hebert DN (2008) The cotranslational maturation program for the type II membrane glycoprotein influenza neuraminidase. J Biol Chem 283 (49):33826–33837. https://doi.org/10.1074/jbc.M806897200

Wei J, Gaut JR, Hendershot LM (1995) In vitro dissociation of BiP-peptide complexes requires a conformational change in BiP after ATP binding but does not require ATP hydrolysis. J Biol Chem 270(44):26677–26682. https://doi.org/10.1074/jbc.270.44.26677

Wijeyesakere SJ, Gagnon JK, Arora K, Brooks CL III, Raghavan M (2015) Regulation of calreticulin–major histocompatibility complex (MHC) class I interactions by ATP. Proc Natl Acad Sci USA 112(41):E5608–E5617. https://doi.org/10.1073/pnas.1510132112

Yan M, Li J, Sha B (2011) Structural analysis of the Sil1–Bip complex reveals the mechanism for Sil1 to function as a nucleotide-exchange factor. Biochem J 438(3):447–455. https://doi.org/10.1042/BJ20110500

Yang J, Nune M, Zong Y, Zhou L, Liu Q (2015) Close and allosteric opening of the polypeptide-binding site in a human Hsp70 chaperone BiP. Structure (London, England: 1993) 23 (12):2191–2203. https://doi.org/10.1016/j.str.2015.10.012

Zapun A, Darby NJ, Tessier DC, Michalak M, Bergeron JJM, Thomas DY (1998) enhanced catalysis of ribonuclease B folding by the interaction of calnexin or calreticulin with ERp57. J Biol Chem 273(11):6009–6012. https://doi.org/10.1074/jbc.273.11.6009

Chapter 4
Proteins Interacting with STIM1 and Store-Operated Ca^{2+} Entry

Wen-An Wang and Nicolas Demaurex

Abstract The endoplasmic reticulum (ER) Ca^{2+} sensor stromal interaction molecule 1 (STIM1) interacts with ORAI Ca^{2+} channels at the plasma membrane to regulate immune and muscle cell function. The conformational changes underlying STIM1 activation, translocation, and ORAI1 trapping and gating, are stringently regulated by post-translational modifications and accessory proteins. Here, we review the recent progress in the identification and characterization of ER and cytosolic proteins interacting with STIM1 to control its activation and deactivation during store-operated Ca^{2+} entry (SOCE).

Keywords Endoplasmic reticulum · Ion channels · Protein trafficking · Cell signaling · Immunity

Abbreviations

Aa	Amino acid
AMPAR	α-amino-3-hydroxy-5-methyl-4-isoxazolepropionic acid receptor
ANO8	Anoctamin 8
APC	Adenomatous polyposis coli
BAG3	B-cell lymphoma 2-associated athanogene 3
BMM	Bone marrow macrophage
CAD	Channel activating domain
CC1α1	Coiled-coil
CLCA2	Chloride channel accessory protein 2
CRAC	Ca^{2+}-release-activated channel
CRACR2	Ca^{2+} release-activated Ca^{2+} channel regulator 2A
CTID	C-terminal inhibitory domain
EB1	Microtubule-associated protein RP/EB family member 1

W.-A. Wang · N. Demaurex (✉)
Department of Cellular Physiology and Metabolism, University of Geneva, Geneva, Switzerland
e-mail: nicolas.demaurex@unige.ch

L. B. Agellon, M. Michalak (eds.), *Cellular Biology of the Endoplasmic Reticulum*, Progress in Molecular and Subcellular Biology 59,
https://doi.org/10.1007/978-3-030-67696-4_4

EFHB	EF-hand domain family member B
EM	Electron microscopy
ER	Endoplasmic reticulum
ERGIC	ER/ER-Golgi intermediate compartment
FCDI/SCDI	Fast Ca^{2+}-dependent inactivation/slow Ca^{2+}-dependent inactivation
FRET	Förster resonance energy transfer
GST	Glutathione S-transferase
IP_3	Inositol triphosphate
IRF3	Interferon regulatory factor 3
ITPR	Inositol triphosphate receptor
MBP	Myelin basic protein
MCS	Membrane contact site
MHCI	Major histocompatibility complex class I
NFAT	Nuclear factor of activated T-cells
NFATc1	Nuclear factor of activated T-cells, cytoplasmic 1
NMDAR	*N*-methyl-D-aspartate receptor
NSF	N-ethlmaleimide-sensitive factor
PDIA3	Protein disulfide-isomerase A3
PIP_2	Phosphatidylinositol-4, 5-bisphophate
PLCγ2	Phospholipase C gamma-2
PM	Plasma membrane
PMCA	Plasma membrane Ca^{2+} ATPase
PMP2	Peripheral myelin protein 22
POST	Partner of STIM1
PRX-4	Peroxiredoxin-4
SAM	Sterile alpha motif
SARAF	SOCE-associated regulatory factor
SNARE	Soluble NSF-attachment protein receptor
SOAR	STIM/Orai activating region
α-SNAP	N-ethylmaleimide-sensitive factor attachment protein alpha
σ1R/σ2R	Sigma 1 receptor/sigma 2 receptor
SOCE	Associated regulatory factor
SERCA	Sarco/endoplasmic reticulum Ca^{2+}-ATPase
SOCE	Store-operated Ca^{2+} entry
STIMATE	STIM-activating enhancer
STIM1	Stromal interaction molecule 1
STING	The stimulator of interferon genes
SURF4	Surfeit locus protein 4
TBK1	TANK-binding kinase 1
TIRF	Total internal reflection fluorescence
TLR	Toll-like receptors
TMEM178	A transmembrane protein 178
TRPC	Transient receptor potential channel
TRPML1	Transient receptor potential mucolipin 1

UNC93B1	Un-coordinated 93 homologue B1
VCGGs	Voltage-gated Ca^{2+} channels

4.1 Introduction

The stromal interaction molecule (STIM) family of proteins mediate the ubiquitous store-operated Ca^{2+} entry (SOCE) pathway that couples the Ca^{2+} depletion of the endoplasmic reticulum (ER) to the activation of Ca^{2+}-release-activated channels (CRAC) at the plasma membrane (PM) (Prakriya and Lewis 2015). Following activation of Ca^{2+} mobilizing receptors, activation of phospholipase C leads to the hydrolysis of phosphatidylinositol 4,5-bisphophate (PIP_2) into diacylglycerol and inositol triphosphate (IP_3). IP_3 in turn binds to and activates ER membrane IP_3 receptors (ITPRs) which cause the release of Ca^{2+} from the ER into the cytoplasm. The depletion of ER Ca^{2+} is sensed by STIM, single-pass transmembrane ER proteins bearing two EF hand and a sterile alpha motif (SAM) in their luminal domain. As Ca^{2+} unbinds from their EF-SAM domains, STIM proteins oligomerize and accumulate in ER-PM junctions to trap and gate the ORAI family of Ca^{2+} channels (ORAI1, 2, and 3), enabling Ca^{2+} influx through these CRAC channels to sustain ER refilling and long-lasting Ca^{2+} signals. By bridging the ER and the PM, STIM proteins also generate membrane contact sites (MCSs) acting as platforms for Ca^{2+}-binding proteins, pumps and channels to facilitate Ca^{2+} exchange and local signaling. In humans, a precise regulation of STIM/ORAI coupling is essential for immune and muscle function and patients with loss and gain of function mutations suffer from severe combined immunodeficiency, myopathy, and Stormorken syndrome (Lacruz and Feske 2015).

Recent studies have revealed the sequence of conformational changes occurring within STIM1 during activation (Fig. 4.1) and the structural rearrangements occurring within ORAI1 as STIM1 binding opens the channel pore and increases its selectivity for Ca^{2+} (Gudlur et al. 2018; Hirve et al. 2018) [reviewed in Qiu and Lewis (2019)]. In the resting state, STIM1 luminal EF-SAM domains are Ca^{2+}-bound and separated while the cytosolic channel activating domains (CAD, also known as the STIM/ORAI activating region, SOAR) are bound to a proximal coiled-coil (CC1α1) in a compact, inactive conformation. Ca^{2+} release causes EF-SAM dimerization and the apposition of TM and CC1α1 domains into an extended coiled-coil, releasing CAD/SOAR dimers at the end of a ~ 15 nm elongated structure that can bridge the gap of ER-PM junctions (Fig. 4.1). The reversible switch of STIM1 between a quiescent to an active state is regulated by post-translational modifications (S-gluthionylation, N- and O-glycosylation, and phosphorylation), by accessory proteins (STIMATE, Septin, SARAF, CRACR2A, UNC93B1), by inhibitory translational and splice variants, and by fast and slow Ca^{2+} inactivation mechanisms [reviewed in Qiu and Lewis (2019)]. The structural basis of most of these regulatory

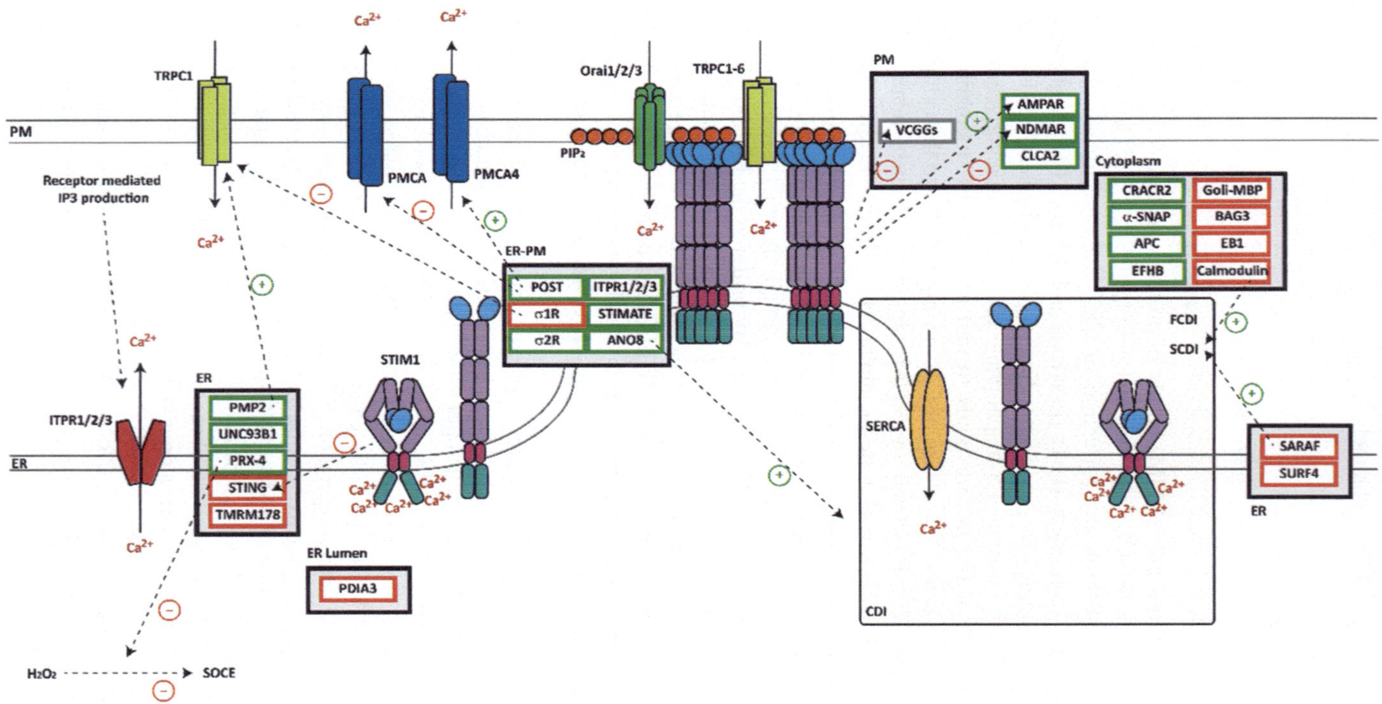

Fig. 4.1 Proteins interacting with STIM1 during activation and termination of SOCE. SOCE is mediated through STIM1. The signal initiates through PM receptor mediated production of IP_3 which in turn activates the release of Ca^{2+} from the ER through the ITPRs. Following this, STIM1 senses the reduced ER Ca^{2+} and undergoes conformational change, oligomerization and translocation to ER-PM membrane contact sites. STIM1 is then able to bind to PM PIP_2, access and activate the PM Ca^{2+} channels, ORAI1/2/3 and TRPC1-6, leading to an influx of Ca^{2+} into the cell. The Ca^{2+} is actively pumped back into the ER by the SERCA to refill ER Ca^{2+} stores and STIM1 inactivation occurs through FCDI/SCDI. The interactors of STIM1 are depicted in this figure separated by location and marked as activators (boxed in green) and inhibitors (boxed in red) of SOCE. VCGGs (boxed in grey) do not impact SOCE. Positive or negative interactions between proteins are indicated by dotted arrows with plus or minus sign. At the PM STIM1 positively regulates AMPARs and negatively regulates VCGGs and NDMAR. At the ER STIM1 negatively modulates STING signaling. PMP2 increases SOCE specifically through activation of TRPC1 and σ1R reduces SOCE

specifically through the inhibition of TRPC1. PRX-4 enhances SOCE by relieving a H_2O_2-mediated inhibition. POST increases SOCE through its negative regulation of PMCAs, which actively extrudes Ca^{2+} from the cell. However POST positively regulates PMCA4. ANO8 is an enhancer of SOCE but also mediates FCDI and SCDI. SARAF and calmodulin are suppressors of SOCE through the positive modulation of SCDI. *AMPAR* α-amino-3-hydroxy-5-methyl-4-isoxazolepropionic acid receptor, *ANO8* Anoctamin 8, *APC* adenomatous polyposis coli, *α-SNAP* N-ethylmaleimide-sensitive factor attachment protein alpha, *BAG3* B-cell lymphoma 2-associated athanogene 3, *CLCA2* chloride channel accessory protein 2, *CRACR2* Ca^{2+} release-activated Ca^{2+} channel regulator 2A, *EB1* microtubule-associated protein RP/EB family member 1, *EFHB* EF-hand domain family member B, *ER* endoplasmic reticulum, *FCDI/SCDI* fast Ca^{2+}-dependent inactivation/slow Ca^{2+}-dependent inactivation, IP_3 inositol triphosphate, *ITPR* inositol triphosphate receptor, *MBP* myelin basic protein, *NMDAR* *N*-methyl-D-aspartate receptor, *PDIA3* protein disulfide-isomerase A3, PIP_2 phosphatidylinositol-4, 5-bisphophate, *PM* plasma membrane, *PMCA* plasma membrane Ca^{2+} ATPase, *PMP2* peripheral myelin protein 22, *POST* partner of STIM1, *PRX-4* peroxiredoxin-4, *SARAF* SOCE-associated regulatory factor, *SERCA* sarco/endoplasmic reticulum Ca^{2+}-ATPase, *SOCE* store-operated Ca^{2+} entry, *σ1R/σ2R* sigma 1 receptor/sigma 2 receptor, *STIMATE* STIM-activating enhancer, *STIM1* stromal interaction molecule 1, *STING* the stimulator of interferon genes, *SURF4* surfeit locus protein 4, *TMEM178* a transmembrane protein 178, *TRPC* transient receptor potential channel, *UNC93B1* un-coordinated 93 homologue B1, *VCGGs* voltage-gated Ca^{2+} channels mechanisms is unclear, but several natural and artificial mutations were shown to impact the activation dynamics of STIM proteins.

This chapter aims to highlight the protein interactors of STIM1 (summarized in Tables 4.1 and 4.2) that modulate the activation and inactivation of the SOCE process as well as the formation of membrane contact sites (MCSs) between the ER and the PM. The different interactors are discussed sequentially based on their primary intracellular location in the ER, PM, and cytosol. This chapter focuses on protein interactions and will not discuss organizer proteins such as extended synaptotagmins, septins, junctate, or junctophilin that contribute to the SOCE process by tethering membranes or by segregating proteins and lipid into specific microdomains. The channels gated by STIM proteins, which besides ORAI include the transient receptor potential cation channels (TRPCs) and the transient receptor potential mucolipin 1 (TRPML1) located in lysosomes, have been extensively reviewed (Lunz et al. 2019; Qiu and Lewis 2019; Lopez et al. 2020; Tedeschi et al. 2019; Kilpatrick et al. 2016). Although these channels are arguably the primary STIM interactors, they are not discussed extensively here. The studies reporting their interactions with STIM proteins are summarized in Table 4.1.

4.1.1 ER Proteins

Many ER luminal and membrane proteins were reported to interact with STIM1 and to modulate its localization, activation, trafficking and inactivation. These proteins can be separated functionally into positive and negative regulators based on their impact on the SOCE process.

4.1.1.1 Positive Regulators of SOCE: ITPRs, STIMATE, UNC93B1, PMP22, ANO8, Prx-4 and POST

ITPRs

Activation of ITPRs by the phospholipase C and IP_3 signaling pathway leads to the release of ER Ca^{2+} and is the main physiological mechanism behind the activation of SOCE (Streb et al. 1983). Therefore, ER Ca^{2+} release through ITPRs and SOCE activation are closely interconnected and studies have documented the synergism between these two processes. A functional interaction was found between ITPRs and TRPCs (Boulay et al. 1999; Kiselyov et al. 1998; Rosado and Sage 2001; Yuan et al. 2003; Jardin et al. 2008a; Tai et al. 2008; Woodard et al. 2010) and between ITPRs and ORAI1 (Woodard et al. 2010; Lur et al. 2011). The presence of ITPRs juxtaposed to STIM-ORAI junctions was therefore proposed to be important for achieving an effective local Ca^{2+} depletion of the ER sufficient to activate SOCE at ER-PM contact sites without causing a global ER Ca^{2+} depletion (Thillaiappan et al. 2017). The interaction between ITPRs and STIM1 further enforced this idea. STIM1 have

Table 4.1 Main interactors of STIM1 in SOCE

Channels	Interaction and differential modulation of calcium currents	Species	Cell type	References
ORAI1 *Role of STIM1*	RNAi-mediated knock-down of STIM1 reduced SOCE	*Drosophila*	S2	Roos et al. (2005), Zhang et al. (2006)
	Overexpression of STIM1 enhanced SOCE	Human	Jurkat T HEK293 SH-SY5Y	
Role of ORAI1	RNAi knockdown of ORAI1 reduced SOCE and CRAC currents	*Drosophila*	S2	Zhang et al. (2006) Vig et al. (2006) Kwon et al. (2017) Wang et al. (2017) Mercer et al. (2006) Peinelt et al. (2006) Takahashi et al. (2007) Lis et al. (2007) DeHaven et al. (2007) Baryshnikov et al. (2009) Inayama et al. (2015)
	ORAI1 overexpression alone or with STIM1 increased SOCE and CRAC currents		S2R+	
	Gene silencing of *ORAI1* reduced SOCE by 35.7%		S2	
	RNAi knockdown of ORAI1 reduced hypoxia-induced SOCE	Mouse	Astrocytes	
	RNAi knockdown of ORAI1 reduced SOCE and CRAC currents	Rat	PASMCs	
	Overexpression of ORAI1 alone did not enhance SOCE	human	HEK293	
	Overexpression of ORAI1 with STIM1 enhanced SOCE and CRAC currents		Jurkat T	
	RNAi knockdown of ORAI1 reduced SOCE		hASMCs OUMS-27	
STIM1 domains	Activation of ORAI1, 2 and 3 requires STIM1 C-terminal	Human	HEK293	Muik et al. (2008, 2009), Frischauf et al. (2009) Park et al. (2009) Yuan et al. (2009) Kawasaki et al. (2009) Wang et al. (2014) Butorac et al. (2019)
	OASF (aa233-450) CAD (aa342-448) SOAR (aa344-442) CCb9 (aa330-444) F394 interacts with N-terminus of ORAI1 (aa73-85) L373S mutation abolishes ORAI1 interaction but not ORAI2 and ORAI3		HEK293 and 293T	

(continued)

Table 4.1 (continued)

Channels	Interaction and differential modulation of calcium currents	Species	Cell type	References
	STIM1 α3 domain is necessary to activate ORAI1 through interaction with ORAI1 TM3			
ORAI1 domains	Interaction with STIM1 requires the N (for activation) and C terminals (for coupling) of Orai1 R91 (for activation)	Human	HEK293	Muik et al. (2008) Yuan et al. (2009) Park et al. (2009) Zheng et al. (2013) Fahrner et al. (2018) Frischauf et al. (2009) Li et al. (2007) McNally et al. (2013) Palty and Isacoff (2016) Zhou et al. (2010, 2016) Takahashi et al. (2007) Niu et al. (2020) Derler et al. (2013, 2018) Yen et al. (2016) Stathopulos et al. (2013) Yen and Lewis (2018)
	ETON, aa73-90 C-terminal and Nexus (aa261-265)			
	Hydrophobic residue L273 (for coupling) and L276		HEK293-H	
ORAI2 *Role of ORAI2*	Gene silencing of *ORAI2* reduced SOCE by 20.3%	Mouse	Astrocytes	Kwon et al. (2017) Wang et al. (2017) Mercer et al. (2006) Alansary et al. (2015) Inayama et al. (2015) Lis et al. (2007), DeHaven et al. (2007) Takahashi et al. (2007) Baryshnikov et al. (2009)
	RNAi knockdown of ORAI2 reduced hypoxia-induced SOCE	Rat	PASMCs	
	Overexpression of ORAI2 alone inhibits SOCE	Human	HEK293 Jurkat T	
	Overexpression of ORAI2 with STIM1 enhanced SOCE and CRAC currents		OUMS-27	
	RNAi knockdown of ORAI2 did not reduce SOCE			
	RNAi knockdown of ORAI2 increased SOCE		hASMCs OUMS-27	
ORAI2 domains	L237 and L244 (moderate)	Human	HEK293	Frischauf et al. (2009)

(continued)

Table 4.1 (continued)

Channels	Interaction and differential modulation of calcium currents	Species	Cell type	References
ORAI3 *Role of ORAI3*	Gene silencing of *ORAI3* reduced SOCE by 26.8%	Mouse	Astrocytes	Kwon et al. (2017) Saliba et al. (2015) Wang et al. (2017) Lis et al. (2007) DeHaven et al. (2007) Alansary et al. (2015) Takahashi et al. (2007) Baryshnikov et al. (2009)
	Gene silencing of *ORAI3* reduced SOCE		Cardiomyocytes	
	RNAi knockdown of ORAI3 reduced hypoxia-induced SOCE	Rat	PASMCs	
	Overexpression of ORAI3 with STIM1 enhanced SOCE and CRAC currents	Human	HEK293	
	Overexpression of ORAI3 alone inhibited SOCE		Jurkat T	
	RNAi knockdown of ORAI3 reduced SOCE			
	RNAi knockdown of ORAI3 did not reduce SOCE		hASMCs	
ORAI3 domains	N terminal ETON region (aa1-57)	Human	HEK293	Fahrner et al. (2018) Derler et al. (2018) Bergsmann et al. (2011) Frischauf et al. (2009) Niu et al. (2020)
	L285 and L292 (moderate) N (aa49-65, R52 and R53) and C (aa268-295, V271, L282, L285, L288 and L292) terminal			
TRPCs *Role of TRPCs*	TRPC1 was the sole contributor to STIM1-induced SOCE	Mouse	VSMCs	Shi et al. (2017a, b) Ng et al. (2012) Almirza et al. (2012) Thebault et al. (2005) Sundivakkam et al. (2012)
	Blocking TRPC1 reduced hypoxia-induced calcium entry		PAMSCs	
	Knockdown of TRPC1 reduced SOCE	Rat	NRK	
	Knockdown of TRPC3 reduced SOCE		PS1	
	Knockout of TRPC4 impaired SOCE	Mouse	Lung endothelial	
Role of TRPCs	STIM1 activity potentiated TRPC1 and TRPC4 channel activity	Human	HEK293 Jurkat T HSG	Huang et al. (2006) Pani et al. (2009) Liu et al. (2000,

(continued)

Table 4.1 (continued)

Channels	Interaction and differential modulation of calcium currents	Species	Cell type	References
	Overexpression of TRPC1 enhanced SOCE and knockdown of TRPC1 reduced SOCE			2007) Kim et al. (2009) Wu et al. (2000) Zagranichnaya et al. (2005) Ong et al. (2007) Liu et al. (2007) Alicia et al. (2008) Sundivakkam et al. (2012) Sours-Brothers et al. (2009) Li et al. (2008) Rosado et al. (2002) Jardin et al. (2008b)
	Knockdown of TRPC1 and TRPC3 reduced SOCE and CRAC currents		RBL	
	Knockdown of TRPC1 attenuated and overexpression of TRPC1 enhanced SOCE		Salivary gland	
	Knockdown of TRPC1 reduced SOCE and CRAC currents		HEK293	
	Knockdown of TRPC1 and TRPC4 reduced SOCE		HLMVECs	
	Knockdown of TRPC1, TRPC4 and TRPC5 reduced SOCE		MCs	
	Blocking TRPC1 with antibody reduced SOCE		VSMCs	
	Blocking TRPC6 with antibody reduced SOCE		Platelets	
STIM1 and TRPC domains	STIM1 binds TRPC1, TRPC2, TRPC4 and TRPC5	Human	HEK293	Worley et al. (2007) Huang et al. (2006) Alicia et al. (2008) Li et al. (2008) Pani et al. (2009) Zeng et al. (2008) Zeng et al. (2008) Lee et al. (2010, 2014) Sundivakkam et al. (2012) Yuan et al. (2007) Jardin et al. (2009)
	STIM1 binding to TRPC1, TRCP2 and TRPC4 requires the ERM domain		Jurkat T	
	STIM1 gating or TRPC1 requires the lysine-rich domain			
	STIM1 K684 and K685 interacts with TRPC1 D639 and D640 via electrostatic interaction			
	TRPC3 D697 and D698			
	TRPC4 E647 and E648			
	TRPC5 D651 and D652			

(continued)

Table 4.1 (continued)

Channels	Interaction and differential modulation of calcium currents	Species	Cell type	References
	TRPC6 E755 and E756			
	STIM1 SOAR domain interacts with the TRPC1 and TRPC5 N terminal coiled coil domain			
	STIM1 regulates TRPC3 via heteromultimerization with TRPC1 and TRPC6 with TRPC4			
	STIM1 binds TRPC6		Platelets	
TRPML1	TRPML1 interacts with STIM1 at ER-lysosome junctions	Mouse	NSC-34	Tedeschi et al. (2019)
	Agonist activation of TRPML1 and TG treatment induced lysosomal calcium efflux			

been co-immunoprecipitated with ITPR1 in bovine thoracic aortic cells (Beliveau et al. 2014) and ITPR3 in Madin-Darby Canine Kidney cells (Santoso et al. 2011), mouse pancreatic tissue and acinar cells (Hong et al. 2011). The interaction between STIM1 and ITPRs were enhanced with induced ITPR-mediated ER Ca^{2+} release and SOCE activation. A small percent of ITPRs was found in HeLa cells to be immobile and cluster in proximity to ER-PM junctions where they may function to regulate SOCE locally (Thillaiappan et al. 2017). Upon activation, ITPRs are recruited and co-localize with STIM1 at ER-PM MCSs, enhancing STIM1 puncta formation and SOCE (Sampieri et al. 2018). The interaction between ITPRs to STIM1 provide a local ER Ca^{2+} depletion in close proximity to the EF hand of STIM1 and this is required for enhanced STIM1 and SOCE activation (Thillaiappan et al. 2017; Sampieri et al. 2018). Inversely, the association of STIM1 to ITPRs potentiated ITPR activity and STIM1 silencing reduced bradykinin and ATP induced ER Ca^{2+} release (Beliveau et al. 2014). It was hypothesized that STIM1-ITPR interaction competed with other ITPR binding and inactivating molecules such as polycystin-2 (Santoso et al. 2011; Beliveau et al. 2014). Alternatively, STIM1 may be recruiting ITPRs to PIP_2-enriched domains within the plasma membrane providing immediate substrate of IP_3 for ITPR activation (Maleth et al. 2014; Thillaiappan et al. 2017).

Table 4.2 Interactors of STIM1

ER proteins	Interaction and differential modulation of calcium currents	Species	Cell type	References
ITPRs	Knockdown of STIM1 reduced ATP and BK induced ER calcium release	Bovine	BAECs	Beliveau et al. (2014)
	STIM1 interacts with IP3R1	Human	HEK293	Sampieri et al. (2018)
	Increased association between STIM1 and IP3R3 reduced ER calcium release	Canine	MDCK	Santoso et al. (2011)
	STIM1 interacts with IP3R3			
	ITPR3 associates with STIM1/ORAI1/TRPC3 complex	Mouse	PACs	Hong et al. (2011)
	ITPR1, ITPR2 and ITPR3 associates with ORAI1			Lur et al. (2011)
STIMATE	STIMATE interacts with STIM1 (aa233-343) at ER-PM junctions	Human	HEK293	Jing et al. (2015)
	Silencing STIMATE reduced SOCE		HeLa, Jurkat T	Quintana et al. (2015)
UNC93B1	3d mutation in UNC93B1 reduced SOCE	Mouse	DC	Maschalidi et al. (2017)
	3d mutation in UNC93B1 disrupted interaction with STIM1		MEFs	
	STIM1 D76A mutant rescued SOCE in cells with UNC93B1 3d mutation			
	UNC93B1 does not bind STIM1 D76A mutant			
PMP22	PMP22 interacts with STIM1	Mouse/Rat	Sciatic nerve	Vanoye et al. (2019)
	Knockout of PMP2 reduced SOC currents		Schwann	
	PMP22 increased SOC currents	Human	HEK293	
ANO8	Ano8 overexpression increase SOCE and STIM1 and ORAI1 at ER-PM junction Ano8 increased STIM1-SARAF interaction Ano8 increased STIM1-ORAI1 complexes, ER-PM junction and size Ano8 acts as a ER-PM tether that requires PIP2	Human	HEK293	Jha et al. (2019)

(continued)

Table 4.2 (continued)

ER proteins	Interaction and differential modulation of calcium currents	Species	Cell type	References
	Ano8 modulates FCDI Ano8 interacts with STIM1			
Prx-4	STIM1 interacts with Prx-4 (CO-IP) in heavy microsomal fractions (no treatment)	Rat	Liver	Tam et al. (2018)
	RNAi knockdown of or Con A inhibition of Prx-4 reduced SOCE in the presence of H2O2		H4IIE	
POST	POST knockdown enhanced PMCA activity with store depletion and reduced SOCE slightly	Human	HEK293 and Jurkat T	Krapivinsky et al. (2011)
	POST overexpression did not alter SOCE or ORAI1 Ca^{2+} influx			
	Exogenous POST co-immunoprecipitates with exogenous STIM1 upon TG treatment		HEK293	
	Endogenous POST co-immunoprecipitates with endogenous STIM1 upon TG treatment		Jurkat T	
	POST enhances PMCA4 activity			Go et al. (2019)
SARAF	SARAF mediated SOCE inactivation (SCDI) following ER calcium refilling	Human	HEK293	Palty et al. (2012)
	SARAF C-domain (aa196-339) inhibited SOCE and interacts with STIM1			
	SARAF inhibited STIM1 through the SOAR domain, regulated by STIM1 CTID (aa448-530)			Jha et al. (2013)
	Reduced STIM1-SARAF interaction with pathological treatment of carbachol			Son et al. (2019)
	STIM1-ORAI1 interaction is required for SARAF-STIM1 interaction			Maleth et al. (2014)
	STIM1 K-domain interaction to PIP2 is required for SARAF-mediated SCDI			

(continued)

Table 4.2 (continued)

ER proteins	Interaction and differential modulation of calcium currents	Species	Cell type	References
	Knockdown of SARAF increased SOCE and STIM1-ORAI1 interaction		MEG01	Albarran et al. (2016a)
	Overexpression of SARAF reduced SOCE and STIM1-ORAI1 interaction			
	SARAF dissociates from STIM1 and binds ORAI1 C-terminus upon SOCE and interacts again with STIM1 for SCDI			
	SARAF at the plasma membrane modulates ORAI1	Mouse/Rat	SH-SY5Y NG115-401L	Albarran et al. (2016b)
	STIM1 regulates plasma membrane expression of SARAF			Albarran et al. (2017)
	SARAF modulates STIM1 and ORAI1 upregulation		Cardiomyocytes	Dai et al. (2018)
	Impairment of STIM1 Y316 phosphorylation increased SARAF interaction and SCDI		NG115-401L	Lopez et al. (2019)
	Knockout of SARAF increased calcium influx during SOCE		Pancreatic acinar	Son et al. (2019)
	SARAF overexpression reduced calcium influx		VSMCs	Yang et al. (2019a)
SURF4	Knockout of SURF4 increased SOCE SURF4 binds to STIM1 ER luminal domain	Mouse/chicken	DT40 B	Fujii et al. (2012)
STING	STIM1 interacts and retains STING at the ER to limit STING signaling at rest	Mouse	MEFs	Srikanth et al. (2019)
	Knockout of STIM1 increased STING-mediated type I IFN response			
	STIM1 interacts with STING N-terminal domain through the transmembrane domains	Human	PBMCs HEK293	
	STING deficient cells increased calcium influx and overexpression of STING reduced SOCE	Mouse	MEFs	
		Human	HEK293 Jurkat T	
	STING expression did not alter SOCE		THP1	

(continued)

Table 4.2 (continued)

ER proteins	Interaction and differential modulation of calcium currents	Species	Cell type	References
TMEM178	TMEM178 deficiency increased SOCE TMEM178 knockout increased SOCE Overexpression of TMEM178 reduced SOCE TMEM178 interacted with STIM1 but not STIM1 D76A mutant TMEM178 interacts with STIM1 via transmembrane domains	Mouse Human	Pre osteoclast BMMs HEK293	Decker et al. (2015) Yang et al. (2019b)
PDIA3	PDIA3 interacts with STIM1 luminal domain and require C49 and C56 Knockout of PDIA3 increased SOCE	Mouse	MEFs	Prins et al. (2011)
	PDIA3 interaction with STIM1 is reduced upon SOCE activation	Human	MKs	Di Buduo et al. (2020)
σR	σ1R overexpression reduced SOCE and σ1R knockdown increased SOCE SK10047 activating σ1R reduced SOCE and BD1047 inhibiting σ1R increased SOCE Exogenous σ1R co-immunopreicipitated with exogenous STIM1 SK10047 increased interaction and BD1047 reduced interaction	Human	HEK293	Srivats et al. (2016)
	σ1R modulates SOCE through TRPC1 Activation of σ2R with NO1 specifically reduced SOCE Overexpression of σ2R increased SOCE and silencing of σ2R decreased SOCE Endogenous σ2R co-immunoprecipitated with endogenous STIM1 Interaction is enhanced with TG and reduced with activation by NO1		SH-SY5Y MDA-MB-231	Sun et al. (2020) Cantonero et al. (2020)

(continued)

Table 4.2 (continued)

ER proteins	Interaction and differential modulation of calcium currents	Species	Cell type	References
PM				
VGCCs	Overexpression of STIM1 reduced calcium influx after membrane depolarization	Rat	Cortical neurons/ A7r5	Park et al. (2010) Wang et al. (2010)
	Overexpression of STIM1 impaired Cav1.2 currents	Human	HEK293	
	STIM1 CAD domain interacts with CaV1.2 C-terminus		Jurkat T	
	STIM1 causes internalization of CaV1.2		SH-SY5Y	
	STIM1 interacts with CaV3.1	Mouse	HL-1	Nguyen et al. (2013)
	STIM1 negatively regulates CaV3,1 activity			
AMPAR	Inhibition of SOCE reduced AMPA-induced calcium influx Inhibition of AMPAR reduced SOCE AMPAR interacts with STIM1	Rat	Cortical neurons Glia	Gruszczynska-Biegala et al. (2016)
NDMAR2	Inhibition of NMDAR reduced SOCE Inhibition of SOCE increased NMDA-induced calcium influx RNAi knockdown of STIM1 increased NMDA-induced calcium influx Overexpression of STIM1 reduced NMDA-induced calcium influx STIM1 interacts with NMDAR2	Rat	Cortical neurons	Gruszczynska-Biegala et al. (2020)
CLCA2	CLCA2 expression increased SOCE and ER calcium stores RNAi knockdown of CLCA2 reduced SOCE CLCA2 interacts with STIM1 at intracellular vesicles and ER-PM junctions	Human	HEK293	Sharma et al. (2018)
Cytoplasm				
Golli-MBP	Knockout of Golli enhanced SOCE and CRAC currents	Mouse	T-cells	Feng et al. (2006)

(continued)

Table 4.2 (continued)

ER proteins	Interaction and differential modulation of calcium currents	Species	Cell type	References
	Golli binds to STIM1 (C-terminus)	Human	HeLa	Walsh et al. (2010)
	Overexpression of Golli reduced SOCE			
	Reduced SOCE and upregulated Golli in patients with oral lichen planus		Lymphocytes	Ma et al. (2016)
	Knockout of Golli reduced SOCE and overexpression increased SOCE	Mouse	Oligodendrocyte precursor	Paez et al. (2007, 2009)
Calmodulin	STIM1 polybasic C-terminus (aa667-685) binds calmodulin	Human	Purified proteins	Bauer et al. (2008), Bhardwaj et al. (2013) O'Connell et al. (2011)
	Calmodulin interaction with STIM1 and ORAI1 (aa48-91) facilitate CDI		HEK293	Mullins et al. (2009)
	Calcium-dependent interaction of STIM1 SOAR (aa344-444) and calmodulin			Li et al. (2017)
	Calmodulin responsible for SCDI and dissociates STIM1-ORAI1 complex			
	STIM1 L374/V375 and L390/F391 important for calmodulin binding and SCDI			Bhardwaj et al. (2020)
CRACR2	CRACR2A interacts with STIM1 (C-terminus) and ORAI1 (N-terminus) at low calcium	Human	HeLa	Srikanth et al. (2010)
	RNAi knockdown of CRACR2A and B proteins reduced SOCE		HEK293	
	CRACR2A regulates SOCE in a calcium-dependent manner through its EF hand		Jurkat T	
EFHB	RNAi knockdown of EFHB reduced SOCE	Human	HEK293	Albarran et al. (2018)
	EFBH interacts with STIM1 upon SOCE		HeLa	
	EFHB required for STIM1-ORAI1 interaction upon SOCE			

(continued)

Table 4.2 (continued)

ER proteins	Interaction and differential modulation of calcium currents	Species	Cell type	References
	Overexpression of EFHB increased STIM1-ORAI1 interaction and SOCE			
	Knockdown of EFHB reduced ER calcium depletion induced STIM1-SARAF dissociation			
α-SNAP	RNAi knockdown of α-SNAP reduced SOCE	Drosophila	Kc	Miao et al. (2013)
	α-SNAP binds STIM1 and ORAI1 at ER-PM junctions	Human	HEK293	
	α-SNAP N-terminal hydrophobic loop (F27, F28, L31, L32) regulates SOCE		Jurkat T	
	α-SNAP binds STIM1 CAD			
BAG3	BAG3 interacts with STIM1 during mitosis	Mouse	T cells	Yu et al. (2019)
	SOCE is inhibited during mitosis	Hamster	CHO	
	STIM1 phosphorylation is not required during mitosis for SOCE inhibition	Human	HeLa	
	STIM1 interacts with BAG3 and localizes to autophagosomes in mitosis		HEK293	
	BAG3-induced STIM1 down regulation during mitosis		Jurkat T	
EB1	STIM1 C-terminus (aa392-652) binds to EB1 C-terminus	Human	HEK293	Grigorievet al. (2008)
	STIM1-EB1 interaction is important for microtubule growth-dependent ER remodeling			
	SxIP motif, TRIP sequence within STIM1 (aa642-645) is important for EB1 binding			Honnappaet al. (2009)
	STIM1-EB1 binding reduced during SOCE			
	Phosphorylation of STIM1 at S575, S608 and S621 are important for EB1 dissociation			Pozo-Guisado et al. (2013)
	STIM1 constitutively bound to EB1 reduced SOCE			

(continued)

Table 4.2 (continued)

ER proteins	Interaction and differential modulation of calcium currents	Species	Cell type	References
	EB1 binding to STIM1 prevents ER-PM translocation of STIM1		HeLa	Chang et al. (2018)
APC	RNAi knockdown of APC reduced SOCE APC knockdown results in STIM1 puncta away from the plasma membrane APC binds STIM1 C-terminus (aa650-685) STIM1 dissociates from EB1 before association to APC	Human	HEK293	Asanov et al. (2013)

STIMATE

TMEM110 or STIM-activating enhancer (STIMATE) is a newly described multi-pass membrane protein that localizes to the ER (Jing et al. 2015; Quintana et al. 2015). STIMATE was first identified as an interactor of STIM1 and enhancer of SOCE through an *in situ* proximity biotin labelling screen, a bimolecular fluorescence complementation secondary screen and another genome-wide screen for regulating nuclear factor of activated T-cells (NFAT), whose translocation depends on SOCE activity (Jing et al. 2015). Co-localization and Förster resonance energy transfer (FRET) studies revealed that STIMATE and STIM1 interact at rest and that, upon ER Ca^{2+} store depletion, a portion of STIMATE moves to ER-PM junctions (Jing et al. 2015). Overexpression of STIMATE enhanced STIM1 puncta size and formation and induced constitutive SOCE activity, while silencing and knockout of STIMATE reduced STIM1 puncta formation and SOCE in HeLa, HEK293 and Jurkat T cells (Quintana et al. 2015; Jing et al. 2015). Co-immunoprecipitation, FRET and glutathione S-transferase (GST) pull-down studies revealed that the STIMATE C-terminal region (aa214-294) interacted with the C-terminus of STIM1, specifically the CC1 domain (aa233-343) (Jing et al. 2015). Therefore, STIMATE appears to act as an enhancer of STIM1 by facilitating the disruption of the CC1-SOAR interaction that maintains STIM in an inactive conformation, thereby promoting STIM-ORAI1 association and SOCE activation (Jing et al. 2015). Furthermore, using electron microscopy (EM) and total internal reflection fluorescence (TIRF) microscopy and cortical ER labelling, STIMATE knockout studies revealed the importance of STIMATE in maintaining and remodeling cortical ER structure, which allows for efficient ER-PM MCS formation during SOCE (Jing et al. 2015; Quintana et al. 2015). Therefore, STIMATE functions as an enhancer of SOCE both by directly interacting with STIM1 to promote its efficient activation and by regulating ER-PM MCS formation.

UNC93B1

Un-coordinated 93 homologue B1 (UNC93B1) is the mammalian homologue of unc-93, a protein initially discovered in *Caenorhabdiis elegans* as a subunit of a two-pore potassium channel involved in coordinated muscle contraction (Levin and Horvitz 1992). In mammals, UNC93B1 is required for signaling of endosomal Toll-like receptors (TLRs) and plays a major role in the innate immune response against foreign microorganisms and in exogenous antigen presentation. UNC93B1 is a multi-spanning ER membrane protein and a single amino acid H412R mutation, which underlies the 'triple D' (3d) phenotype in mice, disrupted toll-like receptor 3, 7 and 9 signaling, antigen cross presentation via the major histocompatibility complex class I (MHCI) and exogenous antigen presentation via MHCII (Tabeta et al. 2006). UNC93B1 functions as an ER protein chaperone for TLR3, 5, 7, 8, 9 and 11 and is important for TLR stability, localization/trafficking and therefore for TLR-mediated immune signaling and response against foreign microbes and viruses (Tabeta et al. 2006; Huh et al. 2014; Itoh et al. 2011; Pifer et al. 2011; Lee et al. 2013; Pelka et al. 2018; Kim et al. 2008). We and others showed that UNC93B1 interacts with STIM1 to regulate phagosomal maturation and antigen cross presentation by enhancing Ca^{2+} signals associated with phagocytosis (Maschalidi et al. 2017). UNC93B1-STIM1 interaction was demonstrated by co-immunoprecipitation, and proximity ligation assays with protein overexpression and this interaction was disrupted by the UNC93B1 H412R 3d mutation (Maschalidi et al. 2017). UNC93B1 interaction with the ER-luminal domain of STIM1 promoted STIM1 oligomerization and ER-PM translocation, and loss of the STIM1 interaction in UNC93B1 3d/3d dendritic cells was associated with reduced SOCE and antigen cross-presentation that could be rescued by a constitutively active form of STIM1 (Maschalidi et al. 2017). STIM1 activity is important for generating local Ca^{2+} elevations (hotspots) at STIM1-mediated ER-phagosome MCSs during phagocytosis in dendritic cells (Nunes-Hasler et al. 2017). These local Ca^{2+} signals are required for phagosome maturation, proteolysis and lysosomal fusion, all steps that are crucial for antigen processing and cross-presentation (Nunes-Hasler et al. 2017). Loss of UNC93B1-STIM1 interaction in UNC93B1 3d/3d dendritic cells reduced ER-phagosome Ca^{2+} hotspots, phagosome maturation and proteolysis (Nunes-Hasler et al. 2017). These findings indicated that UNC93B1 regulates STIM1 activity and that this modulation is important for antigen cross-presentation in innate immunity. UNC93B1 appears to act as a chaperone to regulate STIM1 activation and may be involved in the coupling of STIM1 and TLR signaling during innate immune responses. The STIM1 activation step(s) regulated by UNC93B1 as well as the impact of TLR signaling on the regulation of STIM proteins by UNC93B1 remain to be established.

PMP22

Peripheral myelin protein 22 (PMP22) is a transmembrane protein highly expressed in Schwann cells of the peripheral nervous system whose mutations or gene duplications are linked to the inherited neurodegenerative disease Charcot-Marie-Tooth. PMP22 overexpression in HEK293 cells was associated with increased Gd^{3+}-sensitive, divalent-free induced currents following passive Ca^{2+} depletion of the ER by including EGTA in the pipette solution. The currents associated with PMP22 had a permeability profile similar to currents carried by TRPC1, and were absent in 60% of Schwann cells from $pmp^{-/-}$ mice, implicating a role for PMP22 in modulating TRPC1 channels (Vanoye et al. 2019). Cells expressing an ER-bound PMP22 L16P mutant had increased currents, suggesting that the modulation occurs in the ER and not at the plasma membrane (Vanoye et al. 2019). Consistently, Schwann cells from mice heterozygous for PMP22 L16P had increased intracellular Ca^{2+} levels and SOCE activity (Vanoye et al. 2019). PMP22 co-immunoprecipitated with STIM1 in mouse sciatic nerve suggesting that it might modulate SOCE through interaction with STIM1 at the ER (Vanoye et al. 2019), but how PMP22-STIM1 interaction at the ER could modulate TRPC1 channel activity is unclear.

ANO8

The TMEM16 or Anoctamin (ANO) family of proteins consists of 10 members that are similar in sequence to the yeast ER-PM MCS tether Ist2. Overexpression of ANO8 with STIM1 and ORAI1 increased CRAC current density and slow Ca^{2+}-dependent inactivation, an effect associated with increased STIM1-STIM1 and STIM1-ORAI1 interactions at ER-PM junctions (Jha et al. 2019). Co-immunoprecipitation and FRET studies revealed a direct interaction between ANO8 and STIM1, which increased following store-depletion as ANO8 migrated from the ER to the plasma membrane together with STIM1 (Jha et al. 2019). When constitutively active STIM1 mutants were expressed, knockdown of ANO8 reduced STIM1 clustering as well as CRAC current density and inactivation rates (Jha et al. 2019). ANO8 has a RRR motif (^{948}RPRRP^{952}) for PIP_2 binding and depletion of plasma membrane PI(4,5)P_2 with the FRB/FKBP system prevented the ANO8-mediated increase in STIM1-ORAI1 current, indicating that ANO8 acts as an ER-PM tether that recruits the SOCE machinery to ER-PM junctions. The increased calcium dependent inactivation (CDI) imparted by ANO8 was independent of the slow calcium dependent inactivation (SCDI) regulator SARAF (discussed further later) and was prevented by SERCA pumps inhibition with cyclopiazonic acid, indicating that the increased current inactivation was mediated by facilitated ER Ca^{2+} refilling. ANO8 thus appears to not only interact with STIM1 but to regulate SOCE activation and inactivation via its ability to assemble Ca^{2+} handling proteins at ER-PM junctions, highlighting the importance of MCS in Ca^{2+} signaling.

Prx-4

Cells tightly regulate their redox homeostasis and physiological production of reactive oxygen species (ROS) impacts a variety of cellular processes including proteostasis, immunity and cell death. The level of ROS production also affects Ca^{2+} signaling and was shown to modulate STIM and ORAI proteins via the reversible modification of cysteine residues [reviewed in Nunes and Demaurex (2014)]. Peroxiredoxin-4 (prx-4), an antioxidant enzyme that catalyzes the reduction of hydrogen peroxide to water, was shown to co-immunoprecipitate with endogenous STIM1 in heavy microsomal fractions from rat liver (Tam et al. 2018). Knockdown or inhibition of Prx-4 with Conoidin A in rat liver cells had no effect on Ca^{2+} fluxes but rendered SOCE susceptible to inhibition by H_2O_2 (Tam et al. 2018), suggesting that Prx-4 prevents a tonic inhibition by endogenous ROS. Further investigation is required to elucidate the mechanism of SOCE regulation by prx-4, in particular its effect on disulphide bridge formation between the two luminal cysteine residues of STIM1 (C49 and Y56) and whether this regulation relates to PDIA3 (discussed further below).

POST

TMEM20 or partner of STIM1 (POST) was first identified through tandem affinity purification and mass spectrometry in Jurkat cells (Krapivinsky et al. 2011). POST is an ER and plasma membrane located protein that co-immunoprecipitated with ORAI1 independently of ER Ca^{2+} content and STIM1 upon ER Ca^{2+} store depletion (Krapivinsky et al. 2011). Confocal images suggests movement of POST with STIM1 to ER-PM MCSs and POST-ORAI1 co-localization, however knockdown of POST only slightly reduced SOCE and overexpression of POST did not alter SOCE or ORAI1 dependent influx or current (Krapivinsky et al. 2011). Furthermore, analysis of POST interacting partners revealed a role for POST in facilitating STIM1 interaction with SERCA2A, PMCA, Na/Ka-ATPase, importin-β1 and exportin-1 upon ER Ca^{2+} depletion (Krapivinsky et al. 2011).

Interestingly the interaction between POST and PMCA, a Ca^{2+} export pump, revealed an inhibitory role of the STIM1-POST complex upon store depletion, as overexpression of POST reduced PMCA activity (Krapivinsky et al. 2011). Inhibition of Ca^{2+} extrusion by POST sustains high Ca^{2+} influx and may be relevant in muscular dystrophy characterized by upregulated levels of PMCA (Krapivinsky et al. 2011; Cully et al. 2012). Lastly, in Jurkat T cells Ca^{2+} dependent activation of NFAT was increased by overexpression of PMCA4, whose activity was dependent upon STIM1 and POST binding (Go et al. 2019). POST binding to the preferred PMCA4 splice variant PMCA4b enhanced PMCA4 activation, reducing the localized Ca^{2+} concentration at the membrane thereby indirectly increasing the global Ca^{2+} entry that promotes T cell activation (Go et al. 2019).

4.1.1.2 Negative Regulators of SOCE: SARAF, SURF4, STING, TMEM178 and PDIA3

SARAF

TMEM66 or SOCE-associated regulatory factor (SARAF) is a single membrane spanning protein of the ER that associates with STIM to facilitate the slow Ca^{2+}-dependent inactivation of SOCE (Palty et al. 2012). SARAF was identified during a functional expression screen for candidates that modulate mitochondrial Ca^{2+} homeostasis and serendipitously identified as a SOCE regulator as SARAF overexpression and knockdown impacted ER and cytosolic Ca^{2+} (Palty et al. 2012). Patch-clamp recordings revealed that SARAF expression increased SOCE inhibition (Palty et al. 2012) and that this effect was dependent on Ca^{2+} entry and on ER Ca^{2+} refilling, suggesting the role of SARAF in SCDI (Palty et al. 2012). Subsequent studies showed that SARAF mediates I_{CRAC} SCDI through its interaction with STIM1 and ORAI1 (Palty et al. 2012; Jha et al. 2013; Albarran et al. 2016a). The SARAF cytoplasmic domain (aa196-339) co-immunoprecipitated with STIM1, while the ER luminal domain (aa31-172) regulated the inhibitory activity with mutation of a conserved glutamate (E148) dominantly inactivating the SARAF phenotype (Palty et al. 2012). Co-immunoprecipitation studies revealed that SARAF binds the SOAR domain of STIM1 (aa344-442) and that deletion of a conserved STIM1 C-terminal inhibitory domain (CTID, aa448-530) disrupted the SARAF-STIM1 interaction (Jha et al. 2013), preventing the recruitment of SARAF to ER-PM MCSs and I_{CRAC} SCDI (Jha et al. 2013). The STIM1 CTID domain thus appears to direct SARAF access to SOAR to inhibit ORAI1 channel function.

SARAF binds to STIM1 at rest but dissociates upon ER Ca^{2+} depletion, and upon ER Ca^{2+} restoration, SARAF again interacts with STIM1 to mediate SCDI (Jha et al. 2013). SARAF-ORAI1 interaction was detected through proximal ligation assay and co-immunoprecipitation studies 30 s after TG-induced ER Ca^{2+} depletion, which coincided with SARAF-STIM1 dissociation (Albarran et al. 2016a). Binding of SARAF to ORAI1 occurs through the C-terminal region of ORAI1 and was independent of STIM1 expression (Albarran et al. 2016a) and this association was even required for SARAF-STIM1 interaction (Maleth et al. 2014). Furthermore, SARAF modulated ORAI1 channel activity, as overexpression of SARAF increased ATP-induced Ca^{2+} influx and silencing SARAF expression reduced this Ca^{2+} influx (Albarran et al. 2016a). This STIM1 independent regulation of ORAI1 was also demonstrated in neuroblastoma cells and may be a secondary function of SARAF in regulating the activation of arachidonate-regulated Ca^{2+} channels at the plasma membrane (Albarran et al. 2016b).

SARAF-mediated SCDI of SOCE prevents Ca^{2+} overfilling and spontaneous STIM1 activation and plays an important pathophysiological role (Palty et al. 2012). SARAF overexpression prevented cardiac hypertrophy induced by pressure overload in mice or induced by angiotensin II in cultured cardiomyocyte through the inhibition of SOCE (Dai et al. 2018). Through the downregulation of SOCE activity,

Ca^{2+} signaling mediated proliferation and migration of vascular smooth muscle cells, SARAF overexpression prevented balloon injury-induced neointimal hyperplasia in rats (Yang et al. 2019a). A reduced level of SARAF expression was observed in pancreatic tissue of patients with pancreatitis and in mice with induced pancreatitis, SARAF knockout increased disease severity through increased toxic Ca^{2+} influx (Son et al. 2019). The number of regulatory factors further implicates the importance of the SARAF-STIM1 interaction. The STIM1 polybasic lysine-rich (K) domain was important for SARAF-mediated SCDI through its interaction with plasma membrane PIP_2-rich microdomains and movement of the STIM1-ORAI1 complex from PIP_2-poor to a PIP_2-rich domain was important in facilitating the dynamic interactions between first STIM1 and ORAI1 during SOCE and then STIM1 and SARAF for SCDI (Maleth et al. 2014). The EF-hand domain family member B (EFHB) was identified as another interactor of STIM1 (is discussed in more detail subsequently) and is a regulator of SARAF-STIM1 interaction. Silencing EFHB expression prevented SARAF-STIM1 dissociation after TG-induced activation of SOCE (Albarran et al. 2018). Furthermore, phosphorylation of STIM1 at Y316 was important for SARAF-STIM1 dissociation and the Y316F phosphorylation mutant of STIM1 had increased association with SARAF after TG-induced SOCE (Lopez et al. 2019). To conclude, SARAF is an important interactor of STIM1 and regulator of Ca^{2+} influx and this interaction is in turn fine-tuned by a variety of other factors and post-translational modifications.

SURF4

The surfeit locus protein 4 (SURF4) is an integral membrane protein in the ER/ER-Golgi intermediate compartment (ERGIC) and has been implicated in the early secretion pathway and export of soluble proteins (Mitrovic et al. 2008; Saegusa et al. 2018; Yin et al. 2018; Emmer et al. 2018). SURF4 was identified as a STIM1 binding partner through an affinity purification of STIM1 protein complexes (Fujii et al. 2012). SURF4 co-immunoprecipitated with STIM1 and this interaction was lost when the luminal domain, but not the SAM domain, of STIM1 was deleted (Fujii et al. 2012). Co-immunoprecipitation studies showed that the SURF4-STIM1 interaction was unchanged during TG stimulation of SOCE, however confocal analysis revealed that SURF4 does not form STIM1-like clusters nor does it appear to associate with oligomerized STIM1 (Fujii et al. 2012). Functionally, in the absence of SURF4, Ca^{2+} influx and the number and rate of STIM1 puncta formation were increased during SOCE (Fujii et al. 2012). Therefore, SURF4 is a modulator of SOCE through STIM1 interaction. The mechanism of this modulation is still unknown; however, it may be interesting to investigate the involvement of SURF4 as a regulator of ER protein secretion and ER/ERGIC structure and stability (Mitrovic et al. 2008; Emmer et al. 2018).

STING

Tmem173 or the stimulator of interferon (IFN) genes (STING) is an ER and endosomal multi-pass transmembrane protein that is important for induction of pathways responsible for IFN activity and inflammation in response to DNA pathogens (Ishikawa and Barber 2008). STING upon activation by DNA sensors within the cytosol (ie. IFI16, cGAMP, etc.), translocates from the ER to ERGIC, where it recruits TANK-binding kinase 1 (TBK1) and interferon regulatory factor 3 (IRF3). IRF3 after phosphorylation by TBK1, homodimerizes and translocates into the nucleus to activate transcription of *IFN* genes (Ishikawa and Barber 2008). The STING-STIM1 interaction was uncovered when it was discovered that murine and human cells, as well as patients lacking STIM1 expression, had increased IFN responses (Srikanth et al. 2019). STIM1 deficient cells showed an increase *ifnb1* mRNA at rest and with stimulation, an increase in STING dimers, phosphorylated TBK1, IRF3 dimers and IRF3 nuclear localization (Srikanth et al. 2019). Furthermore, the increase in IFN response in STIM1 deficient cells was rescued through the co-deletion of both *STIM1* and *Tmem173* genes (Srikanth et al. 2019). The association between STIM1 and the STING/TBK1/IRF3/IFN pathway is mediated through the interaction of STING and STIM1. STING and STIM1 both co-localized at the ER at rest and co-immunoprecipitation studies revealed direct STING-STIM1 binding, specifically between the N-terminal domain of STING and the N-terminus (aa1-249) or the transmembrane domain of STIM1 (Srikanth et al. 2019). Furthermore, the STING-STIM1 interaction was reduced upon TG-induced SOCE or cGAMP-induction of the STING pathway (Srikanth et al. 2019). The binding of STING by STIM1 was important for the retention of STING at the ER to prevent spontaneous STING activation and ERGIC translocation, as STIM1 deficiency revealed an increase in ERGIC localization of STING at rest and after stimulation in confocal images (Srikanth et al. 2019). The STING-STIM1 interaction is interesting because STIM1 is the modulator of the activity of STING-induced IFN response.

On the other hand, STING was shown to modulate SOCE in certain cell types. In mouse embryonic fibroblasts and human Jurkat T cells, STING deficiency increased SOCE but in THP1 human monocytes, there was no effect (Srikanth et al. 2019). It would be interesting to see if STING activity can modulate SOCE and if this is related to other immune functions of STIM1 and SOCE. Furthermore, STIM1 activation during SOCE by TG-induced ER Ca^{2+} depletion reduced STING-STIM1 interaction and therefore, SOCE activation may indirectly increase basal level of STING activity. In addition, another protein TMEM203, promoted STING translocation from the ER and competed with STIM1 for STING binding (Li et al. 2019). TMEM203 was shown to be an interactor of ITPR and SERCA2 as well as of STIM1 through co-immunoprecipitation and to be a regulator of ER Ca^{2+} levels (Li et al. 2019). Altogether, these data link Ca^{2+} signaling mediated by STIM1 to the IFN response mediated by STING; however, more studies are needed to establish the mechanism of this relationship.

TMEM178

TMEM178 is a newly described multi-pass integral membrane protein of the ER that is a downstream gene target of nuclear factor of activated T-cells, cytoplasmic 1 (NFATc1), activated by the phospholipase C gamma-2 (PLCγ2) Ca^{2+} signaling pathway during osteoclastogenesis. SOCE is an important process during osteoclastogenesis and knockout of TMEM178 in osteoclast precursors increased Ca^{2+} fluxes and SOCE (Decker et al. 2015; Yang et al. 2019b). TMEM178 is also highly expressed in bone marrow macrophages (BMMs) and TMEM178 knockout in BMMs increased ER Ca^{2+} release and SOCE (Yang et al. 2019b). Furthermore, TMEM178 co-immunoprecipitated STIM1 and this interaction was reduced following ER Ca^{2+} depletion induced by TG (Decker et al. 2015). The interaction is dependent on ER Ca^{2+}, as the STIM1 D76A mutant does not bind TMEM178 and TMEM178 expression does not reduce STIM1 D76A mediated constitutive SOCE activity (Yang et al. 2019b). Based on mutational and co-immunoprecipitation studies, STIM1 transmembrane domain binds TMEM178 transmembrane region 2 and 3 with residues G225 of STIM1 and L212/M216 of TMEM178 modulating this interaction and the activation of SOCE (Yang et al. 2019b). The TMEM178-STIM1 interaction inhibited STIM1 ER-PM puncta formation and prevented ORAI1 activation by STIM1 (Yang et al. 2019b). Therefore, the TMEM178-STIM1 interaction negatively regulates SOCE and due to the expression profile of TMEM178 this negative regulation likely plays a role during osteoclastogenesis and during the activation of macrophages.

PDIA3

Protein disulfide-isomerase A3 (PDIA3) is a thiol oxidoreductase of the ER that in concert with other protein chaperones (i.e. calreticulin and calnexin), functions to facilitate disulphide bond formation and to ensure folding quality of glycoproteins (Frickel et al. 2004). A surface plasmon resonance screen of ER proteins that interacted with the ER luminal domain of STIM1 identified PDIA3 as a binding partner and further FRET analysis confirmed this interaction (Prins et al. 2011). Functional studies revealed that in the absence of PDIA3 there was increased SOCE, increased STIM1 puncta at rest and increased rate of STIM1 puncta formation after TG induction (Prins et al. 2011). Furthermore, the PDIA3-STIM1 interaction is dependent upon two cysteine residues (C49 and C56) upstream of the EF hand within STIM1 and mutation of these two cysteines resulted in reduced SOCE and STIM1 puncta formation (Prins et al. 2011). Therefore, PDIA3 acts to regulate STIM1 activity and Ca^{2+} homeostasis through binding of STIM1. This is further illustrated in patients with myeloproliferative neoplasms due to mutations of the calreticulin gene (which makes up 20–25% of disease) (Di Buduo et al. 2020). In megakaryocytes, thrombopoietin receptor activity causes ITPR-mediated ER Ca^{2+} release and SOCE. With calreticulin mutated megakaryocytes, there is a reduced interaction between calreticulin and PDIA3 and between PDIA3 and STIM1, resulting in spontaneous activation of STIM1 and SOCE that is associated with an increase in cellular proliferation, a hallmark of myeloproliferative neoplasms (Di Buduo et al. 2020). Further investigation is necessary for depicting the modulation of STIM1 by PDIA3 and calreticulin.

Sigma Receptors

The sigma1 receptor is an ER membrane protein with two transmembrane domains activated by a wide variety of ligands. Ligand binding of sigma1 receptors as well as Ca^{2+} store depletion causes its release from ER chaperone GRP78 and movement to interact with and activate plasma membrane proteins, such as ion channels (Srivats et al. 2016). In HEK cells overexpression or activation of sigma1 receptors with agonists decreased SOCE whereas inhibition with antagonists increased SOCE (Srivats et al. 2016). Furthermore, STIM1 co-immunoprecipitated with sigma1 receptors and this interaction was increased by agonists and decreased by antagonists of sigma1 receptors (Srivats et al. 2016). Upon ER Ca^{2+} store depletion the sigma1 receptor translocated to ER-PM MCSs along with STIM1 and colocalized with ORAI1 and STIM1 whose interaction was reduced by overexpression of sigma1 receptors (Srivats et al. 2016). Sigma1 receptors also mediate Ca^{2+} influx through TRPC1 (Sun et al. 2020) as the reduced sigma1 receptor-STIM1 interaction observed upon SOCE induction allowed STIM1 to activate plasma membrane TRPC1 (Sun et al. 2020). This inhibitory function of sigma1 receptor on TRPC1 activity through STIM1 regulation was important in preventing cell death induced by neurotoxins in SH-SY5Y cells (Sun et al. 2020).

The sigma2 receptor is lesser known and has been described to play a role in cholesterol transport (Cantonero et al. 2020). Sigma2 receptor overexpression increased SOCE and silencing of sigma2 receptors reduced SOCE in MDA-MB-231 breast cancer cells lines (Cantonero et al. 2020). Sigma2 receptors co-immunoprecipitated with STIM1 but not ORAI1 during SOCE induction (Cantonero et al. 2020). NO1, a ligand of sigma2 receptor, reduced SOCE and sigma2 receptor-STIM1 interaction and attenuated SOCE-dependent cell migration and proliferation in the triple negative breast cancer cell line, MDA-MB-231 (Cantonero et al. 2020).

4.1.2 Plasma Membrane Proteins

4.1.2.1 Alternate STIM1 Targets: Voltage- and Ligand-Gated Ca^{2+} Channels: VGCCs, AMPAR and NMDAR

While there has been an extensive amount of studies on the mechanism of SOCE in non-excitable cells, studies focussing on SOCE in excitable cells such as neurons, muscle cells and certain endocrine cells are scarce. Ca^{2+} entry in excitable cells is typically mediated by voltage-gated and ligand-gated Ca^{2+} channels, which drive processes such as heart muscle contraction, synaptic transmission and hormone release, however these channels and the SOCE machinery co-exist in both excitable and non-excitable cells. How these various Ca^{2+} entry pathways interact and facilitate differential physiological processes is not fully understood. Here we will review the intricate interactions of STIM1 with voltage-gated Ca^{2+} channels (VGCCs) and the ligand-gated channels α-amino-3-hydroxy-5-methyl-4-isoxazolepropionic acid receptor (AMPAR) and *N*-methyl-D-aspartate receptor (NMDAR).

VGCCs

VGCCs are Ca^{2+} channels that are activated by membrane depolarization and function to trigger neuron excitation, synaptic release of neurotransmitters or hormones and muscle contraction. L-type Ca^{2+} channels are composed of CaV1 family of channel subunits and the CaV1.2 subunits specifically are expressed in cardiac and smooth muscle myocytes, neuronal cell bodies and dendrites as well as endocrine cells. STIM1 was found to directly associate with CaV1.2 channels and to negatively modulate CaV1.2 channel activity during SOCE (Wang et al. 2010; Park et al. 2010). In rat cortical neurons, the main mechanism of Ca^{2+} entry occurs through VGCCs activated by membrane depolarization and TG-induced ER-Ca^{2+} release does not generate a substantial Ca^{2+} influx (Park et al. 2010). Unexpectedly, the induction of the SOCE machinery by TG during membrane depolarization reduced Ca^{2+} influx through VGCCs and this reduction was also seen in cells overexpressing STIM1 (Park et al. 2010). In vascular smooth muscle cells, membrane depolarization to activate CaV1.2 channels were reduced when ER-Ca^{2+} stores were previously depleted to induce SOCE (Wang et al. 2010). Similar negative regulation of CaV1.2 channel activity by ER Ca^{2+} store depletion was also seen in HEK293 cells and overexpression of constitutively active STIM1 (EF hand D76A mutant) further increased this inhibitory effect under both store replete and deplete conditions (Wang et al. 2010). Moreover, membrane depolarization induced Ca^{2+} entry was absent in T-lymphocytes expressing CaV1.2 and Ca^{2+} influx was mediated predominantly through induction of SOCE (Park et al. 2010). Furthermore, when STIM1 expression was silenced in T-lymphocytes expressing CaV1.2, Ca^{2+} influx was observed after membrane depolarization (Park et al. 2010). Altogether, these data suggest that STIM1 expression and activation inhibits CaV1.2 Ca^{2+} conductance.

When ER-Ca^{2+} store depletion was induced, CaV1.2 translocated to STIM1-ORAI1 complexes on the plasma membrane and co-immunoprecipitation studies revealed a direct interaction between STIM1 and CaV1.2 subunits, specifically between the STIM1 CAD domain and the CaV1.2 C-terminus (aa1809-1908) (Wang et al. 2010; Park et al. 2010). The direct binding between STIM1 and CaV1.2 lead to two different mechanisms of negative regulation; the CAD domain of STIM1 directly inhibited channel activity (Wang et al. 2010) and subsequently induced the internalization and degradation of CaV1.2, as shown with reduced

CaV1.2 membrane expression (Park et al. 2010). Therefore, STIM1 functions to reciprocally regulate ORAI1 and CaV1.2 Ca^{2+} channel activity revealing insight into how these two Ca^{2+} entry pathways co-exist and function in cells. The importance of the negative regulatory effect of STIM1 on VGCCs is further illustrated in Ca^{2+}-induced long-term potentiation in postsynaptic dendritic spines (Dittmer et al. 2017). Induction of postsynaptic Ca^{2+} influx through L-type Ca^{2+} channels and glutamate receptor NMDAR, trigger Ca^{2+} induced Ca^{2+} release from the ER and activates STIM1 (Dittmer et al. 2017). Movement of STIM1 to ER-PM MCSs allow STIM1 to feedback and inhibit CaV1.2 channels, which promotes CaV1.2-dependent ER content growth within dendritic spines and shuts down CaV1.2 Ca^{2+} influx dependent signaling and gene expression (Dittmer et al. 2017). CaV1.2-STIM1 interaction is therefore important in the formation of neuronal plasticity.

In cardiomyocytes the fluctuation of Ca^{2+}, through VGCCs, Ca^{2+}-induced Ca^{2+} release and SOCE, are tightly regulated events. In mouse cardiac cells, knockdown of STIM1 increased ryanodine receptor Ca^{2+} release from the sarcoplasmic reticulum (SR) and produced irregular Ca^{2+} oscillations and contractions (Nguyen et al. 2013). Furthermore, there was an increase in current density of T-type Ca^{2+} channels and increased cell surface expression of T-type Ca^{2+} channels when STIM1 was knocked down (Nguyen et al. 2013). The negative regulation of STIM1 on T-type Ca^{2+} channels was mediated though the direct interaction between STIM1 and CaV3.1 subunits. Therefore, STIM1 negatively regulates T-type Ca^{2+} channels to allow SOCE and SR Ca^{2+} refilling during contraction of cardiomyocytes. This may be mediated in a similar manner to STIM1 and CaV1.2 interaction, where channel activity is directly blocked through binding and reduced CaV1.2 cell surface expression through STIM1-mediated internalization and degradation.

AMPAR

The ionotropic glutamate receptors AMPARs are tetramers of glutamate subunits (GluA1-4) and are permeable to K^+, Na^+ and Ca^{2+}, depending on the subunit expression and composition. AMPARs are commonly expressed in neurons and cells of the central nervous system and is important for early synaptic currents. In rat cortical neurons Ca^{2+} influx through AMPARs was reduced by SOCE inhibitors, suggesting a role for SOCE in AMPAR-induced Ca^{2+} signaling (Gruszczynska-Biegala et al. 2016). Additionally, when SOCE was induced through Ca^{2+} store depletion by TG, antagonists of AMPAR reduced Ca^{2+} influx in neurons and glial cells (Gruszczynska-Biegala et al. 2016). Altogether these data suggest a coordination between AMPAR signaling and the SOCE machinery in the modulation of Ca^{2+} in neuronal cells. The interdependence of these two Ca^{2+} entry pathways may be mediated by the direct interaction between STIM1 and AMPAR, as co-immunpreicipitation studies revealed STIM1 binding to GluA1 and GluA2 subunits (Gruszczynska-Biegala et al. 2016).

NMDAR

NMDARs are also ionotropic glutamate receptors; they are heterotetramers composed typically of two GluN1 and two GluN2 subunits. NMDARs are expressed in nerve cells and are important for late synaptic transmission and synaptic plasticity, a critical cellular mechanism underlying learning and memory. When SOCE was initiated by TG-induced Ca^{2+} store depletion, inhibition of NMDARs reduced Ca^{2+} influx in rat cortical neurons (Gruszczynska-Biegala et al. 2020). It appears that NMDAR activity contributes to neuronal SOCE. However, when NMDAR-mediated Ca^{2+} influx was measured, SOCE inhibitors increased Ca^{2+} entry, suggesting a negative modulation of SOCE on NMDAR Ca^{2+} signaling (Gruszczynska-Biegala et al. 2020). In concert with this observation, the knockdown of STIM1 also increased NMDAR Ca^{2+} entry and overexpression of STIM1 had the opposite effect (Gruszczynska-Biegala et al. 2020). The negative modulation of NMDAR-mediated Ca^{2+} influx by STIM1 was suggested to be through direct interaction. Co-immunoprecipitation, proximity ligation and co-localization studies revealed binding of GluN2A and GluN2B by STIM1 (Gruszczynska-Biegala et al. 2020).

4.1.2.2 Plasma Membrane Protein CLCA2

CLCA2

The chloride channel accessory (CLCA) proteins CLCA1 and CLCA2 are transmembrane metalloproteases that activate the Ca^{2+}-activated chloride channel TMEM16A or ANO1 when overexpressed. While CLCA1 directly interacts and activates TMEM16A, CLCA2 modulates TMEM16A currents by promoting SOCE (Sharma et al. 2018). Specifically CLCA2 overexpression increased ER Ca^{2+} release and SOCE while knockdown of CLCA2 reduced ER Ca^{2+} release and SOCE (Sharma et al. 2018). Co-immunoprecipitation studies showed that CLCA2 interacted with both STIM1 and ORAI1 after ER Ca^{2+} store depletion and CLCA2 co-localized with STIM1 and ORAI1 at the plasma membrane (Sharma et al. 2018). CLCA2 does not interact with TMEM16A or affect TMEM16A cell surface expression but modulates TMEM16A current activity by increasing the amount of Ca^{2+} released from ER stores through its positive regulation of SOCE, the CLCA2-STIM1 interaction promoting ER Ca^{2+} store refilling to provide sufficient Ca^{2+} for activation of TMEM16A chloride currents.

4.1.3 Cytoplasmic Proteins

4.1.3.1 Signaling Proteins

Golli-MBP

The golli protein is a second class of proteins encoded by the myelin basic protein (*MBP*) gene and unlike the classic MBPs that are structural components of myelin in the nervous system, golli proteins are also expressed in immune cells and plays a role in the regulation of SOCE. Golli proteins are highly expressed in T-cells, are distributed in the cytoplasm and associate increasingly with the plasma membrane upon T cell activation (Feng et al. 2006). Mice deficient in golli protein had highly proliferative T cells and this was due to increased SOCE activity and activation of proliferative pathways (Feng et al. 2006). In oligodendrocyte precursors, golli proteins also play a role in proliferation through the regulation of SOCE at the plasma membrane (Paez et al. 2007, 2009). Contrary to T cells, golli overexpression increased the proliferation of oligodendrocyte precursors through enhanced SOCE and knockout of golli reduced SOCE and oligodendrocyte precursor proliferation (Paez et al. 2009). This discrepancy may be due to the different golli isoforms expressed, the main isoform expressed in T cells being the golli-BG21 and the main isoform manipulated in oligodendrocyte precursors the golli-J37. The modulation of SOCE by golli-BG21 is attributed to its interaction with the C-terminus of STIM1 and co-localization studies revealed that this interaction occurred at the plasma membrane following ER Ca^{2+} store depletion (Walsh et al. 2010). Furthermore, patients with oral lichen planus, a disease of the oral mucosa mediated by T cells, have increased levels of golli and STIM1 proteins and reduced intracellular Ca^{2+}, implicating golli-STIM1 and SOCE regulation in this pathology (Ma et al. 2016). Therefore, golli-BG21 negatively regulates SOCE and this is mediated through interaction with STIM1. Golli-BG21 is the more widely expressed isoform, however whether golli-J37 also modulates SOCE through STIM1 interaction remains to be seen and the differential effect of SOCE by the two isoforms remains to be elucidated.

Calmodulin

Calmodulin is a widely expressed cytoplasmic protein that in eukaryotes plays a critical role in transducing Ca^{2+} signals into pathways involved in cell growth, proliferation, apoptosis and movement. Calmodulin contains four EF-hand domains and when bound to Ca^{2+} changes to an activate conformation which allows it to interact and activate subsequent protein targets such as kinases. Calmodulin activity is controlled by Ca^{2+} elevations and participates in the Ca^{2+}-dependent inactivation (CDI) of SOCE as Ca^{2+} bound calmodulin binds to STIM1 and ORAI1 to mediate the CDI of CRAC channels (Bauer et al. 2008; Mullins et al. 2009; Bhardwaj et al.

2020). Purified Ca^{2+}/calmodulin constructs bind the polybasic region of STIM1 (aa667-685) with high affinity (Bauer et al. 2008; O'Connell et al. 2011) and co-immunoprecipitation studies demonstrated an interaction between Ca^{2+}/calmodulin and the SOAR domain (aa344-444) of STIM1 (Li et al. 2017). The binding of Ca^{2+}/calmodulin to the polybasic region of STIM1 may be an important interaction that disrupts STIM1-PIP_2 binding and therefore cause STIM1 dissociation from the plasma membrane (Bhardwaj et al. 2013). The binding of Ca^{2+}/calmodulin to both the inactivation domain of STIM1 (aa470-491) and to the N-terminus (aa68-91) of ORAI1 facilitated the fast calcium dependent inactivation (FCDI) of ORAI1 channels (Mullins et al. 2009), whereas interaction between Ca^{2+}/calmodulin and two hydrophobic regions (L374/V375 and L390/F391) within the SOAR domain of STIM1 (Bhardwaj et al. 2020) dissociated the STIM1-ORAI1 complex and modulated SCDI. These data indicate that calmodulin is a negative feedback regulator of the SOCE machinery by controlling the CDI of ORAI1 channels.

CRACR2

The Ca^{2+} release-activated Ca^{2+} channel regulator 2A (CRACR2A) is an EF-hand containing cytoplasmic protein that was first identified as a binding partner of ORAI1 during a large-scale affinity purification analysis and later found to be a direct interactor of both ORAI1 and STIM1 through co-immunoprecipitation (Srikanth et al. 2010). CRACR2A binds to the N-terminus of ORAI1 and to the C-terminus of STIM1 and these interactions were enhanced under low Ca^{2+} conditions (Srikanth et al. 2010). Silencing of CRACR2A or of its homologue CRACR2B reduced SOCE and STIM1-ORAI1 clustering at ER-PM MCSs (Srikanth et al. 2010). Furthermore, expressing an EF-hand mutant of CRACR2A increased SOCE, caused spontaneous STIM1-ORAI1 ER-PM MCS clustering and mediated cell death in Jurkat T-cells (Srikanth et al. 2010). This study suggests that CRACR2A and perhaps CRACR2B maintains STIM1-ORAI1 stability at ER-PM MCSs in a Ca^{2+}-dependent manner but the underlying mechanism requires further investigation.

EFHB

The EF-hand domain family member B (EFHB) is a widely expressed cytosolic protein with two EF-hand domains whose expression levels are associated with SOCE (Albarran et al. 2018). EFHB co-immunoprecipitated with STIM1 and ORAI1 and these interactions increased upon ER-Ca^{2+} release with TG (Albarran et al. 2018). EFHB knocked-down reduced TG-induced STIM1-ORAI1 interaction while EFHB overexpression had the opposite effect (Albarran et al. 2018). The EFHB-STIM1 and SARAF-STIM1 interactions were inversely related as the amount of SARAF co-immunoprecipitated with STIM1 first decreased then recovered after TG treatment while the amount of co-immunoprecipitated EFHB increased and then decreased with a similar time-course (Albarran et al. 2018). In addition, the amount of EFHB co-immunoprecipitated with STIM1 at early time points was reduced by SARAF overexpression, while the amount of co-immunoprecipitated SARAFSTIM1 remained high with EFHB silencing (Albarran et al. 2018). As SARAF did not co-immunoprecipitate EFHB itself, further investigation into the exact EFHB binding sites of STIM1 may reveal a competitive binding mechanism. These findings implicate the EF hand containing protein EFHB as another cytosolic Ca^{2+} sensor that regulates SOCE and SARAF-mediated CDI.

A-SNAP

The N-ethylmaleimide-sensitive factor attachment protein alpha (α-SNAP) plays an important role in intracellular trafficking by acting as an adaptor protein between N-ethlmaleimide-sensitive factor (NSF) and soluble NSF-attachment protein receptor (SNARE) protein complexes to facilitate SNARE complex disassembly and vesicle-target membrane fusion. A genome-wide RNAi screen identified α-SNAP as a candidate protein that modulates SOCE and further validation showed that knock-down of α-SNAP reduced SOCE in HEK-293, Jurkat T and U2OS cells (Miao et al. 2013). α-SNAP was shown to directly interact with both STIM1 and ORAI1 through co-immunoprecipitation and when cells were stimulated with TG, α-SNAP co-localized at clusters in STIM1 and ORAI1 co-expressing cells (Miao et al. 2013). Specifically the LOOP domain of α-SNAP, which consists of key residues F27, F28, L31 and F32 and is located in the N-terminus, is responsible for modulation of SOCE and interaction with STIM1 and ORAI1 (Miao et al. 2013). This is distinctive from the α-SNAP C-terminus region responsible for SNARE complex disassembly. Furthermore, an α-SNAP-L294A mutant dysfunctional in promoting NSF ATPase activity still interacted with STIM1 and ORAI1 and was able to reconstitute SOCE in α-SNAP knockdown cells (Miao et al. 2013). Overall, these data suggest that α-SNAP modulates SOCE independently of its effects on intracellular SNARE-mediated trafficking. The α-SNAP-STIM1 interaction occurred through the CAD domain of STIM1 but α-SNAP-dependent SOCE potentiation was only observed when ORAI1 was expressed with full length STIM1 and not with STIM1 CAD domain, suggesting that the modulation occurs prior to ORAI1 channel activation (Miao et al. 2013). Furthermore, the STIM1/ORAI1 ratio at ER-plasma membrane clusters was reduced with α-SNAP knockdown and increased with α-SNAP overexpression, indicating that α-SNAP modulate STIM1-ORAI1 trapping for efficient SOCE (Miao et al. 2013). Recently, it was shown that the interaction between α-SNAP and the C-terminus of ORAI1 was important for the assembly of ORAI1 and knockdown of α-SNAP increased the proportion of ORAI1 hexamers while decreasing the channel Ca^{2+} permeability and increasing its Na^{+} permeability (Li et al. 2016). These data place α-SNAP as a component of the SOCE machinery that regulates STIM1-ORAI1 interaction and assembly for optimal Ca^{2+} influx during SOCE.

BAG3

Inactivation of Ca^{2+} influx pathways, including SOCE, is important for efficient cell division during both mitosis and meiosis. Accordingly, TG-induced SOCE was downregulated in HEK 293, Jurkat T and $CD4^{+}$ primary T cells undergoing mitosis and upregulation of SOCE during mitosis by overexpressing ORAI1 and STIM1-CAD resulted in mitotic catastrophe and cell death (Yu et al. 2019). The reduction of SOCE during mitosis was not due to altered ORAI1 expression nor to STIM1 phosphorylation (Yu et al. 2019). The Bcl2-associated athanogene 3 (BAG3) is a co-chaperone that modulates a variety of adaptive pathways, such as autophagy, in response to cellular stress (Rosati et al. 2011). Mass spectrometry analysis showed increase BAG3 expression during mitosis and pulldown experiments revealed STIM1-BAG3 interaction during mitosis (Yu et al. 2019). This interaction was suggested to be responsible for the targeting of endogenous STIM1 to LC3 positive autophagosomes for degradation, thereby contributing to a 50% decrease in STIM1 expression during mitosis (Yu et al. 2019). Whether or not this interaction and regulation of STIM1 expression is sufficient in inhibiting SOCE during mitosis is still unclear because other factors such as ER-PM MCS remodeling and prevention of STIM1 cluster formation may also be at play. Nevertheless, BAG3 was identified as a new modulator of STIM1 expression and SOCE activity and may be important for cellular development and cellular stress responses.

4.1.3.2 Cytoskeleton Proteins: Controlling STIM1 Trafficking

EB1

The microtubule-associated protein RP/EB family member 1 (EB1) is a microtubule plus-end-tracking protein that recruit other microtubule plus-end-tracking proteins and modulates microtubule growth dynamics (Tirnauer and Bierer 2000). Through GST pulldown and mass spectrometry, STIM1 was identified as an abundant binding partner of both EB1 and EB3 (Grigoriev et al. 2008). Further co-immunoprecipitation and co-localization studies confirmed direct interaction of STIM1 C-terminus with EB1 C-terminus and STIM1 localization at EB1-positive microtubule ends (Grigoriev et al. 2008). STIM1 possesses a TRIP sequence similar to the SxIP motif responsible for localizing proteins to microtubule plus-ends in an EB1 dependent manner (Honnappa et al. 2009). Phosphorylation sites in proximity to the TRIP motif were identified to regulate EB1 binding and STIM1 dynamics. The phosphorylation of STIM1 at S575, S608, and S621, by ERK1/2 increased with TG stimulation mutations at these sites reduced SOCE and sequential Ca^{2+} influx levels (Pozo-Guisado et al. 2013). Furthermore constitutive phosphorylation of these serine residues reduced EB1 binding and co-localization at microtubule plus-end (Pozo-Guisado et al. 2013). Both the constitutive phosphorylated STIM1 mutant and the TRIP STIM1 mutant disrupted EB1 binding and increased STIM1 multimerization (Pozo-Guisado et al. 2013). The STIM1-EB1 interaction is responsible for specific

ER subcellular localization and phosphorylation of Ser575, S608, and S621 disrupted the interaction and facilitated STIM1 multimerization and translocation to ER-PM puncta, and thereby SOCE activation (Pozo-Guisado et al. 2013; Chang et al. 2018). The STIM1-EB1 interaction illustrates a mechanism by which STIM1 enables growing microtubules to remodel the ER while at the same time limiting SOCE.

APC

The adenomatous polyposis coli (APC) protein is a multifaceted tumor suppressor that plays a role in Wnt signaling, cell adhesion and regulation of the actin and microtubule network. It is an interactor of EB1 and is a microtubule plus-end binding protein. It was observed that knockdown of APC in cells resulted in a reduced ER STIM1 tubular structure and formation of puncta away from the plasma membrane upon TG induction (Asanov et al. 2013). Furthermore, knockdown of APC in cells strongly reduced TG-mediated SOCE and ORAI1 currents (Asanov et al. 2013). Co-localization and co-immunoprecipitation studies revealed that STIM1-EB1 binding occurred at rest and prior to APC-STIM1 binding during TG-induced SOCE (Asanov et al. 2013). The specific APC binding domain of STIM1 was identified in the C-terminus, downstream of the TRIP motif for EB1 binding, at aa650-685 (Asanov et al. 2013). The interaction of STIM1 with EB1 and APC illustrates a mechanism by which cytoskeletal proteins regulate STIM1 translocation from the ER to ER-PM junctions during Ca^{2+} signaling. In this framework, STIM1 binds EB1 at rest and associates with APC during ER Ca^{2+} depletion to facilitate STIM1 translocation to ER-PM junctions.

4.2 Conclusions

The discovery of STIM and ORAI family of proteins 15 years ago initiated a wave of research to identify new molecules interacting with these two essential players of the ubiquitous store-operated Ca^{2+} entry process. In this chapter we attempted to review the proteins of the plasma membrane, ER, and cytosol that were reported to physically interact with STIM1 either as a target of its channel trapping and gating activity, or as a regulator of the complex conformational changes and trafficking events that underlie STIM1-mediated activation, deactivation, and inactivation of SOCE. The field is still in an exponential growing phase and we apologize for any study that we might have omitted in this chapter.

4.3 Summary Points

- More than 20 proteins were reported to interact with the ER Ca^{2+} sensor STIM1 since this protein was identified in 2005 as the main regulator of store-operated Ca^{2+} entry (SOCE)
- In addition to interacting with the Ca^{2+}-permeable ORAI and TRPC channels, STIM was reported to interact with several families of ion channels and to regulate their activity at the plasma membrane.
- Several cytosolic and ER proteins interact with STIM1 and modulate SOCE by regulating the conformational changes underlying STIM1 activation and deactivation, the translocation of STIM1 to membrane contact sites, and the interactions of STIM1 with ion channels and transporters
- By controlling the activity and dynamics of STIM1 and of its target channels, these regulatory proteins can shape the spatiotemporal patterns of Ca^{2+} signals that regulate critical cellular functions

4.4 Future Issues

- Establish whether STIM1 interacting proteins determine specific I_{CRAC} features since STIM1 and ORAI1 are sufficient to reconstitute SOCE (SARAF).
- Establish whether STIM1 interacting proteins are mechanistically relevant for functions beyond SOCE
- Establish whether STIM1 interacting proteins participate in signalling networks relevant for immune or muscle cell function (IPTR, STING, UNC93B1)
- Exploit the additional pharmacological space conferred by STIM1 interacting proteins to design new drugs targeting immune and muscle cell activity
- Establish the molecular mechanism of STIM1 interacting proteins whose mode of action is unclear (UNC93B1, PMP22, PRX-4, SURF4, STING, PDIA3, VCGGs, AMPAR, NMDAR2, golli-MBP, CRACR2)
- Perform replication studies to validate STIM1 interacting proteins reported by a single study (UNC93B1, PMP22, ANO8, PRX-4, STING, NDMAR2, CLCA2, CRACR2, EFHB, a-SNAP, BAG3, APC)

Acknowledgments ND research is financed by SNF grant N° 31000-189042.

Conflict-of-Interest Disclosure Statement None.

References

Alansary D, Bogeski I, Niemeyer BA (2015) Facilitation of Orai3 targeting and store-operated function by Orai1. Biochim Biophys Acta 1853(7):1541–1550. https://doi.org/10.1016/j.bbamcr.2015.03.007

Albarran L, Lopez JJ, Amor NB, Martin-Cano FE, Berna-Erro A, Smani T, Salido GM, Rosado JA (2016a) Dynamic interaction of SARAF with STIM1 and Orai1 to modulate store-operated calcium entry. Sci Rep 6:24452. https://doi.org/10.1038/srep24452

Albarran L, Lopez JJ, Woodard GE, Salido GM, Rosado JA (2016b) Store-operated Ca^{2+} entry-associated regulatory factor (SARAF) plays an important role in the regulation of arachidonate-regulated Ca^{2+} (ARC) channels. J Biol Chem 291(13):6982–6988. https://doi.org/10.1074/jbc.M115.704940

Albarran L, Regodon S, Salido GM, Lopez JJ, Rosado JA (2017) Role of STIM1 in the surface expression of SARAF. Channels (Austin) 11(1):84–88. https://doi.org/10.1080/19336950.2016.1212141

Albarran L, Lopez JJ, Jardin I, Sanchez-Collado J, Berna-Erro A, Smani T, Camello PJ, Salido GM, Rosado JA (2018) EFHB is a novel cytosolic Ca^{2+} sensor that modulates STIM1-SARAF interaction. Cell Physiol Biochem 51(3):1164–1178. https://doi.org/10.1159/000495494

Alicia S, Angelica Z, Carlos S, Alfonso S, Vaca L (2008) STIM1 converts TRPC1 from a receptor-operated to a store-operated channel: moving TRPC1 in and out of lipid rafts. Cell Calcium 44 (5):479–491. https://doi.org/10.1016/j.ceca.2008.03.001

Almirza WH, Peters PH, van Zoelen EJ, Theuvenet AP (2012) Role of Trpc channels, Stim1 and Orai1 in PGF(2alpha)-induced calcium signaling in NRK fibroblasts. Cell Calcium 51 (1):12–21. https://doi.org/10.1016/j.ceca.2011.10.001

Asanov A, Sherry R, Sampieri A, Vaca L (2013) A relay mechanism between EB1 and APC facilitate STIM1 puncta assembly at endoplasmic reticulum-plasma membrane junctions. Cell Calcium 54(3):246–256. https://doi.org/10.1016/j.ceca.2013.06.008

Baryshnikov SG, Pulina MV, Zulian A, Linde CI, Golovina VA (2009) Orai1, a critical component of store-operated Ca^{2+} entry, is functionally associated with Na^{+}/Ca^{2+} exchanger and plasma membrane Ca^{2+} pump in proliferating human arterial myocytes. Am J Physiol Cell Physiol 297 (5):C1103–C1112. https://doi.org/10.1152/ajpcell.00283.2009

Bauer MC, O'Connell D, Cahill DJ, Linse S (2008) Calmodulin binding to the polybasic C-termini of STIM proteins involved in store-operated calcium entry. Biochemistry 47(23):6089–6091. https://doi.org/10.1021/bi800496a

Beliveau E, Lessard V, Guillemette G (2014) STIM1 positively regulates the Ca^{2+} release activity of the inositol 1,4,5-trisphosphate receptor in bovine aortic endothelial cells. PLoS One 9(12): e114718. https://doi.org/10.1371/journal.pone.0114718

Bergsmann J, Derler I, Muik M, Frischauf I, Fahrner M, Pollheimer P, Schwarzinger C, Gruber HJ, Groschner K, Romanin C (2011) Molecular determinants within N terminus of Orai3 protein that control channel activation and gating. J Biol Chem 286(36):31565–31575. https://doi.org/10.1074/jbc.M111.227546

Bhardwaj R, Muller HM, Nickel W, Seedorf M (2013) Oligomerization and Ca^{2+}/calmodulin control binding of the ER Ca^{2+}-sensors STIM1 and STIM2 to plasma membrane lipids. Biosci Rep 33(5):e00077. https://doi.org/10.1042/BSR20130089

Bhardwaj R, Augustynek BS, Ercan-Herbst E, Kandasamy P, Seedorf M, Peinelt C, Hediger MA (2020) Ca(2+)/Calmodulin binding to STIM1 hydrophobic residues facilitates slow Ca(2+)-dependent inactivation of the Orai1 channel. Cell Physiol Biochem 54(2):252–270. https://doi.org/10.33594/000000218

Boulay G, Brown DM, Qin N, Jiang M, Dietrich A, Zhu MX, Chen Z, Birnbaumer M, Mikoshiba K, Birnbaumer L (1999) Modulation of Ca(2+) entry by polypeptides of the inositol 1,4, 5-trisphosphate receptor (IP3R) that bind transient receptor potential (TRP): evidence for roles of TRP and IP3R in store depletion-activated Ca(2+) entry. Proc Natl Acad Sci U S A 96 (26):14955–14960. https://doi.org/10.1073/pnas.96.26.14955

Butorac C, Muik M, Derler I, Stadlbauer M, Lunz V, Krizova A, Lindinger S, Schober R, Frischauf I, Bhardwaj R, Hediger MA, Groschner K, Romanin C (2019) A novel STIM1-Orai1 gating interface essential for CRAC channel activation. Cell Calcium 79:57–67. https://doi.org/10.1016/j.ceca.2019.02.009

Cantonero C, Camello PJ, Abate C, Berardi F, Salido GM, Rosado JA, Redondo PC (2020) NO1, a new sigma 2 receptor/TMEM97 fluorescent ligand, downregulates SOCE and promotes apoptosis in the triple negative breast cancer cell lines. Cancers (Basel) 12(2):257. https://doi.org/10.3390/cancers12020257

Chang CL, Chen YJ, Quintanilla CG, Hsieh TS, Liou J (2018) EB1 binding restricts STIM1 translocation to ER-PM junctions and regulates store-operated Ca(2+) entry. J Cell Biol 217 (6):2047–2058. https://doi.org/10.1083/jcb.201711151

Cully TR, Edwards JN, Friedrich O, Stephenson DG, Murphy RM, Launikonis BS (2012) Changes in plasma membrane Ca-ATPase and stromal interacting molecule 1 expression levels for Ca(2 +) signaling in dystrophic mdx mouse muscle. Am J Physiol Cell Physiol 303(5):C567–C576. https://doi.org/10.1152/ajpcell.00144.2012

Dai F, Zhang Y, Wang Q, Li YY, Ma S, Yang D (2018) Overexpression of SARAF ameliorates pressure overload-induced cardiac hypertrophy through suppressing STIM1-Orai1 in mice. Cell Physiol Biochem 47(2):817–826. https://doi.org/10.1159/000490036

Decker CE, Yang Z, Rimer R, Park-Min KH, Macaubas C, Mellins ED, Novack DV, Faccio R (2015) Tmem178 acts in a novel negative feedback loop targeting NFATc1 to regulate bone mass. Proc Natl Acad Sci U S A 112(51):15654–15659. https://doi.org/10.1073/pnas.1511285112

DeHaven WI, Smyth JT, Boyles RR, Putney JW Jr (2007) Calcium inhibition and calcium potentiation of Orai1, Orai2, and Orai3 calcium release-activated calcium channels. J Biol Chem 282(24):17548–17556. https://doi.org/10.1074/jbc.M611374200

Derler I, Plenk P, Fahrner M, Muik M, Jardin I, Schindl R, Gruber HJ, Groschner K, Romanin C (2013) The extended transmembrane Orai1 N-terminal (ETON) region combines binding interface and gate for Orai1 activation by STIM1. J Biol Chem 288(40):29025–29034. https://doi.org/10.1074/jbc.M113.501510

Derler I, Butorac C, Krizova A, Stadlbauer M, Muik M, Fahrner M, Frischauf I, Romanin C (2018) Authentic CRAC channel activity requires STIM1 and the conserved portion of the Orai N terminus. J Biol Chem 293(4):1259–1270. https://doi.org/10.1074/jbc.M117.812206

Di Buduo CA, Abbonante V, Marty C, Moccia F, Rumi E, Pietra D, Soprano PM, Lim D, Cattaneo D, Iurlo A, Gianelli U, Barosi G, Rosti V, Plo I, Cazzola M, Balduini A (2020) Defective interaction of mutant calreticulin and SOCE in megakaryocytes from patients with myeloproliferative neoplasms. Blood 135(2):133–144. https://doi.org/10.1182/blood.2019001103

Dittmer PJ, Wild AR, Dell'Acqua ML, Sather WA (2017) STIM1 Ca(2+) sensor control of L-type Ca(2+)-channel-dependent dendritic spine structural plasticity and nuclear signaling. Cell Rep 19(2):321–334. https://doi.org/10.1016/j.celrep.2017.03.056

Emmer BT, Hesketh GG, Kotnik E, Tang VT, Lascuna PJ, Xiang J, Gingras AC, Chen XW, Ginsburg D (2018) The cargo receptor SURF4 promotes the efficient cellular secretion of PCSK9. elife 7:e38839. https://doi.org/10.7554/eLife.38839

Fahrner M, Pandey SK, Muik M, Traxler L, Butorac C, Stadlbauer M, Zayats V, Krizova A, Plenk P, Frischauf I, Schindl R, Gruber HJ, Hinterdorfer P, Ettrich R, Romanin C, Derler I (2018) Communication between N terminus and loop2 tunes Orai activation. J Biol Chem 293 (4):1271–1285. https://doi.org/10.1074/jbc.M117.812693

Feng JM, Hu YK, Xie LH, Colwell CS, Shao XM, Sun XP, Chen B, Tang H, Campagnoni AT (2006) Golli protein negatively regulates store depletion-induced calcium influx in T cells. Immunity 24(6):717–727. https://doi.org/10.1016/j.immuni.2006.04.007

Frickel EM, Frei P, Bouvier M, Stafford WF, Helenius A, Glockshuber R, Ellgaard L (2004) ERp57 is a multifunctional thiol-disulfide oxidoreductase. J Biol Chem 279(18):18277–18287. https://doi.org/10.1074/jbc.M314089200

Frischauf I, Muik M, Derler I, Bergsmann J, Fahrner M, Schindl R, Groschner K, Romanin C (2009) Molecular determinants of the coupling between STIM1 and Orai channels: differential activation of Orai1-3 channels by a STIM1 coiled-coil mutant. J Biol Chem 284 (32):21696–21706. https://doi.org/10.1074/jbc.M109.018408

Fujii Y, Shiota M, Ohkawa Y, Baba A, Wanibuchi H, Kinashi T, Kurosaki T, Baba Y (2012) Surf4 modulates STIM1-dependent calcium entry. Biochem Biophys Res Commun 422(4):615–620. https://doi.org/10.1016/j.bbrc.2012.05.037

Go CK, Hooper R, Aronson MR, Schultz B, Cangoz T, Nemani N, Zhang Y, Madesh M, Soboloff J (2019) The Ca(2+) export pump PMCA clears near-membrane Ca(2+) to facilitate store-operated Ca(2+) entry and NFAT activation. Sci Signal 12(602):eaaw2627. https://doi.org/10.1126/scisignal.aaw2627

Grigoriev I, Gouveia SM, van der Vaart B, Demmers J, Smyth JT, Honnappa S, Splinter D, Steinmetz MO, Putney JW Jr, Hoogenraad CC, Akhmanova A (2008) STIM1 is a MT-plus-end-tracking protein involved in remodeling of the ER. Curr Biol 18(3):177–182. https://doi.org/10.1016/j.cub.2007.12.050

Gruszczynska-Biegala J, Sladowska M, Kuznicki J (2016) AMPA receptors are involved in store-operated calcium entry and interact with STIM proteins in rat primary cortical neurons. Front Cell Neurosci 10:251. https://doi.org/10.3389/fncel.2016.00251

Gruszczynska-Biegala J, Strucinska K, Maciag F, Majewski L, Sladowska M, Kuznicki J (2020) STIM protein-NMDA2 receptor interaction decreases NMDA-dependent calcium levels in cortical neurons. Cell 9(1):160. https://doi.org/10.3390/cells9010160

Gudlur A, Zeraik AE, Hirve N, Rajanikanth V, Bobkov AA, Ma G, Zheng S, Wang Y, Zhou Y, Komives EA, Hogan PG (2018) Calcium sensing by the STIM1 ER-luminal domain. Nat Commun 9(1):4536. https://doi.org/10.1038/s41467-018-06816-8

Hirve N, Rajanikanth V, Hogan PG, Gudlur A (2018) Coiled-coil formation conveys a STIM1 signal from ER lumen to cytoplasm. Cell Rep 22(1):72–83. https://doi.org/10.1016/j.celrep.2017.12.030

Hong JH, Li Q, Kim MS, Shin DM, Feske S, Birnbaumer L, Cheng KT, Ambudkar IS, Muallem S (2011) Polarized but differential localization and recruitment of STIM1, Orai1 and TRPC channels in secretory cells. Traffic 12(2):232–245. https://doi.org/10.1111/j.1600-0854.2010.01138.x

Honnappa S, Gouveia SM, Weisbrich A, Damberger FF, Bhavesh NS, Jawhari H, Grigoriev I, van Rijssel FJ, Buey RM, Lawera A, Jelesarov I, Winkler FK, Wuthrich K, Akhmanova A, Steinmetz MO (2009) An EB1-binding motif acts as a microtubule tip localization signal. Cell 138(2):366–376. https://doi.org/10.1016/j.cell.2009.04.065

Huang GN, Zeng W, Kim JY, Yuan JP, Han L, Muallem S, Worley PF (2006) STIM1 carboxyl-terminus activates native SOC, I(crac) and TRPC1 channels. Nat Cell Biol 8(9):1003–1010. https://doi.org/10.1038/ncb1454

Huh JW, Shibata T, Hwang M, Kwon EH, Jang MS, Fukui R, Kanno A, Jung DJ, Jang MH, Miyake K, Kim YM (2014) UNC93B1 is essential for the plasma membrane localization and signaling of Toll-like receptor 5. Proc Natl Acad Sci U S A 111(19):7072–7077. https://doi.org/10.1073/pnas.1322838111

Inayama M, Suzuki Y, Yamada S, Kurita T, Yamamura H, Ohya S, Giles WR, Imaizumi Y (2015) Orai1-Orai2 complex is involved in store-operated calcium entry in chondrocyte cell lines. Cell Calcium 57(5–6):337–347. https://doi.org/10.1016/j.ceca.2015.02.005

Ishikawa H, Barber GN (2008) STING is an endoplasmic reticulum adaptor that facilitates innate immune signalling. Nature 455(7213):674–678. https://doi.org/10.1038/nature07317

Itoh H, Tatematsu M, Watanabe A, Iwano K, Funami K, Seya T, Matsumoto M (2011) UNC93B1 physically associates with human TLR8 and regulates TLR8-mediated signaling. PLoS One 6 (12):e28500. https://doi.org/10.1371/journal.pone.0028500

Jardin I, Lopez JJ, Salido GM, Rosado JA (2008a) Functional relevance of the de novo coupling between hTRPC1 and type II IP3 receptor in store-operated Ca^{2+} entry in human platelets. Cell Signal 20(4):737–747. https://doi.org/10.1016/j.cellsig.2007.12.010

Jardin I, Redondo PC, Salido GM, Rosado JA (2008b) Phosphatidylinositol 4,5-bisphosphate enhances store-operated calcium entry through hTRPC6 channel in human platelets. Biochim Biophys Acta 1783(1):84–97. https://doi.org/10.1016/j.bbamcr.2007.07.007

Jardin I, Gomez LJ, Salido GM, Rosado JA (2009) Dynamic interaction of hTRPC6 with the Orai1-STIM1 complex or hTRPC3 mediates its role in capacitative or non-capacitative Ca(2+) entry pathways. Biochem J 420(2):267–276. https://doi.org/10.1042/BJ20082179

Jha A, Ahuja M, Maleth J, Moreno CM, Yuan JP, Kim MS, Muallem S (2013) The STIM1 CTID domain determines access of SARAF to SOAR to regulate Orai1 channel function. J Cell Biol 202(1):71–79. https://doi.org/10.1083/jcb.201301148

Jha A, Chung WY, Vachel L, Maleth J, Lake S, Zhang G, Ahuja M, Muallem S (2019) Anoctamin 8 tethers endoplasmic reticulum and plasma membrane for assembly of Ca(2+) signaling complexes at the ER/PM compartment. EMBO J 38(12):e101452. https://doi.org/10.15252/embj.2018101452

Jing J, He L, Sun A, Quintana A, Ding Y, Ma G, Tan P, Liang X, Zheng X, Chen L, Shi X, Zhang SL, Zhong L, Huang Y, Dong MQ, Walker CL, Hogan PG, Wang Y, Zhou Y (2015) Proteomic mapping of ER-PM junctions identifies STIMATE as a regulator of Ca(2)(+) influx. Nat Cell Biol 17(10):1339–1347. https://doi.org/10.1038/ncb3234

Kawasaki T, Lange I, Feske S (2009) A minimal regulatory domain in the C terminus of STIM1 binds to and activates ORAI1 CRAC channels. Biochem Biophys Res Commun 385(1):49–54. https://doi.org/10.1016/j.bbrc.2009.05.020

Kilpatrick BS, Yates E, Grimm C, Schapira AH, Patel S (2016) Endo-lysosomal TRP mucolipin-1 channels trigger global ER Ca^{2+} release and Ca^{2+} influx. J Cell Sci 129(20):3859–3867. https://doi.org/10.1242/jcs.190322

Kim YM, Brinkmann MM, Paquet ME, Ploegh HL (2008) UNC93B1 delivers nucleotide-sensing toll-like receptors to endolysosomes. Nature 452(7184):234–238. https://doi.org/10.1038/nature06726

Kim MS, Zeng W, Yuan JP, Shin DM, Worley PF, Muallem S (2009) Native store-operated Ca^{2+} influx requires the channel function of Orai1 and TRPC1. J Biol Chem 284(15):9733–9741. https://doi.org/10.1074/jbc.M808097200

Kiselyov K, Xu X, Mozhayeva G, Kuo T, Pessah I, Mignery G, Zhu X, Birnbaumer L, Muallem S (1998) Functional interaction between InsP3 receptors and store-operated Htrp3 channels. Nature 396(6710):478–482. https://doi.org/10.1038/24890

Krapivinsky G, Krapivinsky L, Stotz SC, Manasian Y, Clapham DE (2011) POST, partner of stromal interaction molecule 1 (STIM1), targets STIM1 to multiple transporters. Proc Natl Acad Sci U S A 108(48):19234–19239. https://doi.org/10.1073/pnas.1117231108

Kwon J, An H, Sa M, Won J, Shin JI, Lee CJ (2017) Orai1 and Orai3 in combination with Stim1 mediate the majority of store-operated calcium entry in astrocytes. Exp Neurobiol 26(1):42–54. https://doi.org/10.5607/en.2017.26.1.42

Lacruz RS, Feske S (2015) Diseases caused by mutations in ORAI1 and STIM1. Ann N Y Acad Sci 1356:45–79. https://doi.org/10.1111/nyas.12938

Lee KP, Yuan JP, So I, Worley PF, Muallem S (2010) STIM1-dependent and STIM1-independent function of transient receptor potential canonical (TRPC) channels tunes their store-operated mode. J Biol Chem 285(49):38666–38673. https://doi.org/10.1074/jbc.M110.155036

Lee BL, Moon JE, Shu JH, Yuan L, Newman ZR, Schekman R, Barton GM (2013) UNC93B1 mediates differential trafficking of endosomal TLRs. elife 2:e00291. https://doi.org/10.7554/eLife.00291

Lee KP, Choi S, Hong JH, Ahuja M, Graham S, Ma R, So I, Shin DM, Muallem S, Yuan JP (2014) Molecular determinants mediating gating of Transient Receptor Potential Canonical (TRPC) channels by stromal interaction molecule 1 (STIM1). J Biol Chem 289(10):6372–6382. https://doi.org/10.1074/jbc.M113.546556

Levin JZ, Horvitz HR (1992) The Caenorhabditis elegans unc-93 gene encodes a putative transmembrane protein that regulates muscle contraction. J Cell Biol 117(1):143–155. https://doi.org/10.1083/jcb.117.1.143

Li Z, Lu J, Xu P, Xie X, Chen L, Xu T (2007) Mapping the interacting domains of STIM1 and Orai1 in Ca^{2+} release-activated Ca^{2+} channel activation. J Biol Chem 282(40):29448–29456. https://doi.org/10.1074/jbc.M703573200

Li J, Sukumar P, Milligan CJ, Kumar B, Ma ZY, Munsch CM, Jiang LH, Porter KE, Beech DJ (2008) Interactions, functions, and independence of plasma membrane STIM1 and TRPC1 in vascular smooth muscle cells. Circ Res 103(8):e97–e104. https://doi.org/10.1161/CIRCRESAHA.108.182931

Li P, Miao Y, Dani A, Vig M (2016) alpha-SNAP regulates dynamic, on-site assembly and calcium selectivity of Orai1 channels. Mol Biol Cell 27(16):2542–2553. https://doi.org/10.1091/mbc.E16-03-0163

Li X, Wu G, Yang Y, Fu S, Liu X, Kang H, Yang X, Su XC, Shen Y (2017) Calmodulin dissociates the STIM1-Orai1 complex and STIM1 oligomers. Nat Commun 8(1):1042. https://doi.org/10.1038/s41467-017-01135-w

Li Y, James SJ, Wyllie DH, Wynne C, Czibula A, Bukhari A, Pye K, Bte Mustafah SM, Fajka-Boja R, Szabo E, Angyal A, Hegedus Z, Kovacs L, Hill AVS, Jefferies CA, Wilson HL, Yongliang Z, Kiss-Toth E (2019) TMEM203 is a binding partner and regulator of STING-mediated inflammatory signaling in macrophages. Proc Natl Acad Sci U S A 116(33):16479–16488. https://doi.org/10.1073/pnas.1901090116

Lis A, Peinelt C, Beck A, Parvez S, Monteilh-Zoller M, Fleig A, Penner R (2007) CRACM1, CRACM2, and CRACM3 are store-operated Ca^{2+} channels with distinct functional properties. Curr Biol 17(9):794–800. https://doi.org/10.1016/j.cub.2007.03.065

Liu X, Wang W, Singh BB, Lockwich T, Jadlowiec J, O'Connell B, Wellner R, Zhu MX, Ambudkar IS (2000) Trp1, a candidate protein for the store-operated Ca(2+) influx mechanism in salivary gland cells. J Biol Chem 275(5):3403–3411. https://doi.org/10.1074/jbc.275.5.3403

Liu X, Cheng KT, Bandyopadhyay BC, Pani B, Dietrich A, Paria BC, Swaim WD, Beech D, Yildrim E, Singh BB, Birnbaumer L, Ambudkar IS (2007) Attenuation of store-operated Ca^{2+} current impairs salivary gland fluid secretion in TRPC1(−/−) mice. Proc Natl Acad Sci U S A 104(44):17542–17547. https://doi.org/10.1073/pnas.0701254104

Lopez E, Frischauf I, Jardin I, Derler I, Muik M, Cantonero C, Salido GM, Smani T, Rosado JA, Redondo PC (2019) STIM1 phosphorylation at Y(316) modulates its interaction with SARAF and the activation of SOCE and I CRAC. J Cell Sci 132(10):jcs226019. https://doi.org/10.1242/jcs.226019

Lopez JJ, Jardin I, Sanchez-Collado J, Salido GM, Smani T, Rosado JA (2020) TRPC channels in the SOCE scenario. Cell 9(1):126. https://doi.org/10.3390/cells9010126

Lunz V, Romanin C, Frischauf I (2019) STIM1 activation of Orai1. Cell Calcium 77:29–38. https://doi.org/10.1016/j.ceca.2018.11.009

Lur G, Sherwood MW, Ebisui E, Haynes L, Feske S, Sutton R, Burgoyne RD, Mikoshiba K, Petersen OH, Tepikin AV (2011) InsP(3)receptors and Orai channels in pancreatic acinar cells: co-localization and its consequences. Biochem J 436(2):231–239. https://doi.org/10.1042/BJ20110083

Ma JM, Wang R, Xu JY, Fan Y (2016) Intracellular Ca^{2+} and related proteins in patients with oral lichen planus. Immunol Res 64(2):531–539. https://doi.org/10.1007/s12026-015-8728-z

Maleth J, Choi S, Muallem S, Ahuja M (2014) Translocation between PI(4,5)P2-poor and PI(4,5)P2-rich microdomains during store depletion determines STIM1 conformation and Orai1 gating. Nat Commun 5:5843. https://doi.org/10.1038/ncomms6843

Maschalidi S, Nunes-Hasler P, Nascimento CR, Sallent I, Lannoy V, Garfa-Traore M, Cagnard N, Sepulveda FE, Vargas P, Lennon-Dumenil AM, van Endert P, Capiod T, Demaurex N, Darrasse-Jeze G, Manoury B (2017) UNC93B1 interacts with the calcium sensor STIM1 for efficient antigen cross-presentation in dendritic cells. Nat Commun 8(1):1640. https://doi.org/10.1038/s41467-017-01601-5

McNally BA, Somasundaram A, Jairaman A, Yamashita M, Prakriya M (2013) The C- and N-terminal STIM1 binding sites on Orai1 are required for both trapping and gating CRAC channels. J Physiol 591(11):2833–2850. https://doi.org/10.1113/jphysiol.2012.250456

Mercer JC, Dehaven WI, Smyth JT, Wedel B, Boyles RR, Bird GS, Putney JW Jr (2006) Large store-operated calcium selective currents due to co-expression of Orai1 or Orai2 with the intracellular calcium sensor, Stim1. J Biol Chem 281(34):24979–24990. https://doi.org/10.1074/jbc.M604589200

Miao Y, Miner C, Zhang L, Hanson PI, Dani A, Vig M (2013) An essential and NSF independent role for alpha-SNAP in store-operated calcium entry. elife 2:e00802. https://doi.org/10.7554/eLife.00802

Mitrovic S, Ben-Tekaya H, Koegler E, Gruenberg J, Hauri HP (2008) The cargo receptors Surf4, endoplasmic reticulum-Golgi intermediate compartment (ERGIC)-53, and p25 are required to maintain the architecture of ERGIC and Golgi. Mol Biol Cell 19(5):1976–1990. https://doi.org/10.1091/mbc.E07-10-0989

Muik M, Frischauf I, Derler I, Fahrner M, Bergsmann J, Eder P, Schindl R, Hesch C, Polzinger B, Fritsch R, Kahr H, Madl J, Gruber H, Groschner K, Romanin C (2008) Dynamic coupling of the putative coiled-coil domain of ORAI1 with STIM1 mediates ORAI1 channel activation. J Biol Chem 283(12):8014–8022. https://doi.org/10.1074/jbc.M708898200

Muik M, Fahrner M, Derler I, Schindl R, Bergsmann J, Frischauf I, Groschner K, Romanin C (2009) A cytosolic homomerization and a modulatory domain within STIM1 C terminus determine coupling to ORAI1 channels. J Biol Chem 284(13):8421–8426. https://doi.org/10.1074/jbc.C800229200

Mullins FM, Park CY, Dolmetsch RE, Lewis RS (2009) STIM1 and calmodulin interact with Orai1 to induce Ca^{2+}-dependent inactivation of CRAC channels. Proc Natl Acad Sci U S A 106 (36):15495–15500. https://doi.org/10.1073/pnas.0906781106

Ng LC, O'Neill KG, French D, Airey JA, Singer CA, Tian H, Shen XM, Hume JR (2012) TRPC1 and Orai1 interact with STIM1 and mediate capacitative Ca(2+) entry caused by acute hypoxia in mouse pulmonary arterial smooth muscle cells. Am J Physiol Cell Physiol 303(11):C1156–C1172. https://doi.org/10.1152/ajpcell.00065.2012

Nguyen N, Biet M, Simard E, Beliveau E, Francoeur N, Guillemette G, Dumaine R, Grandbois M, Boulay G (2013) STIM1 participates in the contractile rhythmicity of HL-1 cells by moderating T-type Ca(2+) channel activity. Biochim Biophys Acta 1833(6):1294–1303. https://doi.org/10.1016/j.bbamcr.2013.02.027

Niu L, Wu F, Li K, Li J, Zhang SL, Hu J, Wang Q (2020) STIM1 interacts with termini of Orai channels in a sequential manner. J Cell Sci 133:jcs239491. https://doi.org/10.1242/jcs.239491

Nunes P, Demaurex N (2014) Redox regulation of store-operated Ca^{2+} entry. Antioxid Redox Signal 21(6):915–932. https://doi.org/10.1089/ars.2013.5615

Nunes-Hasler P, Maschalidi S, Lippens C, Castelbou C, Bouvet S, Guido D, Bermont F, Bassoy EY, Page N, Merkler D, Hugues S, Martinvalet D, Manoury B, Demaurex N (2017) STIM1 promotes migration, phagosomal maturation and antigen cross-presentation in dendritic cells. Nat Commun 8(1):1852. https://doi.org/10.1038/s41467-017-01600-6

O'Connell DJ, Bauer M, Linse S, Cahill DJ (2011) Probing calmodulin protein-protein interactions using high-content protein arrays. Methods Mol Biol 785:289–303. https://doi.org/10.1007/978-1-61779-286-1_20

Ong HL, Cheng KT, Liu X, Bandyopadhyay BC, Paria BC, Soboloff J, Pani B, Gwack Y, Srikanth S, Singh BB, Gill DL, Ambudkar IS (2007) Dynamic assembly of TRPC1-STIM1-Orai1 ternary complex is involved in store-operated calcium influx. Evidence for similarities in store-operated and calcium release-activated calcium channel components. J Biol Chem 282 (12):9105–9116. https://doi.org/10.1074/jbc.M608942200

Paez PM, Spreuer V, Handley V, Feng JM, Campagnoni C, Campagnoni AT (2007) Increased expression of golli myelin basic proteins enhances calcium influx into oligodendroglial cells. J Neurosci 27(46):12690–12699. https://doi.org/10.1523/JNEUROSCI.2381-07.2007

Paez PM, Fulton DJ, Spreuer V, Handley V, Campagnoni CW, Campagnoni AT (2009) Regulation of store-operated and voltage-operated Ca^{2+} channels in the proliferation and death of oligodendrocyte precursor cells by golli proteins. ASN Neuro 1(1):e00003. https://doi.org/10.1042/AN20090003

Palty R, Isacoff EY (2016) Cooperative binding of stromal interaction molecule 1 (STIM1) to the N and C termini of calcium release-activated calcium modulator 1 (Orai1). J Biol Chem 291 (1):334–341. https://doi.org/10.1074/jbc.M115.685289

Palty R, Raveh A, Kaminsky I, Meller R, Reuveny E (2012) SARAF inactivates the store operated calcium entry machinery to prevent excess calcium refilling. Cell 149(2):425–438. https://doi.org/10.1016/j.cell.2012.01.055

Pani B, Ong HL, Brazer SC, Liu X, Rauser K, Singh BB, Ambudkar IS (2009) Activation of TRPC1 by STIM1 in ER-PM microdomains involves release of the channel from its scaffold caveolin-1. Proc Natl Acad Sci U S A 106(47):20087–20092. https://doi.org/10.1073/pnas.0905002106

Park CY, Hoover PJ, Mullins FM, Bachhawat P, Covington ED, Raunser S, Walz T, Garcia KC, Dolmetsch RE, Lewis RS (2009) STIM1 clusters and activates CRAC channels via direct binding of a cytosolic domain to Orai1. Cell 136(5):876–890. https://doi.org/10.1016/j.cell.2009.02.014

Park CY, Shcheglovitov A, Dolmetsch R (2010) The CRAC channel activator STIM1 binds and inhibits L-type voltage-gated calcium channels. Science 330(6000):101–105. https://doi.org/10.1126/science.1191027

Peinelt C, Vig M, Koomoa DL, Beck A, Nadler MJ, Koblan-Huberson M, Lis A, Fleig A, Penner R, Kinet JP (2006) Amplification of CRAC current by STIM1 and CRACM1 (Orai1). Nat Cell Biol 8(7):771–773. https://doi.org/10.1038/ncb1435

Pelka K, Bertheloot D, Reimer E, Phulphagar K, Schmidt SV, Christ A, Stahl R, Watson N, Miyake K, Hacohen N, Haas A, Brinkmann MM, Marshak-Rothstein A, Meissner F, Latz E (2018) The chaperone UNC93B1 regulates toll-like receptor stability independently of endosomal TLR transport. Immunity 48(5):911–922 e917. https://doi.org/10.1016/j.immuni.2018.04.011

Pifer R, Benson A, Sturge CR, Yarovinsky F (2011) UNC93B1 is essential for TLR11 activation and IL-12-dependent host resistance to Toxoplasma gondii. J Biol Chem 286(5):3307–3314. https://doi.org/10.1074/jbc.M110.171025

Pozo-Guisado E, Casas-Rua V, Tomas-Martin P, Lopez-Guerrero AM, Alvarez-Barrientos A, Martin-Romero FJ (2013) Phosphorylation of STIM1 at ERK1/2 target sites regulates interaction with the microtubule plus-end binding protein EB1. J Cell Sci 126(Pt 14):3170–3180. https://doi.org/10.1242/jcs.125054

Prakriya M, Lewis RS (2015) Store-operated calcium channels. Physiol Rev 95(4):1383–1436. https://doi.org/10.1152/physrev.00020.2014

Prins D, Groenendyk J, Touret N, Michalak M (2011) Modulation of STIM1 and capacitative Ca^{2+} entry by the endoplasmic reticulum luminal oxidoreductase ERp57. EMBO Rep 12 (11):1182–1188. https://doi.org/10.1038/embor.2011.173

Qiu R, Lewis RS (2019) Structural features of STIM and Orai underlying store-operated calcium entry. Curr Opin Cell Biol 57:90–98. https://doi.org/10.1016/j.ceb.2018.12.012

Quintana A, Rajanikanth V, Farber-Katz S, Gudlur A, Zhang C, Jing J, Zhou Y, Rao A, Hogan PG (2015) TMEM110 regulates the maintenance and remodeling of mammalian ER-plasma membrane junctions competent for STIM-ORAI signaling. Proc Natl Acad Sci U S A 112(51): E7083–E7092. https://doi.org/10.1073/pnas.1521924112

Roos J, DiGregorio PJ, Yeromin AV, Ohlsen K, Lioudyno M, Zhang S, Safrina O, Kozak JA, Wagner SL, Cahalan MD, Velicelebi G, Stauderman KA (2005) STIM1, an essential and conserved component of store-operated Ca^{2+} channel function. J Cell Biol 169(3):435–445. https://doi.org/10.1083/jcb.200502019

Rosado JA, Sage SO (2001) Activation of store-mediated calcium entry by secretion-like coupling between the inositol 1,4,5-trisphosphate receptor type II and human transient receptor potential (hTrp1) channels in human platelets. Biochem J 356(Pt 1):191–198. https://doi.org/10.1042/0264-6021:3560191

Rosado JA, Brownlow SL, Sage SO (2002) Endogenously expressed Trp1 is involved in store-mediated Ca^{2+} entry by conformational coupling in human platelets. J Biol Chem 277 (44):42157–42163. https://doi.org/10.1074/jbc.M207320200

Rosati A, Graziano V, De Laurenzi V, Pascale M, Turco MC (2011) BAG3: a multifaceted protein that regulates major cell pathways. Cell Death Dis 2:e141. https://doi.org/10.1038/cddis.2011.24

Saegusa K, Sato M, Morooka N, Hara T, Sato K (2018) SFT-4/Surf4 control ER export of soluble cargo proteins and participate in ER exit site organization. J Cell Biol 217(6):2073–2085. https://doi.org/10.1083/jcb.201708115

Saliba Y, Keck M, Marchand A, Atassi F, Ouille A, Cazorla O, Trebak M, Pavoine C, Lacampagne A, Hulot JS, Fares N, Fauconnier J, Lompre AM (2015) Emergence of Orai3 activity during cardiac hypertrophy. Cardiovasc Res 105(3):248–259. https://doi.org/10.1093/cvr/cvu207

Sampieri A, Santoyo K, Asanov A, Vaca L (2018) Association of the IP3R to STIM1 provides a reduced intraluminal calcium microenvironment, resulting in enhanced store-operated calcium entry. Sci Rep 8(1):13252. https://doi.org/10.1038/s41598-018-31621-0

Santoso NG, Cebotaru L, Guggino WB (2011) Polycystin-1, 2, and STIM1 interact with IP(3)R to modulate ER Ca release through the PI3K/Akt pathway. Cell Physiol Biochem 27(6):715–726. https://doi.org/10.1159/000330080

Sharma A, Ramena G, Yin Y, Premkumar L, Elble RC (2018) CLCA2 is a positive regulator of store-operated calcium entry and TMEM16A. PLoS One 13(5):e0196512. https://doi.org/10.1371/journal.pone.0196512

Shi J, Miralles F, Birnbaumer L, Large WA, Albert AP (2017a) Store-operated interactions between plasmalemmal STIM1 and TRPC1 proteins stimulate PLCbeta1 to induce TRPC1 channel activation in vascular smooth muscle cells. J Physiol 595(4):1039–1058. https://doi.org/10.1113/JP273302

Shi J, Miralles F, Kinet JP, Birnbaumer L, Large WA, Albert AP (2017b) Evidence that Orai1 does not contribute to store-operated TRPC1 channels in vascular smooth muscle cells. Channels (Austin) 11(4):329–339. https://doi.org/10.1080/19336950.2017.1303025

Son A, Ahuja M, Schwartz DM, Varga A, Swaim W, Kang N, Maleth J, Shin DM, Muallem S (2019) Ca(2+) influx channel inhibitor SARAF protects mice from acute pancreatitis. Gastroenterology 157(6):1660–1672 e1662. https://doi.org/10.1053/j.gastro.2019.08.042

Sours-Brothers S, Ding M, Graham S, Ma R (2009) Interaction between TRPC1/TRPC4 assembly and STIM1 contributes to store-operated Ca^{2+} entry in mesangial cells. Exp Biol Med (Maywood) 234(6):673–682. https://doi.org/10.3181/0809-RM-279

Srikanth S, Jung HJ, Kim KD, Souda P, Whitelegge J, Gwack Y (2010) A novel EF-hand protein, CRACR2A, is a cytosolic Ca^{2+} sensor that stabilizes CRAC channels in T cells. Nat Cell Biol 12 (5):436–446. https://doi.org/10.1038/ncb2045

Srikanth S, Woo JS, Wu B, El-Sherbiny YM, Leung J, Chupradit K, Rice L, Seo GJ, Calmettes G, Ramakrishna C, Cantin E, An DS, Sun R, Wu TT, Jung JU, Savic S, Gwack Y (2019) The Ca(2+) sensor STIM1 regulates the type I interferon response by retaining the signaling adaptor STING at the endoplasmic reticulum. Nat Immunol 20(2):152–162. https://doi.org/10.1038/s41590-018-0287-8

Srivats S, Balasuriya D, Pasche M, Vistal G, Edwardson JM, Taylor CW, Murrell-Lagnado RD (2016) Sigma1 receptors inhibit store-operated Ca^{2+} entry by attenuating coupling of STIM1 to Orai1. J Cell Biol 213(1):65–79. https://doi.org/10.1083/jcb.201506022

Stathopulos PB, Schindl R, Fahrner M, Zheng L, Gasmi-Seabrook GM, Muik M, Romanin C, Ikura M (2013) STIM1/Orai1 coiled-coil interplay in the regulation of store-operated calcium entry. Nat Commun 4:2963. https://doi.org/10.1038/ncomms3963

Streb H, Irvine RF, Berridge MJ, Schulz I (1983) Release of Ca^{2+} from a nonmitochondrial intracellular store in pancreatic acinar cells by inositol-1,4,5-trisphosphate. Nature 306 (5938):67–69. https://doi.org/10.1038/306067a0

Sun Y, Sukumaran P, Singh BB (2020) Sigma1 receptor inhibits TRPC1-mediated Ca(2+) entry that promotes dopaminergic cell death. Cell Mol Neurobiol. https://doi.org/10.1007/s10571-020-00892-5

Sundivakkam PC, Freichel M, Singh V, Yuan JP, Vogel SM, Flockerzi V, Malik AB, Tiruppathi C (2012) The Ca(2+) sensor stromal interaction molecule 1 (STIM1) is necessary and sufficient for the store-operated Ca(2+) entry function of transient receptor potential canonical (TRPC) 1 and 4 channels in endothelial cells. Mol Pharmacol 81(4):510–526. https://doi.org/10.1124/mol.111.074658

Tabeta K, Hoebe K, Janssen EM, Du X, Georgel P, Crozat K, Mudd S, Mann N, Sovath S, Goode J, Shamel L, Herskovits AA, Portnoy DA, Cooke M, Tarantino LM, Wiltshire T, Steinberg BE, Grinstein S, Beutler B (2006) The Unc93b1 mutation 3d disrupts exogenous antigen presentation and signaling via Toll-like receptors 3, 7 and 9. Nat Immunol 7(2):156–164. https://doi.org/10.1038/ni1297

Tai K, Hamaide MC, Debaix H, Gailly P, Wibo M, Morel N (2008) Agonist-evoked calcium entry in vascular smooth muscle cells requires IP3 receptor-mediated activation of TRPC1. Eur J Pharmacol 583(1):135–147. https://doi.org/10.1016/j.ejphar.2008.01.007

Takahashi Y, Murakami M, Watanabe H, Hasegawa H, Ohba T, Munehisa Y, Nobori K, Ono K, Iijima T, Ito H (2007) Essential role of the N-terminus of murine Orai1 in store-operated Ca^{2+} entry. Biochem Biophys Res Commun 356(1):45–52. https://doi.org/10.1016/j.bbrc.2007.02.107

Tam KC, Ali E, Hua J, Chataway T, Barritt GJ (2018) Evidence for the interaction of peroxiredoxin-4 with the store-operated calcium channel activator STIM1 in liver cells. Cell Calcium 74:14–28. https://doi.org/10.1016/j.ceca.2018.05.002

Tedeschi V, Petrozziello T, Sisalli MJ, Boscia F, Canzoniero LMT, Secondo A (2019) The activation of Mucolipin TRP channel 1 (TRPML1) protects motor neurons from L-BMAA neurotoxicity by promoting autophagic clearance. Sci Rep 9(1):10743. https://doi.org/10.1038/s41598-019-46708-5

Thebault S, Zholos A, Enfissi A, Slomianny C, Dewailly E, Roudbaraki M, Parys J, Prevarskaya N (2005) Receptor-operated Ca^{2+} entry mediated by TRPC3/TRPC6 proteins in rat prostate smooth muscle (PS1) cell line. J Cell Physiol 204(1):320–328. https://doi.org/10.1002/jcp.20301

Thillaiappan NB, Chavda AP, Tovey SC, Prole DL, Taylor CW (2017) Ca(2+) signals initiate at immobile IP3 receptors adjacent to ER-plasma membrane junctions. Nat Commun 8(1):1505. https://doi.org/10.1038/s41467-017-01644-8

Tirnauer JS, Bierer BE (2000) EB1 proteins regulate microtubule dynamics, cell polarity, and chromosome stability. J Cell Biol 149(4):761–766. https://doi.org/10.1083/jcb.149.4.761

Vanoye CG, Sakakura M, Follis RM, Trevisan AJ, Narayan M, Li J, Sanders CR, Carter BD (2019) Peripheral myelin protein 22 modulates store-operated calcium channel activity, providing insights into Charcot-Marie-Tooth disease etiology. J Biol Chem 294(32):12054–12065. https://doi.org/10.1074/jbc.RA118.006248

Vig M, Peinelt C, Beck A, Koomoa DL, Rabah D, Koblan-Huberson M, Kraft S, Turner H, Fleig A, Penner R, Kinet JP (2006) CRACM1 is a plasma membrane protein essential for store-operated Ca^{2+} entry. Science 312(5777):1220–1223. https://doi.org/10.1126/science.1127883

Walsh CM, Doherty MK, Tepikin AV, Burgoyne RD (2010) Evidence for an interaction between Golli and STIM1 in store-operated calcium entry. Biochem J 430(3):453–460. https://doi.org/10.1042/BJ20100650

Wang Y, Deng X, Mancarella S, Hendron E, Eguchi S, Soboloff J, Tang XD, Gill DL (2010) The calcium store sensor, STIM1, reciprocally controls Orai and CaV1.2 channels. Science 330 (6000):105–109. https://doi.org/10.1126/science.1191086

Wang X, Wang Y, Zhou Y, Hendron E, Mancarella S, Andrake MD, Rothberg BS, Soboloff J, Gill DL (2014) Distinct Orai-coupling domains in STIM1 and STIM2 define the Orai-activating site. Nat Commun 5:3183. https://doi.org/10.1038/ncomms4183

Wang J, Xu C, Zheng Q, Yang K, Lai N, Wang T, Tang H, Lu W (2017) Orai1, 2, 3 and STIM1 promote store-operated calcium entry in pulmonary arterial smooth muscle cells. Cell Death Discov 3:17074. https://doi.org/10.1038/cddiscovery.2017.74

Woodard GE, Lopez JJ, Jardin I, Salido GM, Rosado JA (2010) TRPC3 regulates agonist-stimulated Ca^{2+} mobilization by mediating the interaction between type I inositol 1,4,5-trisphosphate receptor, RACK1, and Orai1. J Biol Chem 285(11):8045–8053. https://doi.org/10.1074/jbc.M109.033605

Worley PF, Zeng W, Huang GN, Yuan JP, Kim JY, Lee MG, Muallem S (2007) TRPC channels as STIM1-regulated store-operated channels. Cell Calcium 42(2):205–211. https://doi.org/10.1016/j.ceca.2007.03.004

Wu X, Babnigg G, Villereal ML (2000) Functional significance of human trp1 and trp3 in store-operated Ca(2+) entry in HEK-293 cells. Am J Physiol Cell Physiol 278(3):C526–C536. https://doi.org/10.1152/ajpcell.2000.278.3.C526

Yang J, Li S, Wang Q, Yang D (2019a) Transmembrane protein 66 attenuates neointimal hyperplasia after carotid artery injury by SOCE inactivation. Mol Med Rep 20(2):1436–1442. https://doi.org/10.3892/mmr.2019.10328

Yang Z, Yan H, Dai W, Jing J, Yang Y, Mahajan S, Zhou Y, Li W, Macaubas C, Mellins ED, Shih CC, Fitzpatrick JAJ, Faccio R (2019b) Tmem178 negatively regulates store-operated calcium entry in myeloid cells via association with STIM1. J Autoimmun 101:94–108. https://doi.org/10.1016/j.jaut.2019.04.015

Yen M, Lewis RS (2018) Physiological CRAC channel activation and pore properties require STIM1 binding to all six Orai1 subunits. J Gen Physiol 150(10):1373–1385. https://doi.org/10.1085/jgp.201711985

Yen M, Lokteva LA, Lewis RS (2016) Functional analysis of Orai1 concatemers supports a hexameric stoichiometry for the CRAC channel. Biophys J 111(9):1897–1907. https://doi.org/10.1016/j.bpj.2016.09.020

Yin Y, Garcia MR, Novak AJ, Saunders AM, Ank RS, Nam AS, Fisher LW (2018) Surf4 (Erv29p) binds amino-terminal tripeptide motifs of soluble cargo proteins with different affinities, enabling prioritization of their exit from the endoplasmic reticulum. PLoS Biol 16(8): e2005140. https://doi.org/10.1371/journal.pbio.2005140

Yu F, Hubrack SZ, Chakraborty S, Sun L, Alcantara-Adap E, Kulkarni R, Billing AM, Graumann J, Taylor CW, Machaca K (2019) Remodeling of ER-plasma membrane contact sites but not STIM1 phosphorylation inhibits Ca(2+) influx in mitosis. Proc Natl Acad Sci U S A 116 (21):10392–10401. https://doi.org/10.1073/pnas.1821399116

Yuan JP, Kiselyov K, Shin DM, Chen J, Shcheynikov N, Kang SH, Dehoff MH, Schwarz MK, Seeburg PH, Muallem S, Worley PF (2003) Homer binds TRPC family channels and is required for gating of TRPC1 by IP3 receptors. Cell 114(6):777–789. https://doi.org/10.1016/s0092-8674(03)00716-5

Yuan JP, Zeng W, Huang GN, Worley PF, Muallem S (2007) STIM1 heteromultimerizes TRPC channels to determine their function as store-operated channels. Nat Cell Biol 9(6):636–645. https://doi.org/10.1038/ncb1590

Yuan JP, Zeng W, Dorwart MR, Choi YJ, Worley PF, Muallem S (2009) SOAR and the polybasic STIM1 domains gate and regulate Orai channels. Nat Cell Biol 11(3):337–343. https://doi.org/10.1038/ncb1842

Zagranichnaya TK, Wu X, Villereal ML (2005) Endogenous TRPC1, TRPC3, and TRPC7 proteins combine to form native store-operated channels in HEK-293 cells. J Biol Chem 280 (33):29559–29569. https://doi.org/10.1074/jbc.M505842200

Zeng W, Yuan JP, Kim MS, Choi YJ, Huang GN, Worley PF, Muallem S (2008) STIM1 gates TRPC channels, but not Orai1, by electrostatic interaction. Mol Cell 32(3):439–448. https://doi.org/10.1016/j.molcel.2008.09.020

Zhang SL, Yeromin AV, Zhang XH, Yu Y, Safrina O, Penna A, Roos J, Stauderman KA, Cahalan MD (2006) Genome-wide RNAi screen of Ca(2+) influx identifies genes that regulate Ca(2+) release-activated Ca(2+) channel activity. Proc Natl Acad Sci U S A 103(24):9357–9362. https://doi.org/10.1073/pnas.0603161103

Zheng H, Zhou MH, Hu C, Kuo E, Peng X, Hu J, Kuo L, Zhang SL (2013) Differential roles of the C and N termini of Orai1 protein in interacting with stromal interaction molecule 1 (STIM1) for Ca^{2+} release-activated Ca^{2+} (CRAC) channel activation. J Biol Chem 288(16):11263–11272. https://doi.org/10.1074/jbc.M113.450254

Zhou Y, Meraner P, Kwon HT, Machnes D, Oh-hora M, Zimmer J, Huang Y, Stura A, Rao A, Hogan PG (2010) STIM1 gates the store-operated calcium channel ORAI1 in vitro. Nat Struct Mol Biol 17(1):112–116. https://doi.org/10.1038/nsmb.1724

Zhou Y, Cai X, Loktionova NA, Wang X, Nwokonko RM, Wang X, Wang Y, Rothberg BS, Trebak M, Gill DL (2016) The STIM1-binding site nexus remotely controls Orai1 channel gating. Nat Commun 7:13725. https://doi.org/10.1038/ncomms13725

Chapter 5
Endoplasmic Reticulum (ER) and ER-Phagy

Marisa Loi, Alessandro Marazza, and Maurizio Molinari

Abstract The endoplasmic reticulum (ER) is a biosynthetic organelle in eukaryotic cells. Its capacity to produce proteins, lipids and oligosaccharides responds to physiologic and pathologic demand. The transcriptional and translational unfolded protein response (UPR) programs increase ER size and activity. In contrast, ER-phagy programs in all their flavors select ER subdomains for lysosomal clearance. These programs are activated by nutrient deprivation, accumulation of excess ER (recov-ER-phagy), production of misfolded proteins that cannot be degraded by ER-associated degradation and that are removed from cells by the so-called ER-to-lysosome-associated degradation (ERLAD). Selection of ER subdomains to be cleared from cells relies on ER-phagy receptors, a class of membrane-bound proteins displaying cytosolic domains that engage the cytosolic ubiquitin-like protein LC3. Mechanistically, ER clearance proceeds via macro-ER-phagy, micro-ER-phagy and LC3-regulated vesicular delivery.

M. Loi
Università della Svizzera italiana, Lugano, Switzerland

Institute for Research in Biomedicine, Bellinzona, Switzerland
e-mail: Marisa.Loi@irb.usi.ch

A. Marazza
Università della Svizzera italiana, Lugano, Switzerland

Institute for Research in Biomedicine, Bellinzona, Switzerland

Graduate School for Cellular and Biomedical Sciences, University of Bern, Bern, Switzerland
e-mail: Alessandro.Marazza@irb.usi.ch

M. Molinari (✉)
Università della Svizzera italiana, Lugano, Switzerland

Institute for Research in Biomedicine, Bellinzona, Switzerland

Ecole Polytechnique Fédérale de Lausanne, Lausanne, Switzerland
e-mail: Maurizio.Molinari@irb.usi.ch

L. B. Agellon, M. Michalak (eds.), *Cellular Biology of the Endoplasmic Reticulum*,
Progress in Molecular and Subcellular Biology 59,
https://doi.org/10.1007/978-3-030-67696-4_5

Keywords Endoplasmic reticulum · Macro-ER-phagy and micro-ER-phagy · Vesicular transport · ERLAD · ER-phagy receptors · Autophagy · LC3

Abbreviations

AIM	ATG8-interacting motif
ATG	Autophagy related
ATL	Atlastin
ATP	Adenosine triphosphate
ATZ	α1-antitrypsin Z
CANX	Calnexin
CCPG1	Cell-cycle progression gene 1
CHMP4B	Charged multivesicular body protein 4b
ER	Endoplasmic reticulum
ERAD	Endoplasmic reticulum-associated degradation
ERLAD	Endoplasmic reticulum-to-lysosome associated degradation
ESCRT	Endosomal sorting complexes required for transport
FIP200	Focal adhesion kinase family interacting protein of 200 kDa
FIR	FIP200-interacting region
GABARAP	γ-aminobutyric acid receptor-associated protein
GIM	GABARAP-interacting motif
GnRHR	Gonadotropin-releasing hormone receptor
GTP	Guanosine triphosphate
IDR	Intrinsically disordered region
LC3	Light chain 3
LIR	LC3-interacting region
NPC1	Niemann–Pick type C protein-1
RHD	Reticulon homology domain
RTN	Reticulon
SEC	Secretory protein
SNARE	Soluble NSF attachment protein receptor
TEX264	Testis expressed gene 264
UPR	Unfolded protein response
VAMP	Vesicle associated membrane protein

5.1 The Endoplasmic Reticulum (ER)

The ER is a large and dynamic organelle in eukaryotic cells composed of a complex interconnected network of cisternae and tubules, which are continuous with the nuclear membrane (Schwarz and Blower 2016). Two main regions are distinguishable by the presence or absence of ribosomes (i.e., rough and smooth ER, respectively) (Schwarz and Blower 2016). The rough ER is site of synthesis of secretory and membrane proteins, which compose approximatively 40% of the cell's proteome. The smooth ER has predominant roles in lipid production, hormone biogenesis and detoxification (Schwarz and Blower 2016). The dynamic nature of the intra- and extracellular environments requires effective morphological and functional adaptation of the ER, including changes in volume and activity controlled by anabolic and catabolic pathways. The anabolic events depend on transcriptional and translational processes and are governed by the Unfolded Protein Response (UPR) (Karagoz et al. 2019), which will be treated in a dedicated chapter of this book. Here, we cover the fundamental role of catabolic processes that consist in lysosomal-driven ER turnover in regulation of ER size and function (Wilkinson 2019; Loi et al. 2018). These processes rely on individual or combined activation of a handful of LC3-binding ER-resident membrane proteins named ER-phagy receptors (Fig. 5.1), on shedding of ER subdomains present in excess or that contain toxic material (e.g., proteasome-resistant misfolded proteins), and their delivery to endolysosomal degradative compartments (Loi et al. 2018; Fregno and Molinari 2019; Wilkinson 2019).

5.2 ER-Phagy

5.2.1 The ER Inside the Lysosomes: Historical Evidences

ER-phagy describes a collection of pathways that controls the selective delivery of ER subdomains to the lysosomal compartment for destruction. ER delivery to degradative compartments relies on a series of mechanistically distinct pathways defined as macro-ER-phagy (when the ER is captured by double membrane autophagosomes that eventually fuse with the yeast vacuole or the metazoan's lysosome), micro-ER-phagy (when ER portions are directly engulfed by vacuoles/endolysosomes) or vesicular delivery (when ER derived vesicles directly fuse with vacuoles/endolysosomes to release their content for clearance) (Fig. 5.2).

The first evidence that the ER is delivered within lysosomal compartments for clearance originate from morphometric studies (de Duve 1963). Rough ER membranes were observed inside vesicular structure during the formation of storage granules in insects' fat body (Locke and Collins 1965). Subsequently, ER fragments were observed inside autophagic vacuoles in rat hepatocytes after cessations of phenobarbital treatment or in guinea pig pancreatic cells after subcutaneous cobalt injection (Bolender and Weibel 1973; Tooze et al. 1990).

In 2005, delivery of ER fragments to the vacuole for destruction was observed under starvation conditions in yeast (Hamasaki et al. 2005). Subsequently, the group of Peter Walter coined the term ER-phagy (originated by the contraction of the words "ER" and "autophagy") to describe the selective delivery of ER membrane and content in the yeast vacuole following UPR-induced ER membrane expansion. This type of ER turnover does not depend on canonical autophagy proteins and is characterized by the formation of ER whorls that are directly engulfed by the vacuole in a process that is topologically equivalent to micro-autophagy (Schafer et al. 2020).

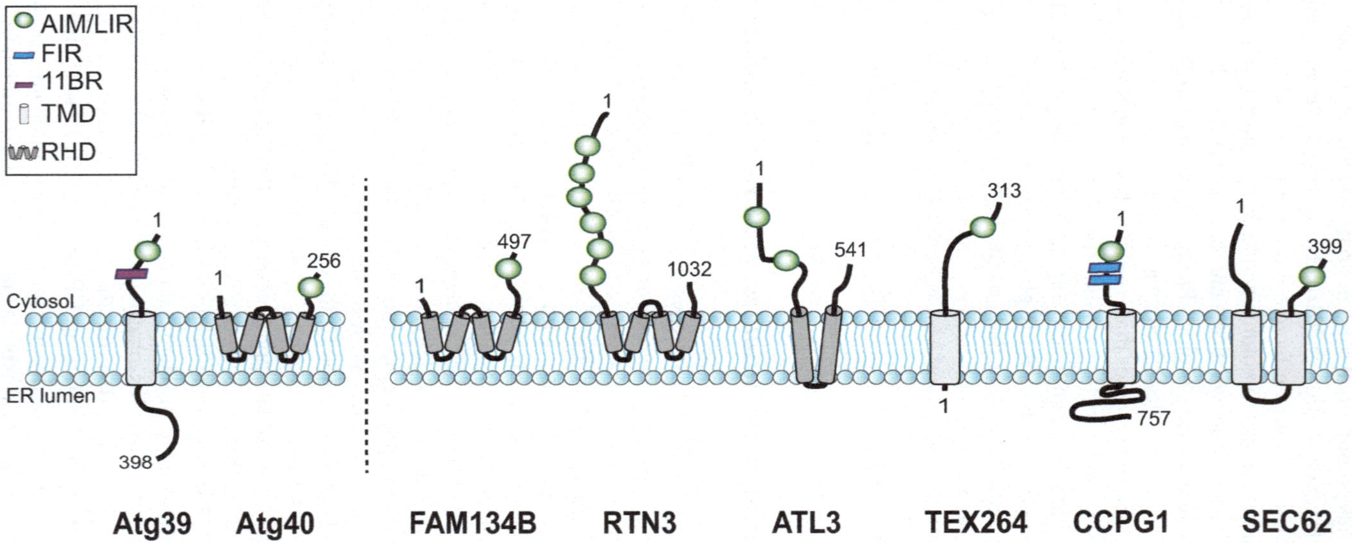

Fig. 5.1 Topology of the ER-phagy receptors. Characteristic features of the two yeast (Atg39, Atg40) and of the six mammalian receptors (FAM134B, RTN3, ATL3, TEX264, CCPG1 and SEC62) are illustrated. Number of residues in protein is shown. *AIM* Atg8-interacting motif, *LIR* LC3-interacting region, *FIR* FIP200-interacting region, *11BR* Atg11-binding region, *TMD* transmembrane domain, *RHD* reticulon-homology domain

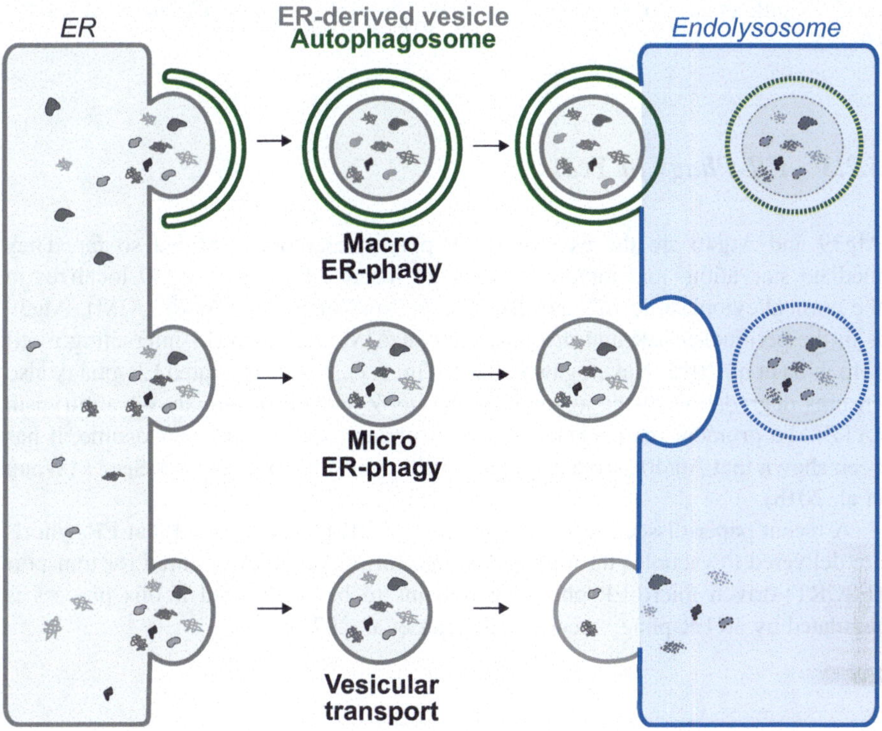

Fig. 5.2 Delivery of ER content to the endolysosomal/vacuolar compartment. *Macro-ER-phagy*, ER-subdomains are engulfed by double membrane autophagosomes that eventually fuse with the endolysosomal/vacuolar compartment. *Micro-ER-phagy*, ER-derived vesicles are directly engulfed by the endolysosomal/vacuolar compartment through membrane invagination. *LC3-dependent vesicular transport*, ER-derived vesicles directly fuse with the endolysosomal/vacuolar compartment to release their content

Consistently, UPR-induced ER turnover has been defined as micro-ER-phagy to distinguish it from macro-ER-phagy pathways induced by the overexpression of integral membrane proteins (Lipatova and Segev 2015) and by starvation (Nakatogawa and Mochida 2015).

5.2.2 The Discovery of ER-Phagy Receptors

The discovery in yeast and mammalian cells of ER-resident membrane proteins that directly bind to Atg8/LC3/GABARAP and regulate ER turnover, the ER-phagy receptors, has been fundamental to understand ER-phagy at the molecular level. So far, two yeast (Atg39 and Atg40) (Mochida et al. 2015) and six mammalian ER-phagy receptors have been described (FAM134B, RTN3L, SEC62, CCPG1, ATL3, and TEX264) (Khaminets et al. 2015; Fumagalli et al. 2016; Grumati et al.

2017; Smith et al. 2018; An et al. 2019; Chen et al. 2019; Chino et al. 2019) (Fig. 5.1).

5.2.3 ER-Phagy in Yeast

Atg39 and Atg40 are the two yeast ER-phagy receptors identified so far. They mediate starvation- and rapamycin-induced macro-ER-phagy. Atg40 localizes in the cortical/cytoplasmic ER and displays an Atg8-interacting motif (AIM), Atg39 is in the perinuclear ER and displays both an AIM and an Atg11-interacting motif (Mochida et al. 2015; Nakatogawa 2020) (Fig. 5.1). Atg40-mediated ER-phagy also ensures degradation of the aggregation-prone, Z variant of human α-1 antitrypsin (ATZ). To promote the package of select-subdomains into autophagosome, it has been shown that Atg40 associates with COPII coat sub-complex Lst1-Sec23 (Wang et al. 2018).

A recent paper dissecting ER stress-induced ER-phagy reported that ER-whorls are delivered to vacuoles through endosomal sorting complex required for transport (ESCRT)-driven micro-ER-phagy. It remains to be established if this process is regulated by an ER-phagy receptor (Schafer et al. 2020).

5.2.4 The Many Ways to Deliver the Mammalian ER to the Endolysosomal Compartments for Clearance

5.2.4.1 Starvation-Induced Macro-ER-Phagy

Nutrient deprivation activates several autophagic pathways, which are responsible for the non-selective degradation of cellular components to provide building blocks required for cell survival and maintain cellular homeostasis (Erbil-Bilir et al. 2017).

Shortage of nutrients is a convenient way also to study macro-ER-phagy as it promiscuously activates ER-phagy receptors to induce ER fragmentation and clearance via macro-ER-phagy, where ER fragments are engulfed by double membrane autophagosomes that eventually fuse with the endolysosomal/vacuolar compartment (Fig. 5.2).

FAM134B The first mammalian ER-phagy receptor reportedly activated by starvation is Family with sequence similarity 134 B (FAM134B; also known as reticulophagy regulator 1, RETREG1) (Khaminets et al. 2015). It primarily resides at the edges of ER sheets where, thanks to its reticulon-homology domain (RHD), it promotes membrane curvature (Fig. 5.1). At the cytosolic, C-terminal domain, FAM134B contains a LC3-interacting region (LIR). Upon oligomerization and association with LC3 and GABARAP proteins, FAM134B promotes ER membrane fragmentation in response to starvation (Khaminets et al. 2015). Its activity as

ER-phagy receptor is regulated by the phosphorylation state of its RHD. Under ER stress conditions and starvation, the kinase CAMK2B phosphorylates FAM134B at S^{151} promoting receptor oligomerization and ER fragmentation (Jiang et al. 2020). Moreover, an N-terminal truncated isoform of FAM134B (FAM134B-2) was identified. This isoform, predicted to be a type I membrane protein, partially lacks the RHD domain and is transcriptionally induced by C/EBPβ under starvation conditions. FAM134B is essential to regulate the lysosomal degradation of ER-retained secretory proteins in nutrient depleted liver cells (Kohno et al. 2019). Ablation of endogenous FAM134B or expression of sensory neuropathy-causing loss-of-function FAM134B mutations promote ER expansion, revealing a role of FAM134B in the constitutive maintenance of basal ER size (Khaminets et al. 2015).

Reticulon 3 (RTN3) RTN3 is also activated and oligomerizes upon nutrient-deprivation. It acts as an ER-phagy receptor involved in the lysosomal clearance of ER-tubules (Grumati et al. 2017). RTN3 belongs to the RHD-containing proteins (RTN1–4) which are highly curving-membrane proteins and drives ER tubulation. Only the long isoform of RTN3 displays six LIRs at the N-terminal domain, essential for its role as ER-phagy receptor (Fig. 5.1).

Despite their importance, RHD and the LIR motif of both FAM134B and RTN3 are probably not sufficient to induce ER fragmentation. This is suggested by the observation that atlastins 1–3 (ATL1–3) are essential for FAM134B mediated macro-ER-phagy. ATLs are ER resident GTPase that drive homotypic ER membrane fusion and are involved in ER branching and scission. During starvation-induced ER-phagy, ATL2 binds to FAM134B and remodels the ER membrane to separate FAM134B-decorated ER fragments for autophagosomal engulfment. Moreover, SEC24C, essential to sort cargo into COPII vesicles, is essential for FAM134B and RTN3 mediated macro-ER-phagy in Torin 2 treated cells (Cui et al. 2019).

Atlastin 3 (ATL3) A unique member of the ATLs protein family is ATL3, which acts as an ER-phagy receptor involved in the turnover of ER tubules during starvation. In addition to the GTPase activity, ATL3 contains two distinct GIM motif through which it selectively binds GABARAP but not LC3 proteins (Fig. 5.1). Depletion of RTN3 significantly reduces the lysosomal clearance of tubular membrane proteins, consistently with its localization, ATL3 interacts with the ER-phagy receptor RTN3 but not with the sheet ER-phagy receptor FAM134B, suggesting that both ATL3 and RTN3 cooperatively act in ER clearance during starvation conditions.

Testis Expressed Gene 264 (TEX264) Another ER-phagy receptor which activity is significantly enhanced under starvation conditions is TEX264. It is a widely expressed single pass transmembrane protein with an extensive cytosolic domain containing a LIR, essential to interact with LC3A, GABARAP and GABARAPL2 (Fig. 5.1). During starvation, TEX264 accumulates at three-way junctions already positive for LC3/GABARAP proteins and is subsequently degraded in an autophagy dependent manner. The long cytosolic C-terminal domain contains an intrinsically disordered region (IDR), suggested to act as a "spacer" to bridge the distance between the ER and autophagosomal membrane allowing the LIR of TEX264 to bind LC3 proteins on the growing autophagosomal membrane. This IDR domain is essential for TEX264 function in ER turnover and, interestingly, IDRs where identified also in FAM134B, CCPG1, RTN3 and SEC62 (Chino et al. 2019). Quantitative mass spectrometry analyses revealed that during 10 h amino acid starvation, the average decrease of ER protein is about 7–8% and that TEX264 is responsible for the autophagic degradation of about 50% of it. Thus, TEX264 in the major ER-phagy receptor for starvation induced macro-ER-phagy. It is important to keep in mind that (An et al. 2019).

5.2.4.2 ER Stress-Induced Macro-ER-Phagy

Cell Cycle Progression Gene 1 (CCPG1) CCPG1 acts as a macro-ER-phagy receptor in cells exposed to ER stress (Smith et al. 2018) (Fig. 5.2). It is a type II single-pass transmembrane protein primarily localized to the perinuclear ER and in small foci at the cell periphery. In addition to its N-terminal LIR domain essential to engage LC3/GABARAP proteins, CCPG1 binds to FIP200 via two additional FIP200-interacting regions, defined as FIRs, which contain amino acid sequences similar to the Atg11-binding motif found in the yeast receptor Atg39 (Fig. 5.1). These regions potentially link the receptors with the autophagy initiation complex. Under chemicals-induced ER stress, CCPG1 in transcriptionally induced, and both LIR and FIRs are essential to drive the lysosomal clearance of the peripheral ER. In addition, cells depleted for CCPG1 retained greater amount of peripheral/tubular ER during nutrient starvation, indicating that CCPG1 regulates ER turnover during both ER stress and starvation. CCPG1 plays an important role in maintaining proteostasis in mouse pancreas, protecting against accumulation of luminal proteins and consequent UPR elevation, thus sustaining tissue health.

5.2.4.3 Micro-ER-Phagy to Recover from ER Stress

SEC62 The translocon component SEC62 acts as an ER-phagy receptor to regulate the endolysosomal clearance of excess ER proteins and membranes produced during the UPR. This process, defined as recov-ER-phagy, regulates the endolysosomal clearance of selected ER subdomains, enriched in ER chaperones and folding enzymes but devoid of ER-associated degradation (ERAD) factors, highlighting the selectivity of the process. Moreover, SEC62-mediated recov-ER-phagy can be induced by the overexpression of recombinant SEC62 or by downregulating SEC63, another component of the translocon complex. This suggest that to remodel the ER, SEC62 dissociates from the translocation machinery. At the C-terminal domain, SEC62 harbors a LIR, which is essential for its function as ER-phagy receptor but is dispensable for its role in protein import into the ER (Fumagalli et al. 2016) (Fig. 5.1). Genetic characterization of recov-ER-phagy revealed that this process requires the LC3 lipidation machinery, but neither the autophagosome biogenesis machinery nor Soluble NSF Attachment Protein Receptors (SNAREs) involved in membrane fusion events. Rather, ultrastructural analysis confirmed that SEC62-decorated ER structures are directly engulfed by endolysosomes via micro-ER-phagy, driven by the ESCRT-III component CHMP4B and the accessory AAA+ ATPase VPS4A (Loi et al. 2019; Loi and Molinari 2020) (Fig. 5.2).

5.2.4.4 Other LC3-Binding Proteins Involved in ER Turnover

The lysosomal ER turnover is not exclusively regulated by ER-resident proteins. An example is the cytosolic autophagy receptor p62 (also known as sequestome1), which is a LIR-containing protein that delivers ubiquitinated polypeptides to the autophagic machinery for clearance. p62 is involved also in the removal of excess ER domains displaying heavily ubiquitinated proteins from mouse liver cells upon withdrawal of the hepatic mitogen 1,4-bis[2-(3,5-dichloropyridyloxy)] benzene (TCPOBOP) (Yang et al. 2016). Moreover, p62 mediates the autophagic clearance of the ER in collaboration with the ER membrane E3 ligase TRIM13. In this context, p62 acts as an N-recognin in the N-degron pathway and regulates the lysosomal ER remodeling in normal and stressed cells and controls the clearance of the pathogenic misfolded proteins α1-antitrypsin mutant Z (Ji et al. 2019). Another case is the cytosolic, ubiquitin binding autophagy receptor optineurin, which interacts with IRE1α at the ER membrane and sequesters this UPR stress sensor into the autophagy pathway in order to terminate ER signaling (Tschurtschenthaler et al. 2017). The mitophagy receptor BNIP3 contains a LIR at its N-terminal region and endogenous BNIP3 localizes at the mitochondria, but during hypoxia conditions, small amounts can relocate at the ER membrane. The over-expression of an ER-targeted version of BNIP3 leads to a LIR-dependent ER clearance. If also endogenous BNIP3 regulates lysosomal ER remodeling is not known (Hanna et al. 2012). The last case is CALCOCO1, a protein that localizes to both the cytoplasm and the nucleus and is proposed to have a role in ER-phagy induced by mTOR inhibition (Stefely et al. 2020).

5.2.4.5 ER-to-Lysosome-Associated Degradation (ERLAD)

The ER does not contain degradative devices and the majority of terminally misfolded proteins generated in this compartment are dislocated across the ER membrane to be cleared by cytosolic 26S proteasomes through a series of pathways defined as ER-associated degradation (ERAD). However, misfolded proteins that form aggregates or polymers, that display chimico/physical features that prevent engagement of ERAD factors or that prevent retro-translocation across the ER membrane are degraded via a different route. These ERAD-resistant misfolded proteins are delivered for clearance from the ER to the endolysosomal compartment by a collection of mechanistically distinct pathways defined as ER-to-Lysosome-associated Degradation (ERLAD) (Fregno and Molinari 2019).

ERAD-resistant polypeptides that fail to be efficiently degraded by cytosolic proteasomes and instead rely on ERLAD pathways comprise the β-subunits of thyrotrophic hormone, several mutant forms of serpin, mutant of the gonadotropin releasing hormone receptor (GnRHR), endogenous and ectopic expressed procollagen, mutant dysferlin, the mutant Niemann–Pick type C protein-1 (NPC1), thyroglobulin, bestrophin-1 (BEST1), pancreatic enzymes, progressive ankylosis protein, proinsulin aggregates and others (Fregno and Molinari 2019).

The receptors responsible for the delivery of a handful of these ERAD-resistant misfolded proteins to endolysosomal compartment for clearance were recently identified (Fregno and Molinari 2019). FAM134B regulates the removal of several ERAD-resistant misfolded proteins. Endolysosomal disposal of diseases relevant ATZ polymers from the ER lumen relies on FAM134B-mediated vesicular transport (Fig. 5.2). With the help of the lectin chaperone calnexin (CANX), luminal ATZ polymers segregate in select ER compartments, decorated with the receptor FAM134B, that eventually undergo scission, generating single membrane ER-derived vesicles containing ATZ polymers but not the membrane protein CANX. These vesicles subsequently release their content within endolysosome upon a membrane fusion event mediated by the ER-resident STX17 and the endolysosomal SNARE VAMP8. Genetic and morphometric analysis confirmed that this receptor-mediated vesicular transport does not rely on the intervention of an autophagosome (Fregno et al. 2018). Moreover, ATZ polymers are removed by the ER lumen through p62-dependent selective autophagy, but the molecular mechanisms regulating this catabolic process are poorly described (Ji et al. 2019).

FAM134B also controls the lysosomal disposal of endogenous misfolded NPC1, which, when mutated, it accumulates in the ER lumen and causes Niemann-Pick type C disease. Two complementary pathways degrade mutated NPC1. On one side, it is cleared via ERLAD through FAM134B mediated macro-ER-phagy, and in part, it is degraded by MARCH6-dependent ERAD.

Collagens are the most abundant proteins in animals and a fraction of the newly synthetized collagen is cleared by the lysosomes as a consequence of inefficient secretion or because it fails to attain its native conformation. Misfolded endogenous procollagen is recognized by the co-receptor CANX, which interacts with the ER-phagy receptor FAM134B. Subsequently FAM134B interacts with LC3 on the autophagosomal membrane, which captures the ER subdomains containing procollagen and CANX for lysosomal clearance (Forrester et al. 2019) (Fig. 5.2). In addition to this macroautophagic pathway, a work reported that specialized ER-exit sites containing ectopic procollagen are engulfed by lysosomes through a mechanistically poorly understood microautophagic-like process (Omari et al. 2018).

In vivo, CCPG1-mediates macro-ER-phagy maintaining ER proteostasis and health in pancreatic acinar cells, protecting against elevated ER stress and tissue injury (Smith et al. 2018).

The proinsulin *Akita* mutant, which forms insoluble aggregates in the ER lumen that entrap wild-type proinsulin, is degraded by the tubular ER-phagy receptor

RTN3. This catabolic event does not require the LIR of the receptor, suggesting a possible involvement of a of a co-receptor (Cunningham et al. 2019).

5.2.5 *ER-Phagy in Health and Disease*

The ER is subjected to constant dynamic environmental changes that requires an adequate cellular homeostatic mechanism, preventing the onset of debilitating diseases. It is well established how ER-phagy induces ER remodeling during recovery from acute ER stress (Fumagalli et al. 2016; Loi et al. 2019). Furthermore, increasing pieces of evidence establish that ER turnover is implicated in delivery of ERAD-resistant misfolded polypeptides to lysosomal compartments for their disposal (Fregno and Molinari 2019). To selectively recognize the specific cargo destined for lysosomal degradation, a series of previously cited ER-phagy has been recently identified. However, mutations linked to ER-phagy receptors can arise a wide spectrum of human disorders. Genetic mutations associated with either *FAM134B* and *ATL3* can develop hereditary sensory and autonomic neuropathies (HSAN) (Kurth et al. 2009; Kornak et al. 2014), a pathological condition where afferent fibers progressively degenerate and cause sensory loss ranging from simple local numbness to complete absence in perceiving pain (Schwartzlow and Kazamel 2019). Furthermore, mutations in *FAM134B* gene are associated with cancers. Indeed, overexpression of FAM134B in esophageal squamous carcinoma has been identified as a common event (Tang et al. 2007). In addition, FAM134B has been associated with the development of colorectal malignancies (Kasem et al. 2014). FAM134B ablation has been observed to facilitate viral production and consequent development of viral infections (Chiramel et al. 2016). For this reason, an interesting approach to counteract viral production within cell is to regulate FAM134B-mediated ER-phagy to limit intracellular viral propagation. FAM134B is able to impair both Dengue and ZIKA virus replication (Lennemann and Coyne 2017). Moreover, flaviviruses have NS3 proteases capable of cleaving specifically a single site within the RHD present in FAM134B, preventing the formation of viral protein-containing ER-derived vesicles and blocking ER-phagy (Lennemann and Coyne 2017).

Another ER-phagy receptor, SEC62, is linked to cancer (Linxweiler et al. 2017). In fact, SEC62 overexpression has been found in several oncogenic pathologies, including prostate cancer, lung cancer, head and neck squamous cell carcinoma, cervical lesions and hepatocellular carcinoma (Wemmert et al. 2016). Enhanced ER-phagy due to SEC62 overexpression may be an explanation why cancer cells are more resistant to ER stress (Bergmann et al. 2017).

Modulation of CCPG1 receptor, important for disposal of pancreatic ER luminal aggregates, may be fundamental to suppress ER stress under pancreatic inflammatory responses or to enhance ER stress to remove tumor cell in pancreatic cancers (Smith et al. 2018). However, CCPG1 has not yet been associated with human disorders (Hubner and Dikic 2020). It has been reported that RTN3-ablated

Alzheimer's mice models promotes amyloid deposition (Shi et al. 2014), suggesting a correlation with Alzheimer's disease and impairments in RTN3 function (Zou et al. 2018). Finally, RTN3 was observed to remove via ER-phagy aggregates derived from Akita proinsulin mutants, which is responsible to develop an autosomal-dominant diabetic syndrome defines as mutant INS-gene-induced diabetes of the youth (MIDY) (Cunningham et al. 2019). Interestingly, RTN3-mediated ER-phagy is an LC3-independent event, suggesting possible other proteins that can interact with RTN3 (Cunningham et al. 2019).

5.3 Summary Points

- The ER is a plastic organelle whose size and content are adapted to physiological and pathological needs.
- ER turnover relies on delivery of fragmented portions to the endolysosomal compartment via mechanistically distinct pathways (macro- and micro-ER-phagy or LC3-dependent vesicular transport).
- The lysosomal-driven ER turnover is regulated by specialized ER-phagy receptors, ER-resident membrane proteins that engage Atg8/LC3s/GABARAPs through their AIM/LIR.
- Two yeast (Atg39 and Atg40) and six mammalian (FAM134B, SEC62, CCPG1, RTN3, ATL3 and TEX264) ER-phagy receptors have been identified so far.
- ERLAD is a collection of pathways that regulate the endolysosomal clearance of ERAD-resistant misfolded proteins.

5.4 Future Issues

- Identification of signals that activate individual ER-phagy receptors and mechanistic characterization of the triggered responses.
- The six different mammalian ER-phagy receptors can be divided into intermembrane (FAM134B, RTN3, ATL3) and transmembrane (SEC62, CCPG1, TEX264) proteins. Their activation and action mechanisms in delimiting and shedding from the bulk ER, the ER sub-domains to be cleared from cells, remain to be established.
- Involvement of ER-phagy in disease and in pathogen infection is an interesting field for future studies.
- Chemical and pharmacologic modulation of ER-phagy pathways is expected to have medical relevance.

Acknowledgments M.M. is supported by AlphaONE Foundation, Foundation for Research on Neurodegenerative Diseases, Swiss National Science Foundation (SNF) and Comel and Gelu Foundations.

Conflict-of-Interest Disclosure Statement The authors declare that they have no conflict of interest.

References

An H, Ordureau A, Paulo JA, Shoemaker CJ, Denic V, Harper JW (2019) TEX264 Is an Endoplasmic Reticulum-Resident ATG8-Interacting Protein Critical for ER Remodeling during Nutrient Stress. Mol Cell 74(5):891–908 e10. https://doi.org/10.1016/j.molcel.2019.03.034

Bergmann TJ, Fumagalli F, Loi M, Molinari M (2017) Role of SEC62 in ER maintenance: a link with ER stress tolerance in SEC62-overexpressing tumors? Mol Cell Oncol 4(2):e1264351. https://doi.org/10.1080/23723556.2016.1264351

Bolender RP, Weibel ER (1973) A morphometric study of the removal of phenobarbital-induced membranes from hepatocytes after cessation of threatment. J Cell Biol 56(3):746–761. https://doi.org/10.1083/jcb.56.3.746

Chen Q, Xiao Y, Chai P, Zheng P, Teng J, Chen J (2019) ATL3 is a tubular ER-phagy receptor for GABARAP-mediated selective autophagy. Curr Biol 29(5):846–855 e6. https://doi.org/10.1016/j.cub.2019.01.041

Chino H, Hatta T, Natsume T, Mizushima N (2019) Intrinsically disordered protein TEX264 mediates ER-phagy. Mol Cell 74(5):909–921 e6. https://doi.org/10.1016/j.molcel.2019.03.033

Chiramel AI, Dougherty JD, Nair V, Robertson SJ, Best SM (2016) FAM134B, the selective autophagy receptor for endoplasmic reticulum turnover, inhibits replication of ebola virus strains makona and mayinga. J Infect Dis 214(suppl 3):S319–S325. https://doi.org/10.1093/infdis/jiw270

Cui Y, Parashar S, Zahoor M, Needham PG, Mari M, Zhu M, Chen S, Ho HC, Reggiori F, Farhan H, Brodsky JL, Ferro-Novick S (2019) A COPII subunit acts with an autophagy receptor to target endoplasmic reticulum for degradation. Science 365(6448):53–60. https://doi.org/10.1126/science.aau9263

Cunningham CN, Williams JM, Knupp J, Arunagiri A, Arvan P, Tsai B (2019) Cells deploy a two-pronged strategy to rectify misfolded proinsulin aggregates. Mol Cell 75(3):442–456. e4. https://doi.org/10.1016/j.molcel.2019.05.011

de Duve C (1963) The lysosome. Sci Am 208:64–72. https://doi.org/10.1038/scientificamerican0563-64

Erbil-Bilir S, Gozuacik D, Kutlu O (2017) Autophagy as a physiological response of the body to starvation. In: Preedy V, Patel VB (eds) Handbook of famine, starvation, and nutrient deprivation: from biology to policy. Springer International Publishing, Cham, pp 1–15

Forrester A, De Leonibus C, Grumati P, Fasana E, Piemontese M, Staiano L, Fregno I, Raimondi A, Marazza A, Bruno G, Iavazzo M, Intartaglia D, Seczynska M, van Anken E, Conte I, De Matteis MA, Dikic I, Molinari M, Settembre C (2019) A selective ER-phagy exerts procollagen quality control via a Calnexin-FAM134B complex. EMBO J 38(2):e99847. https://doi.org/10.15252/embj.201899847

Fregno I, Molinari M (2019) Proteasomal and lysosomal clearance of faulty secretory proteins: ER-associated degradation (ERAD) and ER-to-lysosome-associated degradation (ERLAD) pathways. Crit Rev Biochem Mol Biol 54(2):153–163. https://doi.org/10.1080/10409238.2019.1610351

Fregno I, Fasana E, Bergmann TJ, Raimondi A, Loi M, Solda T, Galli C, D'Antuono R, Morone D, Danieli A, Paganetti P, van Anken E, Molinari M (2018) ER-to-lysosome-associated degradation of proteasome-resistant ATZ polymers occurs via receptor-mediated vesicular transport. EMBO J 37(17):e99259. https://doi.org/10.15252/embj.201899259

Fumagalli F, Noack J, Bergmann TJ, Cebollero E, Pisoni GB, Fasana E, Fregno I, Galli C, Loi M, Solda T, D'Antuono R, Raimondi A, Jung M, Melnyk A, Schorr S, Schreiber A, Simonelli L,

Varani L, Wilson-Zbinden C, Zerbe O, Hofmann K, Peter M, Quadroni M, Zimmermann R, Molinari M (2016) Translocon component Sec62 acts in endoplasmic reticulum turnover during stress recovery. Nat Cell Biol 18(11):1173–1184. https://doi.org/10.1038/ncb3423

Grumati P, Morozzi G, Holper S, Mari M, Harwardt MI, Yan R, Muller S, Reggiori F, Heilemann M, Dikic I (2017) Full length RTN3 regulates turnover of tubular endoplasmic reticulum via selective autophagy. elife 6:e25555. https://doi.org/10.7554/eLife.25555

Hamasaki M, Noda T, Baba M, Ohsumi Y (2005) Starvation triggers the delivery of the endoplasmic reticulum to the vacuole via autophagy in yeast. Traffic 6(1):56–65. https://doi.org/10.1111/j.1600-0854.2004.00245.x

Hanna RA, Quinsay MN, Orogo AM, Giang K, Rikka S, Gustafsson AB (2012) Microtubule-associated protein 1 light chain 3 (LC3) interacts with Bnip3 protein to selectively remove endoplasmic reticulum and mitochondria via autophagy. J Biol Chem 287(23):19094–19104. https://doi.org/10.1074/jbc.M111.322933

Hubner CA, Dikic I (2020) ER-phagy and human diseases. Cell Death Differ 27(3):833–842. https://doi.org/10.1038/s41418-019-0444-0

Ji CH, Kim HY, Heo AJ, Lee SH, Lee MJ, Kim SB, Srinivasrao G, Mun SR, Cha-Molstad H, Ciechanover A, Choi CY, Lee HG, Kim BY, Kwon YT (2019) The N-degron pathway mediates ER-phagy. Mol Cell 75(5):1058–1072. e9. https://doi.org/10.1016/j.molcel.2019.06.028

Jiang X, Wang X, Ding X, Du M, Li B, Weng X, Zhang J, Li L, Tian R, Zhu Q, Chen S, Wang L, Liu W, Fang L, Neculai D, Sun Q (2020) FAM134B oligomerization drives endoplasmic reticulum membrane scission for ER-phagy. EMBO J 39:e102608. https://doi.org/10.15252/embj.2019102608

Karagoz GE, Acosta-Alvear D, Walter P (2019) The unfolded protein response: detecting and responding to fluctuations in the protein-folding capacity of the endoplasmic reticulum. Cold Spring Harb Perspect Biol 11(9):a033886. https://doi.org/10.1101/cshperspect.a033886

Kasem K, Gopalan V, Salajegheh A, Lu CT, Smith RA, Lam AK (2014) The roles of JK-1 (FAM134B) expressions in colorectal cancer. Exp Cell Res 326(1):166–173. https://doi.org/10.1016/j.yexcr.2014.06.013

Khaminets A, Heinrich T, Mari M, Grumati P, Huebner AK, Akutsu M, Liebmann L, Stolz A, Nietzsche S, Koch N, Mauthe M, Katona I, Qualmann B, Weis J, Reggiori F, Kurth I, Hubner CA, Dikic I (2015) Regulation of endoplasmic reticulum turnover by selective autophagy. Nature 522(7556):354–358. https://doi.org/10.1038/nature14498

Kohno S, Shiozaki Y, Keenan AL, Miyazaki-Anzai S, Miyazaki M (2019) An N-terminal-truncated isoform of FAM134B (FAM134B-2) regulates starvation-induced hepatic selective ER-phagy. Life Sci Alliance 2(3):e201900340. https://doi.org/10.26508/lsa.201900340

Kornak U, Mademan I, Schinke M, Voigt M, Krawitz P, Hecht J, Barvencik F, Schinke T, Giesselmann S, Beil FT, Pou-Serradell A, Vilchez JJ, Beetz C, Deconinck T, Timmerman V, Kaether C, De Jonghe P, Hubner CA, Gal A, Amling M, Mundlos S, Baets J, Kurth I (2014) Sensory neuropathy with bone destruction due to a mutation in the membrane-shaping atlastin GTPase 3. Brain 137(Pt 3):683–692. https://doi.org/10.1093/brain/awt357

Kurth I, Pamminger T, Hennings JC, Soehendra D, Huebner AK, Rotthier A, Baets J, Senderek J, Topaloglu H, Farrell SA, Nurnberg G, Nurnberg P, De Jonghe P, Gal A, Kaether C, Timmerman V, Hubner CA (2009) Mutations in FAM134B, encoding a newly identified Golgi protein, cause severe sensory and autonomic neuropathy. Nat Genet 41(11):1179–1181. https://doi.org/10.1038/ng.464

Lennemann NJ, Coyne CB (2017) Dengue and Zika viruses subvert reticulophagy by NS2B3-mediated cleavage of FAM134B. Autophagy 13(2):322–332. https://doi.org/10.1080/15548627.2016.1265192

Linxweiler M, Schick B, Zimmermann R (2017) Let's talk about Secs: Sec61, Sec62 and Sec63 in signal transduction, oncology and personalized medicine. Signal Transduct Target Ther 2:17002. https://doi.org/10.1038/sigtrans.2017.2

Lipatova Z, Segev N (2015) A role for macro-ER-phagy in ER quality control. PLoS Genet 11(7): e1005390. https://doi.org/10.1371/journal.pgen.1005390

Locke M, Collins JV (1965) The structure and formation of protein granules in the fat body of an insect. J Cell Biol 26(3):857–884

Loi M, Molinari M (2020) Mechanistic insights in recov-ER-phagy: micro-ER-phagy to recover from stress. Autophagy 16(2):385–386. https://doi.org/10.1080/15548627.2019.1709767

Loi M, Fregno I, Guerra C, Molinari M (2018) Eat it right: ER-phagy and recovER-phagy. Biochem Soc Trans 46(3):699–706. https://doi.org/10.1042/BST20170354

Loi M, Raimondi A, Morone D, Molinari M (2019) ESCRT-III-driven piecemeal micro-ER-phagy remodels the ER during recovery from ER stress. Nat Commun 10(1):5058. https://doi.org/10.1038/s41467-019-12991-z

Mochida K, Oikawa Y, Kimura Y, Kirisako H, Hirano H, Ohsumi Y, Nakatogawa H (2015) Receptor-mediated selective autophagy degrades the endoplasmic reticulum and the nucleus. Nature 522(7556):359–362. https://doi.org/10.1038/nature14506

Nakatogawa H (2020) Autophagic degradation of the endoplasmic reticulum. Proc Jpn Acad Ser B Phys Biol Sci 96(1):1–9. https://doi.org/10.2183/pjab.96.001

Nakatogawa H, Mochida K (2015) Reticulophagy and nucleophagy: new findings and unsolved issues. Autophagy 11(12):2377–2378. https://doi.org/10.1080/15548627.2015.1106665

Omari S, Makareeva E, Roberts-Pilgrim A, Mirigian L, Jarnik M, Ott C, Lippincott-Schwartz J, Leikin S (2018) Noncanonical autophagy at ER exit sites regulates procollagen turnover. Proc Natl Acad Sci U S A 115(43):E10099–E10108. https://doi.org/10.1073/pnas.1814552115

Schafer JA, Schessner JP, Bircham PW, Tsuji T, Funaya C, Pajonk O, Schaeff K, Ruffini G, Papagiannidis D, Knop M, Fujimoto T, Schuck S (2020) ESCRT machinery mediates selective microautophagy of endoplasmic reticulum in yeast. EMBO J 39(2):e102586. https://doi.org/10.15252/embj.2019102586

Schwartzlow C, Kazamel M (2019) Hereditary sensory and autonomic neuropathies: adding more to the classification. Curr Neurol Neurosci Rep 19(8):52. https://doi.org/10.1007/s11910-019-0974-3

Schwarz DS, Blower MD (2016) The endoplasmic reticulum: structure, function and response to cellular signaling. Cell Mol Life Sci 73(1):79–94. https://doi.org/10.1007/s00018-015-2052-6

Shi Q, Ge Y, Sharoar MG, He W, Xiang R, Zhang Z, Hu X, Yan R (2014) Impact of RTN3 deficiency on expression of BACE1 and amyloid deposition. J Neurosci 34(42):13954–13962. https://doi.org/10.1523/JNEUROSCI.1588-14.2014

Smith MD, Harley ME, Kemp AJ, Wills J, Lee M, Arends M, von Kriegsheim A, Behrends C, Wilkinson S (2018) CCPG1 is a non-canonical autophagy cargo receptor essential for ER-phagy and pancreatic ER proteostasis. Dev Cell 44(2):217–232 e11. https://doi.org/10.1016/j.devcel.2017.11.024

Stefely JA, Zhang Y, Freiberger EC, Kwiecien NW, Thomas HE, Davis AM, Lowry ND, Vincent CE, Shishkova E, Clark NA, Medvedovic M, Coon JJ, Pagliarini DJ, Mercer CA (2020) Mass spectrometry proteomics reveals a function for mammalian CALCOCO1 in MTOR-regulated selective autophagy. Autophagy 6(12):2219–2237. https://doi.org/10.1080/15548627.2020.1719746

Tang WK, Chui CH, Fatima S, Kok SH, Pak KC, Ou TM, Hui KS, Wong MM, Wong J, Law S, Tsao SW, Lam KY, Beh PS, Srivastava G, Chan AS, Ho KP, Tang JC (2007) Oncogenic properties of a novel gene JK-1 located in chromosome 5p and its overexpression in human esophageal squamous cell carcinoma. Int J Mol Med 19(6):915–923

Tooze J, Hollinshead M, Ludwig T, Howell K, Hoflack B, Kern H (1990) In exocrine pancreas, the basolateral endocytic pathway converges with the autophagic pathway immediately after the early endosome. J Cell Biol 111(2):329–345. https://doi.org/10.1083/jcb.111.2.329

Tschurtschenthaler M, Adolph TE, Ashcroft JW, Niederreiter L, Bharti R, Saveljeva S, Bhattacharyya J, Flak MB, Shih DQ, Fuhler GM, Parkes M, Kohno K, Iwawaki T, Janneke van der Woude C, Harding HP, Smith AM, Peppelenbosch MP, Targan SR, Ron D, Rosenstiel P, Blumberg RS, Kaser A (2017) Defective ATG16L1-mediated removal of IRE1alpha drives Crohn's disease-like ileitis. J Exp Med 214(2):401–422. https://doi.org/10.1084/jem.20160791

Wang B, Joo JH, Mount R, Teubner BJW, Krenzer A, Ward AL, Ichhaporia VP, Adams EJ, Khoriaty R, Peters ST, Pruett-Miller SM, Zakharenko SS, Ginsburg D, Kundu M (2018) The COPII cargo adapter SEC24C is essential for neuronal homeostasis. J Clin Invest 128 (8):3319–3332. https://doi.org/10.1172/JCI98194

Wemmert S, Lindner Y, Linxweiler J, Wagenpfeil S, Bohle R, Niewald M, Schick B (2016) Initial evidence for Sec62 as a prognostic marker in advanced head and neck squamous cell carcinoma. Oncol Lett 11(3):1661–1670. https://doi.org/10.3892/ol.2016.4135

Wilkinson S (2019) ER-phagy: shaping up and destressing the endoplasmic reticulum. FEBS J 286 (14):2645–2663. https://doi.org/10.1111/febs.14932

Yang H, Ni HM, Guo F, Ding Y, Shi YH, Lahiri P, Frohlich LF, Rulicke T, Smole C, Schmidt VC, Zatloukal K, Cui Y, Komatsu M, Fan J, Ding WX (2016) Sequestosome 1/p62 protein is associated with autophagic removal of excess hepatic endoplasmic reticulum in mice. J Biol Chem 291(36):18663–18674. https://doi.org/10.1074/jbc.M116.739821

Zou Y, He W, Wang K, Han H, Xiao T, Chen X, Zhou B, Tan J, Xia K, Tang B, Chen C, Shen L, Yan R, Zhang Z (2018) Identification of rare RTN3 variants in Alzheimer's disease in Han Chinese. Hum Genet 137(2):141–150. https://doi.org/10.1007/s00439-018-1868-1

Chapter 6
Defects in Protein Folding and/or Quality Control Cause Protein Aggregation in the Endoplasmic Reticulum

Juthakorn Poothong, Insook Jang, and Randal J. Kaufman

Abstract Protein aggregation is now a common hallmark of numerous human diseases, most of which involve cytosolic aggregates including Aβ (AD) and α-synuclein (PD) in Alzheimer's disease and Parkinson's disease. However, it is also evident that protein aggregation can also occur in the lumen of the endoplasmic reticulum (ER) that leads to specific diseases due to loss of protein function or detrimental effects on the host cell, the former is inherited in a recessive manner where the latter are dominantly inherited. However, the mechanisms of protein aggregation, disaggregation and degradation in the ER are not well understood. Here we provide an overview of factors that cause protein aggregation in the ER and how the ER handles aggregated proteins. Protein aggregation in the ER can result from intrinsic properties of the protein (hydrophobic residues in the ER), oxidative stress or nutrient depletion. The ER has quality control mechanisms [chaperone functions, ER-associated protein degradation (ERAD) and autophagy] to ensure only correctly folded proteins exit the ER and enter the *cis*-Golgi compartment. Perturbation of protein folding in the ER activates the unfolded protein response (UPR) that evolved to increase ER protein folding capacity and efficiency and degrade misfolded proteins. Accumulation of misfolded proteins in the ER to a level that exceeds the ER-chaperone folding capacity is a major factor that exacerbates protein aggregation. The most significant ER resident protein that prevents protein aggregation in the ER is the heat shock protein 70 (HSP70) homologue, BiP/GRP78, which is a peptide-dependent ATPase that binds unfolded/misfolded proteins and releases them upon ATP binding. Since exogenous factors can also reduce protein misfolding and aggregation in the ER, such as chemical chaperones and antioxidants, these treatments have potential therapeutic benefit for ER protein aggregation-associated diseases.

Keywords UPR · BiP/GRP78 · CANX/CALR · Proinsulin · Clotting factor VIII · CFTR · Amyloid

---

J. Poothong · I. Jang · R. J. Kaufman (✉)
Degenerative Diseases Program, Sanford Burnham Prebys Medical Discovery Institute, La Jolla, CA, USA
e-mail: rkaufman@sbpdiscovery.org

L. B. Agellon, M. Michalak (eds.), *Cellular Biology of the Endoplasmic Reticulum*, Progress in Molecular and Subcellular Biology 59,
https://doi.org/10.1007/978-3-030-67696-4_6

6.1 Many Proteins Can Aggregate in the ER

One-third of newly synthesized polypeptides are translocated across the ER membrane where they attain a correct three-dimensional (3D) structure necessary for function. Proteins enter the ER in an unfolded state where they are modified by glycosylation, disulfide bond formation, *cys-trans* prolyl isomerization and other amino acid modifications to obtain their native structures. The primary structure of a protein is the linear chain of amino acids having high entropy and free energy. After forming secondary and tertiary structures, proteins fold into their native conformation (quaternary structure) which decreases their free energy providing protein stability. Correct folding brings all functional groups into close proximity creating sites required for function. Proteins that fail to attain their native conformation due to genetic mutation(s), transcription or translation defects, nutrient deprivation, altered redox status and/or oxidative stress etc., expose hydrophobic amino acids, that are typically buried in the core of a protein, on their surface which predispose to aggregation. Aggregation initiates with the formation of small oligomers that expand into high order ß-structured oligomers and eventually mature amyloid fibrils.

6.2 Protein Aggregation and Amyloid Fibril Formation Initiate Through a Nucleation Event

Misfolded proteins expose hydrophobic patches which can initiate aggregation, however, the molecular mechanism of protein aggregation is not well characterized. The factors dictating the transition between an unfolded protein to a native folded protein or to an aggregate depend on the thermodynamic and kinetic stabilities of intermediates on the pathway for productive folding. The most classical model to explain protein aggregation and amyloid formation is nucleation-dependent polymerization (NDP) (Iadanza et al. 2018) (Fig. 6.1). NDP has three different phases: (1) A Nucleation phase, where a misfolded monomer self-associates to form a small oligomer which forms a nucleus for fibril formation. This initiation step is time and peptide concentration dependent; (2) An Elongation phase, where existing oligomers associate with additional monomers to extend the amyloid fibril. This extension process is a consequence of intermolecular cross ß-strand interactions. At some point in this step, protein intermediates undergo a transition to a ß-sheet prior to fibril assembly (Iannuzzi et al. 2017). Once fibrils with cross ß-sheet structure are formed, they may fragment generating new ends that can recruit additional monomers enhancing fibril growth; and (3) A Saturation phase, where the fibrils further associate to produce well-organized amyloid plaques. Amyloid fibrils are typically unbranched fibrils ~5 to 10 nm in diameter built from 2–3 filaments of ~3 nm protofibrils (Khurana et al. 2003). This nucleation process has been described for various protein aggregates including human islet amyloid polypeptide (IAPP, amylin), Aß and α-synuclein (Ghosh et al. 2015; Tenidis et al. 2000). In some

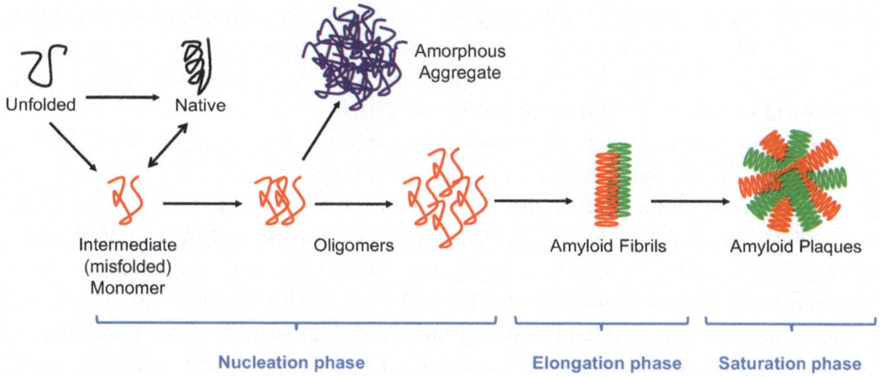

Fig. 6.1 Nucleation dependent polymerization (NDP) for protein aggregation and amyloid fibril formation. Newly synthesized polypeptides enter into the ER where they obtain their functional native structure. Various genetic and environmental defects (i.e. disruption of ER homeostasis, nutrient deprivation, and oxidative stress, etc.) disrupt protein folding and cause exposure of hydrophobic amino acids on the protein surface that can initiate protein oligomerization. NDP is a classic model to explain protein aggregation and amyloid fibril formation: First, a misfolded monomer self-interacts to form small oligomer in the Nucleation phase. Next, these small oligomers associate with additional monomers to extend the amyloid fibril in the Elongation phase. In this step, proteins transition into a ß-sheet structure at some point in advance to fibril assembly. Finally, the fibrils further associate and produce well-organized amyloid plaque in the Saturation phase

cases, amyloid fibril formation can be initiated by intrinsically disordered regions of a protein or even native globular proteins, such as the murine protein tyrosine phosphatase-Bas-like (PDZ2) (Sicorello et al. 2009), human blood clotting factor VIII (FVIII) (Poothong et al. 2020) and insulin (Iannuzzi et al. 2017).

6.3 Intrinsic and Extrinsic Factors Affect Protein Aggregation and Amyloid Fibril Formation

6.3.1 Increased ß-Sheet Propensity Contributes to Protein Aggregation

Although proteins with propensity to form amyloid fibrils have diverse amino acid sequences, they all share common structural features, a ß-sheet or hydrophobic propensity. Hydrophobic amino acids are more likely to initiate aggregation and amyloid fibril formation. Amino acid residues responsible for amyloid fibril formation have been identified in a number of cytosolic proteins, as well as ER-localized proteins including a 79 amino acid region in clotting factor VIII (FVIII) (residues 253–331) (Poothong et al. 2020) and short amino acid sequences FGAIIL (residues 23–27) and NFGAIL (residues 22–27) in IAPP (Tenidis et al. 2000). These residues exhibit high propensity for β-structure and are sufficient for amyloid fibril formation.

On the other hand, amyloid fibrils can also be formed from intrinsically unstructured regions as well as from α-helices in rare cases, such as transthyretin. This process is believed to involve a structural transition (α-helix to ß-structure) during fibril assembly (Jiménez et al. 2002; Iannuzzi et al. 2017).

Algorithms can predict aggregation-prone regions. For example, the TANGO algorithm predicted the increased aggregation propensity of Aß (1–42) over Aß (1–40): Aß (1–42) which has higher toxicity than Aß (1–40) (Fernandez-Escamilla et al. 2004). In the A1 domain of FVIII, TANGO plots showed that a F309S point mutation eliminates a region of β-aggregation propensity and significantly reduces aggregation compared to the wild-type FVIII in cells (Poothong et al. 2020). The TANGO analysis was applied to other proteins, such as insulin. From prediction of full-length proinsulin (insulin precursor), the α-helix from amino acids VEALYL (residues 12–17) within the B-chain that is required for homodimerization in the ER (Sun et al. 2020), and from SLYQLENY (residues 77–84) (Ivanova et al. 2009) within the A-chain exhibit β-sheet propensity, which may explain the structural transition to form a β-stacked structure between the B- and A- chains (Jiménez et al. 2002; Muzaffar and Ahmad 2011; Iannuzzi et al. 2017; Ivanova et al. 2009). In addition, TANGO demonstrated the significance of hydrophobic regions within co-chaperones ERdj4 and ERdj5 that engage misfolded client proteins for transfer to BiP (Behnke et al. 2016).

6.3.2 Protein Aggregation Is Dependent on Protein Concentration

Another important factor that influences amyloid fibril formation is peptide concentration. Overload of newly synthesized polypeptides in the ER can overwhelm the chaperone-mediated folding capacity, causing protein misfolding and aggregation. An equilibrium between the protein folding chaperones and the client protein folding load needs to be maintained to preserve protein folding homeostasis in the ER. At the initiation step, a certain amount of monomer is required to initiate an oligomer nucleus. Addition of monomers to the existing oligomers can accelerate protofibril formation. Lastly, protofibrils might not convert to mature fibrils at low peptide concentrations. This parameter is evident for cytosolic amyloidogenesis including Aß (Meisl et al. 2016), tau (Shammas et al. 2015), α-synuclein (Iljina et al. 2016) as well as ER amyloidogenesis of FVIII (Poothong et al. 2020) and (pro)insulin (Hong et al. 2006). Inhibition of proinsulin synthesis by cycloheximide significantly reduced proinsulin multimer formation (Jang et al. 2019). Conversely, brefeldin A, a treatment that blocks protein trafficking from the ER to the Golgi increased proinsulin oligomerization, possibly because increased proinsulin accumulation in the ER (Jang et al. 2019).

6.3.3 Oxidative Stress Causes Protein Aggregation

Oxidative stress is defined as an imbalance between the generation and elimination of reactive oxygen species (ROS). ROS are oxygen free radicals such as hydroxyl radical (OH) and superoxide anion (O_2^-). Non-radical derivatives of oxygen, such as hydrogen peroxide (H_2O_2), are also considered ROS, since their detoxification generates a free radical. Excessive production of cellular ROS can damage proteins, nucleic acids and lipids that lead to inflammatory processes as well as protein misfolding.

In the ER lumen, ROS is generated as a byproduct of disulfide bond formation during oxidative protein folding. Disulfide bond formation is a reversible process catalyzed between two cysteine residues by thiol-disulfide isomerases, which stabilize the three-dimensional structure of a protein. Protein disulfide isomerase (PDIA1) and ER oxidoreductase 1 (Ero1) are two major thiol-disulfide isomerases that control thiol redox status and regulate disulfide bond formation. During the process, PDI accepts two electrons from thiol groups of cysteine residues from a substrate that results in a disulfide bond in the substrate with reduction of PDI. Re-oxidization of PDI by Ero1 allows another cycle of reduction/oxidation by PDI for other substrates. After oxidizing PDI, Ero1 transfers electrons received from PDI to molecular oxygen (O_2) to produce H_2O_2. PDI can also mediate rearrangement of incorrectly formed disulfide bond(s) by its disulfide isomerase activity which involves a cycle of reduction (breaking of non-native disulfide bonds) and oxidation (introducing a new disulfide bond) (Schwaller et al. 2003) that can be another source of ROS generation.

ROS overload in the ER also affects calcium (Ca^{2+}) homeostasis that is important for cellular and chaperone functions in protein folding. Under normal conditions, Ca^{2+} is pumped into the ER lumen by the sarco/endoplasmic reticulum Ca^{2+}-ATPase (SERCA) and is released through inositol-3-phosphate receptors (IP_3R) and ryanodine receptors (RyR) to maintain an optimal ER Ca^{2+} level at 1–2 mM (GÖrlach et al. 2015). Upon an increase in ROS, the Ca^{2+} concentration in the ER lumen rapidly decreases due to opening of IP_3R and RyR channels (Kania et al. 2015). Unresolved ROS accumulation can disrupt chaperone functions that require Ca^{2+} and compromise protein folding.

6.3.4 ATP Is Required to Prevent Protein Aggregation

Adenosine triphosphate (ATP) is a key energy source required for the most energy-dependent cellular processes such as protein synthesis, folding, translocation into the ER and protein degradation. The function of the chaperone BiP is tightly regulated by ATP levels in the ER. Undoubtedly, ATP is a crucial factor for preventing protein aggregation. Since the ER has no ATP generating system, the ER gets ATP supply from two different sources, glycolysis and oxidative phosphorylation (OxPhos) (Yong et al. 2019). Recent studies suggest that depletion of ER ATP by inhibition of glycolysis with 2-deoxyglucose, but not oxidative phosphorylation, induces FVIII aggregation in the ER (Poothong et al. 2020). In addition, ATP itself can directly act as a biological hydrotrope. It can both prevent and dissolve protein aggregates (Patel et al. 2017; Kang et al. 2019).

6.4 ER Quality Control Ensures Only Properly Folded Proteins Traffic to the Golgi Compartment

In the ER, chaperone machineries play crucial roles in many biological processes such as protein synthesis and translocation into the ER, folding, transport and degradation. Two major chaperone families in the ER are the lectin chaperones CANX/CALR that assist glycoprotein folding and recognize misfolded glycosylated proteins for degradation and the family of Heat Shock Proteins (HSPs). Only properly folded proteins can exit the ER for anterograde traffic to the Golgi apparatus. Proteins that cannot attain a native three-dimensional structure in the ER are targeted for degradation by either ERAD through the 26S proteasome (Ruggiano et al. 2014) or by autophagy to maintain ER protein homeostasis. For these processes, readers are referred to excellent recent reviews (Hwang and Qi 2018; Song et al. 2018).

6.4.1 The Calnexin/Calreticulin (CANX/CALR) Cycle Prevents Aggregation of Misfolded Glycoproteins

Calnexin (CANX) and calreticulin (CALR) are critical ER lectin chaperones responsible for folding and quality control through their interaction with monoglucosylated glycoproteins. CANX and CALR bind to monoglucosylated N-linked glycans on glycoproteins. CANX is an ER type I transmembrane protein and CALR is an ER luminal protein. Both proteins have P-domains on their luminal side that harbor binding sites for the oxidoreductase ERp57/PDIA3 (Araki and Nagata 2011). Interaction with ERp57 catalyzes disulfide bond formation in substrates bound to CANX/CALR. While nascent glycoproteins are being synthesized, a 14-mer oligosaccharide ($Glc_3Man_9GlcNAc_2$; Glc:Glucose; Man:Mannose; GlcNAc:N-acetylglucosamine) is transferred from dolichol diphosphate to nascent polypeptides by oligosaccharyltransferase (Parodi et al. 1972) at the conserved amino acid motif Asn-X-Ser/Thr where X is any amino acid except proline. After attachment to the oligosaccharide, the first external glucose is immediately trimmed by glucosidase I to generate a Glc_2-N-glycan. The second terminal glucose is subsequently trimmed by glucosidase II (Pelletier et al. 2000). Then, monoglucosylated polypeptides can interact with CANX/CALR where it can form disulfide intermediates with ERp57 to further stabilize and assist protein folding while protecting nascent polypeptides from aggregation (Oliver et al. 1997). At this step, if the nascent polypeptide folds correctly, upon release from CANX/CALR it can exit the ER for traffic to the Golgi. However, if released in a misfolded form, the glycopolypeptide interacts with UDP-glucose:glycoprotein glycosyltransferase 1 (UGGT1) that recognizes exposed hydrophobic amino acids as well as the glucose-trimmed oligosaccharide and transfers glucose from UDP-glucose to the core glycan of the misfolded glycoprotein. As a consequence, the monoglucosylated protein re-enters to the CANX/CALR cycle for another attempt for proper folding (Tannous et al. 2015). The selective interaction of UGGT1 with partially folded glycopolypeptides implicates UGGT1 acts as a gatekeeper to only reglycosylated folding intermediates. Subsequent interaction with CANX/CALR can prevent protein aggregation. Indeed, the solubility monoglucosylated α-1-antitrypsin mutant requires interaction with CANX/CALR (Ferris et al. 2013).

6.4.2 BiP/GRP78 Maintains Proteins in a Folding-Competent Conformation, Prevents Aggregation and Promotes Disaggregation

Heat Shock Proteins (HSPs) are classified into several families by their molecular weight and domain organization: Hsp40; Hsp70; Hsp90; Hsp104; Hsp110 etc. The Hsp70 family members exhibit peptide-dependent ATPase activity. BiP/GRP78, the most abundant chaperone in the ER, is the Hsp70 family member localized to the ER. BiP binds to a wide range of client proteins including polypeptides emerging from the ribosome, nascent folding intermediates, unfolded or misfolded proteins and unassembled protein subunits (Clerico et al. 2019). In addition, BiP also acts as a negative regulator for the unfolded protein response (UPR) by binding the UPR signal transducers.

Structurally, BiP is composed of two major domains, an N-terminal nucleotide binding domain (NBD) and a C-terminal substrate binding domain (SBD). The NBD contains the ATPase activity that regulates substrate binding affinity. On the other hand, the SBD comprises 8 β-strands with a helical lid. A hydrophobic linker bridges the two domains and controls allosteric interactions between the NBD and SBD. In the resting state (absence of nucleotide), these two domains are not in close contact and inactive. Upon ATP binding, the SBD lid opens and client proteins are able to bind. There are 7 ERdj co-chaperones (Hsp40 family members that stimulate the ATPase activity of Hsp70 family members) in the ER lumen that interact with client proteins prior to binding to BiP. ERdj proteins bind client proteins for transfer to BiP, while stimulating BiP ATP hydrolysis which induces a conformational change, so the SBD lid closes as the client protein is trapped in the SBD pocket. The Nucleotide Exchange Factors (NEFs), GRP170 and Sil1, subsequently promote the exchange of BiP-bound ADP for ATP. ADP release causes BiP to attain its open conformation to permit client protein release. ATP binding to BiP prepares BiP for another cycle of protein binding (Pobre et al. 2018). Motifs in client proteins that bind BiP usually have short hydrophobic stretches (Clerico et al. 2015; Behnke et al. 2016).

In *Saccharomyces cerevisiae*, overexpression of co-chaperones Sis1 (Hsp40) or co- Sse1 (Hsp110/GRP170) reduces aggregation of carboxypeptidase Y lacking its signal peptide sequence (ΔssCPY). Sis1 is a chaperone that regulates an ATPase activity of Hsp70 targeting misfolded proteins to degradation. Sse1, a NEF for Hsp70 chaperones (Easton et al. 2000), binds to unfolded peptides and prevents their aggregation holding the client in a conformation capable of folding (Hamdan et al. 2017). Therefore, co-expression of Hsp70 co-chaperones appears to enhance chaperone disaggregation activity. Disaggregation activity of ER chaperones is conserved in mammals. Studies from Lindquist et al. demonstrated that Hsp104 can solubilize irreversible protein aggregates. Hsp104 is believed to be the main engine that couples ATP hydrolysis with protein disaggregation with cooperation by BiP. The mechanism is believed to start with binding of the protein aggregate to BiP followed by recruitment of Hsp104 to the complex. The disaggregation activity of Hsp104 is conserved for bacterial ClpB and the plant Hsp101 counterpart. Although there are no Hsp104 homologs in metazoans, it was reported that BiP has disaggregation activity. Overexpression of BiP sufficiently rescued FVIII and rod opsin from aggregation (Athanasiou et al. 2012; Poothong et al. 2020). Although it is not clear yet which domain in rod opsin binds to BiP, a short motif (residues 253–331) in FVIII was identified as both a BiP binding site and aggregation seeding domain. BiP binding to this motif prevented FVIII aggregation and promoted its correct folding. Consistently, inactivation of BiP by Subtilase cytotoxin (SubAB), a protease that specifically cleaves BiP at the hydrophobic linker between the NBD and SBD (Paton et al. 2006), induces aggregation of FVIII (Poothong et al. 2020), proinsulin (Arunagiri et al. 2019) and rod opsin within the ER lumen (Athanasiou et al. 2012). Together, this supports a significant role for BiP in preventing protein aggregation in the ER.

6.5 The Unfolded Protein Response (UPR) Is an Adaptive Response to Prevent Misfolded Protein Accumulation in the ER

To cope with an overload of misfolded ER proteins that disrupt ER function, the UPR is activated to maintain ER homeostasis. This response encompasses mRNA translational attenuation and transcriptional induction of UPR genes and ERAD and autophagy genes to resolve the misfolded protein stress. However, if ER stress is prolonged or severe, cells undergo cell death or apoptosis. The UPR consists of three main pathways: (1) Protein kinase RNA (PKR)-like ER Kinase (PERK); (2) Inositol-Requiring Enzyme 1 (IRE1); and (3) Activating Transcription Factor 6 (ATF6). Normally, these ER stress transducers are bound to BiP which inhibits their activation. Misfolded proteins that accumulate in the ER bind and sequester BiP, thereby releasing BiP from the ER stress sensors triggering UPR activation.

PERK is a type I transmembrane kinase. Under ER stress conditions, PERK undergoes oligomerization, *trans*-autophosphorylation and subsequently phosphorylates its major substrate, eukaryotic translation initiation factor 2α (eIF2α) at serine 51. p-eIF2α attenuates global protein synthesis to immediately and reversibly reduce proteins entering the ER. However, certain mRNAs are selectively translated upon eIF2α phosphorylation, such as Activating Transcription Factor 4 (ATF4). ATF4 induces gene expression of chaperone proteins to improve protein folding capacity, of amino acid biosynthesis and transport machinery, of the anti-oxidative stress response and autophagy. It also induces the pro-apoptotic transcription factor CHOP that promotes cell death if protein misfolding in not resolved.

IRE1, the most conserved UPR signaling pathway across eukaryotes, is a type I ER transmembrane protein containing dual catalytic activities in the cytosolic domain comprising an serine/threonine kinase and a C-terminal endoribonuclease (RNase) (Tirasophon et al. 1998). In response to the ER stress, IRE1 self-association leads to activation of its kinase and RNase activities. Active IRE1 mediates nucleotide specific sequence excision of a 26-nt intron within X-Box-binding Protein (*XBP1*) mRNA (Poothong et al. 2017). This splicing event causes a translational frameshift in *XBP1* mRNA to translate a stable spliced XBP1 (XBP1s) transcription factor (Yoshida et al. 2001; Back et al. 2006). XBP1s induces the expression of genes encoding ER chaperones (BiP, PDIA6, GRP94, etc.), ER protein translocation machinery and ERAD functions (EDEM1–3, etc.) (Shaffer et al. 2004; Hassler et al. 2015). In addition to *XBP1* mRNA cleavage, IRE1 RNase also cleaves ER-localized mRNAs in a process called regulated-Ire1 dependent RNA decay (RIDD) (Hollien and Weissman 2006) and several mRNAs and microRNAs including *insulin* (Han et al. 2009) and *Caspase-2* microRNA (Upton et al. 2012). In addition, IRE1 can initiate an inflammatory response through direct binding to Tumor Necrosis Factor α (TNFα) Receptor Associated Factor 2 (TRAF2). TRAF2 subsequently activates c-Jun N-terminal Kinase (JNK) (Urano et al. 2000; Hetz and Glimcher 2009).

Unlike PERK and IRE1, ATF6 is a type II ER transmembrane protein that contains a basic leucine zipper transcription factor domain in its N-terminus. Under ER stress, ATF6 traffics from the ER to the Golgi apparatus where it is cleaved by Site-1 and Site-2 Proteases (S1P and S2P) releasing the active form of ATF6 (ATF6p50) which moves to the nucleus to induce expression of its target genes. ATF6 upregulates a specific set of UPR target genes, mostly ER chaperones and ERAD functions, including unspliced *XBP1* mRNA (Yoshida et al. 2001).

6.6 Protein Aggregation in the ER Contributes to Human Disease

There are many diseases associated with amyloid fibrillation including neurodegenerative diseases (AD, PD and HD) and metabolic degenerative (Diabetes) diseases. Here, we focus on several diseases associated with protein aggregates or amyloid formation in the ER lumen (liver disease, hemophilia A, type 2 diabetes and cystic fibrosis) as consequences of a genetic disorder and/or cell stress.

6.6.1 α1-Antitrypsin (α1-AT) Aggregation Causes Liver Disease

Human α1-AT deficiency is a common inherited liver disease caused by mutations in the SERPIN peptidase inhibitor A1 (*SERPINA1*) gene on chromosome 14 that encodes α1-AT protein primarily synthesized in hepatocytes. The liver disease is associated with a gain-of-function of α1-AT mutations. Of the more than 100 mutations identified in α1-AT protein, the most common and severe is homozygous Z mutation (E342K) (Perlmutter et al. 2007) which causes α1-AT misfolding and retention in the ER producing polymers leading to hepatic fibrosis and eventually hepatocellular carcinoma (HCC) (Carrell and Lomas 1997; Teckman et al. 2004). The α1-AT mutations also impact lung function via a different mechanism as liver disease. In the lung, defective α1-AT leads to early onset chronic obstructive pulmonary disease because α1-AT is required to inhibit the neutrophil serine elastase to prevent lung tissue fibrosis and deterioration (Anzueto 2015).

Disease development depends on mutations that reduce α1-AT levels in the plasma. On the basis of plasma level and function of α1-AT variants, the disease is categorized into several categories. The M type is due to a major allele (up to 95%) in Caucasian individuals which is characterized by normal plasma levels >20 nmole/L. The S type is a common variant with plasma levels in the range of 8–11 nmole/L. Deficiency or the Z type mutant is more severe with the plasma level of homozygotes ~5–6 nmole/L. The null mutant is a variant that has no detectable circulating α1-AT. The dysfunctional Pittsburgh mutation (M358R) converts α1-AT into an inhibitor of thrombin rather than elastase. The majority of patients are either homozygous or heterozygous for the Z or S variants. Whereas null and dysfunctional mutations are rare.

Human α1-AT is a 52 kDa glycoprotein belonging to the serpin (serine protease inhibitor/Pi) family, with 3 N-linked glycosylation sites. α1-AT is composed of 2 ß-sheets, 9 α-helices and a reactive center loop. This protein is mainly produced by hepatocytes and secreted into blood; however, it is also produced at a low level by other cells, such as neutrophils, macrophages and monocytes. Among the mutants, the Z mutant is most frequently found in patients with liver disease. This mutant results from single amino acid substitution from positively charged lysine to negatively charged glutamate at position 342 at the base of reactive center loop. This mutant distorts the native conformation of α1-AT that allows interaction between the active center loop of one molecule with the ß-sheet of another to form a dimer that can be extended into an α1-AT polymer (Lomas et al. 1992). These polymers can interwind and are predisposed to form insoluble aggregates in the ER of hepatocytes. Accumulation of the aggregates is associated with clinical liver fibrosis (Clark et al. 2018). Although the pathogenesis of the liver disease due to α1-AT aggregation is still unclear, it was shown that accumulation of α1-AT Z mutant induces activation of c-Jun N-terminal kinases 1 and 2 (JNK1, 2) in α1-AT Z mutant transgenic mouse liver (Pastore et al. 2017). Interestingly, accumulation of the α1-AT Z mutant polymer is reduced in *JNK1*$^{-/-}$ and *JNK2*$^{-/-}$ mice due to a decrease in *SERPINA1* transcription (Pastore et al. 2017). *SERPINA1* gene contains a binding site for c-Jun, a target for JNK kinase, suggesting JNK directly regulates α1-AT transcription (Pastore et al. 2017). Specifically, *SERPINA1* expression is also controlled by several cytokines, including interleukin-6 (IL-6) and tumor necrosis factor α (TNFα) (Kalsheker et al. 2002), which are directly regulated by the c-Jun transcription factor (Wang et al. 2013; Guinea-Viniegra et al. 2009).

In addition, accumulation of the α1-AT Z mutant in mouse hepatocytes activates caspase 4 and 12 (Hidvegi et al. 2005; Lindblad et al. 2007) which cleave and activate BAP31, a ER membrane protein that mediates pro-apoptotic signals from the ER to mitochondria (Chandra et al. 2004; Hidvegi et al. 2005). In the mitochondria, caspase-8, −9, and −3 are activated promoting cytochrome c release and apoptosis (Perlmutter 2002; Lindblad et al. 2007), leading to liver injury. Consistent with the mRNA Seq profile, two inflammatory genes (IL-18 and caspase-4) are up-regulated in human induced pluripotent stem cells (hiPSC) derived from a patient with the α1-AT Z mutation (Segeritz et al. 2018). Moreover, induction of the NF_KB transcription factor in an α1-AT Z mouse model (Hidvegi et al. 2005) contributes to the inflammatory response to α1-AT Z accumulation. However, to prevent widespread liver damage, Bad was phosphorylated to prevent its apoptotic function (Lindblad et al. 2007).

Liver disease caused by α1-AT deficiency predisposes to HCC, however, the mechanism involved is unknown. Most likely, it is due to JNK activation in the α1-AT Z transgenic mice, which is also observed in human liver diseases that lead to HCC, such as infection with hepatitis B virus or hepatitis C virus and non-alcoholic steatohepatitis. Phosphorylation of JNK1, but not JNK2, was significantly increased in human HCC samples (Hui et al. 2008). JNK1 can promote hepatocyte proliferation and tumorigenesis by suppressing the transcription factor p21 through c-Myc (Hui et al. 2008). Alternatively, c-Jun can antagonize p53 activity to promote HCC (Eferl et al. 2003). Induction of the oncogenic protein AKR1B10 may be lead to liver fibrosis (Segeritz et al. 2018).

Immunoprecipitation demonstrated that the α1-AT Z mutant interacts with multiple ER chaperones (UGGT1, CANX, GRP94, GEP78 and GRP170) which may cause ER retention. Consistently, sucrose gradient sedimentation suggested that majority of α1-AT Z mutant (approximately 85%) exists in heterogeneous complexes with ER chaperones and the other 15% of the mutant were observed in polymer/aggregated α1-AT devoid of chaperones (Schmidt and Perlmutter 2005). Although data indicate the polymerogenic α1-AT does not cause ER stress, the findings are still under debate. Experiments in HeLa cells with Tet-Off inducible α1-AT Z expression and transgenic mice did not detect altered BiP expression or UPR activation (Hidvegi et al. 2005). Finally, oxidative stress was shown to contribute to liver damage in an α1-AT Z murine model. Higher levels of ROS and a more oxidized cellular redox state was observed in liver tissue from α1-AT Z mice compared to WT mice (Pastore et al. 2017).

6.6.2 FVIII Aggregation Causes Hemophilia A

Hemophilia A is a X-chromosome linked bleeding disorder resulting from deficiency of blood coagulation factor VIII (FVIII), an essential component of the intrinsic blood clotting cascade (Mann 1999). FVIII is synthesized and secreted into the plasma in an inactive form and circulates in a complex with von Willebrand factor (vWF). Upon activation by thrombin specific cleavage, FVIII becomes active and acts as a cofactor to stimulate FIXa mediated cleavage of FX to active FXa. FXa with active FVa stimulates the conversion of prothrombin to thrombin which mediates fibrinogen cleavage to fibrin. Fibrin is finally polymerized resulting in blood clotting.

FVIII is primarily produced in liver sinusoidal endothelial cells and Kupffer cells. FVIII is a large glycoprotein synthesized as a single polypeptide chain with multiple domains comprising A1-A2-B-A3-C1-C2. FVIII has 8 disulfide bonds and 25 potential N-linked glycosylation sites (Orlova et al. 2013). Tagliavacca et al. demonstrated FVIII forms high molecular complex aggregates upon ATP depletion suggesting that ATP is required for FVIII folding and maturation. Recent studies identified FVIII forms higher order amyloid-like FVIII aggregates that co-localize with other ER resident proteins in the ER lumen (Poothong et al. 2020).

Mass Spectrometry analysis of FVIII immunoprecipitates from FVIII expressing cells revealed that FVIII strongly interacts with a number of ER chaperones including BiP, CANX/CALR, UGGT1 and SEL1, of which BiP and CANX/CALR were the most significant interactors. Significantly, inactivation of the CANX/CALR cycle with the α-glucosidase (GS1 and GS2) inhibitor castanospermine (CST) and inhibition of BiP function with the specific BiP protease SubAB both caused FVIII aggregation within the ER implicating both are required to maintain FVIII solubility. Cryo-EM analysis identified aggregated FVIII forms fibrils of ~5 nm in diameter. The aggregates stained with the amyloid specific dye Thioflavin S and were retained upon filtration through cellulose acetate filters, both characteristic of β sheet polymers. Although FVIII has 8 disulfide bonds, FVIII aggregates were not due to disulfide cross-linking, although the aggregates were dissolved by treatment with 1% SDS (Tagliavacca et al. 2000). Collectively, these findings indicate that FVIII forms amyloid aggregates in the ER lumen.

How does FVIII aggregate in the ER? *In vitro* studies demonstrated deletion of the carboxy terminal region of the FVIII A1 domain increased FVIII secretion ~ten-fold (Marquette et al. 1995). More specifically, a short 79 amino acid sequence (residues 253–331) was identified as a minimal aggregation prone motif. TANGO analysis predicted this region has high propensity to form cross β-aggregates that was confirmed by *in vitro* by expression different mutants. Mutations that reduce cross β-sheet propensity reduced FVIII aggregation in the ER and improved FVIII secretion. This region was also characterized as a potential BiP binding site (Marquette et al. 1995; Poothong et al. 2020). As protein concentration is a key that drives protein aggregation in the ER, increased FVIII expression above a threshold caused FVIII aggregation in the ER. As FVIII accumulates in the ER, it may sequester BiP and reduce the ability of BiP to prevent FVIII aggregation. Consistent with this, BiP overexpression reduced FVIII aggregation (Poothong et al. 2020). It was demonstrated that FVIII expression induces oxidative stress both in culture and upon gene delivery to hepatocytes in mice (Malhotra et al. 2008). Strikingly, treatment with antioxidants reduced FVIII aggregation and increased FVIII secretion in cell culture and in mice (Malhotra et al. 2008).

As noted above, BiP function in the ER requires ATP. FVIII folding is also tightly regulated by ATP level in the ER, mostly supplied from oxidative phosphorylation or glycolysis, as depletion of ATP enhanced FVIII aggregation (Yong et al. 2019). Although inhibition of glucose metabolism by 2′-deoxyglucose induced FVIII aggregation (Tagliavacca et al. 2000; Poothong et al. 2020), the mechanism has not been defined. Significantly, FVIII aggregation induced by 2-deoxyglucose was reversible upon repletion of cells with glucose, and the disaggregated FVIII was able to refold and be secreted as a functional clotting factor. This is the first evidence for reversible protein amyloid formation that can produce a functional protein in mammals.

Incomplete glycosylation can cause protein misfolding and degradation. The majority of glycosylation in FVIII occurs in the B-domain which harbors 18 potential N-linked glycans. However, deletion of the B domain (~1000 amino acids) produces a molecule that is secreted as a functional protein, although it still aggregates in the ER. In contrast, partial deletion of 75% of the B-domain in a construct that retains 6 glycosylation sites in the B domain produced a molecule that did not aggregate and was efficiently secreted as a functional molecule. It is proposed that the 6 glycans in the partial B domain deleted FVIII may engage the CANX/CALR cycle to prevent aggregation.

Due to accumulation of misfolded FVIII in the ER lumen, FVIII was the first native protein shown to activate the UPR *in vitro* and *in vivo*. Expression of FVIII in cell lines and in mice showed induction of UPR markers BiP, ATF4, CHOP, phosphorylated eIF2α and spliced XBP1 mRNA (Poothong et al. 2020). Similarly, hydrodynamic tail vein delivery of plasmid DNA encoding FVIII (Malhotra et al. 2008) or adeno-associated viral vectors (AAV) carrying FVIII induced UPR activation in murine hepatocytes (Zolotukhin et al. 2016; Lange et al. 2016).

6.6.3 Proinsulin High Molecular Weight (HMW) Aggregation Is an Early Step in β-Cell Failure in Type 2 Diabetes (T2D)

Insulin is a small peptide hormone secreted by pancreatic β cells that regulates blood glucose levels. Insulin is synthesized as a precursor preproinsulin that is translocated into the ER. This precursor is processed to proinsulin upon signal peptide cleavage by signal peptidase. Proinsulin contains B and A chains connected by an unstructured C-peptide. In the ER, proinsulin attains a 3-dimensional structure with three disulfide bonds. In the *trans*-Golgi the C-peptide is cleaved away by the prohormone convertases (PC1/3 and 2) to produce mature insulin. This cleavage occurs at specific sites of dibasic amino acids flanking the C-peptide (B/C and C/A junctions). Mature insulin is stored as Zn^{2+} stabilized hexamers in secretory granules that fuse with the plasma membrane to release monomeric insulin upon glucose stimulation. Glucose concentrations (>25 mM) induce insulin secretion and increase insulin biosynthesis ~ten-fold above baseline (Itoh and Okamoto 1980). T2D results from peripheral insulin resistance. Under these conditions, β cells compensate by increasing proinsulin synthesis. A chronic increase in proinsulin production overwhelms the ER folding capacity that may cause ER stress and β-cell failure toward development of T2D.

Proper proinsulin folding requires formation of three disulfide bonds. These disulfide bonds are composed of two intermolecular disulfide bonds Cys(B7)-Cys(A7) and Cys(B19)-Cys(A20) that bridge the B and A chains, and one intramolecular disulfide bond within the A-chain Cys(A6)-Cys(A11). Mispairing any of these disulfide bonds causes proinsulin misfolding and retention in the ER lumen. Among the three disulfide bonds, the Cys(B19)-Cys(A20) linkage is believed to have the most substantial impact on proinsulin structural stability (Chang et al. 2004). However, common mispairing by formation of Cys(A11)-Cys(A20), Cys(A7)-Cys(A20), and Cys(B19)-Cys(A11) disulfide bonds can disrupt the critical Cys(B19)-Cys(A20) pairing (Haataja et al. 2016).

Mutations in proinsulin gene can cause proinsulin misfolding leading to early onset diabetes in patients called mutant-*INS* gene diabetes of youth (MIDY) (Liu et al. 2010). These patients carry one mutant *INS* and one wild-type proinsulin allele (Liu et al. 2015). More than 30 missense mutations in human *INS* have been identified. Most proinsulin mutants cannot fold properly and are trapped in the ER. MIDY mutant proinsulins are inherited in a dominant manner because the mutant protein traps the wild-type proinsulin in the ER due to intermolecular disulfide bonding between the mutant and wildtype protein (Arunagiri et al. 2018). To compensate, β-cells increase proinsulin biosynthesis (both WT and mutant proinsulin) that exacerbates proinsulin misfolding and intermolecular covalent interactions between the mutant and wildtype protein in the ER which causes ER stress and eventually β cell failure.

Akita proinsulin C(A7)Y is the best characterized MIDY mutant where cysteine 7 in the A chain is mutated to tyrosine and misfolds due to inability to form the Cys

(B7)-Cys(A7) disulfide bond, leaving Cys(B7) unpaired (Wang et al. 1999; Izumi et al. 2003). Recent evidence shows this unpaired cysteine can form an intermolecular disulfide bond with another misfolded *Akita* or WT proinsulin molecule to generate high molecular weight (HMW) complexes (Cunningham et al. 2017). Indeed, two regions in proinsulin have propensity to form β-sheets in both the B-chain VEALYL (residues 12–17) (Sun et al. 2020) and the A-chain SLYQLENY (residues 77–84) (Ivanova et al. 2009) that are thought to be aggregation prone/amyloidogenic, however, there is no direct evidence indicating the involvement of these motifs in proinsulin aggregation. Recent studies in human T2D patient samples and pre diabetic (*ob/ob*) mouse islets revealed significant increases in proinsulin HMW complexes (Arunagiri et al. 2019; Tran et al. 2020) with T2D disease progression.

In the ER, several chaperone proteins have been identified as interactors with both WT and misfolded *Akita* mutant proinsulin. The most prominent proinsulin interacting partners are BiP, PDIA6, GRP170, and ERdjs (ERdj 3, 4 and 6). Among these, PDIA6, ERdj6, and GRP170 specifically interact to a greater degree with misfolded proinsulin (Gorasia et al. 2016; Cunningham et al. 2017, 2019). Physical interaction with chaperones including GRP170, ERdj4, and Hrd1, which function in ERAD suggests that misfolded proinsulin can be eliminated by the proteasome degradation pathway (Fritz et al. 2013; Cunningham et al. 2017; Hu et al. 2019). Significantly, overexpression of GRP170 enhances and restores WT proinsulin secretion (Cunningham et al. 2017). Peroxiredoxin-4 (PRDX4) and PDI are oxidoreductases that may contribute to proinsulin disulfide bond formation. Since defective proinsulin disulfide bond formation can cause oxidative stress, PDIA1 deletion increased ROS production by menadione induced oxidative stress in islets from β cell specific *Pdia1* deleted mice compared to wildtype mice (Jang et al. 2019). Significantly, proinsulin HMW complexes accumulated in these *Pdia1*-deleted β cells. Importantly, overexpression of human PDI by adenovirus delivery to mouse embryonic fibroblasts (MEFs) that express human proinsulin increased amount of reduced proinsulin, suggesting that PDIA1 may act as a reductase to limit accumulation of proinsulin HMW complexes (Jang et al. 2019). A recent study in T2D human islets revealed that PRDX4 is another proinsulin interacting protein that is also required to prevent proinsulin aggregation. In T2D human islets PRDX4 is inactivated by sulfonylation due to excessive oxidative stress which stimulates proinsulin aggregation (Tran et al. 2020).

One of the most abundant proinsulin interacting proteins is BiP (Tran et al. 2020). It was reported that BiP may prevent proinsulin HMW complex formation. Acute loss function of BiP by SubAB protease treatment induced proinsulin aggregation in both mouse and human islets (Arunagiri et al. 2019). Increased co-localization of proinsulin with CANX under SubAB treatment indicates these complexes are localized to the ER (Arunagiri et al. 2019). The significance of BiP in promoting proinsulin folding was also demonstrated *in vivo*. Transgenic mice with β cell specific BiP overexpression were protected from high-fat-diet (HFD) induced diabetes (Teodoro-Morrison et al. 2013). Similarly, BiP overexpression in a β cell line (INS-1) improved glucose-stimulated insulin secretion suggesting these β cells fold proinsulin more efficiently (Zhang et al. 2009). TANGO analysis predicted that amino acids in close proximity to tyrosine (Y) at residue 16 in the B-chain exhibit β sheet propensity, characteristic of a BiP binding site.

Recent biochemical and genetic evidence in humans and mouse models revealed that the UPR (IRE1α, PERK and ATF6) is activated upon accumulation of mutant proinsulin (Hartley et al. 2010). Increased UPR protein markers BiP, ERdj6/p58^{IPK} and CHOP were increased in human pancreata sections from T2D patients (Laybutt et al. 2007) and XBP1s increased in pancreata from *ob/ob* mice (Laybutt et al. 2007). IRE1 appears to be involved in proinsulin proper folding as loss of IRE1/XBP1 splicing in β cell specific *Ire1* deleted mice decreased proinsulin and insulin contents which correlated with decreased proinsulin oxidative folding in insulinoma 6 cells (Tsuchiya et al. 2018). Significantly, mice with *Ire1*-deletion in β cells were defective in glucose-stimulated transcription of many genes that function in protein translocation into the ER, due to the absence of XBP1s and caused preproinsulin to accumulate in the cytosol (Hassler et al. 2015). The impact of PERK in proinsulin aggregation was also demonstrated. β cell specific *Perk*-deleted mice accumulated proinsulin in a distended ER (Harding et al. 2001). PERK inhibition overnight in a β cell line increased proinsulin translation as well as proinsulin aggregation (Arunagiri et al. 2019)' likely due to a lack of eIF2α phosphorylation which attenuates protein synthesis. In contrast, IRE1 hyperactivation in β cell lines caused *insulin* mRNA degradation that would reduce proinsulin influx into the ER and presumably reduce proinsulin aggregation (Han et al. 2009). These findings support the idea that the rate of proinsulin synthesis correlates with proinsulin HMW complex formation (Back and Kaufman 2012; Jang et al. 2019). Although, there are no direct data that connect ATF6 to proinsulin aggregation, lack of ATF6 reduced insulin content and produced a distended ER (Usui et al. 2012). As noted above, misfolded proinsulin that accumulates in the ER is likely degraded by ERAD. It was shown that up-regulation of downstream targets of XBP1 and ATF6 transcription factors that activate genes encoding functions in ERAD (EDEM, Hrd1 and Sel1) in β cells expressing a mutant proinsulin suggest a UPR requirement to maintain proinsulin folding homeostasis (Shaffer et al. 2004; Allen et al. 2004; Hartley et al. 2010).

6.6.4 Cystic Fibrosis Is Due to Mutations that Cause Misfolding of CFTR

Cystic Fibrosis, a lethal autosomal recessive disease, is caused by a failure of ER protein quality control. Mutations in the Cystic Fibrosis Transmembrane conductance Regulator (*CFTR*) gene are the main cause for the disease. *CFTR* gene encodes a transmembrane protein, a cAMP-regulated chloride channel expressed at the apical membrane of epithelial cells. Therefore, mutations in *CFTR* cause abnormal chloride regulation in epithelial cells (Yadav et al. 2019; Sheppard and Welsh 1999; Kusum et al. 2019). Although more than 1000 *CFTR* gene mutations related to cystic fibrosis are known, the deletion of phenylalanine at position 508 (ΔF508) is the most common and responsible for about 66% of cystic fibrosis resulting from CFTR mutations (Bobadilla et al. 2002).

CFTR is a ~ 170 kDa protein comprised of two membrane-spanning domains (MSD1 and MSD2, six transmembrane segments for each), two nucleotide-binding domains (NBD1 and NBD2), and a central regulatory (R) domain. MSD1 and MSD2 are connected by four cytoplasmic loops (CLs 1–4). The NBDs bind and hydrolyze ATP. Serine residues in the R domain are phosphorylated by protein kinase A (PKA) and protein kinase C (PKC), and the modifications regulate channel gating of the NBDs. The interaction between the NBDs and CLs are crucial for the proper folding and function of CFTR as a chloride channel (Hunt et al. 2013; Riordan et al. 1989).

Biosynthesis of CFTR begins as it is translocated across the ER membrane. CFTR folding occurs at the same time when transmembrane segments are inserted into the ER membrane (Kim and Skach 2012). First, the MSD1 transmembrane domain diffuses laterally out of the Sec61 translocon for integration into the ER membrane which stimulates by NBD1 synthesis. NBD1 transiently binds to two molecular chaperones, Hdj2 (Hsp40) and Hsc70 stabilize its conformation, providing time to synthesize the R domain. Binding of the R domain to the N-terminal of CFTR reduces Hsp40 interaction. Then, MSD2 is synthesized and incorporated into the ER membrane. This also stabilizes the NBD1-R-domain interaction and releases most of the Hsp40 from CFTR. Finally, NBD2 is synthesized and CFTR acquires its native folded structure in an ATP dependent manner (Lu et al. 1998; Ostedgaard et al. 1997).

Hsc70/Hsp70 interacts with exposed hydrophobic residues near the core of the native protein and prevent protein misfolding. Once CFTR reaches the ER, Hsc70/Hsp70 bind to the cytosolic domains to stabilize their conformation and scrutinize their folding status. This step is critical to distinguish between wildtype and the ΔF508 mutant. Deletion of F508 destabilizes NBD1 folding and assembly by inhibiting interaction between NBD1 and MSD2 (Du et al. 2005; Du and Lukacs 2009). As a result, the unfolded ΔF508 mutant is removed by ERAD. Once wildtype CFTR passes this checkpoint, it is de-glycosylated and re-glycosylated for interaction with CANX. When CFTR attains its native conformation it exits the ER for traffic to the Golgi compartment. However, if wildtype CFTR is retained too long in the CANX cycle, it is also targeted to ERAD. Only 20–40% of nascent CFTR obtains a native structure in non-polarized cells, while the remaining is degraded. In polarized epithelial cells, the folding efficiency increases up to 60–100%, however, degradation is maintained at 40% (Varga et al. 2004). Removal of CFTR protein by degradation causes loss of function in epithelial cells.

In cystic fibrosis, retained mutant CFTR in the ER may induce ER stress and UPR activation. Overexpression of ΔF508 mutant, but not wildtype CFTR induced the UPR in various cell types. For instance, ATF6 and BiP/GRP78 were activated in human epithelial cells transiently expressing ΔF508 mutant while knockdown of ATF6 by siRNA increased plasma membrane localization of ΔF508 mutant CFTR (Kerbiriou et al. 2007). In patients with cystic fibrosis, chronic airway inflammation was observed due to increased susceptibility to infections, however, studies also demonstrated that ER stress related inflammation mediated by NFκB activation occurred even without any bacterial or viral infection, indicating that the ER stress response may trigger sterile inflammation in cystic fibrosis lungs (Knorre et al. 2002). On the other hand, induction of ER stress and UPR activation inhibited endogenous ΔF508 mutant and wildtype CFTR biogenesis at the transcriptional level by binding of ATF6 to a minimal CFTR promoter to cause histone deacetylation and methylation of the promoter region (Bartoszewski et al. 2008).

Many strategies have been employed to rescue CFTR folding and function. Although they need to be further confirmed, low temperature, chemical chaperones, and second-site suppressor mutations in the NBD1 or at the NBD1-MSD2 interface were able to partially rescue surface expression of the ΔF508 mutant up to 15% of wildtype CFTR. Sodium 4-phenylbutyrate (PBA) is a chemical chaperone which are small molecules that can buffer protein aggregation much like molecular chaperones which are proteins. Chemical chaperones bind to exposed hydrophobic regions in unfolded proteins and prevents aggregation. PBA also acts as an inhibitor of histone deacetylase (HDAC) and activates gene transcription. In bronchial epithelial cells, PBA improved trafficking of ΔF508 mutant and wildtype CFTR by inducing Erp29, a novel ER resident molecular chaperone (Suaud et al. 2011). In addition, small molecules have been identified by high-throughput screening although they have limited efficiencies at rescue. Among them, VX-809 improved ΔF508 mutant CFTR processing in the ER and chloride secretion in human bronchial epithelial cells from a patient with homozygous ΔF508 mutation to ~14% of normal bronchial epithelial cells (Van Goor et al. 2011). Recently there was a major breakthrough for cystic fibrosis treatment. The FDA approved Trikafta for treatment of patients with heterozygous or homozygous ΔF508 CFTR mutations which can improve lung function up to 15% of normal. Trikafta is composed of three different drugs: Elexacaftor and Texacaftor work together to correct CFTR folding so more ΔF508 protein gets to the cell surface; and Ivacaftor helps the ΔF508 protein channel stay open longer at the cell surface.

6.7 Therapeutic Strategies May Prevent ER Protein Aggregation-Associated Diseases

Protein misfolding in the ER is a common event in cells due intrinsic defects in the protein, such as gene mutation, or extrinsic factors that influence protein homeostasis in the ER, such as cell stress, defects in chaperone function or the UPR or defects in ERAD and/or autophagy. Misfolded proteins with exposed hydrophobic regions predispose to aggregation that in some cases can form amyloid-like fibrils. As chaperones are critical for ER homeostasis, enhancing chaperone activity or treating cells with chemical chaperones may offer a therapeutic strategy for diseases associated with protein misfolding and aggregation.

6.7.1 *Chemical Chaperones Buffer Protein Folding*

Chemical chaperones are small molecules that limit protein misfolding in the cell. The most well-known chemical chaperones are sodium 4-phenylbutyrate (PBA); Tauroursodeoxycholic acid (TUDCA), UDCA conjugated with the amino acid taurine. The proposed mechanism for these chemical chaperones is binding to exposed hydrophobic surfaces on misfolded proteins to protect them from aggregation. In addition to chaperone property, PBA is also a HDAC inhibitor that can indirectly enhance protein folding capacity in the ER by inducing transcription of BiP (Wright et al. 2004). Long-term administration of PBA significantly reduced α-synuclein aggregates in the brain and prevented degeneration of cognitive brain function (Inden et al. 2007; Zhou et al. 2011). Moreover, PBA and TUDCA have potential to benefit several diseases that result from protein misfolding in the ER, including T2D and liver disease. Oral administration of either PBA or TUDCA relieved ER stress and significantly reduced blood glucose to normal in an obese mouse model (*ob/ob* mice) (Özcan et al. 2006). The potential therapeutic activity of PBA was also evaluated in α1-AT deficiency. PBA treatment increased secretion of the Z allele of α1-AT in cell culture and in transgenic mice oral PBA administration increased blood levels of human α1-AT Z close to normal (Burrows et al. 2000). The impact of PBA and TUDCA on coagulation factor secretion was demonstrated for recombinant mutant factor VII-160R (rFVII-160R), an ER retention variant of FVII. PBA, not TUDCA, treatment decreased ER stress and was able to increase 2.5-fold FVII secretion in cell culture (Andersen et al. 2019). Preventing exposure of hydrophobic misfolded protein by chemical chaperones to delay aggregation might be an alternative therapeutic approach for protein aggregation associated diseases. Amyotrophic Lateral Sclerosis (ALS) or Lou Gehrig's disease is apparently caused by protein aggregation in motor neurons leading to loss of motor neuron functions. Significantly, a recent placebo-controlled phase 2 clinical study demonstrated that a combination of PBA and TUDCA in ALS patients delayed decline in motor functions for 6 wks over 6 mon, approximately 25% more than the placebo (Paganoni et al. 2020). This likely portends that chemical chaperones have potential to treat many protein aggregation-associated diseases.

6.7.2 *Antioxidants Can Improve ER Protein Folding*

Protein misfolding in the ER can cause oxidative stress. It was shown that antioxidant treatment not only reduced oxidative stress, but also improved protein folding and secretion to prevent ER stress-induced apoptosis (Malhotra et al. 2008; Han et al. 2015). Many studies demonstrated that antioxidant treatment can prevent or delay the course of disease. One common antioxidant is butylated hydroxyanisole (BHA). BHA treatment not only attenuated ROS production and UPR activation in cells and in mice that express human FVIII in hepatocytes, but also improved FVIII secretion

(Malhotra et al. 2008). Later studies demonstrated that betaine, an antioxidant as well as chemical chaperone, was even more effective than BHA to improve FVIII secretion *in vitro* and *in vivo* (Roth et al. 2012). Deletion of the ERdj family member $P58^{IPK}$ induces oxidative stress and β cell failure to cause diabetes in mice. Feeding $p58^{IPK-/-}$ mice with BHA restored β cell function by reducing oxidative stress and improving proinsulin biosynthesis and folding (Han et al. 2015). Thus, an antioxidant can replace function of a molecular chaperone to preserve β cell function. Reducing ROS production by antioxidants may provide an alternative strategy to prevent protein misfolding and aggregation.

6.7.3 Preemptive UPR Activation Reduces ER Protein Misfolding

Many insults exhibit a biphasic dose-response, called hormesis, where a low dose of a stressor elicits an opposite effect to that observed by a high dose of the same stressor. This is characteristic of ER stress, where cells respond and adapt to low level ER stress, but induce apoptosis in response to a higher level of ER stress. Significantly, mild ER stress can induce adaptation to a higher dose of ER stress through activation of the UPR (Rutkowski and Kaufman 2007). In fact, induction of ER stress by one insult can protect cells from a toxic response to a different insult that causes ER stress. For example, genetic mutation to induce mild ER protein misfolding can protect cells from subsequent lethal pharmacological ER stress by induction of ER chaperone expression (Rutkowski et al. 2006). This observation led to the notion that preemptive activation of the UPR may protect cells from subsequent ER stress. For example, AAV mediated delivery of active ATF6p50 into the heart induced expression of catalase that improved cardiac function in a mouse model of ischemic heart disease and reduced accumulation of intracellular α1-AT Z mutant (Jin et al. 2017; Smith et al. 2011). In addition, activation of either ATF6 or IRE1/XBP1 reduced secretion of amyloidogenic variant immunoglobulin light chain (ALLC) (Cooley et al. 2014) and aggregation-prone variant rhodopsin (Chiang et al. 2012). However, the molecular mechanism of XBP1s and ATF6 in modulating ER stress may be distinct. Transcriptional and proteomic profiling revealed that XBP1s regulates the entire ER proteostasis pathway including protein trafficking, folding and degradation, etc., which is broader than ATF6 that selectively regulates expression only subset of UPR genes involved in protein folding and degradation (Shoulders et al. 2013).

Recently, small molecules that selectively induce ATF6 activation were identified (Plate et al. 2016). Compound 147 covalently modifies thiol residues in PDI family members that leads to ATF6 activation (Paxman et al. 2018). Treatment with compound 147 reduced secretion of aggregation prone mutant transthyretin (TTR; $^{FL}TTR^{A25T}$) by approximately 45% relative to wild-type TTR without effecting general protein secretion (Plate et al. 2016) and reduced secretion and aggregation of amyloidogenic immunoglobulin light chain (Plate et al. 2016). Administration of compound 147 *in vivo* activated ATF6 to improve cardiac performance in an ischemic model of heart disease in mice (Blackwood et al. 2019). These findings suggest that activating the UPR by genetic manipulation or small molecules may provide therapeutic opportunities for protein aggregation-associated diseases.

6.8 Conclusions

The ER is a highly versatile protein folding, assembly and processing factory that is equipped with chaperones and folding enzymes to facilitate protein folding. Importantly, the ER has strict surveillance systems to ensure only correctly folded proteins exit the ER for anterograde trafficking through the secretory pathway. Although protein aggregation is not unique to the ER, when ER protein folding is compromised, misfolded proteins may aggregate. Here, we reviewed several proteins which have distinct folding problems in the ER (FVIII, α1-AT, proinsulin and CFTR) that misfold and/or aggregate in the ER lumen. In most cases, protein aggregation is initiated from non-native conformations where hydrophobic residues are exposed at the protein surface. Protein aggregation in the ER appears to be concentration dependent. Therefore, ER retention of misfolded proteins, through interaction with ER chaperones, can increase the misfolded protein concentration in the ER that predisposes to aggregation. An important outstanding question is whether these aggregates are directly toxic to the cell or whether they act indirectly to interfere with cell function. To avoid the accumulation of protein aggregates, cells have evolved protein quality control, the UPR, ERAD and autophagy to limit accumulation of aggregated proteins in the ER. BiP is multifaceted protein that plays multiple roles to maintain ER homeostasis (protein translocation, protein folding, protein degradation, protein complex assembly and repression of UPR activation). Accumulated misfolded proteins bind BiP to sequester it away from the UPR transducers (IRE1α, PERK and ATF6), thus permitting their activation. All data to date indicate that BiP prevents toxicity from misfolded protein accumulation. Some chemical chaperones (PBA and TUDCA) and antioxidants (BHA) were demonstrated to reduce protein aggregation and associated toxicity for aggregation prone proteins (FVIII and α1-AT). Understanding fundamental mechanisms of protein misfolding/aggregation, factors affecting protein misfolding/aggregation and how the ER responds to misfolded/aggregated proteins may offer novel treatments for protein aggregation associated diseases.

6.9 Summary Points

- Disruption of ER homeostasis can cause protein misfolding and lead to protein aggregation in the ER.

- Protein aggregation in the ER is concentration dependent and the capacity of ER chaperones, such as BiP, is essential to prevent the formation of protein aggregates.
- Molecular and chemical chaperones, antioxidants, and preemptive UPR activation may be useful therapeutic approaches for ER protein aggregate-associated diseases.

6.10 Future Issues

- What are underlying molecular mechanisms of protein misfolding, aggregation and clearance of protein aggregates in the ER?
- What molecules are directly involved in regulating and resolving protein aggregation in the ER? Do they form a complex?
- Are protein aggregates in the ER a fundamental cause for the onset of disease and what signaling pathways are required for propagating their toxicity?

Acknowledgements This work was supported by NIH/NCI grants CA198103, DK103185, DK110973, DK113171, AG062190 to RJK and the Sanford Burnham Prebys NCI Cancer Center Grant P30 CA030199.

Funding NIH/NCI Grants R01CA198103, R01DK113171, R01DK103185, R24DK110973, R01AG062190 and the Sanford Burnham Prebys NCI Cancer Center Grant P30 CA030199.

References

Allen JR, Nguyen LX, Sargent KE, Lipson KL, Hackett A, Urano F (2004) High ER stress in beta-cells stimulates intracellular degradation of misfolded insulin. Biochem Biophys Res Commun 324:166–170

Andersen E, Chollet ME, Baroni M, Pinotti M, Bernardi F, Skarpen E, Sandset PM, Skretting G (2019) The effect of the chemical chaperone 4-phenylbutyrate on secretion and activity of the p. Q160R missense variant of coagulation factor FVII. Cell Biosci 9:69

Anzueto A (2015) Alpha-1 antitrypsin deficiency-associated chronic obstructive pulmonary disease: a family perspective. COPD 12:462–467

Araki K, Nagata K (2011) Protein folding and quality control in the ER. Cold Spring Harb Perspect Biol 3:a007526

Arunagiri A, Haataja L, Cunningham CN, Shrestha N, Tsai B, Qi L, Liu M, Arvan P (2018) Misfolded proinsulin in the endoplasmic reticulum during development of beta cell failure in diabetes. Ann N Y Acad Sci 1418:5–19

Arunagiri A, Haataja L, Pottekat A, Pamenan F, Kim S, Zeltser LM, Paton AW, Paton JC, Tsai B, Itkin-Ansari P, Kaufman RJ, Liu M, Arvan P (2019) Proinsulin misfolding is an early event in the progression to type 2 diabetes. elife 8:e44532

Athanasiou D, Kosmaoglou M, Kanuga N, Novoselov SS, Paton AW, Paton JC, Chapple JP, Cheetham ME (2012) BiP prevents rod opsin aggregation. Mol Biol Cell 23:3522–3531

Back SH, Kaufman RJ (2012) Endoplasmic reticulum stress and type 2 diabetes. Annu Rev Biochem 81:767–793

Back SH, Lee K, Vink E, Kaufman RJ (2006) Cytoplasmic IRE1alpha-mediated XBP1 mRNA splicing in the absence of nuclear processing and endoplasmic reticulum stress. J Biol Chem 281:18691–18706

Bartoszewski R, Rab A, Twitty G, Stevenson L, Fortenberry J, Piotrowski A, Dumanski J, Bebok Z (2008) The mechanism of cystic fibrosis transmembrane conductance regulator transcriptional repression during the unfolded protein response. J Biol Chem 283:12154–12165

Behnke J, Mann MJ, Scruggs FL, Feige MJ, Hendershot LM (2016) Members of the Hsp70 family recognize distinct types of sequences to execute ER quality control. Mol Cell 63:739–752

Blackwood EA, Azizi K, Thuerauf DJ, Paxman RJ, Plate L, Kelly JW, Wiseman RL, Glembotski CC (2019) Pharmacologic ATF6 activation confers global protection in widespread disease models by reprograming cellular proteostasis. Nat Commun 10:187

Bobadilla JL, Macek M Jr, Fine JP, Farrell PM (2002) Cystic fibrosis: a worldwide analysis of CFTR mutations--correlation with incidence data and application to screening. Hum Mutat 19:575–606

Burrows JAJ, Willis LK, Perlmutter DH (2000) Chemical chaperones mediate increased secretion of mutant α1-antitrypsin (α1-AT) Z: a potential pharmacological strategy for prevention of liver injury and emphysema in α1-AT deficiency. Proc Natl Acad Sci 97:1796

Carrell RW, Lomas DA (1997) Conformational disease. Lancet 350:134–138

Chandra D, Choy G, Deng X, Bhatia B, Daniel P, Tang DG (2004) Association of active caspase 8 with the mitochondrial membrane during apoptosis: potential roles in cleaving BAP31 and caspase 3 and mediating mitochondrion-endoplasmic reticulum cross talk in etoposide-induced cell death. Mol Cell Biol 24:6592–6607

Chang S-G, Choi K-D, Jang S-H, Shin H-C (2004) Role of disulfide bonds in the structure and activity of human insulin. Mol Cells 16:323–330

Chiang W-C, Hiramatsu N, Messah C, Kroeger H, Lin JH (2012) Selective activation of ATF6 and PERK endoplasmic reticulum stress signaling pathways prevent mutant rhodopsin accumulation. Invest Ophthalmol Vis Sci 53:7159–7166

Clark VC, Marek G, Liu C, Collinsworth A, Shuster J, Kurtz T, Nolte J, Brantly M (2018) Clinical and histologic features of adults with alpha-1 antitrypsin deficiency in a non-cirrhotic cohort. J Hepatol 69:1357–1364

Clerico EM, Tilitsky JM, Meng W, Gierasch LM (2015) How Hsp70 molecular machines interact with their substrates to mediate diverse physiological functions. J Mol Biol 427:1575–1588

Clerico EM, Meng W, Pozhidaeva A, Bhasne K, Petridis C, Gierasch LM (2019) Hsp70 molecular chaperones: multifunctional allosteric holding and unfolding machines. Biochem J 476:1653–1677

Cooley CB, Ryno LM, Plate L, Morgan GJ, Hulleman JD, Kelly JW, Wiseman RL (2014) Unfolded protein response activation reduces secretion and extracellular aggregation of amyloidogenic immunoglobulin light chain. Proc Natl Acad Sci U S A 111:13046

Cunningham CN, He K, Arunagiri A, Paton AW, Paton JC, Arvan P, Tsai B (2017) Chaperone-driven degradation of a misfolded proinsulin mutant in parallel with restoration of wild-type insulin secretion. Diabetes 66:741–753

Cunningham CN, Williams JM, Knupp J, Arunagiri A, Arvan P, Tsai B (2019) Cells deploy a two-pronged strategy to rectify misfolded proinsulin aggregates. Mol Cell 75:442–456.e4

Du K, Lukacs GL (2009) Cooperative assembly and misfolding of CFTR domains in vivo. Mol Biol Cell 20:1903–1915

Du K, Sharma M, Lukacs GL (2005) The DeltaF508 cystic fibrosis mutation impairs domain-domain interactions and arrests post-translational folding of CFTR. Nat Struct Mol Biol 12:17–25

Easton DP, Kaneko Y, Subjeck JR (2000) The hsp110 and Grp1 70 stress proteins: newly recognized relatives of the Hsp70s. Cell Stress Chaperones 5:276–290

Eferl R, Ricci R, Kenner L, Zenz R, David J-P, Rath M, Wagner EF (2003) Liver tumor development: c-Jun antagonizes the proapoptotic activity of p53. Cell 112:181–192

Fernandez-Escamilla AM, Rousseau F, Schymkowitz J, Serrano L (2004) Prediction of sequence-dependent and mutational effects on the aggregation of peptides and proteins. Nat Biotechnol 22:1302–1306

Ferris SP, Jaber NS, Molinari M, Arvan P, Kaufman RJ (2013) UDP-glucose:glycoprotein glucosyltransferase (UGGT1) promotes substrate solubility in the endoplasmic reticulum. Mol Biol Cell 24:2597–2608

Fritz JM, Dong M, Apsley KS, Martin EP, Na C-L, Sitaraman S, Weaver TE (2013) Deficiency of the BiP cochaperone ERdj4 causes constitutive endoplasmic reticulum stress and metabolic defects. Mol Biol Cell 25:431–440

Ghosh D, Singh PK, Sahay S, Jha NN, Jacob RS, Sen S, Kumar A, Riek R, Maji SK (2015) Structure based aggregation studies reveal the presence of helix-rich intermediate during α-Synuclein aggregation. Sci Rep 5:9228–9228

Gorasia DG, Dudek NL, Safavi-Hemami H, Perez RA, Schittenhelm RB, Saunders PM, Wee S, Mangum JE, Hubbard MJ, Purcell AW (2016) A prominent role of PDIA6 in processing of misfolded proinsulin. Biochim Biophys Acta (BBA) Proteins Proteom 1864:715–723

GÖrlach A, Bertram K, Hudecova S, Krizanova O (2015) Calcium and ROS: a mutual interplay. Redox Biol 6:260–271

Guinea-Viniegra J, Zenz R, Scheuch H, Hnisz D, Holcmann M, Bakiri L, Schonthaler HB, Sibilia M, Wagner EF (2009) TNFα shedding and epidermal inflammation are controlled by Jun proteins. Genes Dev 23:2663–2674

Haataja L, Manickam N, Soliman A, Tsai B, Liu M, Arvan P (2016) Disulfide mispairing during proinsulin folding in the endoplasmic reticulum. Diabetes 65:1050

Hamdan N, Kritsiligkou P, Grant CM (2017) ER stress causes widespread protein aggregation and prion formation. J Cell Biol 216:2295–2304

Han D, Lerner AG, Vande Walle L, Upton J-P, Xu W, Hagen A, Backes BJ, Oakes SA, Papa FR (2009) IRE1alpha kinase activation modes control alternate endoribonuclease outputs to determine divergent cell fates. Cell 138:562–575

Han J, Song B, Kim J, Kodali VK, Pottekat A, Wang M, Hassler J, Wang S, Pennathur S, Back SH, Katze MG, Kaufman RJ (2015) Antioxidants complement the requirement for protein chaperone function to maintain β-cell function and glucose homeostasis. Diabetes 64:2892–2904

Harding HP, Zeng H, Zhang Y, Jungries R, Chung P, Plesken H, Sabatini DD, Ron D (2001) Diabetes mellitus and exocrine pancreatic dysfunction in perk-/- mice reveals a role for translational control in secretory cell survival. Mol Cell 7:1153–1163

Hartley T, Siva M, Lai E, Teodoro T, Zhang L, Volchuk A (2010) Endoplasmic reticulum stress response in an INS-1 pancreatic beta-cell line with inducible expression of a folding-deficient proinsulin. BMC Cell Biol 11:59

Hassler JR, Scheuner DL, Wang S, Han J, Kodali VK, Li P, Nguyen J, George JS, Davis C, Wu SP, Bai Y, Sartor M, Cavalcoli J, Malhi H, Baudouin G, Zhang Y, Yates JR III, Itkin-Ansari P, Volkmann N, Kaufman RJ (2015) The IRE1alpha/XBP1s pathway is essential for the glucose response and protection of beta cells. PLoS Biol 13:e1002277

Hetz C, Glimcher LH (2009) Fine-tuning of the unfolded protein response: assembling the IRE1alpha interactome. Mol Cell 35:551–561

Hidvegi T, Schmidt B, Hale P, Perlmutter D (2005) Accumulation of mutant 1-antitrypsin Z in the endoplasmic reticulum activates caspases-4 and -12, NF B, and BAP31 but not the unfolded protein response. J Biol Chem 280:39002–39015

Hollien J, Weissman JS (2006) Decay of endoplasmic reticulum-localized mRNAs during the unfolded protein response. Science 313:104–107

Hong DP, Ahmad A, Fink AL (2006) Fibrillation of human insulin A and B chains. Biochemistry 45:9342–9353

Hu Y, Gao Y, Zhang M, Deng K-Y, Singh R, Tian Q, Gong Y, Pan Z, Liu Q, Boisclair YR, Long Q (2019) Endoplasmic reticulum–associated degradation (ERAD) has a critical role in supporting glucose-stimulated insulin secretion in pancreatic β-cells. Diabetes 68:733

Hui L, Zatloukal K, Scheuch H, Stepniak E, Wagner EF (2008) Proliferation of human HCC cells and chemically induced mouse liver cancers requires JNK1-dependent p21 downregulation. J Clin Invest 118:3943–3953

Hunt JF, Wang C, Ford RC (2013) Cystic fibrosis transmembrane conductance regulator (ABCC7) structure. Cold Spring Harb Perspect Med 3:a009514

Hwang J, Qi L (2018) Quality control in the endoplasmic reticulum: crosstalk between ERAD and UPR pathways. Trends Biochem Sci 43:593–605

Iadanza MG, Jackson MP, Hewitt EW, Ranson NA, Radford SE (2018) A new era for understanding amyloid structures and disease. Nat Rev Mol Cell Biol 19:755–773

Iannuzzi C, Borriello M, Portaccio M, Irace G, Sirangelo I (2017) Insights into insulin fibril assembly at physiological and acidic pH and related amyloid intrinsic fluorescence. Int J Mol Sci 18:2551

Iljina M, Garcia GA, Horrocks MH, Tosatto L, Choi ML, Ganzinger KA, Abramov AY, Gandhi S, Wood NW, Cremades N, Dobson CM, Knowles TPJ, Klenerman D (2016) Kinetic model of the aggregation of alpha-synuclein provides insights into prion-like spreading. Proc Natl Acad Sci 113:E1206

Inden M, Kitamura Y, Takeuchi H, Yanagida T, Takata K, Kobayashi Y, Taniguchi T, Yoshimoto K, Kaneko M, Okuma Y, Taira T, Ariga H, Shimohama S (2007) Neurodegeneration of mouse nigrostriatal dopaminergic system induced by repeated oral administration of rotenone is prevented by 4-phenylbutyrate, a chemical chaperone. J Neurochem 101:1491–1504

Itoh N, Okamoto H (1980) Translational control of proinsulin synthesis by glucose. Nature 283:100–102

Ivanova MI, Sievers SA, Sawaya MR, Wall JS, Eisenberg D (2009) Molecular basis for insulin fibril assembly. Proc Natl Acad Sci U S A 106:18990–18995

Izumi T, Yokota-Hashimoto H, Zhao S, Wang J, Halban PA, Takeuchi T (2003) Dominant negative pathogenesis by mutant proinsulin in the akita diabetic mouse. Diabetes 52:409

Jang I, Pottekat A, Poothong J, Yong J, Lagunas-Acosta J, Charbono A, Chen Z, Scheuner DL, Liu M, Itkin-Ansari P, Arvan P, Kaufman RJ (2019) PDIA1/P4HB is required for efficient proinsulin maturation and ß cell health in response to diet induced obesity. elife 8:e44528

Jiménez JL, Nettleton EJ, Bouchard M, Robinson CV, Dobson CM, Saibil HR (2002) The protofilament structure of insulin amyloid fibrils. Proc Natl Acad Sci U S A 99:9196

Jin JK, Blackwood EA, Azizi K, Thuerauf DJ, Fahem AG, Hofmann C, Kaufman RJ, Doroudgar S, Glembotski CC (2017) ATF6 decreases myocardial ischemia/reperfusion damage and links ER stress and oxidative stress signaling pathways in the heart. Circ Res 120:862–875

Kalsheker N, Morley S, Morgan K (2002) Gene regulation of the serine proteinase inhibitors alpha (1)-antitrypsin and alpha(1)-antichymotrypsin. Biochem Soc Trans 30(2):93–98

Kang J, Lim L, Song J (2019) ATP binds and inhibits the neurodegeneration-associated fibrillization of the FUS RRM domain. Commun Biol 2:223

Kania E, PajĄK B, Orzechowski A (2015) Calcium homeostasis and ER stress in control of autophagy in cancer cells. Biomed Res Int 2015:352794

Kerbiriou M, Le Drévo MA, FÉrec C, Trouvé P (2007) Coupling cystic fibrosis to endoplasmic reticulum stress: differential role of Grp78 and ATF6. Biochim Biophys Acta 1772:1236–1249

Khurana R, Ionescu-Zanetti C, Pope M, Li J, Nielson L, Ramírez-Alvarado M, Regan L, Fink AL, Carter SA (2003) A general model for amyloid fibril assembly based on morphological studies using atomic force microscopy. Biophys J 85:1135–1144

Kim SJ, Skach WR (2012) Mechanisms of CFTR folding at the endoplasmic reticulum. Front Pharmacol 3:201

Knorre A, Wagner M, Schaefer HE, Colledge WH, Pahl HL (2002) DeltaF508-CFTR causes constitutive NF-kappaB activation through an ER-overload response in cystic fibrosis lungs. Biol Chem 383:271–282

Kusum Y, Anurag Y, Priyanka V, Veda PP, Upendra ND (2019) Protein misfolding diseases and therapeutic approaches. Curr Protein Pept Sci 20:1226–1245

Lange AM, Altynova ES, Nguyen GN, Sabatino DE (2016) Overexpression of factor VIII after AAV delivery is transiently associated with cellular stress in hemophilia A mice. Mol Ther Methods Clin Dev 3:16064

Laybutt DR, Preston AM, Akerfeldt MC, Kench JG, Busch AK, Biankin AV, Biden TJ (2007) Endoplasmic reticulum stress contributes to beta cell apoptosis in type 2 diabetes. Diabetologia 50:752–763

Lindblad D, Blomenkamp K, Teckman J (2007) Alpha-1-antitrypsin mutant Z protein content in individual hepatocytes correlates with cell death in a mouse model. Hepatology 46:1228–1235

Liu M, Hodish I, Haataja L, Lara-Lemus R, Rajpal G, Wright J, Arvan P (2010) Proinsulin misfolding and diabetes: mutant INS gene-induced diabetes of youth. Trends Endocrinol Metab 21:652–659

Liu M, Sun J, Cui J, Chen W, Guo H, Barbetti F, Arvan P (2015) INS-gene mutations: from genetics and beta cell biology to clinical disease. Mol Asp Med 42:3–18

Lomas DA, Li-Evans D, Finch JT, Carrell RW (1992) The mechanism of Z α1-antitrypsin accumulation in the liver. Nature 357:605–607

Lu Y, Xiong X, Helm A, Kimani K, Bragin A, Skach WR (1998) Co- and posttranslational translocation mechanisms direct cystic fibrosis transmembrane conductance regulator N terminus transmembrane assembly. J Biol Chem 273:568–576

Malhotra JD, Miao H, Zhang K, Wolfson A, Pennathur S, Pipe SW, Kaufman RJ (2008) Antioxidants reduce endoplasmic reticulum stress and improve protein secretion. Proc Natl Acad Sci U S A 105:18525–18530

Mann KG (1999) Biochemistry and physiology of blood coagulation. Thromb Haemost 82:165–174

Marquette KA, Pittman DD, Kaufman RJ (1995) A 110-amino acid region within the A1-domain of coagulation factor VIII inhibits secretion from mammalian cells. J Biol Chem 270:10297–10303

Meisl G, Yang X, Frohm B, Knowles TP, Linse S (2016) Quantitative analysis of intrinsic and extrinsic factors in the aggregation mechanism of Alzheimer-associated Abeta-peptide. Sci Rep 6:18728

Muzaffar M, Ahmad A (2011) The mechanism of enhanced insulin amyloid fibril formation by NaCl is better explained by a conformational change model. PLoS One 6:e27906

Oliver JD, Van Der Wal FJ, Bulleid NJ, High S (1997) Interaction of the thiol-dependent reductase ERp57 with nascent glycoproteins. Science 275:86–88

Orlova NA, Kovnir SV, Vorobiev II, Gabibov AG, Vorobiev AI (2013) Blood clotting factor VIII: from evolution to therapy. Acta Nat 5:19–39

Ostedgaard LS, Rich DP, Deberg LG, Welsh MJ (1997) Association of domains within the cystic fibrosis transmembrane conductance regulator. Biochemistry 36:1287–1294

Özcan U, Yilmaz E, Özcan L, Furuhashi M, Vaillancourt E, Smith RO, GÖrgün CZ, Hotamisligil GS (2006) Chemical chaperones reduce ER stress and restore glucose homeostasis in a mouse model of type 2 diabetes. Science 313:1137

Paganoni S, Macklin EA, Hendrix S, Berry JD, Elliott MA, Maiser S, Karam C, Caress JB, Owegi MA, Quick A, Wymer J, Goutman SA, Heitzman D, Heiman-Patterson T, Jackson CE, Quinn C, Rothstein JD, Kasarskis EJ, Katz J, Jenkins L, Ladha S, Miller TM, Scelsa SN, Vu TH, Fournier CN, Glass JD, Johnson KM, Swenson A, Goyal NA, Pattee GL, Andres PL, Babu S, Chase M, Dagostino D, Dickson SP, Ellison N, Hall M, Hendrix K, Kittle G, McGovern M, Ostrow J, Pothier L, Randall R, Shefner JM, Sherman AV, Tustison E, Vigneswaran P, Walker J, Yu H, Chan J, Wittes J, Cohen J, Klee J, Leslie K, Tanzi RE, Gilbert W, Yeramian PD, Schoenfeld D, Cudkowicz ME (2020) Trial of sodium phenylbutyrate–taurursodiol for amyotrophic lateral sclerosis. N Engl J Med 383:919–930

Parodi AJ, Behrens NH, Leloir LF, Carminatti H (1972) The role of polyprenol-bound saccharides as intermediates in glycoprotein synthesis in liver. Proc Natl Acad Sci U S A 69:3268–3272

Pastore N, Attanasio S, Granese B, Castello R, Teckman J, Wilson AA, Ballabio A, Brunetti-Pierri N (2017) Activation of the c-Jun N-terminal kinase pathway aggravates proteotoxicity of hepatic mutant Z alpha1-antitrypsin. Hepatology 65:1865–1874

Patel A, Malinovska L, Saha S, Wang J, Alberti S, Krishnan Y, Hyman AA (2017) ATP as a biological hydrotrope. Science 356:753

Paton AW, Beddoe T, Thorpe CM, Whisstock JC, Wilce MC, Rossjohn J, Talbot UM, Paton JC (2006) AB5 subtilase cytotoxin inactivates the endoplasmic reticulum chaperone BiP. Nature 443:548–552

Paxman R, Plate L, Blackwood EA, Glembotski C, Powers ET, Wiseman RL, Kelly JW (2018) Pharmacologic ATF6 activating compounds are metabolically activated to selectively modify endoplasmic reticulum proteins. elife 7:e37168

Pelletier M, Marcil A, Sevigny G, Jakob CA, Tessier D, Chevet E, Menard R, Bergeron J, Thomas D (2000) The heterodimeric structure of glucosidase II is required for its activity, solubility, and localization in vivo. Glycobiology 10:815–827

Perlmutter DH (2002) Liver injury in alpha1-antitrypsin deficiency: an aggregated protein induces mitochondrial injury. J Clin Invest 110:1579–1583

Perlmutter DH, Brodsky JL, Balistreri WF, Trapnell BC (2007) Molecular pathogenesis of alpha-1-antitrypsin deficiency-associated liver disease: a meeting review. Hepatology 45:1313–1323

Plate L, Cooley CB, Chen JJ, Paxman RJ, Gallagher CM, Madoux F, Genereux JC, Dobbs W, Garza D, Spicer TP, Scampavia L, Brown SJ, Rosen H, Powers ET, Walter P, Hodder P, Wiseman RL, Kelly JW (2016) Small molecule proteostasis regulators that reprogram the ER to reduce extracellular protein aggregation. elife 5:e15550

Pobre K, Poet G, Hendershot L (2018) The endoplasmic reticulum (ER) chaperone BiP is a master regulator of ER functions: Getting by with a little help from ERdj friends. J Biol Chem 294:2098–2108

Poothong J, Sopha P, Kaufman RJ, Tirasophon W (2017) IRE1α nucleotide sequence cleavage specificity in the unfolded protein response. FEBS Lett 591:406–414

Poothong J, Pottekat A, Siriin M, Campos AR, Paton AW, Paton JC, Lagunas-Acosta J, Chen Z, Swift M, Volkmann N, Hanein D, Yong J, Kaufman RJ (2020) Factor VIII exhibits chaperone-dependent and glucose-regulated reversible amyloid formation in the endoplasmic reticulum. Blood 135(21):1899–1911

Riordan JR, Rommens JM, Kerem B, Alon N, Rozmahel R, Grzelczak Z, Zielenski J, Lok S, Plavsic N, Chou JL et al (1989) Identification of the cystic fibrosis gene: cloning and characterization of complementary DNA. Science 245:1066–1073

Roth SD, Schüttrumpf J, Milanov P, Abriss D, Ungerer C, Quade-Lyssy P, Simpson JC, Pepperkok R, Seifried E, Tonn T (2012) Chemical chaperones improve protein secretion and rescue mutant factor VIII in mice with hemophilia A. PLoS One 7:e44505

Ruggiano A, Foresti O, Carvalho P (2014) ER-associated degradation: protein quality control and beyond. J Cell Biol 204:869–879

Rutkowski DT, Kaufman RJ (2007) That which does not kill me makes me stronger: adapting to chronic ER stress. Trends Biochem Sci 32:469–476

Rutkowski DT, Arnold SM, Miller CN, Wu J, Li J, Gunnison KM, Mori K, Sadighi Akha AA, Raden D, Kaufman RJ (2006) Adaptation to ER stress is mediated by differential stabilities of pro-survival and pro-apoptotic mRNAs and proteins. PLoS Biol 4:e374

Schmidt BZ, Perlmutter DH (2005) Grp78, Grp94, and Grp170 interact with α1-antitrypsin mutants that are retained in the endoplasmic reticulum. Am J Physiol Gastrointest Liver Physiol 289: G444–G455

Schwaller M, Wilkinson B, Gilbert HF (2003) Reduction-reoxidation cycles contribute to catalysis of disulfide isomerization by protein-disulfide isomerase. J Biol Chem 278:7154–7159

Segeritz C-P, Rashid ST, De Brito MC, Serra MP, Ordonez A, Morell CM, Kaserman JE, Madrigal P, Hannan NRF, Gatto L, Tan L, Wilson AA, Lilley K, Marciniak SJ, Gooptu B,

Lomas DA, Vallier L (2018) hiPSC hepatocyte model demonstrates the role of unfolded protein response and inflammatory networks in α_1-antitrypsin deficiency. J Hepatol 69: 851–860
Shaffer AL, Shapiro-Shelef M, Iwakoshi NN, Lee A-H, Qian S-B, Zhao H, Yu X, Yang L, Tan BK, Rosenwald A, Hurt EM, Petroulakis E, Sonenberg N, Yewdell JW, Calame K, Glimcher LH, Staudt LM (2004) XBP1, downstream of Blimp-1, expands the secretory apparatus and other organelles, and increases protein synthesis in plasma cell differentiation. Immunity 21:81–93
Shammas SL, Garcia GA, Kumar S, Kjaergaard M, Horrocks MH, Shivji N, Mandelkow E, Knowles TP, Mandelkow E, Klenerman D (2015) A mechanistic model of tau amyloid aggregation based on direct observation of oligomers. Nat Commun 6:7025
Sheppard DN, Welsh MJ (1999) Structure and function of the CFTR chloride channel. Physiol Rev 79:S23–S45
Shoulders MD, Ryno LM, Genereux JC, Moresco JJ, Tu PG, Wu C, Yates JR III, Su AI, Kelly JW, Wiseman RL (2013) Stress-independent activation of XBP1s and/or ATF6 reveals three functionally diverse ER proteostasis environments. Cell Rep 3:1279–1292
Sicorello A, Torrassa S, Soldi G, Gianni S, Travaglini-Allocatelli C, Taddei N, Relini A, Chiti F (2009) Agitation and high ionic strength induce amyloidogenesis of a folded PDZ domain in native conditions. Biophys J 96:2289–2298
Smith SE, Granell S, Salcedo-Sicilia L, Baldini G, Egea G, Teckman JH, Baldini G (2011) Activating transcription factor 6 limits intracellular accumulation of mutant α(1)-antitrypsin Z and mitochondrial damage in hepatoma cells. J Biol Chem 286:41563–41577
Song S, Tan J, Miao Y, Zhang Q (2018) Crosstalk of ER stress-mediated autophagy and ER-phagy: involvement of UPR and the core autophagy machinery. J Cell Physiol 233:3867–3874
Suaud L, Miller K, Alvey L, Yan W, Robay A, Kebler C, Kreindler JL, Guttentag S, Hubbard MJ, Rubenstein RC (2011) ERp29 regulates DeltaF508 and wild-type cystic fibrosis transmembrane conductance regulator (CFTR) trafficking to the plasma membrane in cystic fibrosis (CF) and non-CF epithelial cells. J Biol Chem 286:21239–21253
Sun J, Xiong Y, Li X, Haataja L, Chen W, Mir SA, Lv L, Madley R, Larkin D, Anjum A, Dhayalan B, Rege N, Wickramasinghe ND, Weiss MA, Itkin-Ansari P, Kaufman RJ, Ostrov DA, Arvan P, Liu M (2020) Role of proinsulin self-association in mutant INS gene-induced diabetes of youth. Diabetes 69(5):954–964
Tagliavacca L, Wang Q, Kaufman RJ (2000) ATP-dependent dissociation of non-disulfide-linked aggregates of coagulation factor VIII is a rate-limiting step for secretion. Biochemistry 39:1973–1981
Tannous A, Pisoni GB, Hebert DN, Molinari M (2015) N-linked sugar-regulated protein folding and quality control in the ER. Semin Cell Dev Biol 41:79–89
Teckman JH, An J-K, Blomenkamp K, Schmidt B, Perlmutter D (2004) Mitochondrial autophagy and injury in the liver in α1-antitrypsin deficiency. Am J Physiol Gastrointest Liver Physiol 286: G851–G862
Tenidis K, Waldner M, Bernhagen J, Fischle W, Bergmann M, Weber M, Merkle M-L, Voelter W, Brunner H, Kapurniotu A (2000) Identification of a penta- and hexapeptide of islet amyloid polypeptide (IAPP) with amyloidogenic and cytotoxic properties1 1Edited by R. Huber. J Mol Biol 295:1055–1071
Teodoro-Morrison T, Schuiki I, Zhang L, Belsham DD, Volchuk A (2013) GRP78 overproduction in pancreatic beta cells protects against high-fat-diet-induced diabetes in mice. Diabetologia 56:1057–1067
Tirasophon W, Welihinda AA, Kaufman RJ (1998) A stress response pathway from the endoplasmic reticulum to the nucleus requires a novel bifunctional protein kinase/endoribonuclease (Ire1p) in mammalian cells. Genes Dev 12:1812–1824
Tran DT, Pottekat A, Mir SA, Loguercio S, Jang I, Campos AR, Scully KM, Lahmy R, Liu M, Arvan P, Balch WE, Kaufman RJ, Itkin-Ansari P (2020) Unbiased profiling of the human proinsulin biosynthetic interaction network reveals a role for peroxiredoxin 4 in proinsulin folding. Diabetes 69(8):1723–1734

Tsuchiya Y, Saito M, Kadokura H, Miyazaki J-I, Tashiro F, Imagawa Y, Iwawaki T, Kohno K (2018) IRE1–XBP1 pathway regulates oxidative proinsulin folding in pancreatic β cells. J Cell Biol 217(4):1287–1301

Upton J-P, Wang L, Han D, Wang E, Huskey N, Lim L, Truitt M, McManus M, Ruggero D, Goga A, Papa F, Oakes S (2012) IRE1α Cleaves Select microRNAs during ER Stress to Derepress Translation of Proapoptotic Caspase-2. Science 338(6108):818–822

Urano F, Wang X, Bertolotti A, Zhang Y, Chung P, Harding HP, Ron D (2000) Coupling of stress in the ER to activation of JNK protein kinases by transmembrane protein kinase IRE1. Science 287:664

Usui M, Yamaguchi S, Tanji Y, Tominaga R, Ishigaki Y, Fukumoto M, Katagiri H, Mori K, Oka Y, Ishihara H (2012) Atf6alpha-null mice are glucose intolerant due to pancreatic beta-cell failure on a high-fat diet but partially resistant to diet-induced insulin resistance. Metabolism 61:1118–1128

Van Goor F, Hadida S, Grootenhuis PD, Burton B, Stack JH, Straley KS, Decker CJ, Miller M, McCartney J, Olson ER, Wine JJ, Frizzell RA, Ashlock M, Negulescu PA (2011) Correction of the F508del-CFTR protein processing defect in vitro by the investigational drug VX-809. Proc Natl Acad Sci U S A 108:18843–18848

Varga K, Jurkuvenaite A, Wakefield J, Hong JS, Guimbellot JS, Venglarik CJ, Niraj A, Mazur M, Sorscher EJ, Collawn JF, Bebök Z (2004) Efficient intracellular processing of the endogenous cystic fibrosis transmembrane conductance regulator in epithelial cell lines. J Biol Chem 279:22578–22584

Wang J, Takeuchi T, Tanaka S, Kubo SK, Kayo T, Lu D, Takata K, Koizumi A, Izumi T (1999) A mutation in the insulin 2 gene induces diabetes with severe pancreatic beta-cell dysfunction in the Mody mouse. J Clin Invest 103:27–37

Wang A, Al-Kuhlani M, Johnston SC, Ojcius DM, Chou J, Dean D (2013) Transcription factor complex AP-1 mediates inflammation initiated by Chlamydia pneumoniae infection. Cell Microbiol 15:779–794

Wright JM, Zeitlin PL, Cebotaru L, Guggino SE, Guggino WB (2004) Gene expression profile analysis of 4-phenylbutyrate treatment of IB3-1 bronchial epithelial cell line demonstrates a major influence on heat-shock proteins. Physiol Genomics 16:204–211

Yadav K, Yadav A, Vashistha P, Pandey VP, Dwivedi UN (2019) Protein misfolding diseases and therapeutic approaches. Curr Protein Pept Sci 20:1226–1245

Yong J, Bischof H, Burgstaller S, Siirin M, Murphy A, Malli R, Kaufman RJ (2019) Mitochondria supply ATP to the ER through a mechanism antagonized by cytosolic Ca(2). elife 8:e49682

Yoshida H, Matsui T, Yamamoto A, Okada T, Mori K (2001) XBP1 mRNA is induced by ATF6 and spliced by IRE1 in response to ER stress to produce a highly active transcription factor. Cell 107:881–891

Zhang L, Lai E, Teodoro T, Volchuk A (2009) GRP78, but not protein-disulfide isomerase, partially reverses hyperglycemia-induced inhibition of insulin synthesis and secretion in pancreatic-cells. J Biol Chem 284:5289–5298

Zhou W, Bercury K, Cummiskey J, Luong N, Lebin J, Freed CR (2011) Phenylbutyrate up-regulates the DJ-1 protein and protects neurons in cell culture and in animal models of Parkinson disease. J Biol Chem 286:14941–14951

Zolotukhin I, Markusic DM, Palaschak B, Hoffman BE, Srikanthan MA, Herzog RW (2016) Potential for cellular stress response to hepatic factor VIII expression from AAV vector. Mol Ther Methods Clin Dev 3:16063

Chapter 7
Roles of Calreticulin in Protein Folding, Immunity, Calcium Signaling and Cell Transformation

Arunkumar Venkatesan, Leslie S. Satin, and Malini Raghavan

Abstract The endoplasmic reticulum (ER) is an organelle that mediates the proper folding and assembly of proteins destined for the cell surface, the extracellular space and subcellular compartments such as the lysosomes. The ER contains a wide range of molecular chaperones to handle the folding requirements of a diverse set of proteins that traffic through this compartment. The lectin-like chaperones calreticulin and calnexin are an important class of structurally-related chaperones relevant for the folding and assembly of many N-linked glycoproteins. Despite the conserved mechanism of action of these two chaperones in nascent protein recognition and folding, calreticulin has unique functions in cellular calcium signaling and in the immune response. The ER-related functions of calreticulin in the assembly of major histocompatibility complex (MHC) class I molecules are well-studied and provide many insights into the modes of substrate and co-chaperone recognition by calreticulin. Calreticulin is also detectable on the cell surface under some conditions, where it induces the phagocytosis of apoptotic cells. Furthermore, mutations of calreticulin induce cell transformation in myeloproliferative neoplasms (MPN). Studies of the functions of the mutant calreticulin in cell transformation and immunity have provided many insights into the normal biology of calreticulin, which are discussed.

Keywords Calreticulin · Calnexin · PDIA3 · Calcium signaling · Major Histocompatibility Complex (MHC) class I · Phagocytosis of apoptotic cells · Myeloproliferative neoplasms

---

A. Venkatesan · M. Raghavan (✉)
Department of Microbiology and Immunology, Michigan Medicine, University of Michigan, Ann Arbor, MI, USA
e-mail: malinir@umich.edu

L. S. Satin
Department of Pharmacology, Brehm Diabetes Research Center, Michigan Medicine, University of Michigan, Ann Arbor, MI, USA

L. B. Agellon, M. Michalak (eds.), *Cellular Biology of the Endoplasmic Reticulum*, Progress in Molecular and Subcellular Biology 59,
https://doi.org/10.1007/978-3-030-67696-4_7

7.1 Introduction to the Proteins

Calreticulin and calnexin are structurally-related chaperones of the ER that exert major influences on glycoprotein folding, solubility and assembly. Calreticulin is an ER lumenal resident protein, whereas calnexin is membrane-bound. Both calreticulin and calnexin have a structurally-related globular lectin domain, with a β-sandwich structure comprising a large convex β-sheet and a concave β-sheet (Kozlov et al. 2010b; Chouquet et al. 2011; Schrag et al. 2001; Blees et al. 2017) (Fig. 7.1A, B). The domains contain a conserved carbohydrate-binding site responsible for a glycan-dependent binding of glycoprotein substrates and chaperone activity towards such substrates. Both calreticulin and calnexin preferentially bind mono-glucosylated ($Glc_1Man_9GlcNAc_2$) glycoprotein substrates (Hebert et al. 1995; Rodan et al. 1996). In addition to the carbohydrate-binding site, the globular domains of both calreticulin and calnexin contain a high-affinity Ca^{2+}-binding site (Schrag et al. 2001; Kozlov et al. 2010b) [$K_D \sim 17\ \mu M$ (Wijeyesakere et al. 2011)] which is important for the structural stability of the proteins (Li et al. 2001; Conte et al. 2007; Wijeyesakere et al. 2011).

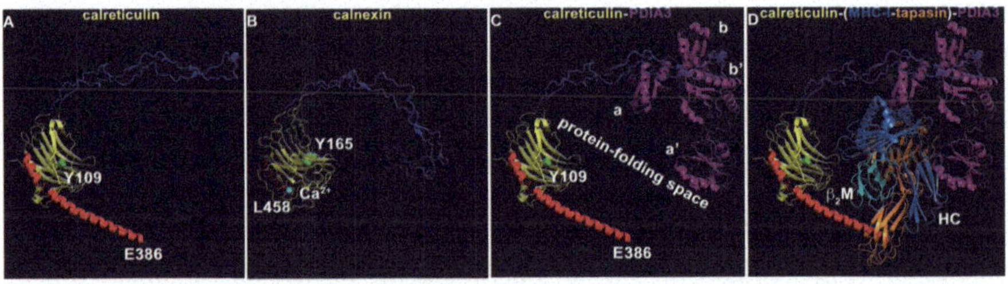

Fig. 7.1 Structures of Calreticulin, Calnexin, Calreticulin-PDIA3, and Calreticulin-MHC class I-tapasin-PDIA3 complexes. Overall domain organizations of calreticulin (**A**) calnexin (**B**) and calreticulin-PDIA3 complexes (**C**), and calreticulin-MHC class I-tapasin-PDIA3 based on cryoelectron microscopy studies (cryo EM) [(Blees et al. 2017) PDB 6eny] and crystal [(Schrag et al. 2001) PDB 1jhn] structures. (**A**) Calreticulin contains a globular lectin-like domain (yellow, residues 18-203 from the N-domain and 301–335 from the C-domain) bearing a recognition site for monoglucosylated glycans (including indicated residue Y109) and a high-affinity calcium-binding site, a P domain (blue, residues 206–300), the tip of which is the site of co-chaperone binding, and an α-helical (residues 336–417) C-domain (red) that contains multiple low affinity calcium-binding sites. (**B**) The ER lumenal domain of calnexin (21–482) contains a globular lectin-like domain (yellow, residues 21–276 from the N-domain and 411–443 from the C-domain), bearing a recognition site for monoglucosylated glycans (including indicated residue Y165) and high-affinity calcium-binding site (cyan), a P domain (blue, residues 277–410) and a C-domain (residues 444–482 (red), not resolved beyond residue 458 in the crystal structure). (**C**) PDIA3 contains four domains (a, b, b′ and a′) and interacts with the calreticulin and calnexin P-domains via its b' domain. (**D**) Selective components of the MHC class I peptide-loading complex including calreticulin, PDIA3, MHC class I heavy chains (hc), β2m, and tapasin. Higher resolution crystal structures were superimposed on the low-resolution map to reconstruct the low-resolution map (Blees et al. 2017). Within this complex, calreticulin interacts with its glycan binding site with the conserved MHC class I hc glycan and Tapasin C95 forms a disulfide-linked conjugate with C57 of the a domain of PDIA3 (within its catalytic thiol-disulfide exchanging motif) (Dong et al. 2009)

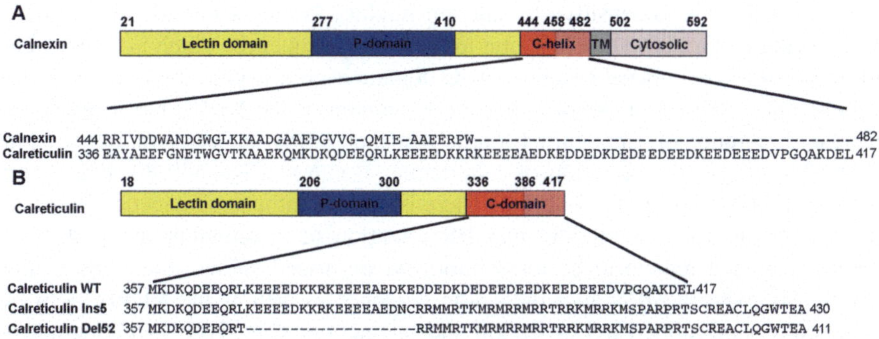

Fig. 7.2 Domain structures and sequence alignments of calreticulin, calnexin and calreticulin MPN mutants. Both calnexin (**a**) and calreticulin (**b**) contains globular lectin domains (yellow), P-domains (blue) and C-terminal helices (red). Calnexin additionally includes transmembrane and cytosolic domains. Sequence alignments of the C-terminal helices of calreticulin and calnexin are shown in (**a**). A longer helix is present in calreticulin. (**b**) Alignments of wild type and MPN mutant calreticulin sequences. Frame-shift mutations change the acidic amino acids of the calreticulin C-terminal domain into basic residues. The most frequently observed mutations are a 52 base-pair deletion (Del52) and a 5 base-pair (Ins5) insertion. Amino acid sequences of altered regions of the calreticulin C-terminal domain are shown

Inserted into the lectin domains of calreticulin and calnexin is an arm-like P-domain, rich in proline residues (Moreau et al. 2016; Schrag et al. 2001). The P-domains of calnexin and calreticulin form extended ß-hairpin structures of different lengths in the two proteins (Ellgaard et al. 2001a, b; Schrag et al. 2001) (Fig. 7.1A, B). The tip of the hairpin loop of the P-domain is shown to interact with several co-chaperones, including PDIA3 (Frickel et al. 2002) (Fig. 7.1C), cyclophilin B (Kozlov et al. 2010a), which have thiol-disulfide and peptidyl-prolyl isomerase activities respectively, and the chaperone PDIA9 (Kozlov et al. 2017). Thus, the P-domains function as plurivalent adaptors for co-chaperones (Kozlov et al. 2017), that could enhance the lectin-domain dependent folding of newly synthesized glycoproteins, via facilitation of disulfide-bond formation, proline isomerization and other interactions important for folding. Calreticulin is also shown to interact with PDIA1 (Baksh et al. 1995; Avezov et al. 2015), an interaction that is calcium sensitive, and which impacts functional activities of both proteins. It remains to be established whether calreticulin-PDIA1 interaction involves the conventional P-domain-dependent mode of calreticulin binding to other PDI family members.

Unique to calreticulin is the presence of a long C-terminal helix (C-domain) that extends beyond its lectin domain (Blees et al. 2017) (Fig. 7.1A, C, D), with a C-terminal region rich in acidic amino acids (Figs. 7.1 and 7.2a). The acidic residues are absent in calnexin's C-domain (Fig. 7.2a). The C-domain of calreticulin is capable of binding multiple calcium ions with low affinity ($K_D \sim 600$ μM) (Baksh and Michalak 1991; Wijeyesakere et al. 2011), contributing to calcium buffering and storage in the ER (Nakamura et al. 2001). The overall structure of the calreticulin

C-terminal domain is sensitive to calcium binding (Villamil Giraldo et al. 2010; Wijeyesakere et al. 2011). In addition to the KDEL sequence, which is responsible for retrieval of ER lumenal proteins, the C-domain itself also contributes to retaining calreticulin in the ER. Loss of either the C-domain or the KDEL motif results in increased secretion of calreticulin (Sonnichsen et al. 1994), and calcium-depletion in cells also induces the secretion of calreticulin (Booth and Koch 1989; Peters and Raghavan 2011; Jeffery et al. 2011). There is thus an intimate connection between the C-domain, calcium binding and ER retention of calreticulin, the molecular details of which remain to be understood. As discussed below, mutations of the acidic C-domain in MPN alter the acidic sequence as well as the KDEL motif of calreticulin, and influence localization and interactions (Fig. 7.2B). Calreticulin is also known to undergo non-degrading retro-translocation from the ER to the cytosol, in a process regulated by the acidic C-domain (Afshar et al. 2005).

7.2 Engagement, Folding and ER Retention of Glycoproteins with Monoglucosylated Glycans

Calreticulin and calnexin function in the folding of N-glycosylated proteins that travel from the ER through the secretory pathway into the plasma membrane, extracellular space or subcellular compartments. As nascent proteins are synthesized and co-translationally inserted into the ER, their asparagine-linked glycosylation sites are modified by the addition of preassembled glycans ($Glc_3Man_9GlcNAc_2$) via oligosaccharyl transferases (Kornfeld and Kornfeld 1985). Further, through the action of glucosidase I and glucosidase II, the two-terminal glucose residues are removed, thereby generating a monoglucosylated glycan ($Glc_1Man_9GlcNAc_2$), which is favored for binding to calreticulin and calnexin (Hebert et al. 1995). For glycoproteins that adopt a native structure, glucosidase II removes the terminal glucose, allowing substrates to exit the calreticulin and calnexin cycle. However, glycoproteins that have not obtained a native structure are re-glucosylated by UDP-glucose: glycoprotein glucosyltransferase (UGGT), which permits reentry into the calreticulin and calnexin cycle. UGGT preferentially re-glycosylates ($Man_9GlcNAc_2$) misfolded/unfolded substrates but not properly folded substrates (Taylor et al. 2004; Calles-Garcia et al. 2017). Proteins that fail to properly fold following repeated interactions with calreticulin/calnexin are eventually shuttled into a degradation pathway.

The P-domain of calreticulin is known to exhibit considerable conformational flexibility measurable through fluorescence resonance energy transfer (FRET) spectroscopic studies, which showed the induction of an open P-domain conformation following engagement of glycoprotein substrates, and a more closed conformation in the presence of glycoprotein substrates and PDIA3 (Wijeyesakere et al. 2013). X-ray crystal structures and small angle x-ray scattering studies also provide insights into conformational variations of the calreticulin P-domain (Moreau et al. 2016). In combination with PDIA3, the P-domain may restrain substrate conformational fluctuations and provide an enclosed space to assist the chaperoning of glycoprotein substrates (Moreau et al. 2016) (Fig. 7.1C, D).

ATP is a key regulator of many chaperone-substrate interactions, and early studies have shown that ATP influences the conformational characteristics of both calnexin and calreticulin (Ihara et al. 1999; Saito et al. 1999; Brockmeier and Williams 2006). Recent studies point to the existence of an ATP binding site in the globular domain of calreticulin and indicate a role for ATP binding in destabilizing glycoprotein substrate interactions with calreticulin (Wijeyesakere et al. 2015).

7.3 Glycan-Independent Protein Binding by Calreticulin

Despite a high degree of conservation in their structures and carbohydrate recognition sites, there are distinct and variable requirements for calreticulin and calnexin in the folding and assembly of individual substrates (Hebert et al. 1997; Molinari et al. 2004). These distinct requirements and functions may arise due to differences in their respective localization (membrane vs. lumenal) or those in their P-domain lengths (Fig. 7.1A, B), which could impose different constraints on their substrate interactions. It is also possible that both proteins recognize distinct polypeptide components of the substrate glycoproteins. In addition to glycan based chaperone activity, several studies have indicated that both calreticulin and calnexin have the capacity to inhibit the *in vitro* aggregation of non-glycosylated proteins (Ihara et al. 1999; Saito et al. 1999; Mancino et al. 2002; Rizvi et al. 2004; Thammavongsa et al. 2005; Del Cid et al. 2010), an activity induced by heat shock, calcium depletion, or deletion of acidic C-terminal domain (Rizvi et al. 2004; Thammavongsa et al. 2005). Notably, the secondary structure of calreticulin is thermostable (Wijeyesakere et al. 2011), and the above-mentioned conditions likely provide suitable structural surfaces for inhibiting protein aggregation by exposing hydrophobic residues for non-native protein binding. However, the polypeptide-specific chaperone activities of calreticulin and calnexin are also measured under physiological conditions with non-glycosylated substrates that display instability under those conditions (Del Cid et al. 2010; Jeffery et al. 2011; Brockmeier and Williams 2006). The exact location of the polypeptide-binding site(s) of calreticulin is controversial (Pocanschi et al. 2011; Moreau et al. 2016; Wijeyesakere et al. 2013), although some studies have suggested a location in the vicinity of the glycan-binding site (Wijeyesakere et al. 2013; Chouquet et al. 2011). Despite these *in vitro* measurements of chaperone activities towards many different non-glycosylated proteins, the cellular activity of calreticulin or calnexin in the folding of non-glycosylated substrates remain to be demonstrated. The association kinetics for calreticulin binding to monoglucosylated substrates are higher than towards non-glycosylated substrates (Wijeyesakere et al. 2013), and thus it is likely that monoglucosylated proteins are the preferred substrates of calreticulin in cells. Once recruited, both glycan and polypeptide-based interactions could contribute to substrate stabilization, folding and assembly.

7.4 Calreticulin, Cellular Calcium Homeostasis and Calcium Signaling

The ER is a major storage site for Ca^{2+}. The total ER Ca^{2+} concentration is in the range of 400 μM–1 mM, while the resting cytosolic concentrations are orders of magnitude lower, around 100 nM (Clapham 2007). Much of the ER Ca^{2+} is sequestered via binding to low affinity and rapidly exchanging calcium-binding sites on ER-resident factors such as calreticulin (Mery et al. 1996; Nakamura et al. 2001; Baksh and Michalak 1991) and BiP (Lievremont et al. 1997). Over-expression of calreticulin leads to increased cellular Ca^{2+} content, and its deficiency results in reduced ER Ca^{2+} storage, as well as impairments in agonist stimulated calcium release from the ER (Mery et al. 1996; Mesaeli et al. 1999; Nakamura et al. 2001; Arnaudeau et al. 2002). Compartmental Ca^{2+} maintenance within the cell is determined by specific Ca^{2+} transporters (Fig. 7.3). The sarco/endoplasmic reticulum Ca^{2+}-ATPase (SERCA) pump transports Ca^{2+} from the cytosol to the ER in an ATP-dependent manner (Toyoshima and Inesi 2004). Following stimulation with the agonists, ER Ca^{2+} is released to the cytoplasm via the 1,4,5-triphosphate (IP3R) receptor (Berridge 2016) and the ryanodine receptor (RyR) (Lanner et al. 2010). When the ER Ca^{2+} concentration is reduced, the stromal cell surface molecule 1 (Stim1), an ER Ca^{2+} sensor, undergoes conformational changes that induce its oligomerization and trigger interactions with a plasma membrane (PM) calcium channel called Orai1 to allow for influx of extracellular calcium into the cytoplasm, a process known as store operated Ca^{2+} entry (SOCE) (Lewis 2020). Previous studies have shown that over-expression of calreticulin inhibits SOCE activation in cells, presumably by increasing ER Ca^{2+} and inhibiting Stim1 oligomerization (Bastianutto et al. 1995; Mery et al. 1996; Arnaudeau et al. 2002). Conversely, as discussed below, calreticulin insufficiency under certain pathological conditions, is also suggested to enhance SOCE activation (Di Buduo et al. 2014).

Calcium is a universal second messenger that controls various cellular functions in the cell. As noted above, calreticulin has multiple low affinity calcium-binding sites in its acidic C-terminus (Wijeyesakere et al. 2011; Baksh and Michalak 1991), which are responsible for storing a significant fraction of the cellular Ca^{2+} in the ER lumen (Nakamura et al. 2001; Mery et al. 1996). Calreticulin deficiency is embryonically lethal in mice due to defects in cardiac development (Mesaeli et al. 1999). Calreticulin deficiency results in impaired IP3R-mediated calcium release from the ER, in turn inducing insufficient nuclear translocation of the nuclear factor of activated T cells (NF-AT) in response to calcium stimulation (Mesaeli et al. 1999). The calcium-dependent phosphatase activity of calcineurin is required for the nuclear translocation of NF-AT, and indeed cardiac-specific expression of activated

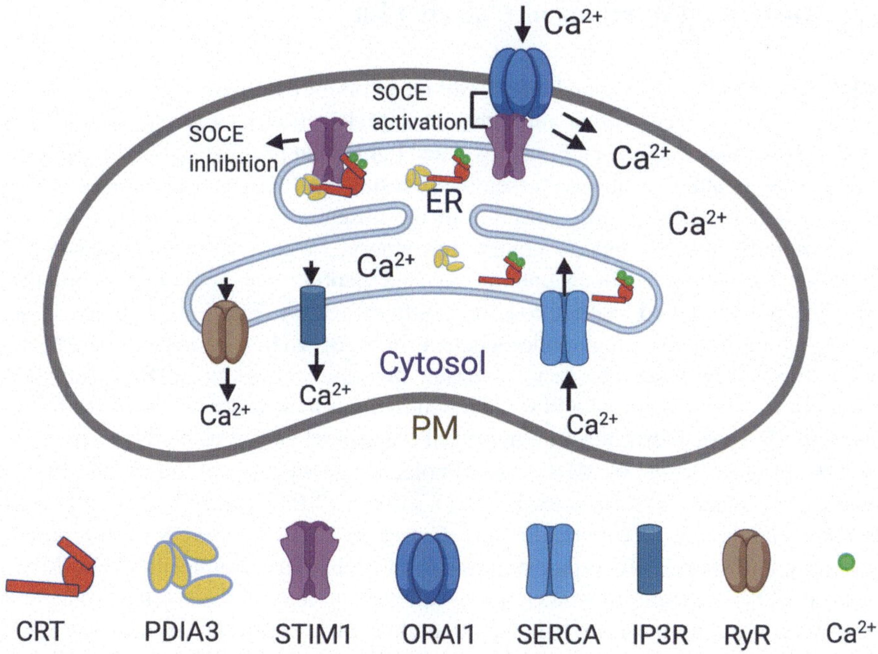

Fig. 7.3 Calcium movement within cells. The SERCA pump transports Ca^{2+} from the cytosol to the ER in an ATP-dependent manner. In the ER, calcium is bound with low affinity to calcium binding proteins such as calreticulin. Following stimulation with agonists, ER Ca^{2+} is released to the cytoplasm via the IP3R receptor and RyR. When the ER Ca^{2+} concentration is reduced, Stim1, an ER Ca^{2+} sensor, undergoes conformational changes that induce oligomerization and trigger interactions with a plasma membrane calcium channel Orai1, to allow for influx of extracellular calcium into the cytoplasm, a process known as store operated Ca^{2+} entry (SOCE). PDIA3 is suggested to regulate SOCE by binding to the lumenal domain of Stim1 and inhibiting SOCE activity (Prins et al. 2011). Recent studies suggest that the number of calreticulin-PDIA3 interactions with Stim1 are reduced in cells expressing MPN mutant calreticulins (which are secreted), findings which suggest that calreticulin-PDIA3 complexes are regulators of SOCE activity (Di Buduo et al. 2020)

calcineurin reverses the cardiac developmental defect in calreticulin-deficient mice and rescues embryonic lethality (Guo et al. 2002). These studies suggested a calreticulin, calcium, activated calcineurin, and NF-AT-dependent signaling pathway that is important for early cardiac development (Guo et al. 2002; Mesaeli et al. 1999; Rauch et al. 2000). Transforming growth factor β (TGF-β) activates the calcineurin-NF-AT pathway via Ca^{2+} to trigger extracellular matrix production. Calreticulin-deficient cells also have impairments in this pathway (Zimmerman et al. 2013).

7.5 Immune Functions of Calreticulin

MHC class I molecules, ligands for T cell receptors of cytotoxic T cells, are one of the best characterized calreticulin substrates. MHC class I molecules comprise a glycosylated heavy chain, a non-glycosylated light chain [β2-microglobulin ((β2m)], and a short peptide, which are assembled within the ER. Heterodimers of heavy chain and β2m are generally unstable in the absence of peptide and are in turn recognized by the ER quality control machinery, to generate monoglucosylated forms of the heavy chain. Monoglucosylated peptide-free heterodimers interact with the MHC class I peptide-loading complex (PLC) (Fig. 7.1D) (Blees et al. 2017), which includes the peptide-deficient forms of MHC-1 molecules, the MHC class I-dedicated assembly factor tapasin, and the calreticulin-PDIA3 complex (Blum et al. 2013). Within the PLC, calreticulin binds to a conserved MHC-1 glycan, and PDIA3 forms a disulfide conjugate with tapasin (Dong et al. 2009) to stabilize peptide-deficient forms of MHC class I molecules. Mutations of either the glycan-binding site residues of calreticulin (Del Cid et al. 2010; Wearsch et al. 2011) or PDIA3 binding site of calreticulin (Del Cid et al. 2010) interfere with the proper assembly and stable cell surface expression of MHC class I molecules. The MHC class I PLC structure illustrates how calreticulin and PDIA3 (chaperone and co-chaperone) are used to achieve assembly of a substrate protein complex (Fig. 7.1D).

While calreticulin is predominantly ER localized, certain conditions induce its cell surface expression. As noted above, calcium depletion in the ER induces the secretory trafficking of calreticulin and its cell surface expression (Booth and Koch 1989; Peters and Raghavan 2011; Jeffery et al. 2011). The polypeptide-binding mode of calreticulin is relevant to its interactions with cell surface proteins (Jeffery et al. 2011). Calreticulin is also expressed on the surface of dying cells, and is shown to promote the phagocytosis of dying cells by antigen presenting cells (Gardai et al. 2005). Certain types of cell death conditions, termed immunogenic cell death, are shown to enhance the induction of protective immune response against antigens derived from the dying cells. Surface calreticulin expression is a marker of immunogenic forms of cell death, inducing enhanced phagocytosis and presentation of antigens derived from the dying cells to T cells (Obeid et al. 2007). Calreticulin binds to phosphatidylserine on the surface of apoptotic cells via a calcium and C-domain-dependent pathway and with phagocytes via lectin and P-domain-dependent interaction that may involve its generic polypeptide-binding site (Wijeyesakere et al. 2016) (Fig. 7.4). Thus, calreticulin is a pro-phagocytic signal, bridging apoptotic cells to phagocytes.

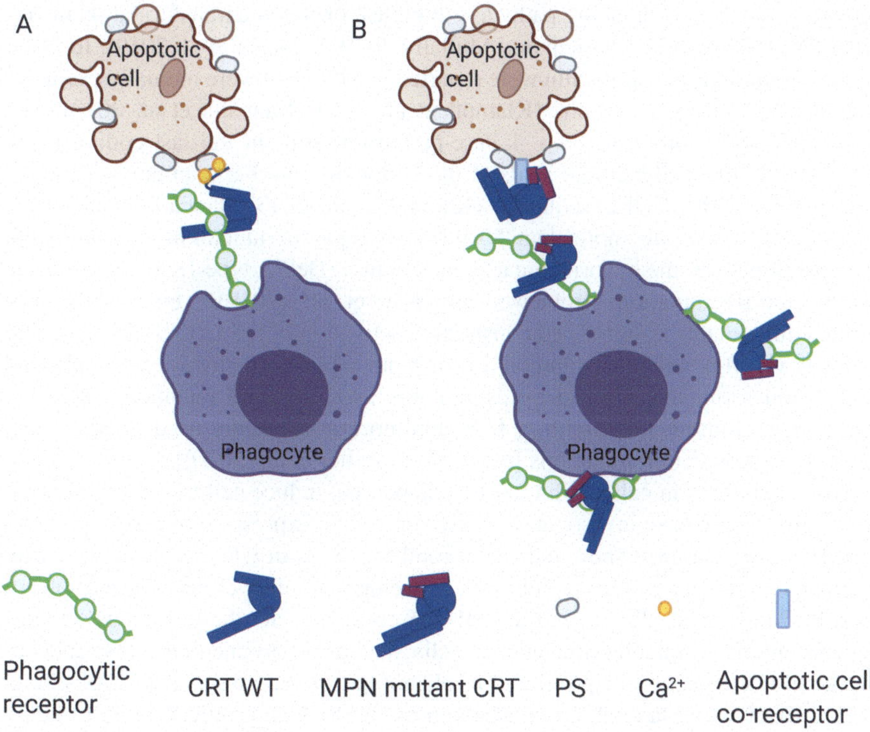

Fig. 7.4 Calreticulin-dependent phagocytosis of apoptotic cells. (**a**) Calreticulin engages surface phosphatidylserine of apoptotic cells via a calcium and C-domain-dependent manner, and phagocyte receptors via a P-domain and lectin domain-dependent manner (Wijeyesakere et al. 2016). (**b**) MPN mutant calreticulins, due to size and charge alterations to the C-domain, are likely to engage the surface of apoptotic cells via distinct mechanisms, to inhibit phagocytosis. Additionally, secreted calreticulin, via binding to phagocyte cell surface receptors, are shown to inhibit the phagocytosis of apoptotic cells and the induction of tumor antigen-specific T cell responses (Liu et al. 2020)

7.6 Calreticulin C-Domain Mutations in Myeloproliferative Neoplasms

Myeloproliferative neoplasms (MPNs) are clonal hematopoietic stem cell disorders that include polycythemia vera (PV), essential thrombocythemia (ET), primary myelofibrosis (MF), post-ET MF and post PV MF (Arber et al. 2016). These conditions induce abnormal production of red blood cells, and platelets and can cause fibrosis in the bone marrow. In normal hematopoiesis, the binding of thrombopoietin to the thrombopoietin receptor/myeloproliferative leukemia virus (Mpl) or erythropoietin to the erythropoietin receptor (EPOR) induces the activation of the Janus kinase 2 (JAK2) and transcriptional activation via STAT (Schieber et al. 2019). Specific somatic mutations of either JAK2 or Mpl are known to constitutively activate downstream signaling pathways to induce oncogenic transformation underlying PV, ET and MF (Marneth and Mullally 2020; Schieber et al. 2019). In 2013, somatic mutations in calreticulin were reported in MPN that were mutually exclusive with the JAK2/Mpl mutations (Klampfl et al. 2013; Nangalia et al. 2013). The calreticulin mutations cause a + 1 base-pair frame-shift in the last coding exon, that converts the acidic amino acids of the C-domain into basic amino acids along with a loss of the KDEL sequence (Klampfl et al. 2013; Nangalia et al. 2013) (Fig. 7.2B). There are more than 50 different types of mutations in calreticulin reported; however, the majority are a 52 bp deletion (Del52, type-I) or 5 bp insertion (Ins5, type-II) mutations [reviewed in How et al. (2019)] (Fig. 7.2B). The calreticulin mutants also induced constitutive activation of the JAK/STAT signaling pathway via Mpl, and thus common downstream signaling pathways are implicated in cell transformation induced by calreticulin, JAK2 or Mpl mutations (Schieber et al. 2019). Indeed, ruxolitinib, a JAK inhibitor, is frequently used to treat MPN induced by calreticulin mutations [reviewed in Asher et al. (2020)].

How mutations in calreticulin, an ER chaperone, induce cell transformation has been a topic of intense investigations in recent years. Expression of both type-I and type-II mutants in mice show induced thrombocytosis, an ET-like phenotype, with expansion of megakaryocytes, and further development of myelofibrosis and splenomegaly (Marty et al. 2016; Li et al. 2018). Consistent with these findings, *in vitro* expression of calreticulin mutants in cells that are cytokine-dependent induces cytokine-independent cell proliferation phenotype in the presence of Mpl (Araki et al. 2016; Chachoua et al. 2016; Marty et al. 2016). Cell proliferation induced by calreticulin mutants specifically requires Mpl, as calreticulin mutants are unable to activate cells in the context of the structurally related receptors (Chachoua et al. 2016; Elf et al. 2016). Several studies have shown induced binding between Mpl and the MPN mutant calreticulins, relative to the wild type proteins (Araki et al. 2016; Elf et al. 2018). A glycan linked to asparagine 117 (N117) in extracellular domain of Mpl is shown to be important for mutant calreticulin binding (Chachoua et al. 2016; Pecquet et al. 2019). While the novel C-domain is indispensable for the oncogenic activation, expression of novel C-domain by itself is not sufficient to induce activation, and large truncations of novel C-domain result in impaired oncogenic activation potential (Elf et al. 2016, 2018). Overall, these finding suggest a model in which both glycan-dependent and C-domain-dependent interactions allow mutant calreticulins to bind and aberrantly activate Mpl. The exact nature of the complexes remains to be elucidated.

Early studies have shown that a number of conditions including heat-shock, reduced pH, truncation of acidic C-domain and calcium depletion induce calreticulin oligomerization coincident with the induction of its polypeptide-specific chaperone activity (Jorgensen et al. 2003; Mancino et al. 2002; Rizvi et al. 2004). Interestingly, the novel C-domain in the MPN calreticulin mutants also induces calreticulin multimerization (Araki et al. 2019; Charonis et al. 2017). Mutant calreticulin multimers are also observed in platelets from patients with calreticulin-mutated MPN (unpublished results). The nature of the mutant calreticulin multimers and their contributions to Mpl activation are of great interest.

Calcium signaling plays an important role in megakaryocyte function and pro-platelet formation (Di Buduo et al. 2014). Given that the domains containing low affinity calcium binding sites of calreticulin undergoes sequence and charge alterations in the MPN mutants (Fig. 7.2B), it is possible that the MPN mutants also induce aberrant calcium signaling in cells as a component of their cell transformation mechanism. Abnormal cytosolic calcium signals have been detected in cultured megakaryocytes from patients expressing type I calreticulin mutations (Pietra et al. 2016). More recent reports demonstrated reduced interactions of mutant calreticulin and PDIA3 with STIM1 in megakaryocytes from patients expressing calreticulin mutants, which was suggested to activate STIM-1 and induce SOCE (Di Buduo et al. 2020). The MPN mutant calreticulins are known to be secreted from cells (Han et al. 2016; Garbati et al. 2016; Arshad and Cresswell 2018), owing to the absence of the KDEL sequence. The resulting calreticulin insufficiency in the ER is implicated in enhanced SOCE activation (Di Buduo et al. 2020). PDIA3 is previously known to interact with STIM1 and regulate its activity, but a direct role for calreticulin or calreticulin-PDIA3 complexes in the regulation of STIM1 activity remains to be demonstrated. While some studies have suggested that the MPN mutants are defective for PDIA3 binding (Elf et al. 2018), the MPN mutations do not affect the PDIA3 binding site of calreticulin, which, as noted above, is located at the tip of the calreticulin P domain (Fig. 7.1C). Thus, any effects of the calreticulin MPN mutations on PDIA3 binding are expected to be indirect rather than direct.

Mutant calreticulins are secreted and detectable *in vitro* in cell supernatants (Garbati et al. 2016; Han et al. 2016; Arshad and Cresswell 2018), in the plasma of mice expressing calreticulin mutants, and in patient serum (Liu et al. 2020; Sollazzo et al. 2016). As noted above, calreticulin is expressed on the surface of apoptotic cells under some cell death conditions via calcium-dependent binding to phosphatidylserine and engages phagocyte receptors to enhance the phagocytosis of apoptotic cells (Fig. 7.4A). This process is thought to be important for priming of $CD8^+$ T cell responses in antigens derived from dying cells such as tumor cells (Obeid et al. 2007). In mice, MPN mutants, which are secreted, are shown to interfere with this process by binding to phagocyte receptors and blocking the calreticulin-dependent uptake of dying cells (Fig. 7.4B), thereby interfering with the priming of immune response (Liu et al. 2020). In this manner, MPN mutants are suggested to interfere with the induction of tumor-specific immune responses. Since the C-terminal domain of calreticulin is relevant to calcium-dependent binding to phosphatidylserine on apoptotic cells (Fig. 7.4A) (Wijeyesakere et al. 2016), it is likely that MPN mutant calreticulin will engage the cell surface of dying cells via different mechanisms than wild type calreticulin, which could also block induction of tumor-specific immune response (Fig. 7.4B).

Notably, due to their enhanced secretion, MPN mutant calreticulin are not efficiently incorporated into the PLC, and MHC class I assembly is suboptimal in the context of mutant calreticulins. Engineering of a KDEL sequence into the mutant calreticulin induced incorporation into the PLC, but nonetheless disrupted the ability of the PLC to induce MHC-I assembly (Arshad and Cresswell 2018), likely due to disruptions in contact between the calreticulin C-terminus and tapasin, a point of protein-protein interactions within the PLC (Fig. 7.1D) (Blees et al. 2017).

7.7 Summary Points and Future Issues

- Calreticulin shares its function with calnexin in maintaining ER homeostasis by folding glycoprotein substrates and inhibiting their misfolding and insolubility.
- Three-dimensional structural studies and other biophysical studies have provided insights into the substrate binding properties of calreticulin and its interactions with co-chaperones. Further studies are needed to delineate the structure of acidic C-domain of wild type calreticulin and the basic C-domains of MPN mutants.
- Apart from protein folding, calreticulin is also important for maintaining ER Ca^{2+} homeostasis.
- Calreticulin C-domain frameshift mutations in MPN alter its localization and induce a binding preference for Mpl. The localization and sequence changes affect cellular calcium signaling, MHC class I assembly, and in mice, inhibit the abilities of dendritic cells to phagocytose dying tumor cells. Further understanding of these processes will allow for the development of more targeted approaches to combat MPN and provide insights into critical cellular functions of this ubiquitous chaperone.

Acknowledgements We thank the many contributors to the field and all present and previous members of the Raghavan lab for their contributions. We thank Dr. Jie Geng for assistance with Fig. 7.1. This work was supported by the National Institute of Allergy and Infectious Diseases of the National Institutes of Health, United States (Grants RO1AI123957 and R01AI044115 to M.R), by the National Institute of Diabetes and Digestive and Kidney Diseases of the National Institutes of Health, United States (Grant RO1 DK46409 to LS), the Juvenile Diabetes Research Foundation (to LS) and the University of Michigan Fast Forward Protein Folding Diseases Initiative (to MR and LS). Work in the Satin Lab is also supported by HIRN 1U01DK120463-01.

References

Afshar N, Black BE, Paschal BM (2005) Retrotranslocation of the chaperone calreticulin from the endoplasmic reticulum lumen to the cytosol. Mol Cell Biol 25(20):8844–8853. https://doi.org/10.1128/MCB.25.20.8844-8853.2005

Araki M, Yang Y, Masubuchi N, Hironaka Y, Takei H, Morishita S, Mizukami Y, Kan S, Shirane S, Edahiro Y, Sunami Y, Ohsaka A, Komatsu N (2016) Activation of the thrombopoietin receptor by mutant calreticulin in CALR-mutant myeloproliferative neoplasms. Blood 127 (10):1307–1316. https://doi.org/10.1182/blood-2015-09-671172

Araki M, Yang Y, Imai M, Mizukami Y, Kihara Y, Sunami Y, Masubuchi N, Edahiro Y, Hironaka Y, Osaga S, Ohsaka A, Komatsu N (2019) Homomultimerization of mutant calreticulin is a prerequisite for MPL binding and activation. Leukemia 33(1):122–131. https://doi.org/10.1038/s41375-018-0181-2

Arber DA, Orazi A, Hasserjian R, Thiele J, Borowitz MJ, Le Beau MM, Bloomfield CD, Cazzola M, Vardiman JW (2016) The 2016 revision to the World Health Organization classification of myeloid neoplasms and acute leukemia. Blood 127(20):2391–2405. https://doi.org/10.1182/blood-2016-03-643544

Arnaudeau S, Frieden M, Nakamura K, Castelbou C, Michalak M, Demaurex N (2002) Calreticulin differentially modulates calcium uptake and release in the endoplasmic reticulum and mitochondria. J Biol Chem 277(48):46696–46705. https://doi.org/10.1074/jbc.M202395200

Arshad N, Cresswell P (2018) Tumor-associated calreticulin variants functionally compromise the peptide loading complex and impair its recruitment of MHC-I. J Biol Chem 293 (25):9555–9569. https://doi.org/10.1074/jbc.RA118.002836

Asher S, McLornan DP, Harrison CN (2020) Current and future therapies for myelofibrosis. Blood Rev 42:100715. https://doi.org/10.1016/j.blre.2020.100715

Avezov E, Konno T, Zyryanova A, Chen W, Laine R, Crespillo-Casado A, Melo EP, Ushioda R, Nagata K, Kaminski CF, Harding HP, Ron D (2015) Retarded PDI diffusion and a reductive shift in poise of the calcium depleted endoplasmic reticulum. BMC Biol 13:2. https://doi.org/10.1186/s12915-014-0112-2

Baksh S, Michalak M (1991) Expression of calreticulin in Escherichia coli and identification of its Ca2+ binding domains. J Biol Chem 266(32):21458–21465

Baksh S, Burns K, Andrin C, Michalak M (1995) Interaction of calreticulin with protein disulfide isomerase. J Biol Chem 270(52):31338–31344. https://doi.org/10.1074/jbc.270.52.31338

Bastianutto C, Clementi E, Codazzi F, Podini P, De Giorgi F, Rizzuto R, Meldolesi J, Pozzan T (1995) Overexpression of calreticulin increases the Ca2+ capacity of rapidly exchanging Ca2+ stores and reveals aspects of their lumenal microenvironment and function. J Cell Biol 130 (4):847–855. https://doi.org/10.1083/jcb.130.4.847

Berridge MJ (2016) The inositol trisphosphate/calcium signaling pathway in health and disease. Physiol Rev 96(4):1261–1296. https://doi.org/10.1152/physrev.00006.2016

Blees A, Januliene D, Hofmann T, Koller N, Schmidt C, Trowitzsch S, Moeller A, Tampe R (2017) Structure of the human MHC-I peptide-loading complex. Nature 551(7681):525–528. https://doi.org/10.1038/nature24627

Blum JS, Wearsch PA, Cresswell P (2013) Pathways of antigen processing. Annu Rev Immunol 31:443–473. https://doi.org/10.1146/annurev-immunol-032712-095910

Booth C, Koch GL (1989) Perturbation of cellular calcium induces secretion of luminal ER proteins. Cell 59(4):729–737

Brockmeier A, Williams DB (2006) Potent lectin-independent chaperone function of calnexin under conditions prevalent within the lumen of the endoplasmic reticulum. Biochemistry 45 (42):12906–12916. https://doi.org/10.1021/bi0614378

Calles-Garcia D, Yang M, Soya N, Melero R, Menade M, Ito Y, Vargas J, Lukacs GL, Kollman JM, Kozlov G, Gehring K (2017) Single-particle electron microscopy structure of UDP-glucose: glycoprotein glucosyltransferase suggests a selectivity mechanism for misfolded proteins. J Biol Chem 292(27):11499–11507. https://doi.org/10.1074/jbc.M117.789495

Chachoua I, Pecquet C, El-Khoury M, Nivarthi H, Albu RI, Marty C, Gryshkova V, Defour JP, Vertenoeil G, Ngo A, Koay A, Raslova H, Courtoy PJ, Choong ML, Plo I, Vainchenker W, Kralovics R, Constantinescu SN (2016) Thrombopoietin receptor activation by myeloproliferative neoplasm associated calreticulin mutants. Blood 127(10):1325–1335. https://doi.org/10.1182/blood-2015-11-681932

Charonis AS, Michalak M, Groenendyk J, Agellon LB (2017) Endoplasmic reticulum in health and disease: the 12th International Calreticulin Workshop, Delphi, Greece. J Cell Mol Med 21 (12):3141–3149. https://doi.org/10.1111/jcmm.13413

Chouquet A, Paidassi H, Ling WL, Frachet P, Houen G, Arlaud GJ, Gaboriaud C (2011) X-ray structure of the human calreticulin globular domain reveals a peptide-binding area and suggests a multi-molecular mechanism. PLoS One 6(3):e17886. https://doi.org/10.1371/journal.pone.0017886

Clapham DE (2007) Calcium signaling. Cell 131(6):1047–1058. https://doi.org/10.1016/j.cell.2007.11.028

Conte IL, Keith N, Gutierrez-Gonzalez C, Parodi AJ, Caramelo JJ (2007) The interplay between calcium and the in vitro lectin and chaperone activities of calreticulin. Biochemistry 46 (15):4671–4680. https://doi.org/10.1021/bi6026456

Del Cid N, Jeffery E, Rizvi SM, Stamper E, Peters LR, Brown WC, Provoda C, Raghavan M (2010) Modes of calreticulin recruitment to the major histocompatibility complex class I assembly pathway. J Biol Chem 285(7):4520–4535. https://doi.org/10.1074/jbc.M109.085407

Di Buduo CA, Moccia F, Battiston M, De Marco L, Mazzucato M, Moratti R, Tanzi F, Balduini A (2014) The importance of calcium in the regulation of megakaryocyte function. Haematologica 99(4):769–778. https://doi.org/10.3324/haematol.2013.096859

Di Buduo CA, Abbonante V, Marty C, Moccia F, Rumi E, Pietra D, Soprano PM, Lim D, Cattaneo D, Iurlo A, Gianelli U, Barosi G, Rosti V, Plo I, Cazzola M, Balduini A (2020) Defective interaction of mutant calreticulin and SOCE in megakaryocytes from patients with myeloproliferative neoplasms. Blood 135(2):133–144. https://doi.org/10.1182/blood.2019001103

Dong G, Wearsch PA, Peaper DR, Cresswell P, Reinisch KM (2009) Insights into MHC class I peptide loading from the structure of the tapasin-ERp57 thiol oxidoreductase heterodimer. Immunity 30(1):21–32. https://doi.org/10.1016/j.immuni.2008.10.018

Elf S, Abdelfattah NS, Chen E, Perales-Paton J, Rosen EA, Ko A, Peisker F, Florescu N, Giannini S, Wolach O, Morgan EA, Tothova Z, Losman JA, Schneider RK, Al-Shahrour F, Mullally A (2016) Mutant calreticulin requires both its mutant C-terminus and the thrombopoietin receptor for oncogenic transformation. Cancer Discov 6(4):368–381. https://doi.org/10.1158/2159-8290.CD-15-1434

Elf S, Abdelfattah NS, Baral AJ, Beeson D, Rivera JF, Ko A, Florescu N, Birrane G, Chen E, Mullally A (2018) Defining the requirements for the pathogenic interaction between mutant calreticulin and MPL in MPN. Blood 131(7):782–786. https://doi.org/10.1182/blood-2017-08-800896

Ellgaard L, Riek R, Braun D, Herrmann T, Helenius A, Wuthrich K (2001a) Three-dimensional structure topology of the calreticulin P-domain based on NMR assignment. FEBS Lett 488 (1-2):69–73

Ellgaard L, Riek R, Herrmann T, Guntert P, Braun D, Helenius A, Wuthrich K (2001b) NMR structure of the calreticulin P-domain. Proc Natl Acad Sci U S A 98(6):3133–3138. https://doi.org/10.1073/pnas.051630098

Frickel EM, Riek R, Jelesarov I, Helenius A, Wuthrich K, Ellgaard L (2002) TROSY-NMR reveals interaction between ERp57 and the tip of the calreticulin P-domain. Proc Natl Acad Sci U S A 99(4):1954–1959. https://doi.org/10.1073/pnas.042699099

Garbati MR, Welgan CA, Landefeld SH, Newell LF, Agarwal A, Dunlap JB, Chourasia TK, Lee H, Elferich J, Traer E, Rattray R, Cascio MJ, Press RD, Bagby GC, Tyner JW, Druker BJ, Dao KH (2016) Mutant calreticulin-expressing cells induce monocyte hyperreactivity through a paracrine mechanism. Am J Hematol 91(2):211–219. https://doi.org/10.1002/ajh.24245

Gardai SJ, McPhillips KA, Frasch SC, Janssen WJ, Starefeldt A, Murphy-Ullrich JE, Bratton DL, Oldenborg PA, Michalak M, Henson PM (2005) Cell-surface calreticulin initiates clearance of viable or apoptotic cells through trans-activation of LRP on the phagocyte. Cell 123(2):321–334

Guo L, Nakamura K, Lynch J, Opas M, Olson EN, Agellon LB, Michalak M (2002) Cardiac-specific expression of calcineurin reverses embryonic lethality in calreticulin-deficient mouse. J Biol Chem 277(52):50776–50779. https://doi.org/10.1074/jbc.M209900200

Han L, Schubert C, Kohler J, Schemionek M, Isfort S, Brummendorf TH, Koschmieder S, Chatain N (2016) Calreticulin-mutant proteins induce megakaryocytic signaling to transform hematopoietic cells and undergo accelerated degradation and Golgi-mediated secretion. J Hematol Oncol 9(1):45. https://doi.org/10.1186/s13045-016-0275-0

Hebert DN, Foellmer B, Helenius A (1995) Glucose trimming and reglucosylation determine glycoprotein association with calnexin in the endoplasmic reticulum. Cell 81(3):425–433. https://doi.org/10.1016/0092-8674(95)90395-x

Hebert DN, Zhang JX, Chen W, Foellmer B, Helenius A (1997) The number and location of glycans on influenza hemagglutinin determine folding and association with calnexin and calreticulin. J Cell Biol 139(3):613–623

How J, Hobbs GS, Mullally A (2019) Mutant calreticulin in myeloproliferative neoplasms. Blood 134(25):2242–2248. https://doi.org/10.1182/blood.2019000622

Ihara Y, Cohen-Doyle MF, Saito Y, Williams DB (1999) Calnexin discriminates between protein conformational states and functions as a molecular chaperone in vitro. Mol Cell 4(3):331–341

Jeffery E, Peters LR, Raghavan M (2011) The polypeptide binding conformation of calreticulin facilitates its cell surface expression under conditions of ER stress. J Biol Chem 286 (4):2402–2415. https://doi.org/10.1074/jbc.M110.180877

Jorgensen CS, Ryder LR, Steino A, Hojrup P, Hansen J, Beyer NH, Heegaard NH, Houen G (2003) Dimerization and oligomerization of the chaperone calreticulin. Eur J Biochem 270 (20):4140–4148. https://doi.org/10.1046/j.1432-1033.2003.03808.x

Klampfl T, Gisslinger H, Harutyunyan AS, Nivarthi H, Rumi E, Milosevic JD, Them NC, Berg T, Gisslinger B, Pietra D, Chen D, Vladimer GI, Bagienski K, Milanesi C, Casetti IC, Sant'Antonio E, Ferretti V, Elena C, Schischlik F, Cleary C, Six M, Schalling M, Schonegger A, Bock C, Malcovati L, Pascutto C, Superti-Furga G, Cazzola M, Kralovics R (2013) Somatic mutations of calreticulin in myeloproliferative neoplasms. N Engl J Med 369 (25):2379–2390. https://doi.org/10.1056/NEJMoa1311347

Kornfeld R, Kornfeld S (1985) Assembly of asparagine-linked oligosaccharides. Annu Rev Biochem 54:631–664. https://doi.org/10.1146/annurev.bi.54.070185.003215

Kozlov G, Bastos-Aristizabal S, Maattanen P, Rosenauer A, Zheng F, Killikelly A, Trempe JF, Thomas DY, Gehring K (2010a) Structural basis of cyclophilin B binding by the calnexin/calreticulin P-domain. J Biol Chem 285(46):35551–35557. https://doi.org/10.1074/jbc.M110.160101

Kozlov G, Pocanschi CL, Rosenauer A, Bastos-Aristizabal S, Gorelik A, Williams DB, Gehring K (2010b) Structural basis of carbohydrate recognition by calreticulin. J Biol Chem 285 (49):38612–38620. https://doi.org/10.1074/jbc.M110.168294

Kozlov G, Munoz-Escobar J, Castro K, Gehring K (2017) Mapping the ER interactome: the P domains of calnexin and calreticulin as plurivalent adapters for foldases and chaperones. Structure 25(9):1415–1422 e1413. https://doi.org/10.1016/j.str.2017.07.010

Lanner JT, Georgiou DK, Joshi AD, Hamilton SL (2010) Ryanodine receptors: structure, expression, molecular details, and function in calcium release. Cold Spring Harb Perspect Biol 2(11): a003996. https://doi.org/10.1101/cshperspect.a003996

Lewis RS (2020) Store-operated calcium channels: from function to structure and back again. Cold Spring Harb Perspect Biol 12(5):a035055. https://doi.org/10.1101/cshperspect.a035055

Li Z, Stafford WF, Bouvier M (2001) The metal ion binding properties of calreticulin modulate its conformational flexibility and thermal stability. Biochemistry 40(37):11193–11201. https://doi.org/10.1021/bi010948l

Li J, Prins D, Park HJ, Grinfeld J, Gonzalez-Arias C, Loughran S, Dovey OM, Klampfl T, Bennett C, Hamilton TL, Pask DC, Sneade R, Williams M, Aungier J, Ghevaert C, Vassiliou GS, Kent DG, Green AR (2018) Mutant calreticulin knockin mice develop thrombocytosis and myelofibrosis without a stem cell self-renewal advantage. Blood 131(6):649–661. https://doi.org/10.1182/blood-2017-09-806356

Lievremont JP, Rizzuto R, Hendershot L, Meldolesi J (1997) BiP, a major chaperone protein of the endoplasmic reticulum lumen, plays a direct and important role in the storage of the rapidly exchanging pool of Ca2+. J Biol Chem 272(49):30873–30879

Liu P, Zhao L, Loos F, Marty C, Xie W, Martins I, Lachkar S, Qu B, Waeckel-Enee E, Plo I, Vainchenker W, Perez F, Rodriguez D, Lopez-Otin C, van Endert P, Zitvogel L, Kepp O,

Kroemer G (2020) Immunosuppression by mutated calreticulin released from malignant cells. Mol Cell 77(4):748–760 e749. https://doi.org/10.1016/j.molcel.2019.11.004

Mancino L, Rizvi SM, Lapinski PE, Raghavan M (2002) Calreticulin recognizes misfolded HLA-A2 heavy chains. Proc Natl Acad Sci U S A 99(9):5931–5936

Marneth AE, Mullally A (2020) The molecular genetics of myeloproliferative neoplasms. Cold Spring Harb Perspect Med 10(2):a034876. https://doi.org/10.1101/cshperspect.a034876

Marty C, Pecquet C, Nivarthi H, El-Khoury M, Chachoua I, Tulliez M, Villeval JL, Raslova H, Kralovics R, Constantinescu SN, Plo I, Vainchenker W (2016) Calreticulin mutants in mice induce an MPL-dependent thrombocytosis with frequent progression to myelofibrosis. Blood 127(10):1317–1324. https://doi.org/10.1182/blood-2015-11-679571

Mery L, Mesaeli N, Michalak M, Opas M, Lew DP, Krause KH (1996) Overexpression of calreticulin increases intracellular Ca2+ storage and decreases store-operated Ca2+ influx. J Biol Chem 271(16):9332–9339. https://doi.org/10.1074/jbc.271.16.9332

Mesaeli N, Nakamura K, Zvaritch E, Dickie P, Dziak E, Krause KH, Opas M, MacLennan DH, Michalak M (1999) Calreticulin is essential for cardiac development. J Cell Biol 144 (5):857–868. https://doi.org/10.1083/jcb.144.5.857

Molinari M, Eriksson KK, Calanca V, Galli C, Cresswell P, Michalak M, Helenius A (2004) Contrasting functions of calreticulin and calnexin in glycoprotein folding and ER quality control. Mol Cell 13(1):125–135. https://doi.org/10.1016/s1097-2765(03)00494-5

Moreau C, Cioci G, Iannello M, Laffly E, Chouquet A, Ferreira A, Thielens NM, Gaboriaud C (2016) Structures of parasite calreticulins provide insights into their flexibility and dual carbohydrate/peptide-binding properties. IUCrJ 3(Pt 6):408–419. https://doi.org/10.1107/S2052252516012847

Nakamura K, Zuppini A, Arnaudeau S, Lynch J, Ahsan I, Krause R, Papp S, De Smedt H, Parys JB, Muller-Esterl W, Lew DP, Krause KH, Demaurex N, Opas M, Michalak M (2001) Functional specialization of calreticulin domains. J Cell Biol 154(5):961–972. https://doi.org/10.1083/jcb.200102073

Nangalia J, Massie CE, Baxter EJ, Nice FL, Gundem G, Wedge DC, Avezov E, Li J, Kollmann K, Kent DG, Aziz A, Godfrey AL, Hinton J, Martincorena I, Van Loo P, Jones AV, Guglielmelli P, Tarpey P, Harding HP, Fitzpatrick JD, Goudie CT, Ortmann CA, Loughran SJ, Raine K, Jones DR, Butler AP, Teague JW, O'Meara S, McLaren S, Bianchi M, Silber Y, Dimitropoulou D, Bloxham D, Mudie L, Maddison M, Robinson B, Keohane C, Maclean C, Hill K, Orchard K, Tauro S, Du MQ, Greaves M, Bowen D, Huntly BJ, Harrison CN, Cross NC, Ron D, Vannucchi AM, Papaemmanuil E, Campbell PJ, Green AR (2013) Somatic CALR mutations in myeloproliferative neoplasms with nonmutated JAK2. N Engl J Med 369(25):2391–2405. https://doi.org/10.1056/NEJMoa1312542

Obeid M, Tesniere A, Ghiringhelli F, Fimia GM, Apetoh L, Perfettini JL, Castedo M, Mignot G, Panaretakis T, Casares N, Metivier D, Larochette N, van Endert P, Ciccosanti F, Piacentini M, Zitvogel L, Kroemer G (2007) Calreticulin exposure dictates the immunogenicity of cancer cell death. Nat Med 13(1):54–61

Pecquet C, Chachoua I, Roy A, Balligand T, Vertenoeil G, Leroy E, Albu RI, Defour JP, Nivarthi H, Hug E, Xu E, Ould-Amer Y, Mouton C, Colau D, Vertommen D, Shwe MM, Marty C, Plo I, Vainchenker W, Kralovics R, Constantinescu SN (2019) Calreticulin mutants as oncogenic rogue chaperones for TpoR and traffic-defective pathogenic TpoR mutants. Blood 133(25):2669–2681. https://doi.org/10.1182/blood-2018-09-874578

Peters LR, Raghavan M (2011) Endoplasmic reticulum calcium depletion impacts chaperone secretion, innate immunity, and phagocytic uptake of cells. J Immunol 187(2):919–931. https://doi.org/10.4049/jimmunol.1100690

Pietra D, Rumi E, Ferretti VV, Di Buduo CA, Milanesi C, Cavalloni C, Sant'Antonio E, Abbonante V, Moccia F, Casetti IC, Bellini M, Renna MC, Roncoroni E, Fugazza E, Astori C, Boveri E, Rosti V, Barosi G, Balduini A, Cazzola M (2016) Differential clinical effects of different mutation subtypes in CALR-mutant myeloproliferative neoplasms. Leukemia 30(2):431–438. https://doi.org/10.1038/leu.2015.277

Pocanschi CL, Kozlov G, Brockmeier U, Brockmeier A, Williams DB, Gehring K (2011) Structural and functional relationships between the lectin and arm domains of calreticulin. J Biol Chem 286(31):27266–27277. https://doi.org/10.1074/jbc.M111.258467

Prins D, Groenendyk J, Touret N, Michalak M (2011) Modulation of STIM1 and capacitative Ca2+ entry by the endoplasmic reticulum luminal oxidoreductase ERp57. EMBO Rep 12 (11):1182–1188. https://doi.org/10.1038/embor.2011.173

Rauch F, Prud'homme J, Arabian A, Dedhar S, St-Arnaud R (2000) Heart, brain, and body wall defects in mice lacking calreticulin. Exp Cell Res 256(1):105–111

Rizvi SM, Mancino L, Thammavongsa V, Cantley R, Raghavan M (2004) A polypeptide-binding conformation of calreticulin is induced by heat-shock, calcium depletion, or by mutation of the C-terminal acidic domain. Mol Cell 15:913–923

Rodan AR, Simons JF, Trombetta ES, Helenius A (1996) N-linked oligosaccharides are necessary and sufficient for association of glycosylated forms of bovine RNase with calnexin and calreticulin. EMBO J 15(24):6921–6930

Saito Y, Ihara Y, Leach MR, Cohen-Doyle MF, Williams DB (1999) Calreticulin functions in vitro as a molecular chaperone for both glycosylated and non-glycosylated proteins. EMBO J 18 (23):6718–6729. https://doi.org/10.1093/emboj/18.23.6718

Schieber M, Crispino JD, Stein B (2019) Myelofibrosis in 2019: moving beyond JAK2 inhibition. Blood Cancer J 9(9):74. https://doi.org/10.1038/s41408-019-0236-2

Schrag JD, Bergeron JJ, Li Y, Borisova S, Hahn M, Thomas DY, Cygler M (2001) The structure of calnexin, an ER chaperone involved in quality control of protein folding. Mol Cell 8 (3):633–644. https://doi.org/10.1016/s1097-2765(01)00318-5

Sollazzo D, Forte D, Polverelli N, Perricone M, Romano M, Luatti S, Vianelli N, Cavo M, Palandri F, Catani L (2016) Circulating Calreticulin Is Increased in Myelofibrosis: Correlation with Interleukin-6 Plasma Levels, Bone Marrow Fibrosis, and Splenomegaly. Mediat Inflamm 2016:5860657. https://doi.org/10.1155/2016/5860657

Sonnichsen B, Fullekrug J, Nguyen Van P, Diekmann W, Robinson DG, Mieskes G (1994) Retention and retrieval: both mechanisms cooperate to maintain calreticulin in the endoplasmic reticulum. J Cell Sci 107(Pt 10):2705–2717

Taylor SC, Ferguson AD, Bergeron JJ, Thomas DY (2004) The ER protein folding sensor UDP-glucose glycoprotein-glucosyltransferase modifies substrates distant to local changes in glycoprotein conformation. Nat Struct Mol Biol 11(2):128–134. https://doi.org/10.1038/nsmb715

Thammavongsa V, Mancino L, Raghavan M (2005) Polypeptide substrate recognition by calnexin requires specific conformations of the calnexin protein. J Biol Chem 280(39):33497–33505. https://doi.org/10.1074/jbc.M503648200

Toyoshima C, Inesi G (2004) Structural basis of ion pumping by Ca2+-ATPase of the sarcoplasmic reticulum. Annu Rev Biochem 73:269–292. https://doi.org/10.1146/annurev.biochem.73.011303.073700

Villamil Giraldo AM, Lopez Medus M, Gonzalez Lebrero M, Pagano RS, Labriola CA, Landolfo L, Delfino JM, Parodi AJ, Caramelo JJ (2010) The structure of calreticulin C-terminal domain is modulated by physiological variations of calcium concentration. J Biol Chem 285 (7):4544–4553. https://doi.org/10.1074/jbc.M109.034512

Wearsch PA, Peaper DR, Cresswell P (2011) Essential glycan-dependent interactions optimize MHC class I peptide loading. Proc Natl Acad Sci U S A 108(12):4950–4955. https://doi.org/10.1073/pnas.1102524108

Wijeyesakere SJ, Gafni AA, Raghavan M (2011) Calreticulin is a thermostable protein with distinct structural responses to different divalent cation environments. J Biol Chem 286:8771–8785. https://doi.org/10.1074/jbc.M110.169193

Wijeyesakere SJ, Rizvi SM, Raghavan M (2013) Glycan-dependent and -independent interactions contribute to cellular substrate recruitment by calreticulin. J Biol Chem 288(49):35104–35116. https://doi.org/10.1074/jbc.M113.507921

Wijeyesakere SJ, Gagnon JK, Arora K, Brooks CL III, Raghavan M (2015) Regulation of calreticulin-major histocompatibility complex (MHC) class I interactions by ATP. Proc Natl Acad Sci U S A 112(41):E5608–E5617. https://doi.org/10.1073/pnas.1510132112

Wijeyesakere SJ, Bedi SK, Huynh D, Raghavan M (2016) The C-terminal acidic region of calreticulin mediates phosphatidylserine binding and apoptotic cell phagocytosis. J Immunol 196(9):3896–3909. https://doi.org/10.4049/jimmunol.1502122

Zimmerman KA, Graham LV, Pallero MA, Murphy-Ullrich JE (2013) Calreticulin regulates transforming growth factor-beta-stimulated extracellular matrix production. J Biol Chem 288 (20):14584–14598. https://doi.org/10.1074/jbc.M112.447243

Chapter 8
Impact of Calreticulin and Its Mutants on Endoplasmic Reticulum Function in Health and Disease

Najla Arshad and Peter Cresswell

Abstract The endoplasmic reticulum (ER) performs key cellular functions including protein synthesis, lipid metabolism and signaling. While these functions are spatially isolated in structurally distinct regions of the ER, there is cross-talk between the pathways. One vital player that is involved in ER function is the ER-resident protein calreticulin (CALR). It is a calcium ion-dependent lectin chaperone that primarily assists in glycoprotein synthesis in the ER as part of the protein quality control machinery. CALR also buffers calcium ion release and mediates other glycan-independent protein interactions. Mutations in CALR have been reported in a subset of chronic blood tumors called myeloproliferative neoplasms. The mutations consist of insertions or deletions in the CALR gene that all cause a + 1 bp shift in the reading frame and lead to a dramatic alteration of the amino acid sequence of the C-terminal domain of CALR. This alters CALR function and affects cell homeostasis. This chapter will discuss how CALR and mutant CALR affect ER health and disease.

Keywords Endoplasmic reticulum · Protein folding · Lipid biosynthesis · Calcium signaling · Calreticulin · Myeloproliferative neoplasms

Abbreviations

ATF6	Activating factor 6
CALR	Calreticulin
CALR-FS	Frameshift
CNX	Calnexin

N. Arshad (✉)
Departments of Immunobiology, Yale University School of Medicine, New Haven, CT, USA
e-mail: najla.arshad@yale.edu

P. Cresswell
Departments of Immunobiology, Yale University School of Medicine, New Haven, CT, USA

Department of Cell Biology, Yale University School of Medicine, New Haven, CT, USA

L. B. Agellon, M. Michalak (eds.), *Cellular Biology of the Endoplasmic Reticulum*, Progress in Molecular and Subcellular Biology 59,
https://doi.org/10.1007/978-3-030-67696-4_8

CTD	C-terminal domain
ER	Endoplasmic reticulum
ERAD	ER-Associated degradation
ERES	ER exit sites
ERGIC	ER-Golgi intermediate compartments
ET	Essential thrombocythemia
GPI	Glycosylphosphatidylinositol
IRE1	Inositol-requiring enzyme 1
JAK2	Janus kinase 2
MHC-I	Major histocompatibility complex I
MPL	Myeloproliferative leukemia virus oncogene
MPN	Myeloproliferative neoplasms
MPO	Myeloperoxidase
PERK	PKR-like endoplasmic reticulum kinase
PLC	Peptide loading complex
PMF	Primary myelofibrosis
PV	Polycythemia vera
RAV	Ribosome-associated vesicles
RER	Rough endoplasmic reticulum
SER	Smooth endoplasmic reticulum
SOCE	Store-operated Ca2+ entry
SREBP	Sterol regulatory-element binding protein
STAT	Signal transducer and activator of transcription
STIM1	Stromal interaction molecule 1
TAP	*T*ransporter *a*ssociated with antigen *p*rocessing
UPR	Unfolded protein response

8.1 Introduction

The endoplasmic reticulum is the largest organelle in a cell, accounting for ~50% of the total membrane content even though it is only ~10% of the cell volume. This structurally and functionally complex organelle is present in almost all cell types. It is the major site of lipid synthesis, calcium signaling, and protein synthesis, quality control and secretion (Alberts et al. 2002). It is therefore intuitive that disruption of ER structure or function can be detrimental to the cell and to human health. Among the proteins that are responsible for the myriad functions of the ER, we will focus on the role of calreticulin (CALR) in these processes. Many features of the ER are discussed in detail in other chapters of this book, but we will summarize them here to provide context to the function of CALR in the ER.

8.2 The Endoplasmic Reticulum

Structure The ER is a highly organized yet dynamic membrane network with distinct morphological and functional domains. It fans out through the cell from the nuclear envelope (the lumen of which is continuous with the ER) to the plasma membrane, making contacts with other organelles along the way (Phillips and Voeltz 2016). A structural domain within it is the rough ER (RER), which is comprised of flat membrane cisternae or sheets that are dotted with ribosomes, where proteins are synthesized and inserted into the ER lumen if they are destined for the secretory pathway. Such proteins progressing beyond the ER are packaged into vesicles that bud off from ER exit sites (ERES), which are devoid of ribosomes. The next destination is usually the Golgi complex via intermediary compartments such as the ER-Golgi intermediate compartments (ERGIC) (Schwarz and Blower 2016). The other structural domain is the smooth ER (SER), named so due to the paucity of associated ribosomes. It is made up of branched tubules that make contact sites with other organelles and the plasma membrane and is the main site of lipid biosynthesis (Schwarz and Blower 2016). A newly identified ER sub-compartment consisting of ribosome-associated vesicles (RAVs) was recently reported. RAVs are distinct from the ER and are found to interact with mitochondria via direct membrane contacts (Carter et al. 2020). The structural segregation of the ER correlates with distinct functions. Each region is inhabited by a unique cohort of proteins leading to structural and functional specialization.

Function The ER is involved in critical cellular processes such as lipid biogenesis, synthesis and trafficking of proteins, and calcium storage and release.

8.2.1 Lipid Synthesis and Transport

The smooth ER consists of tubules that are the site of lipid metabolism and make contacts with other organelles such as mitochondria, endosomes and the plasma membrane, tethering them to the ER (Phillips and Voeltz 2016). These sites are the hotbed of sterol and phospholipid synthesis, with lipid transport proteins exchanging lipids between organelle membranes.

Lipids are not only essential for the structure and curvature of the cell membrane but also affect protein function. Lipidation of proteins facilitates membrane localization and membrane fluidity is regulated by lipid composition, which also influences protein function, e.g. receptor clustering that affects cell signaling. Lipid moieties are also required for the biosynthesis of glycan intermediates, which are then used to modify proteins in the ER, regulating their stability and function. These may even be proteins involved in lipid biosynthesis and transport (William James et al. 2019) demonstrating the interplay between cellular processes that maintains homeostasis.

8.2.2 Protein Synthesis and Trafficking

Around a third of the proteome is targeted to the ER. This includes ER-resident proteins, plasma membrane proteins, secreted proteins and those targeted to other cellular organelles such as the lysosome (Stevenson et al. 2016). All these proteins encode a signal sequence, generally N-terminal but occasionally internal, that targets them to the surface of ER sheets, where they are co-translationally or post-translationally translocated into the ER lumen. Ribosomes dot the surface of the ER when co-translational translocation occurs, giving the ER a "rough" appearance when observed by electron microscopy, compared to the smooth ER where lipid synthesis occurs (Schwarz and Blower 2016). Once in the lumen, these proteins are further processed and delivered to their site of function. Processing includes further folding and, in the case of multi-subunit proteins, assembly, which is assisted by chaperones where disulfide bonds may be formed. Posttranslational modifications such as the addition of glyco-phosphatidyl inositol (GPI) anchors or glycans, and trimming by proteases may also occur. See Chap. 11 for more information on ER chaperones.

Glycosylation is an important posttranslational modification that has a role to play in protein folding, stability, and even regulation of protein function. It is also a means of cell-to-cell communication and adhesion. This modification commonly occurs on Asn residues or on Ser/Thr residues, called N-linked and O-linked glycosylation respectively (McCaffrey and Braakman 2016). N-linked glycosylation occurs on a specific motif (Asn-Xaa-Ser/Thr) that is recognized by the oligosaccharyl transferase enzyme, which transfers a 14-sugar oligosaccharide structure onto the protein from a lipid-linked precursor. The sugars, specifically a terminal glucose residue, serve as handles for protein folding mediated by the lectin chaperones calnexin (CNX) and CALR. These in turn associate with ERp57, a protein disulfide isomerase that also assists in protein folding (McCaffrey and Braakman 2016). The glycan structure is also subject to modification by glycosidase and glycosyl transferase enzymes present in both the lectin folding pathway and along the ER and Golgi. The glycans thus also serve as markers for glycoprotein progression along the secretory pathway, which is an indicator of functional maturity and also aids in the recognition of misfolded forms.

Misfolded proteins in the ER are detected and retrotranslocated into the cytosol for degradation by proteasomes, in a process referred to as ER-associated degradation or ERAD (McCaffrey and Braakman 2016). Together this forms the ER quality control machinery, parts of which have been co-opted by other cellular pathways such as those involved in immune function (Blum et al. 2013) and metabolite sensing (Stevenson et al. 2016). Chapter 3 discusses this function in greater detail. Once a protein is deemed folded, it transfers to ER exit sites where the protein cargo is packaged into vesicles for trafficking into the ERGIC and the Golgi complex, en route to the cell surface or destination organelle (McCaffrey and Braakman 2016).

8.2.3 ER Stress

In pathophysiological conditions such as neurodegenerative diseases (e.g. Alzheimer's disease), inflammatory bowel disease, diabetes, exposure to radiation, and various cancers (Schwarz and Blower 2016), normal protein folding is disrupted leading to the accumulation of misfolded proteins. This causes signaling from the ER that initiates an unfolded protein response or UPR. Three main stress sensors of the UPR, ATF6 (activating factor 6) α and β, PERK (PKR-like endoplasmic reticulum kinase) and IRE1 (inositol-requiring enzyme 1) α and β, are activated, leading to the transcription of genes encoding protein chaperones as well as components of ERAD required to restore ER homeostasis (Schwarz and Blower 2016). In response to stress and as part of the UPR, regions of the ER may be engulfed by autophagosomes in a selective type of autophagy called ER-phagy. This is receptor mediated and the engulfed membrane is removed by lysosomal degradation (Strzyz 2020). This topic is covered in detail in Chap. 1, and more on ER stress in various pathological conditions is found in Chaps. 4, 9, 10 and 12.

8.2.4 Calcium Signaling

Calcium ions are versatile second messengers in the cell. They play a role in an exhaustive number of physiological processes including cell adhesion, cell division, immune responses, muscle contraction, pathways mediated by extrinsic signals such as hormonal activity, environmental changes, microbial activity or cytotoxic reagents, to list a few (Carreras-Sureda et al. 2018; Berridge et al. 2000). The extracellular Ca^{2+} concentration is ~1–2 mM while the cytosolic concentration of these ions is ~100 nM, almost 4-log orders lower. Free ER calcium ion levels are estimated to be ~100–800 μM, but can be almost five-fold higher if bound Ca^{2+} is considered. When signaling occurs, bound $Ca^{2+}$is released into the cytosol from the ER, raising the concentration by ten-fold and regulating a host of proteins and enzymes as part of the cellular response. This release stimulates the influx of extracellular Ca^{2+}, in a process called store-operated Ca^{2+} entry (SOCE) (Carreras-Sureda et al. 2018). To restore homeostasis, Ca^{2+} is then exported from the cytoplasm either out of the cell or back into the ER. Proteins in the ER such as CALR serve as Ca^{2+} reserves, as they contain low-affinity, high-capacity Ca^{2+} binding domains. This chapter and Chap. 2 provide more details on this aspect of ER function.

It is clear that several key cellular processes occur on or in the ER and that these are made possible by a host of ER-associated proteins. CALR is a multi-faceted protein that is involved in many of the various ER functions. We will discuss the myriad roles of CALR in ER function how this affects human health and disease.

8.3 Calreticulin: Role in ER Function in Health and Disease

It was nearly half a century ago that that CALR was identified in the sarcoplasmic membrane of skeletal muscle cells as a luminal calcium ion binding protein. Fifteen years later, around 1990, it was cloned by many groups independently and named differently by each but a consensus was reached to name the protein calreticulin (Michalak et al. 1992). Human and mouse CALR share an amino acid sequence identity of 94%, and in fact it is highly conserved evolutionarily, present in plants and animals, from algae to higher plants (Berridge et al. 2000) and from amoebae to mammals. The notable exception is yeast, which contains a protein more closely related to the transmembrane chaperone calnexin, but it is not a homolog of CALR (Parlati et al. 1995). CALR is well expressed in all cells except erythrocytes and is primarily localized to the ER although its presence in other sub-cellular locations and roles in regulating myriad aspects of cell function are constantly emerging. An N-terminal, 180-amino acid, fragment of CALR called vasostatin is found in circulation (Pike et al. 1998).

The CALR knockout mouse is embryonic lethal, indicating that CALR expression is essential for survival (Mesaeli et al. 1999). Alterations of CALR expression due to transcriptional or protein coding mutations have been reported in many cancers (Leone et al. 2013) and neurological disorders (Kotian et al. 2019). These occur along with other driver mutations and may not have a large role in the disease mechanism. However, in 2013, a somatic, frameshift mutant of CALR (CALR-FS) was identified in myeloproliferative neoplasms (MPNs), a type of chronic blood disease (Klampfl et al. 2013; Nangalia et al. 2013). MPNs that are Philadelphia chromosome–negative and myeloproliferative leukemia virus (MPL) oncogene–negative can be further categorized into three classes: polycythemia vera (PV), essential thrombocythemia (ET), and primary myelofibrosis (PMF). An activating mutation in Janus kinase 2 (JAK2V617F) is seen in 95% of PV patients, whereas only 50–60% of patients suffering from ET or PMF have this mutation. Klampf et al. and Nangalia et al. found that the majority of ET or PMF patients with non-mutated JAK2 kinase carry insertions or deletions in exon 9 of the CALR gene. This chapter will focus on the various roles of CALR in the ER, and the pathology associated with its mutants.

CALR Structure CALR is comprised of an N-terminal lectin or glycan binding domain, a proline-rich P-domain, and a C-terminal domain (CTD) comprised mainly of acidic amino acids. The lectin domain interacts with the glycans on nascent glycoproteins and assists in their folding. CALR also associates with other proteins in a glycan-independent way via the P-domain (Kozlov et al. 2017) and the N-terminal domain (Pocanschi et al. 2011). The P-domain also has a high-affinity Ca^{2+}-binding site, which is necessary for lectin function. The CTD serves to sequester calcium ions by virtue of its negatively charged amino acids (Michalak et al. 2009). It also encodes the ER-retention sequence KDEL (Lys-Asp-Glu-Leu) and is involved in phosphatidylserine binding and cell phagocytosis (Wijeyesakere et al. 2016).

In the MPN-associated frame-shift mutant of CALR (CALR-FS), the CTD is altered as a consequence of deletions ($CALR\text{-}FS^{DEL}$) or insertions ($CALR\text{-}FS^{INS}$) in exon 9 of the gene. Over 50 different insertions or deletions have been identified, but deletions of 52 bp (Type 1) or insertions of 5 bp (Type 2) are the most common. All insertions and deletions lead to a 1-bp change that alters the reading frame, generating a novel C-terminus where the ER-retention sequence and the acidic residues are lost (almost completely in Type 1 and by half in Type 2) and replaced by basic residues. This alteration has been shown to influence the function of CALR significantly, playing a key role in disease pathology as discussed below.

CALR Function and Effect of Disease-Associated Mutations When CALR-FS was identified in MPN patients it was found to activate the JAK-STAT signaling pathway. Subsequent studies have not only identified the underlying mechanism responsible but have also uncovered unique functional properties that the mutation confers on CALR. To study the mechanisms and effects of this mutation, groups have used patient samples, human or mouse cell lines, and *in vitro* biochemical techniques. Mutant CALR transgenic mice, used by several groups, successfully recapitulate the disease (Dunbar et al. 2017). Figure 8.1 summarizes the cellular processes that involve CALR (Fig. 8.1a) and the processes that are disrupted in the presence of CALR-FS (Fig. 8.1b).

8.3.1 CALR as a Lectin Chaperone

In concert with CNX and in complex with ERp57, CALR assists in the folding of newly-synthesized glycoproteins. The lectins associate with their glycoprotein substrates via the terminal glucose carried on their newly added N-linked glycans and draw them into the lectin folding cycle (McCaffrey and Braakman 2016). The chaperone roles of CNX and CALR are somewhat redundant, although a level of client-chaperone specificity has been identified. Cells deficient in CALR exhibit an increase in ER retention of collagen and myeloperoxidase, a lysosomal antibacterial enzyme, while overexpression of CALR enhances the expression of epithelial sodium channel subunits, human insulin receptors, and sterol transporters (Jiang et al. 2014).

Glycoproteins associating with the lectin folding pathway for too long are subjected to ERAD. The glycans of folded glycoproteins do not carry the terminal glucose and are therefore released from the folding cycle, preventing prolonged, futile interactions. CALR is also spatially removed from folded proteins as it is retained in the ER due to the presence of the C-terminal KDEL sequence while its folded substrates traverse the secretory pathway. CALR-FS, however, is able to progress along the secretory pathway due to the loss of the ER-retention sequence, potentially influencing the chaperone function of CALR, thereby affecting the function of its client proteins.

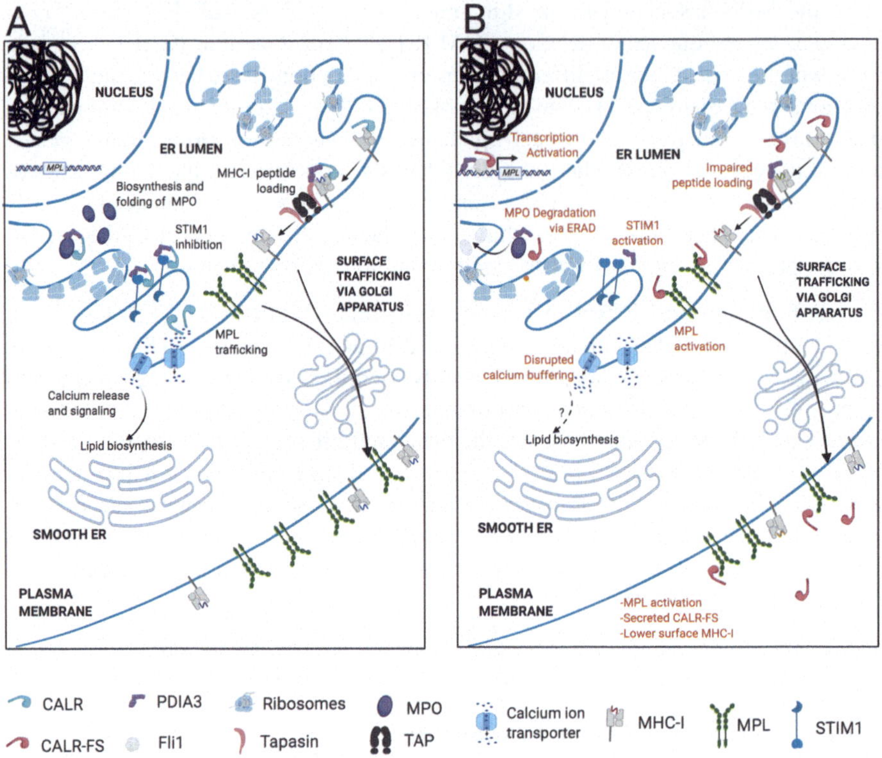

Fig. 8.1 Impact of CALR and CALR-FS on cellular processes. (**a**) Cellular functions mediated by CALR: folding of newly synthesized proteins such as MPO and MPL is mediated by CALR and ERp57 in the ER. The two chaperones interact with STIM1, maintaining it as an inactive monomer in the presence of high calcium ion concentrations in the ER. Ca^{2+} signaling regulates lipid synthesis, among other functions, and its levels are regulated by calcium ion channels and Ca^{2+} binding proteins such as CALR. CALR recruits MHC-I molecules to the PLC, which also contains ERp57, tapasin and TAP. Peptide-associated MHC-I and receptors such as MPL are trafficked to the plasma membrane via the Golgi apparatus. (**b**) Disrupted cellular functions in the presence of CALR-FS: The interaction between ERp57 and CALR-FS is reduced compared to CALR, leading to improper chaperone function and folding of proteins such as MPO and their subsequent degradation via ERAD, as well as dissociation of ERp57 from STIM1 leading to its oligomerization and activation. Reduced Ca^{2+} buffering capacity due to the mutant CALR affects Ca^{2+} signaling in the cell and may also influence lipid biosynthesis. Aberrant interaction of CALR-FS with MPL in the secretory pathway and cell surface causes its activation. Mutant CALR does not facilitate MHC-I interaction with the PLC, resulting in lower surface MHC-I levels. CALR-FS interacts with Fli1 and is recruited to the promoter of *MPL*, enhancing its transcription. CALR-FS is also secreted from the cells and present in the extracellular space

Gain of Function of CALR-FS It was known that activating mutations in MPL, the thrombopoetin receptor, drives a subset of MPNs. Like other cytokine receptors, MPL is a glycoprotein synthesized in the ER that undergoes lectin chaperone-mediated folding. Thus, the interaction between CALR-FS and MPL was investigated by three independent groups. They all demonstrated a glycan-dependent interaction between the two that allowed activation of the receptor independent of its ligand (Araki et al. 2016; Chachoua et al. 2016; Elf et al. 2016; Masubuchi et al. 2020). A mutated CALR-FS that is deficient in glycan binding failed to interact with MPL (Masubuchi et al. 2020), and mutational analysis of MPL's 4 glycosylation sites showed that the interaction of CALR is mediated by one specific glycan. This glycan normally matures to a complex, sialic acid-bearing form in the Golgi, but maturation is blocked in the presence of CALR-FS, even though both are present on the surface. This indicates that the association occurs in the secretory pathway and may persist at the plasma membrane. Even mutant MPL receptors deficient in trafficking are localized to the surface in the presence of CALR-FS, as is MPL containing an ER-retention sequence (Masubuchi et al. 2020). Activation of MPL by its normal ligand involves dimerization of the receptor, and CRT-FS similarly induces dimerization. CRT-FS does not inhibit ligand binding and also requires a hydrophobic region in MPL to mediate activation, indicating the CRT-FS activation mechanism mimics, but is distinct from, that of the ligand. CALR-FS is reported to form a multimer via its mutant CTD that may facilitate ligand-independent MPL activation (Araki et al. 2019).

To assess the importance of CALR-FS localization in associating with and activating MPL, the N-terminal ER-targeting sequence of CALR-FS was deleted. This mutant had no effect on MPL activity and in fact, accumulated in the nucleus due to the positive charges at the C-terminus that resemble a nuclear localization sequence (Masubuchi et al. 2020). Clearly, localization of CALR to the secretory pathway is necessary, but whether it is sufficient was tested by Elf et al. who deleted the ER-retention sequence of wild-type CALR and assessed the effect on MPL. Deleting this sequence allowed the secretion of CALR (Arshad and Cresswell 2018) but it failed to activate MPL (Elf et al. 2016). Replacing the basic amino acids of CALR-FS with glycine resulted in the loss of the MPL-activating gain-of-function phenotype, indicating that the basic nature of the mutant CTD is important. Expression of the CTD of CALR-FS by itself in the ER did not activate MPL, indicating the importance of the N-terminal domain (Elf et al. 2016). Together, these studies show that for the gain-of-function phenotype exhibited by CALR-FS in MPL activation CALR-FS needs: to be present in the secretory pathway, to progress through the secretory pathway, and to forge interactions with MPL via its lectin domain. However, the transformative capacity also requires the frameshift-induced change from a negative to positive electrostatic charge of the CTD, which probably influences interactions of the other domains.

Loss of Function of CALR-FS Given the anomalous chaperone function of CALR-FS, the effect of this mutation on other substrates has been investigated. An important CALR substrate is myeloperoxidase (MPO), a lysosomal glycoprotein enzyme expressed exclusively in myeloid cells that is involved in the generation of reactive oxygen species required for antimicrobial activity. Failure of MPO to mature in the secretory pathway leads to its degradation, and loss of MPO expression or function results in a disease known as MPO deficiency. Although individuals with complete MPO deficiency are very rare, they exhibit an increased incidence of severe inflammatory and infectious diseases (Theocharides et al. 2016). Theocharides et al. found that acquired MPO deficiency in MPN patients is exclusively associated with those who are homozygous for mutant CALR. Neutrophils from these patients have no detectable intracellular CALR expression, phenocopying CALR-knockout cells. Decreased MPO expression in these neutrophils was apparent and peroxidase activity was undetectable. MPO genetically introduced into mouse embryonic fibroblasts lacking CALR expression showed an increased rate of proteasomal degradation compared to wildtype cells that could be rescued by the expression of CALR. It was inferred that CALR-FS in the neutrophils of homozygous patients cannot facilitate MPO folding, leading to its degradation by ERAD and causing MPO deficiency.

Major Histocompatibility Complex I (MHC-I) proteins are comprised of a glycosylated transmembrane heavy chain and a non-glycosylated light chain, β2-microglobulin (β_2m). MHC-I-β_2m heterodimers bind proteasomally generated peptides and display them on the cell surface. This allows the immune system, in particular CD8-positive T lymphocytes, to distinguish self from non-self. This is critically important for resistance to infection and tumor surveillance. To facilitate peptide loading, newly synthesized MHC-I molecules associate in the ER with an assembly of proteins known as the peptide loading complex (PLC) (Blum et al. 2013). The PLC is comprised of the dimeric *T*ransporter associated with *A*ntigen *P*rocessing (TAP) that imports peptides from the cytosol into the ER lumen, associated tapasin, a transmembrane glycoprotein that facilitates peptide exchange by PLC-associated MHC-I molecules, and the chaperone proteins ERp57 and CALR. ERp57 is disulfide-linked to tapasin. CALR interacts with the MHC-I heavy chain via its glycan, recruiting it to the PLC. Within the PLC CALR also interacts with ERp57 via its P domain and with tapasin via its CTD, as deduced by analysis of the cryo-EM structure of the PLC (Blees et al. 2017).

We investigated the effect of CALR-FS on PLC function (Arshad and Cresswell 2018). The loss of CALR in cells led to a decrease in MHC-I localized to the surface that was rescued by the introduction of wildtype CALR but not by CALR-FS. This indicated a loss-of-function of CALR-FS that phenocopied a knockout of CALR. Indeed, very little CALR-FS was detected in the cell as most of it was secreted. Further analysis showed that PLC function was impaired, as CALR-FS was not incorporated into the complex and there was a decrease in PLC-associated MHC-I. This was not due to CRT-FS secretion because C-terminal addition of KDEL, the classic ER-retention sequence, to the mutant C-terminal region led to retention of CALR-FS in the ER. While this restored its association with the PLC it did not rescue surface MHC-I expression.

The interactions between CALR-FS and other proteins in the PLC were examined to understand the compromised PLC function (Arshad and Cresswell 2018). CALR-FS was able to interact with MHC-I via its glycan, but the interaction with ERp57 and tapasin was reduced. Analysis of the PLC structure showed that the acidic C-terminal region of wildtype CALR is accommodated in a positively-charged groove in tapasin. It is likely that the change in electrostatic charge of CALR-FS disrupts this association. Due to these alterations of protein interactions it is highly probable that the overall structural integrity of the PLC is compromised, rendering it functionally ineffective. This results in impaired peptide binding to MHC-I molecules, reducing their stability and a consequent decrease in cell surface expression. A notable result of this is that immune surveillance of MPN tumor cells is likely to be impaired. For more on the effect on immune function, refer to Chap. 5.

8.3.2 CALR Mutants and Non-glycan Mediated Protein Interactions

Our demonstration of a reduced interaction between CALR-FS and ERp57 was confirmed by other groups (Pronier et al. 2018; Elf et al. 2018) and a functional consequence of this on Ca^{2+} signaling was reported by Di Buduo et al. (discussed in the next section). In another study, Poiner et al., elucidated the interactome of CALR and CALR-FS in mouse cells. While a decrease was observed in CALR-FS interactions with ERp57 and Msi2, an RNA binding protein, there were enhanced interactions of CALR-FS with proteins of the UPR, cytoskeleton, and ribosomes. Interestingly, only CALR-FS^{DEL} showed an enhanced interaction with the transcription factor Fli1, a known regulator of *MPL* transcription. CALR was also found to be associated with Fli1, although to a lower extent, and the authors suggest that it functions to chaperone Fli1 to the nucleus. There was an elevation of Fli1 in the cytoplasm and chromatin of cells expressing CALR-FS^{DEL} compared to cells expressing CALR or CALR-FS^{INS}. This was also seen in mononuclear cells from CALR-FS^{DEL}-positive MPN patients when compared to healthy donors. An increase in the mRNA levels of *MPL* was observed, likely due to enhanced recruitment of Fli1, and perhaps CALR-FS^{DEL}, to the *MPL* promoter. An increase in Fli1 at promoters and transcription start sites of other genes it regulates was also observed. Thus CALR-FS^{DEL} may permit irregular transcriptional changes due to an enhanced cytosolic interaction with the *MPL* transcription factor Fli1.

8.3.3 CALR and Ca^{2+} Signaling

The negatively charged C-terminal region of CALR serves as a reserve for cellular Ca^{2+} and accounts for nearly 50% of the ER luminal Ca^{2+} content. Ablation of the *Calr* gene in all tissues in an attempt to make a calreticulin-deficient mouse resulted in an embryonic lethal phenotype. Further investigation revealed that cardiogenesis was compromised due to impaired calcium signaling that hampered activation of Ca^{2+}-dependent transcriptional pathways, resulting in embryonic death or failure of embryonic development. Expression of downstream effector proteins in cardiac cells rescued this phenotype and these mice are termed *Calr*$^{-/-}$-rescued mice (Michalak et al. 2009). The absence of the normal CTD in CALR-FS^{DEL} results in the loss of negatively charged residues that bind Ca^{2+}, while CALR-FS^{INS} retains about half the residues. The effect of this total vs partial loss of the Ca^{2+} buffering capacity of CALR-FS was studied in patients (Pietra et al. 2016). They found that Type 1 deletions were more associated with a myelofibrosis phenotype and a high risk of progression from ET to MF, while Type 2 insertional mutations were primarily associated with an ET phenotype. Given the role they play in MPNs, megakaryocytes were isolated from patients and calcium flux in them was assessed *in vitro*. Upon stimulation of Ca^{2+} release, the magnitude of change in cytosolic Ca^{2+} concentration observed in megakaryocytes from patients with Type 1 CALR mutation was much larger than the change seen in those from patients with Type 2 or JAK kinase mutations or even healthy subjects. A similar exaggerated increase of cytosolic calcium from extracellular sources due to store-operated Ca^{2+} entry was also seen only in cells harboring Type 1 mutation. These observations were in keeping with a hampered ER Ca^{2+} storage capacity, and also indicated a potential dysregulation of the store-operated calcium entry (SOCE) machinery.

Stromal interaction molecule 1 (STIM1) is a part of the SOCE system and plays a sensory and regulatory role. It is expressed in the ER and exists a monomer in the presence of Ca^{2+}. On Ca^{2+} depletion, it oligomerizes and translocates to a region proximal to the cell membrane where it activates membrane calcium channels leading to the influx of Ca^{2+} (Carreras-Sureda et al. 2018). This signaling cascade can also be initiated by MPL signaling. ERp57 regulates STIM1 by associating with it in the ER lumen in its inactive monomeric state and disassociating from it when it is activated. Di Buduo et al. showed that CALR also interacted with STIM1 in resting conditions and dissociated from it under conditions that activated SOCE, in a manner similar to that seen for ERp57 (Di Buduo et al. 2020). Since ERp57 and CALR form a co-chaperone complex, it was suggested that CALR brings ERp57 to STIM1. Thus, a disruption of the interaction between CALR and ERp57 should lead to spontaneous activation of STIM1, causing abnormal Ca^{2+} flux. This was indeed the case for CALR-FS, which has a decreased interaction with ERp57. It causes spontaneous cytosolic Ca^{2+} flux and results in abnormal megakaryocyte proliferation, a hallmark of ET, which can be reversed by a specific SOCE inhibitor. This revealed a new axis of signaling in the development of CALR-positive MPNs.

8.3.4 CALR and Lipid Synthesis

Elevated serum lipids in rescued CALR-KO mice suggested a link between altered lipid metabolism and disrupted ER Ca^{2+} homeostasis in non-cardiac tissues (Wang et al. 2017). Wang et al. investigated this phenomenon and found total cholesterol levels were not affected, but the distribution of cholesterol within the ER was altered. This resulted in the sequestering of a bigger pool of cholesterol from the proteins involved in cholesterol sensing such as Sterol Regulatory-Element Binding Proteins (SREBPs). SREBPs are transcription factors that regulate the expression of genes involved in lipid synthesis and are distributed between the nucleus and cytoplasm. In

CALR-deficient cells, increased nuclear localization of SERBPs was observed, with a concomitant increase in their transcriptional activity. This was reversed when ER Ca^{2+} levels were restored to those seen in CALR expressing cells, revealing the regulatory role of ER Ca^{2+} in lipid homeostasis. While the loss of CALR clearly has an impact on the sterol sensing mechanism, further studies need to be done to assess the impact of CALR-FS on this newly discovered pathway that regulates lipid metabolism.

8.3.5 CALR Mutants and ER Stress

Under conditions where misfolded proteins accumulate and cause ER stress, cells induce the UPR, leading to the upregulation of ERAD associated proteins and ER-phagy. Although CALR is a folding chaperone there is no activation of the UPR in CALR knockout cells. The effect of CALR-FS on the cellular response to ER stress was first studied in K562 cells, a human lymphoblast cell line the proliferation of which is not dependent on cytokines (Salati et al. 2019). There was no change in UPR response induced by hypoxia or thapsigargin, although the PERK pathway was impaired by tunicamycin. In another study, transcriptomic analysis of megakaryocytes from CALR-FS-positive MPN patients showed a robust upregulation of UPR, specifically IRE1 activation (Nam et al. 2019). A third study (Mansier et al. 2019) utilized human cell lines with megakaryocytic features (UT-7 and DAMI) that are dependent on cytokines for proliferation. They showed that expression of Type 1 or Type 2 CALR variants did not induce UPR but instead led to a global decrease in the expression of UPR target genes, similar to what was seen with the overexpression of WT CALR. Inducing ER stress by tunicamycin resulted in same UPR response across all cell lines. It was suggested that normal CALR in the heterozygous cells is sufficient to maintain cell homeostasis. Given the different conclusions reached by these three groups, there is no consensus with regard to the effect of CALR-FS on UPR and further studies are needed to fully understand the ER stress response in cells expressing these mutants.

8.3.6 CALR as a Signaling Molecule

Cellular expression of CALR (Han et al. 2019) and serum CALR levels serve as a prognostic marker for disease outcome. Vasostatin, the N-terminal 180-aa region of CALR found in circulation inhibits angiogenesis and has tumor suppressive properties (Michalak et al. 2009). The presence of cell surface CALR, induced by chemotherapeutic agents for example, serves as a pro-phagocytic signal and leads to immunogenic cell death. Macrophages have been shown to secrete CALR that coats target cells destined for phagocytosis via weak lectin interactions (Feng et al. 2015). In both these cases, the mechanism by which CALR translocates to the surface and is expelled from the cell are unclear; cleavage of the KDEL sequence and a role of CALR phosphorylation have both been suggested (Byrne et al. 2013). Several groups have shown that CALR-FS is naturally secreted (Arshad and Cresswell 2018; Garbati et al. 2016; Mansier et al. 2019) and is on the surface of the cells that make it, but this does not induce phagocytosis (Daitoku et al. 2016). A more recent study showed that it serves as a decoy, blocking the uptake of CALR-exposing cells by dendritic cells, which leads to immunosuppression and disease progression (Liu et al. 2020). Secreted CRT also acts as a chaperone cytokine, or "chaperokine", affecting the differentiation of mouse and human monocytes (Bajor et al. 2011). While CALR-FS has been shown to primarily have an autocrine effect there is some indication of a paracrine activity (Garbati et al. 2016). This finding warrants further investigation.

8.4 Summary Points

- ER as an organelle is the seat of crucial cell functions such as signaling, lipid biosynthesis, and protein synthesis and trafficking.
- Calreticulin is an ER-localized Ca^{2+}-dependent lectin chaperone that is also a Ca^{2+} binding protein and has a direct or indirect role to play in almost all ER functions.
- Mutant CALR was discovered to be the driver of a subset of MPNs by causing ligand-independent activation of MPL, activating the JAK/STAT signaling pathway.
- The mutation is a frameshift, which changes the C-terminal domain of CALR and confers new properties on the protein that affect cellular homeostasis.

8.5 Future Issues

To further our knowledge of the role of CALR in ER health and disease, gaps in our understanding of the regulation and function of the protein need to be filled.

- The mechanism of CALR localization to extra-ER sites such as the cell surface.
- Posttranslational modifications of CALR and their role in CALR function and localization.
- Clarifying the role of CALR-FS on the ER stress response and lipid metabolism.
- Enhancing our understanding of CALR and CALR-FS chaperokine functions and identifying their target receptors.
- Targeting the unique structure and protein interactions of CALR-FS for therapy.
- Screening for mutations in CALR in other diseases and elucidating their contribution to disease pathology.

Acknowledgments The cited work from this laboratory was funded by the National Institutes of Health (R01-AI09726). We thank the members of the Cresswell laboratory for valuable discussions.

Conflict-of-Interest Disclosure Statement The authors declare no conflicts of interest.

References

Alberts B, Johnson A, Lewis J, Raff M, Roberts K, Walter P (2002) Molecular biology of the cell. Garland Science, New York

Araki M, Yang Y, Masubuchi N, Hironaka Y, Takei H, Morishita S, Mizukami Y, Kan S, Shirane S, Edahiro Y, Sunami Y, Ohsaka A, Komatsu N (2016) Activation of the thrombopoietin receptor by mutant calreticulin in CALR-mutant myeloproliferative neoplasms. Blood 127 (10):1307–1316. https://doi.org/10.1182/blood-2015-09-671172

Araki M, Yang Y, Imai M, Mizukami Y, Kihara Y, Sunami Y, Masubuchi N, Edahiro Y, Hironaka Y, Osaga S, Ohsaka A, Komatsu N (2019) Homomultimerization of mutant calreticulin is a prerequisite for MPL binding and activation. Leukemia 33(1):122–131. https://doi.org/10.1038/s41375-018-0181-2

Arshad N, Cresswell P (2018) Tumor-associated calreticulin variants functionally compromise the peptide loading complex and impair its recruitment of MHC-I. J Biol Chem 293 (25):9555–9569. https://doi.org/10.1074/jbc.RA118.002836

Bajor A, Tischer S, Figueiredo C, Wittmann M, Immenschuh S, Blasczyk R, Eiz-Vesper B (2011) Modulatory role of calreticulin as chaperokine for dendritic cell-based immunotherapy. Clin Exp Immunol 165(2):220–234. https://doi.org/10.1111/j.1365-2249.2011.04423.x

Berridge MJ, Lipp P, Bootman MD (2000) The versatility and universality of calcium signalling. Nat Rev. Mol Cell Biol 1(1):11–21. https://doi.org/10.1038/35036035

Blees A, Januliene D, Hofmann T, Koller N, Schmidt C, Trowitzsch S, Moeller A, Tampe R (2017) Structure of the human MHC-I peptide-loading complex. Nature 551(7681):525–528. https://doi.org/10.1038/nature24627

Blum JS, Wearsch PA, Cresswell P (2013) Pathways of antigen processing. Annu Rev Immunol 31:443–473. https://doi.org/10.1146/annurev-immunol-032712-095910

Byrne JC, Ni Gabhann J, Stacey KB, Coffey BM, McCarthy E, Thomas W, Jefferies CA (2013) Bruton's tyrosine kinase is required for apoptotic cell uptake via regulating the phosphorylation and localization of calreticulin. J Immunol 190(10):5207–5215. https://doi.org/10.4049/jimmunol.1300057

Carreras-Sureda A, Pihan P, Hetz C (2018) Calcium signaling at the endoplasmic reticulum: fine-tuning stress responses. Cell Calcium 70:24–31. https://doi.org/10.1016/j.ceca.2017.08.004

Carter SD, Hampton CM, Langlois R, Melero R, Farino ZJ, Calderon MJ, Li W, Wallace CT, Tran NH, Grassucci RA, Siegmund SE, Pemberton J, Morgenstern TJ, Eisenman L, Aguilar JI, Greenberg NL, Levy ES, Yi E, Mitchell WG, Rice WJ, Wigge C, Pilli J, George EW, Aslanoglou D, Courel M, Freyberg RJ, Javitch JA, Wills ZP, Area-Gomez E, Shiva S, Bartolini F, Volchuk A, Murray SA, Aridor M, Fish KN, Walter P, Balla T, Fass D, Wolf SG, Watkins SC, Carazo JM, Jensen GJ, Frank J, Freyberg Z (2020) Ribosome-associated vesicles: A dynamic subcompartment of the endoplasmic reticulum in secretory cells. Sci Adv 6 (14):eaay9572. https://doi.org/10.1126/sciadv.aay9572

Chachoua I, Pecquet C, El-Khoury M, Nivarthi H, Albu RI, Marty C, Gryshkova V, Defour JP, Vertenoeil G, Ngo A, Koay A, Raslova H, Courtoy PJ, Choong ML, Plo I, Vainchenker W, Kralovics R, Constantinescu SN (2016) Thrombopoietin receptor activation by myeloproliferative neoplasm associated calreticulin mutants. Blood 127(10):1325–1335. https://doi.org/10.1182/blood-2015-11-681932

Daitoku S, Takenaka K, Yamauchi T, Yurino A, Jinnouchi F, Nunomura T, Eto T, Kamimura T, Higuchi M, Harada N, Saito N, Miyamoto T, Iwasaki H, Akashi K (2016) Calreticulin mutation does not contribute to disease progression in essential thrombocythemia by inhibiting phagocytosis. Exp Hematol 44(9):817–825 e813. https://doi.org/10.1016/j.exphem.2016.05.001

Di Buduo CA, Abbonante V, Marty C, Moccia F, Rumi E, Pietra D, Soprano PM, Lim D, Cattaneo D, Iurlo A, Gianelli U, Barosi G, Rosti V, Plo I, Cazzola M, Balduini A (2020) Defective interaction of mutant calreticulin and SOCE in megakaryocytes from patients with myeloproliferative neoplasms. Blood 135(2):133–144. https://doi.org/10.1182/blood.2019001103

Dunbar A, Nazir A, Levine R (2017) Overview of transgenic mouse models of myeloproliferative neoplasms (MPNs). Curr Protoc Pharmacol 77:14 40 11–14 40 19. https://doi.org/10.1002/cpph.23

Elf S, Abdelfattah NS, Chen E, Perales-Paton J, Rosen EA, Ko A, Peisker F, Florescu N, Giannini S, Wolach O, Morgan EA, Tothova Z, Losman JA, Schneider RK, Al-Shahrour F, Mullally A (2016) Mutant calreticulin requires both its mutant C-terminus and the thrombopoietin receptor for oncogenic transformation. Cancer Discov 6(4):368–381. https://doi.org/10.1158/2159-8290.CD-15-1434

Elf S, Abdelfattah NS, Baral AJ, Beeson D, Rivera JF, Ko A, Florescu N, Birrane G, Chen E, Mullally A (2018) Defining the requirements for the pathogenic interaction between mutant calreticulin and MPL in MPN. Blood 131(7):782–786. https://doi.org/10.1182/blood-2017-08-800896

Feng M, Chen JY, Weissman-Tsukamoto R, Volkmer JP, Ho PY, McKenna KM, Cheshier S, Zhang M, Guo N, Gip P, Mitra SS, Weissman IL (2015) Macrophages eat cancer cells using their own calreticulin as a guide: roles of TLR and Btk. Proc Natl Acad Sci U S A 112 (7):2145–2150. https://doi.org/10.1073/pnas.1424907112

Garbati MR, Welgan CA, Landefeld SH, Newell LF, Agarwal A, Dunlap JB, Chourasia TK, Lee H, Elferich J, Traer E, Rattray R, Cascio MJ, Press RD, Bagby GC, Tyner JW, Druker BJ, Dao KH (2016) Mutant calreticulin-expressing cells induce monocyte hyperreactivity through a paracrine mechanism. Am J Hematol 91(2):211–219. https://doi.org/10.1002/ajh.24245

Han A, Li C, Zahed T, Wong M, Smith I, Hoedel K, Green D, Boiko AD (2019) Calreticulin is a critical cell survival factor in malignant neoplasms. PLoS Biol 17(9):e3000402. https://doi.org/10.1371/journal.pbio.3000402

Jiang Y, Dey S, Matsunami H (2014) Calreticulin: roles in cell-surface protein expression. Membranes (Basel) 4(3):630–641. https://doi.org/10.3390/membranes4030630

Klampfl T, Gisslinger H, Harutyunyan AS, Nivarthi H, Rumi E, Milosevic JD, Them NC, Berg T, Gisslinger B, Pietra D, Chen D, Vladimer GI, Bagienski K, Milanesi C, Casetti IC, Sant'Antonio E, Ferretti V, Elena C, Schischlik F, Cleary C, Six M, Schalling M, Schonegger A, Bock C, Malcovati L, Pascutto C, Superti-Furga G, Cazzola M, Kralovics R (2013) Somatic mutations of calreticulin in myeloproliferative neoplasms. N Engl J Med 369 (25):2379–2390. https://doi.org/10.1056/NEJMoa1311347

Kotian V, Sarmah D, Kaur H, Kesharwani R, Verma G, Mounica L, Veeresh P, Kalia K, Borah A, Wang X, Dave KR, Yavagal DR, Bhattacharya P (2019) Evolving evidence of calreticulin as a pharmacological target in neurological disorders. ACS Chem Neurosci 10(6):2629–2646. https://doi.org/10.1021/acschemneuro.9b00158

Kozlov G, Munoz-Escobar J, Castro K, Gehring K (2017) Mapping the ER interactome: the P domains of calnexin and calreticulin as plurivalent adapters for foldases and chaperones. Structure 25(9):1415–1422. e1413. https://doi.org/10.1016/j.str.2017.07.010

Leone P, Shin EC, Perosa F, Vacca A, Dammacco F, Racanelli V (2013) MHC class I antigen processing and presenting machinery: organization, function, and defects in tumor cells. J Natl Cancer Inst 105(16):1172–1187. https://doi.org/10.1093/jnci/djt184

Liu P, Zhao L, Loos F, Marty C, Xie W, Martins I, Lachkar S, Qu B, Waeckel-Enee E, Plo I, Vainchenker W, Perez F, Rodriguez D, Lopez-Otin C, van Endert P, Zitvogel L, Kepp O,

Kroemer G (2020) Immunosuppression by mutated calreticulin released from malignant cells. Mol Cell 77(4):748–760 e749. https://doi.org/10.1016/j.molcel.2019.11.004

Mansier O, Prouzet-Mauleon V, Jegou G, Barroso K, Raymundo DP, Chauveau A, Dumas PY, Lagarde V, Turcq B, Pasquet JM, Viallard JF, James C, Praloran V, Voutetakis K, Chatziioannou A, Mahon FX, Chevet E, Lippert E (2019) The expression of myeloproliferative neoplasm-associated calreticulin variants depends on the functionality of ER-associated degradation. Cancers (Basel) 11(12). https://doi.org/10.3390/cancers11121921

Masubuchi N, Araki M, Yang Y, Hayashi E, Imai M, Edahiro Y, Hironaka Y, Mizukami Y, Kihara Y, Takei H, Nudejima M, Koike M, Ohsaka A, Komatsu N (2020) Mutant calreticulin interacts with MPL in the secretion pathway for activation on the cell surface. Leukemia 34 (2):499–509. https://doi.org/10.1038/s41375-019-0564-z

McCaffrey K, Braakman I (2016) Protein quality control at the endoplasmic reticulum. Essays Biochem 60(2):227–235. https://doi.org/10.1042/EBC20160003

Mesaeli N, Nakamura K, Zvaritch E, Dickie P, Dziak E, Krause KH, Opas M, MacLennan DH, Michalak M (1999) Calreticulin is essential for cardiac development. J Cell Biol 144 (5):857–868. https://doi.org/10.1083/jcb.144.5.857

Michalak M, Milner RE, Burns K, Opas M (1992) Calreticulin. Biochem J 285(Pt 3):681–692. https://doi.org/10.1042/bj2850681

Michalak M, Groenendyk J, Szabo E, Gold LI, Opas M (2009) Calreticulin, a multi-process calcium-buffering chaperone of the endoplasmic reticulum. Biochem J 417(3):651–666. https://doi.org/10.1042/BJ20081847

Nam AS, Kim KT, Chaligne R, Izzo F, Ang C, Taylor J, Myers RM, Abu-Zeinah G, Brand R, Omans ND, Alonso A, Sheridan C, Mariani M, Dai X, Harrington E, Pastore A, Cubillos-Ruiz JR, Tam W, Hoffman R, Rabadan R, Scandura JM, Abdel-Wahab O, Smibert P, Landau DA (2019) Somatic mutations and cell identity linked by genotyping of transcriptomes. Nature 571 (7765):355–360. https://doi.org/10.1038/s41586-019-1367-0

Nangalia J, Massie CE, Baxter EJ, Nice FL, Gundem G, Wedge DC, Avezov E, Li J, Kollmann K, Kent DG, Aziz A, Godfrey AL, Hinton J, Martincorena I, Van Loo P, Jones AV, Guglielmelli P, Tarpey P, Harding HP, Fitzpatrick JD, Goudie CT, Ortmann CA, Loughran SJ, Raine K, Jones DR, Butler AP, Teague JW, O'Meara S, McLaren S, Bianchi M, Silber Y, Dimitropoulou D, Bloxham D, Mudie L, Maddison M, Robinson B, Keohane C, Maclean C, Hill K, Orchard K, Tauro S, Du MQ, Greaves M, Bowen D, Huntly BJP, Harrison CN, Cross NCP, Ron D, Vannucchi AM, Papaemmanuil E, Campbell PJ, Green AR (2013) Somatic CALR mutations in myeloproliferative neoplasms with nonmutated JAK2. N Engl J Med 369(25):2391–2405. https://doi.org/10.1056/NEJMoa1312542

Parlati F, Dominguez M, Bergeron JJ, Thomas DY (1995) Saccharomyces cerevisiae CNE1 encodes an endoplasmic reticulum (ER) membrane protein with sequence similarity to calnexin and calreticulin and functions as a constituent of the ER quality control apparatus. J Biol Chem 270(1):244–253. https://doi.org/10.1074/jbc.270.1.244

Phillips MJ, Voeltz GK (2016) Structure and function of ER membrane contact sites with other organelles. Nat Rev. Mol Cell Biol 17(2):69–82. https://doi.org/10.1038/nrm.2015.8

Pietra D, Rumi E, Ferretti VV, Di Buduo CA, Milanesi C, Cavalloni C, Sant'Antonio E, Abbonante V, Moccia F, Casetti IC, Bellini M, Renna MC, Roncoroni E, Fugazza E, Astori C, Boveri E, Rosti V, Barosi G, Balduini A, Cazzola M (2016) Differential clinical effects of different mutation subtypes in CALR-mutant myeloproliferative neoplasms. Leukemia 30(2):431–438. https://doi.org/10.1038/leu.2015.277

Pike SE, Yao L, Jones KD, Cherney B, Appella E, Sakaguchi K, Nakhasi H, Teruya-Feldstein J, Wirth P, Gupta G, Tosato G (1998) Vasostatin, a calreticulin fragment, inhibits angiogenesis and suppresses tumor growth. J Exp Med 188(12):2349–2356. https://doi.org/10.1084/jem.188.12.2349

Pocanschi CL, Kozlov G, Brockmeier U, Brockmeier A, Williams DB, Gehring K (2011) Structural and functional relationships between the lectin and arm domains of calreticulin. J Biol Chem 286(31):27266–27277. https://doi.org/10.1074/jbc.M111.258467

Pronier E, Cifani P, Merlinsky TR, Berman KB, Somasundara AVH, Rampal RK, LaCava J, Wei KE, Pastore F, Maag JL, Park J, Koche R, Kentsis A, Levine RL (2018) Targeting the CALR interactome in myeloproliferative neoplasms. JCI Insight 3(22):e122703. https://doi.org/10.1172/jci.insight.122703

Salati S, Genovese E, Carretta C, Zini R, Bartalucci N, Prudente Z, Pennucci V, Ruberti S, Rossi C, Rontauroli S, Enzo E, Calabresi L, Balliu M, Mannarelli C, Bianchi E, Guglielmelli P, Tagliafico E, Vannucchi AM, Manfredini R (2019) Calreticulin Ins5 and Del52 mutations impair unfolded protein and oxidative stress responses in K562 cells expressing CALR mutants. Sci Rep 9(1):10558. https://doi.org/10.1038/s41598-019-46843-z

Schwarz DS, Blower MD (2016) The endoplasmic reticulum: structure, function and response to cellular signaling. Cell Mol Life Sci 73(1):79–94. https://doi.org/10.1007/s00018-015-2052-6

Stevenson J, Huang EY, Olzmann JA (2016) Endoplasmic reticulum-associated degradation and lipid homeostasis. Annu Rev. Nutr 36:511–542. https://doi.org/10.1146/annurev-nutr-071715-051030

Strzyz P (2020) Foundations of ER-phagy regulation. Nat Rev Mol Cell Biol 21(5):251. https://doi.org/10.1038/s41580-020-0238-8

Theocharides AP, Lundberg P, Lakkaraju AK, Lysenko V, Myburgh R, Aguzzi A, Skoda RC, Manz MG (2016) Homozygous calreticulin mutations in patients with myelofibrosis lead to acquired myeloperoxidase deficiency. Blood 127(25):3253–3259. https://doi.org/10.1182/blood-2016-02-696310

Wang WA, Liu WX, Durnaoglu S, Lee SK, Lian J, Lehner R, Ahnn J, Agellon LB, Michalak M (2017) Loss of calreticulin uncovers a critical role for calcium in regulating cellular lipid homeostasis. Sci Rep 7(1):5941. https://doi.org/10.1038/s41598-017-05734-x

Wijeyesakere SJ, Bedi SK, Huynh D, Raghavan M (2016) The C-terminal acidic region of calreticulin mediates phosphatidylserine binding and apoptotic cell phagocytosis. J Immunol 196(9):3896–3909. https://doi.org/10.4049/jimmunol.1502122

William James A, Ravi C, Srinivasan M, Nachiappan V (2019) Crosstalk between protein N-glycosylation and lipid metabolism in Saccharomyces cerevisiae. Sci Rep 9(1):14485. https://doi.org/10.1038/s41598-019-51054-7

Chapter 9
Cancer Biology of the Endoplasmic Reticulum Lectin Chaperones Calreticulin, Calnexin and PDIA3/ERp57

Shing Tat Theodore Lam and Chinten James Lim

Abstract The lectin chaperones calreticulin (CALR) and calnexin (CANX), together with their co-chaperone PDIA3, are increasingly implicated in studies of human cancers in roles that extend beyond their primary function as quality control facilitators of protein folding within the endoplasmic reticulum (ER). Led by the discovery that cell surface CALR functions as an immunogen that promotes anti-tumour immunity, studies have now expanded to include their potential uses as prognostic markers for cancers, and in regulation of oncogenic signaling that regulate such diverse processes including integrin-dependent cell adhesion and migration, proliferation, cell death and chemotherapeutic resistance. The diversity stems from the increasing recognition that these proteins have an equally diverse spectrum of subcellular and extracellular localization, and which are aberrantly expressed in tumour cells. This review describes key foundational discoveries and highlight recent findings that further our understanding of the plethora of activities mediated by CALR, CANX and PDIA3.

Keywords Calreticulin · Calnexin · PDIA3 · Oncogenic signaling · Immunogenic cell death · Integrins · Drug resistance

Abbreviations

CAM-DR	Cell adhesion mediated drug resistance
CALR	Calreticulin, also known as calregulin, CRT, CRP55, CaBP3 and ERp60
CANX	Calnexin, also known as CNX, IP90 and P90

S. T. T. Lam · C. J. Lim (✉)
Department of Pediatrics, University of British Columbia, Vancouver, BC, Canada

Michael Cuccione Childhood Cancer Research Program, B.C. Children's Hospital Research Institute, Vancouver, BC, Canada
e-mail: cjlim@mail.ubc.ca

L. B. Agellon, M. Michalak (eds.), *Cellular Biology of the Endoplasmic Reticulum*, Progress in Molecular and Subcellular Biology 59,
https://doi.org/10.1007/978-3-030-67696-4_9

DAMP	Damage associated molecular patterns
ER	Endoplasmic reticulum
PDIA3	Protein disulfide isomerase A3, also known as ERp57, ER60, GRP57, GRP58
ERK	Extracellular signal regulated kinase
ICD	Immunogenic cell death
MAPK	Mitogen activated protein kinase
STAT3	Signal transducer and activator of transcription 3

9.1 Introduction

Calreticulin (CALR) and calnexin (CANX) are calcium-binding lectin-like molecular chaperones responsible for the proper folding of newly synthesized glycoproteins. Each can be found in a complex with PDIA3, a thiol oxidoreductase with protein disulfide isomerase activity, to form the CALR/CANX cycle that is a key part of the quality control for new proteins being made and assembled in the lumen of the endoplasmic reticulum (ER) (Adams et al. 2019; Halperin et al. 2014). An increasing revelation gained from studies spanning over two decades have highlighted additional non-classical roles for CALR, CANX and PDIA3 in multiple cellular locales that include the nucleus, the extra-ER cytosol and on the extracellular surface of the plasma membrane (Gold et al. 2010; Hettinghouse et al. 2018). However, their relative low abundance in subcellular compartments outside of the ER presents challenges in studies seeking to elucidate their specific functional significance. Here, we will review some of the implications of these findings in the cellular and pathobiology of cancers.

9.2 Expression of CALR in Human Cancers

Many ER chaperones are known to be aberrantly expressed in a variety of tumour tissues. As the expression of ER chaperones is highly regulated in normal cells, the detection and quantification of these ER chaperones in patients' tumour cells may have potential prognostic significance (Gutierrez and Simmen 2014). Among these trio of proteins, the expression of CALR in various cancers have been the most investigated, owing largely to CALR's enhanced significance and prominence as a pro-phagocytic signal deemed important for innate and adaptive anti-tumour immunity.

In a survey of multiple malignancies including hematologic and solid tumours, cell surface CALR levels were inherently much higher when compared to their normal cell counterparts (Chao et al. 2010). However, this did not necessarily translate to a better disease outcome, as the increased surface CALR on tumour cells was accompanied by enhanced expression of the anti-phagocytic CD47 signal receptor. In the same study, the authors found that increased expression of *CALR* mRNA is associated with an adverse clinical prognosis in neuroblastoma, bladder cancer and mantle cell lymphoma (Chao et al. 2010). It's unclear how the increased *CALR* mRNA may account for increased CALR protein expressed in the various subcellular locales that include the cell surface.

An immunohistological survey of resected pancreatic cancers identified an association between high CALR expression and poor patient survival (Matsukuma et al. 2016). High CALR expression was also associated with a poorer prognosis for patients with nasopharyngeal carcinoma (Han et al. 2019b). CALR was also found to be highly expressed in oral squamous cell carcinoma tumour tissues in comparison to matched non-cancerous tissues (Chiang et al. 2013). This study also showed that stable knockdown of CALR expression led to reduced cell proliferation, colony forming potential and anchorage-independent growth. Immunohistological analysis of oral squamous cell carcinoma tissues also identified an association between elevated CALR expression with advanced staging and reduced term survival in patients (Harada et al. 2017). A recent analysis of gene expression datasets across a large cohort of patients that included breast, kidney and brain cancers, found a high level of consistency that high *CALR* mRNA expression associated with the worst survival outcomes (Han et al. 2019a). However, the authors cautioned that the enhanced transcript expression data have not been correlated with increased protein expression.

Thus, it is interesting that CALR deficiency can also be associated with poorer prognosis in other malignancies. CALR is highly expressed in the normal prostate, and seen to be downregulated in prostate cancer specimens from more advanced stages of disease (Alur et al. 2009). Accordingly, enforced CALR overexpression in prostate cancer cells inhibited cell proliferation in both monolayer and soft agar conditions, concomitant with reduced growth and metastasis in an orthotopic xenograft tumour model (Alur et al. 2009). Additionally, high expression of CALR in non-small cell lung cancer is associated with a most favorable prognosis for patients (Fucikova et al. 2016); this retrospective study found high CALR expression in tumour tissues correlated with dense intratumoral infiltration with immune cell subtypes suggestive of an enhanced adaptive immune response. Similar observations were also made in ovarian cancer carcinoma, where high CALR expression is associated with enhanced immune response and improvement in patients' survival (Kasikova et al. 2019; Stoll et al. 2016). Importantly, this correlation was seen in treatment-naïve primary tumours, suggesting an ongoing and effective anti-tumour immune response prior to the onset of chemotherapy.

It was found that high levels of anti-CALR autoantibodies in serum is a potential biomarker for bladder cancer, although a correlation between survival and serum anti-CALR levels could not be established (Minami et al. 2014). The level of soluble CALR was found to be significantly elevated in the serum of lung cancer patients when compared to healthy subjects (Liu et al. 2012). In the same study, serum CALR was found to be higher for patients receiving chemotherapy compared to those who did not. It is worth noting that a serum CALR survey of patients with chronic lymphocytic leukemia taken at diagnosis revealed a correlation between higher

CALR levels to earlier onset of disease progression that necessitated treatment (Molica et al. 2016). It's a compelling notion that a simple blood test for soluble CALR may provide not only diagnostic and prognostic information, but may also be useful to assess treatment response.

Although it may seem contradictory that CALR can be seen to both promote and suppress oncogenesis, these findings illustrate the complexity and dynamic nature of CALR's role in various biological processes. It seems clear that a simple relationship between CALR expression and tumour aggressiveness cannot be made at this point. It is likely the value of such a relationship will depend not only on the individual tumour types, but also on the consideration of how the CALR protein is aberrantly expressed and distinguished at the subcellular level. In a later section, we will consider how the subcellular biology of CALR may promote either a pro- or anti-tumour effect.

9.3 Expression of PDIA3 and CANX in Human Cancers

Like CALR, PDIA3 is also known to be aberrantly expressed in various types of cancer. High expression of PDIA3 in patients' hepatocellular carcinoma tissues linked to hepatitis infection was associated with poorer prognosis, increased tumour cell proliferation and decreased apoptosis (Liu et al. 2017; Takata et al. 2016). High PDIA3 expression could have prognostic significance in ovarian cancer, being overexpressed in malignant tissues in comparison to normal, and increasing expression was correlated with tumour progression and poor survival (Samanta et al. 2017). Similarly, high expression of *PDIA3* mRNA in clear cell renal carcinoma was associated with poor survival and mechanistically linked to enhanced oncogenic signaling through STAT3 (Liu et al. 2019b). On the other hand, gastric cancer tissues in patients are observed to have a low level of PDIA3 expression when compared to normal mucosa or low grade gastric tumors. However, gastric cancer tissues with the highest PDIA3 levels was associated with favourable staging and survival outcome (Leys et al. 2007; Shimoda et al. 2019). Thus, these findings demonstrate that PDIA3 expression, like CALR, may have both tumour promotion and suppression properties.

Patient cohort studies involving CANX expression lag behind those of CALR and PDIA3. Still, the available studies provide an early consensus that increased CANX expression is generally associated with a poorer prognosis. High expression of *CANX* mRNA was correlated with poorer survival for breast cancer patients (Alam et al. 2019). In a study that evaluated expression of several ER-stress chaperones in colorectal cancer, increased CANX protein expression, but not of CALR, GRP78 or GRP94, was predictive of poorer clinical outcome (Ryan et al. 2016). A proteomics based approach comparing individual-matched pairs of tumour and normal tissue from patients with oral squamous cell carcinoma found significantly elevated CANX expression in tumour cells (Chen et al. 2019). This study also evaluated immunohistological staining of tumour tissues distinguished at the subcellular level, revealing that high plasma membrane localized CANX was correlated with poorer patient survival, while enrichment of cytosolic CANX was not. Furthermore, surgical derived tumour tissues with increased membranous CANX expression was found to contain fewer infiltrated T-cells, suggesting an immune-suppressive role for CANX (Chen et al. 2019). This phenomenon is supported by the finding that CANX expression was inversely correlated with MHC1 presentation in breast cancers (Alam et al. 2019). Given that CANX and CALR are both PDIA3 binding proteins, it is intriguing to consider that overexpression of CANX may reduce the level of CALR expressed, thus accounting for CANX's role in immune suppression.

9.4 Cell Surface CALR Signals Immunogenic Cell Death

The Nomenclature Committee on Cell Death defined immunogenic cell death (ICD) as a "regulated cell death that is sufficient to activate an adaptive immune response" (Galluzzi et al. 2018). In this regard, the discovery that cell surface exposed CALR functions as a damage associated molecular pattern (DAMP) able to stimulate engulfment by professional phagocytes (Gardai et al. 2005) was another landmark moment that promoted investigation of CALR's role in an extra-ER locale. The significance for cancer studies was further exemplified when it was found that certain cytotoxic agents, including anthracyclines that are strongly favored for inclusion in most combination chemotherapies, can efficiently induce surface CALR exposure (Obeid et al. 2007). Mechanistically, drug-induced surface CALR occurs rapidly within several hours and is considered to be pre-apoptotic, since this phenomena preceded surface presentation of phosphaditylserine. It was observed also that drug treatment resulted in PDIA3 being co-presented on the cell surface (Obeid et al. 2007). Other ER-chaperones including GRP78, PDIA1, and GRP94 are also presented on the cell surface (Wiersma et al. 2015), although not all necessarily involve induction of cell stress. Of note, the membrane integral protein CANX has been characterized on the surface of many cell types at low levels (Okazaki et al. 2000), but with yet unclear physiologic roles. Still, surface CALR remains unique in being the only ER-chaperone with DAMP-functionality able to stimulate an immune-response. Hence, there has been much research interest seeking to understand the molecular mechanisms that regulate trafficking of CALR from its predominant ER localization to the cell surface in ICD, and how this phenomenon may be exploited to enhance the anti-tumour immune response.

As with most ER-lumen resident proteins, CALR is translationally directed into the ER-lumen via an N-terminal signal peptide that is subsequently proteolytically removed. CALR's retention and enrichment in the ER is further mediated by the C-terminal KDEL motif, which engages KDEL-receptors for retrograde transport from the Golgi back to the ER (Gold et al. 2010; Wiersma et al. 2015). Despite these targeted enrichment and retention mechanisms, it was estimated that up to 14% of total CALR resides in the extra-ER cytosol in non-stressed cells, and that the cytosolic form is without the signal peptide (Afshar et al. 2005; Shaffer et al.

2005). This highlighted that extra-ER forms of CALR likely originates from the ER-luminal pool.

The advent of CRISPR-Cas9 gene editing methods ushered new abilities to easily generate loss of function cellular models to facilitate our lab's studies of ER to surface trafficking of CALR. In this manner, we showed that surface CALR induced by doxorubicin- or oxaliplatin-treatment of wildtype Jurkat T-lymphoblasts was completely abolished in *CALR*$^{-/-}$ cells (Liu et al. 2016). Importantly, drug-induced increases of surface CALR in *CALR*$^{-/-}$ cells could only be rescued upon re-expression of CALR containing the signal peptide, while expression of the truncated cytosol-targeted variant, which bypasses the ER-lumen, resulted in constitutively high levels of surface CALR detected, irrespective of drug-treatment. This indicated that the bulk of drug-induced surface CALR is derived from the ER-luminal pool, and involved transit through the cytosol. Interestingly, it was shown that inhibition of ER to Golgi transport using brefeldin A, as well as inhibition of actin polymerization using latrunculin B, resulted in reduced drug-induced surface CALR (Garg et al. 2012). Thus, it remains likely that drug stress induced ER to surface translocation of CALR may involve multiple mechanisms.

It is now evident that CALR cannot translocate to the cell surface on its own. It was shown that anthracyclines induced surface presentation of PDIA3 in addition to CALR (Obeid et al. 2007). It was reported that in murine cell models, knockdown of PDIA3 inhibited surface CALR presentation, and likewise, knockdown of CALR inhibited surface PDIA3 presentation (Obeid 2008; Panaretakis et al. 2008), suggesting CALR and PDIA3 are co-required in ER to surface trafficking. Our studies using the human Jurkat derivative *CALR*$^{-/-}$ and *PDIA3*$^{-/-}$ cells revealed that PDIA3 is similarly required for surface CALR presentation, while PDIA3 remains efficiently translocated to the surface in a CALR-independent manner (Liu et al. 2019a). In addition, we found that non-drug treated *PDIA3*$^{-/-}$ cells is devoid of a cytosolic pool of CALR typically observed in wildtype cells, indicating that PDIA3 is a critical mediator of possibly all extra-ER localized CALR. These findings highlight an indispensable role for PDIA3 to facilitate ER to cytosol to surface presentation of CALR that serves the immunogenic role important for ICD. In this regard, CALR-PDIA3 is a sufficient and necessary 'immunogenic complex' able to translocate to the cell surface.

What about CANX? Given that PDIA3 act as a co-chaperone with the ability to bind CALR or CANX to form a stable complex (Leach et al. 2002; Russell et al. 2004), it is likely the subcellular localization of the trio of proteins will overlap. It's interesting to note that surface CANX was originally described as being enriched on thymocytes (Wiest et al. 1995), which are immature T-lymphocytes. Indeed, ultraviolet light-induced apoptosis of Jurkat cells was accompanied by surface CANX expression (Franz et al. 2007). It's not clear if this process requires PDIA3, but we do note that Jurkat cells lacking CALR presented increased levels of surface PDIA3 upon drug treatment (Liu et al. 2019a). Unlike CALR, surface CANX is not known to be immunogenic, but an intriguing postulate would predict that tumour cells expressing higher CANX to CALR ratios may present decreased levels of surface CALR owing to competition for binding to PDIA3, and hence be less immunogenic.

Likewise, increased expression of CALR and/or PDIA3 may enhance ICD, and this balance of CALR, CANX and PDIA3 may account for differences in their prognostic values for cancers as reflected in the preceding section.

9.5 Intersection of α-Integrins, CALR and ICD

Integrins are transmembrane heterodimeric cell adhesion receptor proteins consisting of α- and β-subunits. The human genome encodes 21α and 9β isoforms, which combine to form 21 known αβ-heterodimers. The integrin extracellular domains mediate attachment to other cells or to the extracellular matrix, while the cytosolic 'tails' mediate intracellular signaling by interaction with downstream effector proteins (Hynes 2002; Takada et al. 2007).

Prior to cell surface CALR being identified as a DAMP, CALR was identified as an interacting protein with the membrane proximal GFFKR motif found on the cytoplasmic tail of α-integrins (Rojiani et al. 1991). GFFKR is a highly conserved motif amongst all 21 α-integrin tails, serving a critical structural function by interacting with its β-subunit counterpart to form the 'inner membrane clasp' and stable αβ-heterodimer expression (Fig. 9.1). Subsequently, it was shown that physiological cell adhesion, which entails the conformational activation of integrins, inducibly enhances the interaction of α-integrins with CALR, and that *Calr*$^{-/-}$ mouse embryonic fibroblasts exhibit defects in integrin-mediated adhesion and Ca^{2+} flux (Coppolino et al. 1995, 1997).

Admittedly, it was rather by accident that work conducted in our lab led to a 'rediscovery' of CALR-integrin interaction relevant to cancer studies. We rationalized that a phenomenon known as cell adhesion mediated drug resistance (CAM-DR) involved α-integrin tail interactions with downstream effector proteins, and proceeded to express a tail-truncated version of integrin α4, called α4δ, with

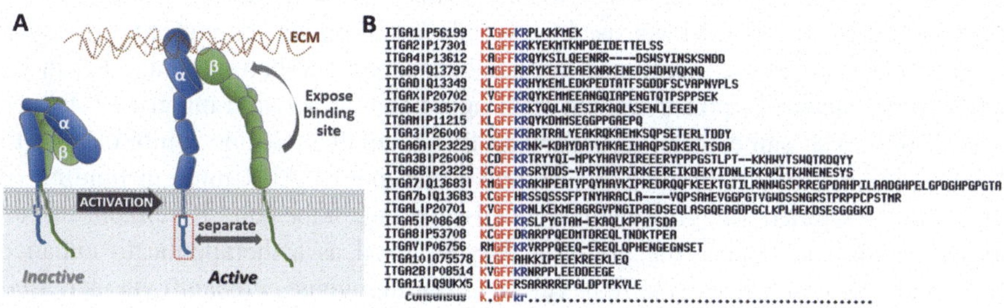

Fig. 9.1 Integrins are αβ-heterodimeric cell adhesion receptors. (**a**) Integrin activation switches the conformation from low- to high-affinity ligand binding state, to include exposure of the substrate binding site and physical separation of the αβ transmembrane and cytosolic domains. (**b**) Cytoplasmic domain sequence alignment of all 21 human α-integrin isoforms. The juxta-membrane GFFKR motif is highly conserved in 21, and fully conserved in 16

only the minimal KxGFFKR cytosolic tail to maintain αβ-heterodimer pairing. Unexpectedly, rather than abrogating CAM-DR, α4δ-expression enhanced chemoresistance in a manner that bypassed the requirement for cell adhesion (Liu et al. 2013). This phenomenon was reproduced by expression of Tacδ, essentially a non-integrin dummy transmembrane receptor with the same KxGFFKR cytosolic tail, while a scrambled version KxRFGFK that does not bind CALR failed to do so. Furthermore, whereas cells expressing wildtype α4 required cell adhesion to stimulate Akt phosphorylation, Ca^{2+}-flux and integrin-CALR interaction, α4δ and Tacδ cells activated these pathways constitutively in an adhesion-independent manner (Liu et al. 2013). We extended these findings to show that drug induced surface CALR expression was attenuated for cells plated to engage integrins, while α4δ and Tacδ expression decreased surface CALR expression in an adhesion-independent manner (Liu et al. 2016). Further to this, treatment of cells with either chemical- or antibody-based agonists of integrin activation resulted in complete suppression of drug-induced surface CALR exposure, while cells with overall reduced α-integrin expression enhanced it (Liu et al. 2019a). The integrin modulatory effect on cell surface CALR is functionally relevant, as cells with activated integrins, stimulated via adhesion or molecular agonists, exhibited reduced phagocytosis by macrophages concomitant with reduced surface CALR levels. In sum, our findings add a new layer to the pro-tumour survival effects mediated by integrins to include suppression of a surface immunogen important in anti-tumour immunity.

Importantly, GFFKR as a highly conserved motif found in the cytosolic tail of α-integrins provision a widely applicable universal mechanism in multiple cancer cell types where various integrin-substrate interactions can potentiate reduced therapy-induced ICD. Mechanistically, it is interesting to speculate how CALR-integrin binding can occur in a manner stimulated by cell adhesion and how this binding reduces surface CALR presented. Structural studies of the integrin α-β-transmembrane helices provide evidence that the GFFKR moiety transitions from a partial membrane-embedded state in inactive integrins, to full exposure within the cytosol when activated (Kim et al. 2011; Lau et al. 2009). In addition, the transmembrane and cytosolic domains of αβ-integrins is physically separated, facilitating access for binding by cytosolic effector proteins. More recently, proximity ligation assays (PLA) was used to provide *in situ* evidence that CALR-integrin interaction occurs at the membrane proximal cytosolic compartment, and the interaction was enhanced upon treatment with agonists that activate integrins (Ohkuro et al. 2018). The same study also showed that a small molecule inhibitor able to disrupt CALR-GFFKR binding *in vitro* also disrupted CALR-integrin binding *in situ*, and blockaded integrin-mediated cell-substrate adhesion. It remains to be shown if this small molecule inhibitor may be used as a therapeutic to enhance both the chemosensitivity of tumour cells and immunogenic cell death via increased surface CALR presentation.

Collectively, the combined studies support a model whereby conformational change accompanying integrin activation exposes the GFFKR moiety of α-integrin within the membrane proximal cytosol available to bind CALR (Fig. 9.2). This pool of CALR is dramatically increased for cells treated with ICD-inducers, which in the

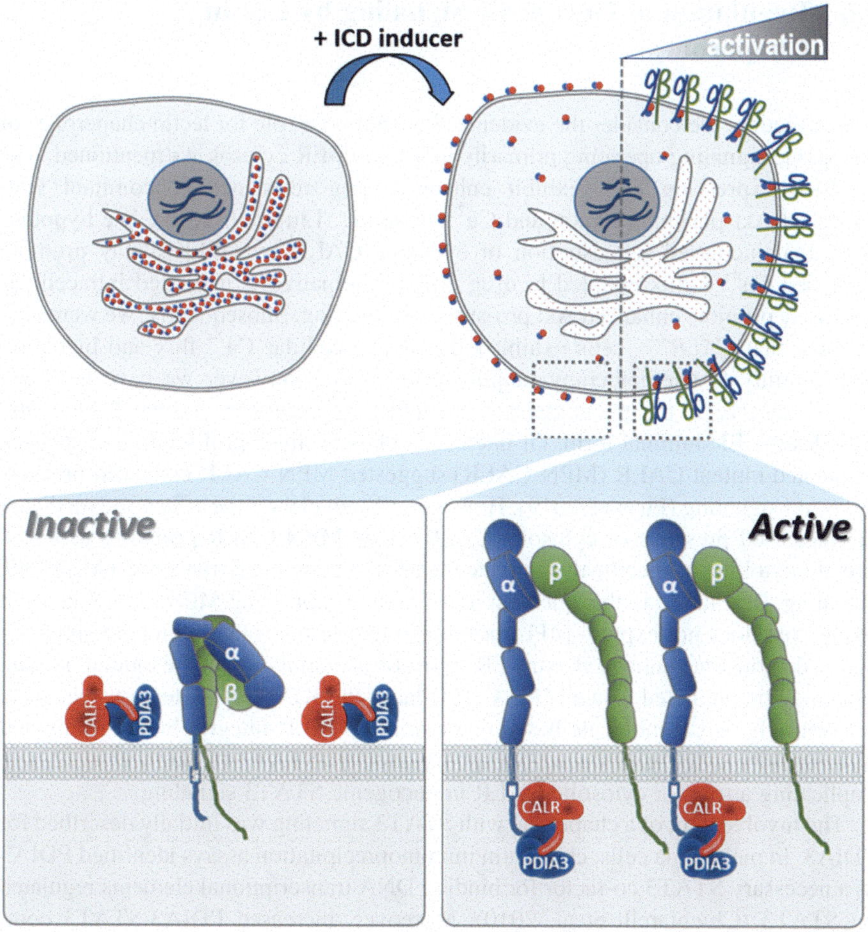

Fig. 9.2 Activated integrins suppress immunogenic cell death by limiting presentation of cell surface CALR. At rest, CALR (red) and PDIA3 (blue) is highly enriched in the ER-lumen. Treatment with ICD-inducers promote ER-stress and release of CALR-PDIA3 complex into the extra-ER cytosolic space. In the absence of activated integrins, the CALR-PDIA3 'immunogenic complex' readily translocates to the cell surface, where CALR acts as a DAMP to promote phagocytosis. When activated, integrins undergo conformational change that exposes the membrane proximal GFFKR motif of α-integrin to bind and sequester CALR-PDIA3 complex in the cytosolic compartment and reducing the levels of surface CALR and PDIA3 presented

absence of activated integrins, translocate to the cell surface to act as the immunogen. When integrins are activated, GFFKR-mediated integrin-CALR binding act to sequester CALR within the cytosol, thereby limiting the levels of surface CALR presented.

9.6 Regulation of Oncogenic Signaling by Lectin Chaperones

In this section, we consider the evidence that support a role for lectin chaperones in oncogenic signaling operating primarily in the extra-ER cytosol. As mentioned, α4δ or Tacδ expressing cells exhibit enhanced drug resistance concomitant with increased Akt phosphorylation and Ca^{2+} signaling (Liu et al. 2013). We hypothesized that increased sequestration of cytosolic CALR by GFFKR may promote increased Ca^{2+} influx coupled to drug efflux; alternatively, increased intracellular Ca^{2+} may promote enhanced Akt pro-survival signaling. Subsequently, we were able to show that *CALR*$^{-/-}$ cells exhibit reduced intracellular Ca^{2+}-flux and increased susceptibility to chemotherapy (unpublished results); however we have not been able to resolve if the phenomena is mediated by cytosolic or total CALR that includes the ER-luminal form. Of interest, studies of myeloproliferative neoplasms associated mutant CALR (MPN-CALR) suggested MPN-CALR enhanced proliferative Akt signaling (Fu et al. 2019). However, it's not clear if the effects are mediated via increased presence of cytosolic CALR, since MPN-CALR primarily acts as a pseudo-cytokine to specifically activate the MPL receptor and transduce JAK/STAT signaling. Resolution to this question must involve expressing MPN-CALR in a cell model that does not express MPL receptor to negate the role of cytokine signaling and to demonstrate enhanced extra-ER cytosolic accumulation of the mutant protein. This may be predicted, since MPN-CALR lacks the KDEL ER-retention sequence. Interestingly, small molecule based disruption of CALR-integrin binding inhibited inflammatory cytokine induced phosphorylation of STAT3 (Ohkuro et al. 2018), implicating a role for cytosolic CALR in oncogenic STAT3 signaling.

The involvement of a chaperone with STAT3 signaling was initially described for PDIA3. In melanoma cells, chromatin immunoprecipitation assays identified PDIA3 as a necessary STAT3 co-factor for binding DNA transcriptional elements regulated by STAT3 (Chichiarelli et al. 2010). Moreover, increased PDIA3-STAT3 complexes was detected in radioresistant laryngeal cancer cells, while PDIA3 depletion results in decreased STAT3 phosphorylation induced by irradiation (Choe et al. 2015). In contrast, *Pdia3*$^{-/-}$ mouse embryonic fibroblasts exhibit increased levels of phosphorylated STAT3 and increased transcription of a STAT3 regulated promoter (Coe et al. 2010). In this report, *Calr*$^{-/-}$ fibroblasts exhibited no differences in phosphorylated STAT3, while STAT3 regulated transcription was enhanced in a similar fashion as *Pdia3*$^{-/-}$ cells. Furthermore, PDIA3 inhibition of STAT3 transcriptional activity was partially reconstituted in *Pdia3*$^{-/-}$ fibroblasts expressing an ER-targeted form of PDIA3, but not with cytosolic PDIA3, suggesting that luminal PDIA3 modulates STAT3. STAT3 transcription activity was not affected in *Canx*$^{-/-}$ cells, but requires a functional CALR-PDIA3 complex, since re-expression of a PDIA3 binding mutant of CALR does not rescue *Calr*$^{-/-}$ cells (Coe et al. 2010). Recently, it was shown that CALR knockdown in human nasopharyngeal carcinoma cell lines inhibited cell migration, invasion and proliferation, as well as decreased total and phosphorylated STAT3 levels (Han et al. 2019b). It's unclear what accounts for the reported differences in activities for CALR and PDIA3; different species background may play a role (mouse vs human), as so is the cell type specific context (different tissues, and normal vs tumour cells). Further studies are required to establish a clearer consensus for the role/s of PDIA3 and CALR in oncogenic STAT3 signaling.

Other prominent pathways potentially regulated by ER-chaperones include MAPK and p53. Knockdown of CALR led to decreased migration and invasion in breast cancer cells that was correlated with significantly reduced expression of genes in MAPK signaling, including *MAP2K6*, *MAPK8*, *ERK2*, and *MAPK14*, and upregulation of several target genes of p53 (Zamanian et al. 2016). This is supported by studies in pancreatic cancer cells, where *CALR* silencing inhibited EGF-induced pathways involved in epidermal to mesenchymal transition, including E-cadherin, β-catenin, EGFR and ERK (Sheng et al. 2017). Coincidentally, the inhibition of *PDIA3* with siRNA in human acute myeloid leukemia cells resulted in decreased proliferation with increased apoptosis, concomitant with significant decreases in phosphorylated p38, JNK and ERK (Ye et al. 2018). Interestingly, *CALR* silencing in several solid tumours led to significant increases in caspase-dependent apoptotic proteins, even though an increase in cell death was observed (Han et al. 2019a). This was attributed to increased expression of total and activated p53, since *CALR* silencing induced cell death was effectively negated upon p53 activity blockade with a dominant negative peptide-based inhibitor.

The accumulated evidence clearly highlight the multitude of oncogenic pathways that are regulated by lectin-ER chaperones stemming from their varied localizations including the cell surface, extra-ER cytosol, nucleus, and of course, the ER-lumen. An interesting hypothesis for their roles in oncogenesis and tumour progression stems from their enrichment within these subcellular compartments leading to dysregulation of pathways not commonly observed in normal, non-tumour cells. A key aspect that remains to be exploited is detailed multiparameter studies that compares the relative enrichment of all three proteins to better understand their relationship as prognostic biomarkers in cancer, and their relative contribution in oncogenic signaling.

9.7 Summary Points

- High expression of CALR and/or PDIA3 in tumour tissues may have positive or negative prognostic significance that vary by tumour type. More limited studies suggest high CANX expression is associated with a poorer prognosis.
- ICD-inducer mediated presentation of surface CALR on tumour cells is dependent on PDIA3 to facilitate translocation of the CALR-PDIA3 as a complex from the ER-lumen to the extracellular cell surface.
- Activated integrins mediate the suppression of surface CALR presentation in ICD. The mechanism involves a conserved cytosolic GFFKR motif on α-integrin that facilitate binding to and predicted sequestration of CALR in the cytosol.

- CALR and PDIA3 can modulate activities of oncogenic pathways including STAT3, Akt, MAPK, p53, and Ca^{2+} flux manifested in alterations of cell adhesion, migration, proliferation, and cell death.

9.8 Future Issues

- It is not known if surface CALR presented on the surface of ICD-inducer treated tumour cells requires physical interaction with a receptor-like protein. LRP1 (LDL receptor-related protein 1) is a known trans-acting CALR receptor expressed on macrophages that may facilitate this role as a cis-acting receptor on tumour cells.
- Thus far, functional integrin suppression of surface CALR have been studied with leukemic blasts that grow viably in suspension. It remains to be shown if cells of solid tumour origin, which are anchorage-dependent to remain viable, are inherently more resistant to ICD-inducing therapies.
- A number of existing biologics target integrin-substrate interactions that may be evaluated as ICD-enhancing therapeutics. Consideration must be given to ensure that these agents act in a manner to blockade integrin function without their activation, which would achieve the opposite desired effect whereby immunogenicity of the tumour is reduced.
- Due to their multi-locale nature, there remains a continuing challenge to resolve the compartment specific functions of CALR, CANX and PDIA3 in a manner distinct from their roles as ER-chaperones. This limitation can be partly addressed by cell models that achieve compartment specific expression of the respective proteins that allows distinction between activities mediated by their extra-ER localization and those originating from the ER-lumen.
- Fundamental reported differences exist for signaling attributed to CALR and PDIA3 expression. Further work will be required to resolve if the differences could be species dependent, tissue cell type dependent, or even differences arising from oncogenic transformation vs normal cells.
- Is surface CANX on tumour cells immunosuppressive? Or is this role facilitated by competitive inhibition of surface CALR involving limited PDIA3?

Acknowledgments Research conducted in CJL's laboratory was supported by grants from the Canadian Institutes of Health Research, Natural Sciences and Engineering Research Council of Canada, Canadian Cancer Society Research Institute and the Leukemia and Lymphoma Society of Canada. We also acknowledge additional funding received from the Michael Cuccione Foundation for Childhood Cancer Research and the BC Children's Hospital Foundation.

Conflict-of-Interest Disclosure Statement The authors declare no existing conflict-of-interest.

References

Adams BM, Oster ME, Hebert DN (2019) Protein quality control in the endoplasmic reticulum. Protein J 38:317–329. https://doi.org/10.1007/s10930-019-09831-w

Afshar N, Black BE, Paschal BM (2005) Retrotranslocation of the chaperone calreticulin from the endoplasmic reticulum lumen to the cytosol. Mol Cell Biol 25:8844–8853. https://doi.org/10.1128/MCB.25.20.8844-8853.2005

Alam A et al (2019) SMAR1 favors immunosurveillance of cancer cells by modulating calnexin and MHC I expression. Neoplasia 21:945–962. https://doi.org/10.1016/j.neo.2019.07.002

Alur M et al (2009) Suppressive roles of calreticulin in prostate cancer growth and metastasis. Am J Pathol 175:882–890. https://doi.org/10.2353/ajpath.2009.080417

Chao MP et al (2010) Calreticulin is the dominant pro-phagocytic signal on multiple human cancers and is counterbalanced by CD47. Sci Transl Med 2:63ra94. https://doi.org/10.1126/scitranslmed.3001375

Chen Y et al (2019) Calnexin impairs the antitumor immunity of CD4(+) and CD8(+) T cells. Cancer Immunol Res 7:123–135. https://doi.org/10.1158/2326-6066.CIR-18-0124

Chiang WF et al (2013) Calreticulin, an endoplasmic reticulum-resident protein, is highly expressed and essential for cell proliferation and migration in oral squamous cell carcinoma. Oral Oncol 49:534–541. https://doi.org/10.1016/j.oraloncology.2013.01.003

Chichiarelli S et al (2010) Role of ERp57 in the signaling and transcriptional activity of STAT3 in a melanoma cell line. Arch Biochem Biophys 494:178–183. https://doi.org/10.1016/j.abb.2009.12.004

Choe MH et al (2015) ERp57 modulates STAT3 activity in radioresistant laryngeal cancer cells and serves as a prognostic marker for laryngeal cancer. Oncotarget 6:2654–2666. https://doi.org/10.18632/oncotarget.3042

Coe H, Jung J, Groenendyk J, Prins D, Michalak M (2010) ERp57 modulates STAT3 signaling from the lumen of the endoplasmic reticulum. J Biol Chem 285:6725–6738. https://doi.org/10.1074/jbc.M109.054015

Coppolino M, Leung-Hagesteijn C, Dedhar S, Wilkins J (1995) Inducible interaction of integrin alpha 2 beta 1 with calreticulin. Dependence on the activation state of the integrin. J Biol Chem 270:23132–23138

Coppolino MG, Woodside MJ, Demaurex N, Grinstein S, St-Arnaud R, Dedhar S (1997) Calreticulin is essential for integrin-mediated calcium signalling and cell adhesion. Nature 386:843–847

Franz S et al (2007) After shrinkage apoptotic cells expose internal membrane-derived epitopes on their plasma membranes. Cell Death Differ 14:733–742. https://doi.org/10.1038/sj.cdd.4402066

Fu C et al (2019) AKT activation is a feature of CALR mutant myeloproliferative neoplasms. Leukemia 33:271–274. https://doi.org/10.1038/s41375-018-0224-8

Fucikova J et al (2016) Calreticulin expression in human non-small cell lung cancers correlates with increased accumulation of antitumor immune cells and favorable prognosis. Cancer Res 76:1746–1756. https://doi.org/10.1158/0008-5472.CAN-15-1142

Galluzzi L et al (2018) Molecular mechanisms of cell death: recommendations of the Nomenclature Committee on Cell Death 2018. Cell Death Differ 25:486–541. https://doi.org/10.1038/s41418-017-0012-4

Gardai SJ et al (2005) Cell-surface calreticulin initiates clearance of viable or apoptotic cells through trans-activation of LRP on the phagocyte. Cell 123:321–334. https://doi.org/10.1016/j.cell.2005.08.032

Garg AD et al (2012) A novel pathway combining calreticulin exposure and ATP secretion in immunogenic cancer cell death. EMBO J 31:1062–1079. https://doi.org/10.1038/emboj.2011.497

Gold LI et al (2010) Calreticulin: non-endoplasmic reticulum functions in physiology and disease. FASEB J 24:665–683

Gutierrez T, Simmen T (2014) Endoplasmic reticulum chaperones and oxidoreductases: critical regulators of tumor cell survival and immunorecognition. Front Oncol 4:291. https://doi.org/10.3389/fonc.2014.00291

Halperin L, Jung J, Michalak M (2014) The many functions of the endoplasmic reticulum chaperones and folding enzymes. IUBMB Life 66:318–326. https://doi.org/10.1002/iub.1272

Han A et al (2019a) Calreticulin is a Critical Cell Survival Factor in Malignant Neoplasms. PLoS Biol 17:e3000402. https://doi.org/10.1371/journal.pbio.3000402

Han Y et al (2019b) High expression of calreticulin indicates poor prognosis and modulates cell migration and invasion via activating Stat3 in nasopharyngeal carcinoma. J Cancer 10:5460–5468. https://doi.org/10.7150/jca.35362

Harada K, Takenawa T, Ferdous T, Kuramitsu Y, Ueyama Y (2017) Calreticulin is a novel independent prognostic factor for oral squamous cell carcinoma. Oncol Lett 13:4857–4862. https://doi.org/10.3892/ol.2017.6062

Hettinghouse A, Liu R, Liu CJ (2018) Multifunctional molecule ERp57: From cancer to neurodegenerative diseases. Pharmacol Ther 181:34–48. https://doi.org/10.1016/j.pharmthera.2017.07.011

Hynes RO (2002) Integrins: bidirectional, allosteric signaling machines. Cell 110:673–687

Kasikova L et al (2019) Calreticulin exposure correlates with robust adaptive antitumor immunity and favorable prognosis in ovarian carcinoma patients. J Immunother Cancer 7:312. https://doi.org/10.1186/s40425-019-0781-z

Kim C, Schmidt T, Cho EG, Ye F, Ulmer TS, Ginsberg MH (2011) Basic amino-acid side chains regulate transmembrane integrin signalling. Nature 481:209–213. https://doi.org/10.1038/nature10697

Lau TL, Kim C, Ginsberg MH, Ulmer TS (2009) The structure of the integrin alphaIIbbeta3 transmembrane complex explains integrin transmembrane signalling. EMBO J 28:1351–1361. https://doi.org/10.1038/emboj.2009.63

Leach MR, Cohen-Doyle MF, Thomas DY, Williams DB (2002) Localization of the lectin, ERp57 binding, and polypeptide binding sites of calnexin and calreticulin. J Biol Chem 277:29686–29697. https://doi.org/10.1074/jbc.M202405200

Leys CM, Nomura S, LaFleur BJ, Ferrone S, Kaminishi M, Montgomery E, Goldenring JR (2007) Expression and prognostic significance of prothymosin-alpha and ERp57 in human gastric cancer. Surgery 141:41–50. https://doi.org/10.1016/j.surg.2006.05.009

Liu R et al (2012) Calreticulin as a potential diagnostic biomarker for lung cancer. Cancer Immunol Immunother 61:855–864. https://doi.org/10.1007/s00262-011-1146-8

Liu CC, Leclair P, Yap SQ, Lim CJ (2013) The membrane-proximal KXGFFKR motif of alpha-integrin mediates chemoresistance. Mol Cell Biol 33:4334–4345. https://doi.org/10.1128/MCB.00580-13

Liu CC, Leclair P, Monajemi M, Sly LM, Reid GS, Lim CJ (2016) alpha-Integrin expression and function modulates presentation of cell surface calreticulin. Cell Death Disease 7:e2268. https://doi.org/10.1038/cddis.2016.176

Liu M, Du L, He Z, Yan L, Shi Y, Shang J, Tang H (2017) Increased ERp57 expression in HBV-related hepatocellular carcinoma: possible correlation and prognosis. Biomed Res Int 2017:1252647. https://doi.org/10.1155/2017/1252647

Liu CC et al (2019a) Integrins and ERp57 coordinate to regulate cell surface calreticulin in immunogenic cell death. Front Oncol 9:411. https://doi.org/10.3389/fonc.2019.00411

Liu Y et al (2019b) Upregulation of ERp57 promotes clear cell renal cell carcinoma progression by initiating a STAT3/ILF3 feedback loop. J Exp Clin Cancer Res 38:439. https://doi.org/10.1186/s13046-019-1453-z

Matsukuma S et al (2016) Calreticulin is highly expressed in pancreatic cancer stem-like cells. Cancer Sci 107:1599–1609. https://doi.org/10.1111/cas.13061

Minami S et al (2014) Detection of tumor-associated antigens in culture supernatants using autoantibodies in sera from patients with bladder cancer. Biomed Res 35:25–35. https://doi.org/10.2220/biomedres.35.25

Molica S et al (2016) Serum levels of soluble calreticulin predict for time to first treatment in early chronic lymphocytic leukaemia. Br J Haematol 175:983–985. https://doi.org/10.1111/bjh.13907

Obeid M (2008) ERP57 membrane translocation dictates the immunogenicity of tumor cell death by controlling the membrane translocation of calreticulin. J Immunol 181:2533–2543

Obeid M et al (2007) Calreticulin exposure dictates the immunogenicity of cancer cell death. Nat Med 13:54–61. https://doi.org/10.1038/nm1523

Ohkuro M et al (2018) Calreticulin and integrin alpha dissociation induces anti-inflammatory programming in animal models of inflammatory bowel disease. Nat Commun 9:1982. https://doi.org/10.1038/s41467-018-04420-4

Okazaki Y, Ohno H, Takase K, Ochiai T, Saito T (2000) Cell surface expression of calnexin, a molecular chaperone in the endoplasmic reticulum. J Biol Chem 275:35751–35758. https://doi.org/10.1074/jbc.M007476200

Panaretakis T et al (2008) The co-translocation of ERp57 and calreticulin determines the immunogenicity of cell death. Cell Death Differ 15:1499–1509. https://doi.org/10.1038/cdd.2008.67

Rojiani MV, Finlay BB, Gray V, Dedhar S (1991) In vitro interaction of a polypeptide homologous to human Ro/SS-A antigen (calreticulin) with a highly conserved amino acid sequence in the cytoplasmic domain of integrin alpha subunits. Biochemistry 30:9859–9866

Russell SJ et al (2004) The primary substrate binding site in the b' domain of ERp57 is adapted for endoplasmic reticulum lectin association. J Biol Chem 279:18861–18869. https://doi.org/10.1074/jbc.M400575200

Ryan D et al (2016) Calnexin, an ER stress-induced protein, is a prognostic marker and potential therapeutic target in colorectal cancer. J Transl Med 14:196. https://doi.org/10.1186/s12967-016-0948-z

Samanta S et al (2017) Expression of protein disulfide isomerase family members correlates with tumor progression and patient survival in ovarian cancer. Oncotarget 8:103543–103556. https://doi.org/10.18632/oncotarget.21569

Shaffer KL, Sharma A, Snapp EL, Hegde RS (2005) Regulation of protein compartmentalization expands the diversity of protein function. Dev Cell 9:545–554. https://doi.org/10.1016/j.devcel.2005.09.001

Sheng W et al (2017) Calreticulin promotes EGF-induced EMT in pancreatic cancer cells via Integrin/EGFR-ERK/MAPK signaling pathway. Cell Death Dis 8:e3147. https://doi.org/10.1038/cddis.2017.547

Shimoda T et al (2019) Expression of protein disulfide isomerase A3 and its clinicopathological association in gastric cancer. Oncol Rep 41:2265–2272. https://doi.org/10.3892/or.2019.6999

Stoll G, Iribarren K, Michels J, Leary A, Zitvogel L, Cremer I, Kroemer G (2016) Calreticulin expression: Interaction with the immune infiltrate and impact on survival in patients with ovarian and non-small cell lung cancer. Onco Targets Ther 5:e1177692. https://doi.org/10.1080/2162402X.2016.1177692

Takada Y, Ye X, Simon S (2007) The integrins. Genome Biol 8:215. https://doi.org/10.1186/gb-2007-8-5-215

Takata H et al (2016) Increased expression of PDIA3 and its association with cancer cell proliferation and poor prognosis in hepatocellular carcinoma. Oncol Lett 12:4896–4904. https://doi.org/10.3892/ol.2016.5304

Wiersma VR, Michalak M, Abdullah TM, Bremer E, Eggleton P (2015) Mechanisms of translocation of ER chaperones to the cell surface and immunomodulatory roles in cancer and autoimmunity. Front Oncol 5:7. https://doi.org/10.3389/fonc.2015.00007

Wiest DL, Burgess WH, McKean D, Kearse KP, Singer A (1995) The molecular chaperone calnexin is expressed on the surface of immature thymocytes in association with clonotype-independent CD3 complexes. EMBO J 14:3425–3433

Ye Q, Fu P, Dou J, Wang N (2018) Downregulation of PDIA3 inhibits proliferation and invasion of human acute myeloid leukemia cells. Onco Targets Ther 11:2925–2935. https://doi.org/10.2147/OTT.S162407

Zamanian M, Qader Hamadneh LA, Veerakumarasivam A, Abdul Rahman S, Shohaimi S, Rosli R (2016) Calreticulin mediates an invasive breast cancer phenotype through the transcriptional dysregulation of p53 and MAPK pathways. Cancer Cell Int 16:56. https://doi.org/10.1186/s12935-016-0329-y

Chapter 10
Maintenance of Endoplasmic Reticulum Protein Homeostasis in Cancer: Friend or Foe

Mari McMahon, Afshin Samali, and Eric Chevet

Abstract The endoplasmic reticulum, as the site of synthesis for proteins in the secretory pathway has evolved select machineries to ensure the correct folding and modification of proteins. However, sometimes these quality control mechanisms fail and proteins are misfolded. Other factors, such as nutrient deprivation, hypoxia or an increased demand on protein synthesis can also cause the accumulation of unfolded or misfolded proteins in the endoplasmic reticulum. There are mechanisms that recognise and deal with this accumulation of protein through degradation and/or export. Many diseases are associated with aberrant quality control mechanisms, and among these, cancer has emerged as a group of diseases that rely on endoplasmic reticulum homeostasis to sustain development and growth. The knowledge of how protein quality control operates in cancer has identified opportunities for these pathways to be pharmacologically targeted, which could lead to newer or more effective treatments in the future.

Keywords Endoplasmic reticulum · proteostasis · quality control · cancer

M. McMahon
INSERM U1242, University of Rennes 1, Rennes, France

Centre Eugène Marquis, Rennes, France

Apoptosis Research Centre, NUI Galway, Galway, Ireland

A. Samali
Apoptosis Research Centre, NUI Galway, Galway, Ireland

E. Chevet (✉)
INSERM U1242, University of Rennes 1, Rennes, France

Centre Eugène Marquis, Rennes, France
e-mail: eric.chevet@inserm.fr

L. B. Agellon, M. Michalak (eds.), *Cellular Biology of the Endoplasmic Reticulum*,
Progress in Molecular and Subcellular Biology 59,
https://doi.org/10.1007/978-3-030-67696-4_10

Abbreviations

17-AAG	17-*N*-allylamino-17-demethoxygeldanamycin
4u8c	4-methyl umbelliferone 8-carbaldehyde
ATF4	Activating transcription factor 4
ATF6	Activating transcription factor 6
CHOP	C/EBP homologous protein
COPII	Coat protein complex II
eIF2α	Eukaryotic translation initiation factor 2 subunit 1
ER	Endoplasmic reticulum
ERAD	Endoplasmic reticulum associated degradation
ERN1	Endoplasmic reticulum to nucleus signalling 1
ERSU	Endoplasmic reticulum stress surveillance pathway
EZH2	Enhancer of zeste homolog 2
GADD34	Growth arrest and DNA damage-inducible protein GADD34
GRP78	Glucose-regulated protein, 78 kDa
HIF1α	Hypoxia-inducible factor 1-alpha
HRas	Harvey rat sarcoma viral oncogene homolog
HRD1	HMG-CoA reductase degradation 1 homolog
HSP90	Heat shock protein
ID1	Inhibitor of DNA binding 1
IRE1	Inositol-requiring enzyme 1
ISRIB	Integrated stress response inhibitor
JNK	c-Jun N-terminal kinase
LAMP3	Lysosomal associated membrane protein 3
Nrf2	Nuclear factor, erythroid 2 like 2
OS-9	Osteosarcoma amplified 9, endoplasmic reticulum lectin
PDIA5	Protein disulphide isomerase family A member 5
PER1	Period circadian regulator 1
PERK	Protein kinase R-like ER kinase
PP1	Protein phosphatase 1
RIDD	Regulated IRE1-dependent decay
RtcB	RNA 2′,3′-cyclic phosphate and 5’-OH ligase
S1P	Site-1 protease
S2P	Site-2 protease
SCD1	Stearoyl-CoA desaturase
Sec24	Secretory protein 24
Sec61	Homolog of yeast SecY/61
SEL1L	Suppressor of Lin-12-like protein 1
SPARC	Secreted protein acidic and rich in cysteine
TRAF2	TNF Receptor associated factor 2
UPR	Unfolded protein response
XBP1	X-box binding protein 1
XTP3-B	Endoplasmic reticulum lectin 1

10.1 Introduction

As described in chapter one of this book, protein quality control mechanisms operate in the cell to assist in the correct protein folding of newly-synthesised proteins, as well as to assist in the degradation of misfolded proteins or protein aggregates. The endoplasmic reticulum (ER) is an integral part of the protein quality control network of the cell, as it is the site of protein folding and maturation for approximately one-third of the proteome, consisting mainly of proteins destined for the secretory pathway. In the ER, chaperone proteins interact with the nascent protein to ensure the correct final structure and other enzymes make post-translational modifications to the protein before it can travel through the rest of the secretory pathway. One of the functions of the ER in protein quality control is the unfolded protein response (UPR), a signalling pathway that senses disturbances in the protein folding capacity of the ER in stressful conditions and activates a co-ordinated transcriptional programme that aims to alleviate ER stress/restore ER homeostasis or commit to apoptosis if the stress is irredeemable. Another important role of the ER in protein quality control is ER-associated degradation (ERAD), where misfolded proteins are removed from the ER for degradation by the ubiquitin-proteasome system. There are many cross-talk mechanisms between ERAD and the UPR, and pharmacological targeting of one can affect the other. Protein quality control mechanisms operate as part of protein homeostasis (proteostasis) which ensures that the proteome meets the cells' requirements to respond to changes in its environment. In harsh or challenging environments e.g. in cancer or other diseases, proteostasis is altered due to intrinsic factors such as increased protein synthesis demand, high mutation rates or extrinsic factors such as hypoxia or nutrient deprivation, which leads to a disturbance in proteostasis, and an activation of protein quality control mechanisms in response to stress (Madden et al. 2019; Urra et al. 2017). The cell uses a combination of strategies to restore proteostasis, from translational attenuation through the UPR to relieve the pressure on the ER by blocking the introduction of any more protein to the already stressed ER, the actions of chaperone proteins in the ER to ensure correct folding and detection of misfolded proteins and finally by clearance of folded or misfolded proteins from the ER.

10.2 The Unfolded Protein Response

One of the mechanisms by which the ER responds to alterations in proteostasis is the UPR, a co-ordinated response of three ER membrane-bound proteins to a build up of improperly folded proteins in the ER, which converges on a number of transcription factors that aim to rescue the cell from the stress. The UPR consists of activating transcription factor 6 (ATF6), protein kinase R-like ER kinase (PERK) and inositol-requiring enzyme 1 (IRE1). The UPR itself is controlled by a chaperone protein in the ER, GRP78, which is usually bound to the UPR sensors, however when unfolded proteins accumulate in the ER, GRP78 dissociates from the UPR sensors, allowing their activation through phosphorylation in the case of PERK and IRE1, and through cleavage of the protein in the case of ATF6 (Hetz et al. 2015). Recently, there is also evidence that misfolded proteins interact directly with the PERK and IRE1, which also results in their activation (Adams et al. 2019). The UPR is primarily an adaptive mechanism that aims to restore proteostasis in the ER, but if this cannot be achieved, the UPR will eventually lead to apoptosis. Apart from this adaptive and apoptotic signalling, there is increasing evidence that the UPR also acts as a signalling mechanism to influence cellular behaviour (e.g. migration/invasion ability, communication with neighbour cells, inter-organelle relationships), especially in cells that have a high protein synthesis demand, regardless of disease state.

ATF6α (hereafter referred to as ATF6) is an ER membrane glycoprotein which contains a bZIP domain in the cytosolic portion of the protein. ATF6 has two isoforms, ATF6α and ATF6β, but ATF6β displays poor transcriptional activation activity and so ATF6α is better characterised. Upon activation of the UPR, GRP78 dissociates from ATF6, exposing Golgi localisation sequences in the luminal domain, and the protein disulphide isomerase PDIA5 modulates disulphide bonds in the protein that together direct ATF6 to the Golgi apparatus via COPII vesicles (Higa et al. 2014). In the Golgi, ATF6 is cleaved by the Golgi-resident proteases S1P and S2P to release the cytosolic fragment. The now active ATF6 transcription factor then travels to the nucleus, where it is involved in the regulation of genes from the UPR and ERAD pathways in an attempt to relieve the ER stress (Hillary and FitzGerald 2018).

IRE1α (hereafter referred to as IRE1) is the most evolutionarily conserved arm of the UPR and has dual functions in its cytoplasmic region, namely a serine/threonine kinase domain and ribonuclease domain. Like ATF6, IRE1 also has two isoforms; IRE1α and IRE1β, but again IRE1α has been the subject of a larger number of studies. This might be due to the fact that while the expression of IRE1α is ubiquitous, that of IRE1β is mostly restricted to the gut and to a lesser extend, to the airway epithelia (Almanza et al. 2019). IRE1 is also activated by dissociation of GRP78, which results in the formation of IRE1 homodimers or oligomers, followed by trans-autophosphorylation and activation of the IRE1 nuclease domain. X-box binding protein 1 (XBP1) mRNA is the major substrate for IRE1 RNase activity, as IRE1 cleaves a 26-nucleotide intron from the mRNA. The RNA fragments flanking the intron are then ligated by the tRNA ligase RtcB. This causes a frameshift in the *XBP1* mRNA open reading frame and its translation product, the protein XBP1s is more stable than the XBP1u form produced from the non-ligated mRNA. XBP1s is a transcription factor that also regulates genes encoding ER chaperones and degradation enzymes, as well as membrane and lipid synthesis genes. Another function of IRE1, that has been attributed to the kinase domain, is an association with the signal transducer TNF receptor-associated factor-2 (TRAF2), which signals to the JNK pathway through the activation of the kinase ASK1, promoting apoptosis (Rubio-Patiño et al. 2018). As well as XBP1 mRNA, the IRE1 RNase also degrades other mRNAs and microRNAs through the process of regulated IRE1-dependent decay

(RIDD), which may also help to ease the burden of protein accumulation in the ER (Maurel et al. 2014).

Although the luminal domains of IRE1 and PERK share weak amino-acid sequence homology (<12% homology overall), PERK activated in a somewhat similar manner to IRE1, by dissociation from GRP78, followed by homodimerization and trans-autophosphorylation. Unlike IRE1, PERK has exclusively kinase activity in its cytoplasmic region and phosphorylates several substrates including the eukaryotic translation initiation factor 2α (eIF2α) and the transcription factor Nrf2. The phosphorylation of eIF2α on serine 51 halts translation in the cell by blocking the formation of the ternary complex thereby attenuating almost all protein synthesis in attempt to alleviate ER stress. However, some mRNAs that contain short upstream open reading frames in their 5′-UTRs are still translated in these conditions. The transcription factor ATF4 is translated which launches a signalling cascade including the C/EBP homologous protein (CHOP), which is involved in apoptosis and autophagy signalling and GADD34, which as a subunit of the PP1 phosphatase that dephosphorylates eIF2α, is part of a negative feedback loop that shuts down PERK signalling (Bu and Diehl 2016) (Fig. 10.1).

10.3 ER-Associated Degradation

Although the ER has evolved optimal conditions for protein folding, proteins are not always folded correctly which results in an accumulation of misfolded protein or protein aggregates. The ER does not have a clearance system of its own so proteins must be retro-translocated to the cytosol for clearance by the ubiquitin proteasome system, in a process known as ER-associated degradation (ERAD). This separation of synthesis and degradation pathways is also a means of protection against inadvertent degradation of protein during synthesis and modification.

The ERAD process broadly consists of substrate recognition, ubiquitination and translocation from the ER and finally degradation by the proteasome. Substrate recognition differs depending on the topology of the protein and on the nature/location of the misfolded segments; ERAD-L, ERAD-M and ERAD-C deal with proteins misfolded in their luminal, membrane or cytoplasmic domains respectively. These pathways are well-defined in yeast, where much of the initial discoveries about ERAD were made, but in the more complex mammalian systems, there are overlapping components between the systems that have yet to be fully described. For ERAD-L, mannose residues of misfolded glycoproteins are recognised by three ER degradation-enhancing alpha-mannosidase like proteins (EDEM1, 2 and 3), allowing further recognition by the lectins OS-9 and XTP3-B which help to localise the complexes at the ER membrane in preparation for ER exit (Sun and Brodsky 2019). ERAD-M bypasses the lectin stage and instead interact directly with E3 ligases. ERAD-C also bypasses this route, as substrate recognition occurs in the cytoplasm in conjunction with chaperone proteins (Christianson and Ye 2014). Regardless of the mechanism of substrate recognition, the exact mechanisms of

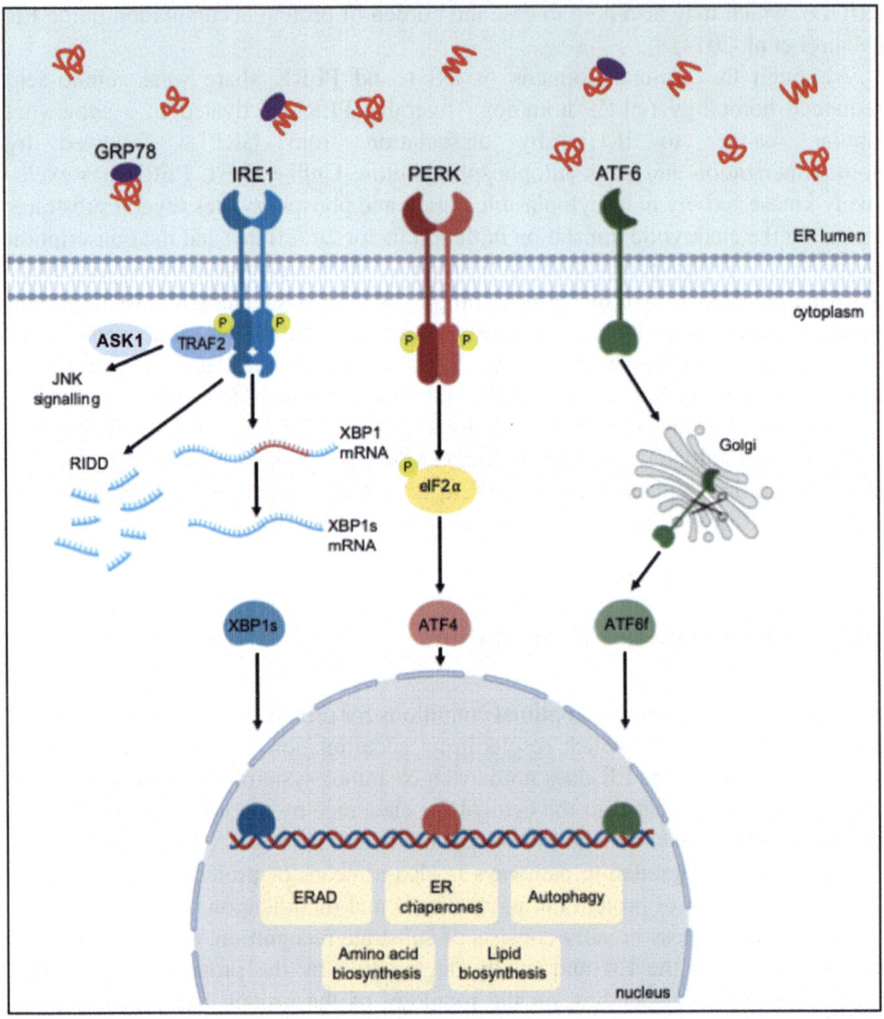

Fig. 10.1 Schematic representation of the conventional signalling component of the unfolded protein response—Unfolded or misfolded proteins accumulate in the ER lumen in unfavourable conditions. An attempt to restore proteostasis is made in the ER, where the GRP78 chaperone protein dissociates from the UPR sensors and instead binds to unfolded proteins. IRE1 and PERK are activated by dimerization/oligomerization and trans-autophosphorylation, and ATF6 by cleavage in the Golgi apparatus by S1P and S2P proteases. Activation of the UPR results in the production of a network of transcription factors; XBP1s in the case of IRE1, ATF4 in the case of PERK and ATF6f in the case of ATF6 which regulate the expression of a wide range of genes in the nucleus aimed at restoring correct protein folding conditions. In addition, IRE1 also activates JNK signalling through interaction with TRAF2 and controls RNA degradation through Regulated IRE1 Dependent Decay (RIDD)

protein retrotranslocation from the ER are not fully resolved. Some of the candidates for the composition of the retrotranslocon are Sec61 and the Derlin family of proteins, with interacting partners from the ubiquitin proteasome system including the HRD1 E3 ligase, the cofactor SEL1L and the AAA+ ATPase p97, which provides the energy required to cross the ER membrane (Kadowaki and Nishitoh 2019). Ubiquitination and retrotranslocation appear to be coupled in mammals, as the ubiquitin ligases that have been identified in the ERAD pathway may need access to misfolded proteins in the cytosol, which requires that transport across the ER membrane should be at least partially completed. Regardless of the mechanism, ubiquitinated proteins in the cytosol are delivered to the proteasome and degraded (Needham et al. 2019). The ERAD and UPR pathways are distinct cellular mechanisms, but ultimately share the same goal; that of restoring the protein folding capacity of the ER. There is some cross-talk between the pathways, as GRP78 participates in both processes and the transcription factors involved in the UPR can upregulate components of the ERAD pathway.

10.4 Surveillance Mechanisms

Other mechanisms also exist that ensure protein quality control in the ER, such as ER stress surveillance (ERSU) and protein bulk reflux from the ER into the cytosol (Igbaria et al. 2019), however these systems have only been described in yeast to date. ER proteins have occasionally been observed in the cytosol, but no explanation for this was explored. Recently, reports have demonstrated that some proteins are refluxed back to the cytosol, but in a folded form, making this process distinct from ERAD, as it is independent of ubiquitin E3 ligases that are obligatory for ERAD to occur. There are some unresolved questions remaining about this pathway, such as the identity of the translocation machinery and if there is a homologous pathway in other eukaryotes (Lajoie and Snapp 2019; Igbaria et al. 2019). The ERSU system is in place to ensure that a dividing cell that is experiencing ER stress does not transmit stressed ER to daughter cells, as new ER has to be transmitted and cannot be synthesised. Although the mechanism has been described in yeast, it is unclear if a similar system operates in mammalian cells (Spang 2018).

10.5 ER Import and Export Systems

There is overlap between protein quality control mechanisms and protein synthesis mechanisms, for example, the chaperone protein GRP78, which participates in the UPR, ERAD and protein synthesis. During protein synthesis, proteins are directed to the ER with the presence of a signal peptide that is recognised at the ribosome. With the aid of the Sec61 translocon, the protein can enter the ER to undergo proper folding and post-translational modifications in the oxidising environment of the

ER. Folding starts as the protein enters the ER, and further modifications to the protein may be made, such as glycosylation or the formation of disulphide bonds. Proteins are then trafficked out of the ER in COPII vesicles with the aid of Sec24 proteins, at distinct ER exit sites to the Golgi apparatus (McCaffrey and Braakman 2016). The ER itself does not operate alone, in the case of the secretory pathway it operates very closely with the Golgi apparatus where proteins are further processed and packaged for export. The ER also maintains contacts with mitochondria and the plasma membrane. Indeed, IRE1 and PERK have both been reported to interact with filamin A in the remodelling of the actin cytoskeleton (Urra et al. 2018; van Vliet et al. 2017). This interconnected network allows the ER to influence cell behaviour on a macro level with the aim of maintaining proteostasis.

10.6 ER Proteostasis Control in Cancer

Many studies in recent years have clearly demonstrated that proteostasis is perturbed in cancer, a fact that is unsurprising given the high protein synthesis demand for oncogenic processes such as angiogenesis and invasion, as well as a hostile tumour micro-environment. Tumour cells can also engage the UPR and other protein quality control mechanisms in response to therapy, and indeed some chemotherapies target components of the proteostasis network e.g. the proteasome inhibitor Bortezomib. The UPR is an attractive target for therapeutic interventions, either alone or in combination with conventional therapies given that it can control the balance between life and death for the cell, but as there is a large amount of cross-talk and some redundancy in the signalling network, careful targeting of specific components is required.

10.6.1 Oncogenes and Tumour Suppressors

The actions of oncogenes on cellular processes can cause disturbances in the proteostatic network because by their very nature, oncogenes drive an increase in proliferation and cell survival thereby changing the proteome of the cell in the process and activating protein quality control systems. Overexpression of the myc family of transcription factors is a common feature to many cancers and the UPR has been linked to aberrant myc signalling in cancer. For example, in a lymphoma model, increased myc signalling had multiple effects on the protein quality control, as forced expression of c-Myc led to an increase in the size of the ER and also affected the UPR by activation of both the PERK and IRE1 arms, with an increase in ATF4 and spliced XBP1 compared to healthy donors. The authors suggest that co-opting the UPR is one of the mechanisms by which myc signalling transforms the cell (Hart et al. 2012). In another study involving high-myc expressing Burkitt's lymphoma, the IRE1 arm of the UPR was shown to regulate lipid metabolism through the regulation of SCD1 by the XBP1s transcription factor. This helps to maintain the ER membrane under ER stress by providing unsaturated lipids, therefore regulating protein quality control at multiple levels by engaging the UPR to aid in driving expansion of the ER (Xie et al. 2018). In breast cancer, myc also engages the IRE1 pathway at multiple levels by regulating the ERN1 (ER to Nucleus signalling 1) gene, encoding IRE1, expression and by co-regulating gene expression in a heterodimer with XBP1s (Zhao et al. 2018). In T-cell acute lymphoblastic leukaemia, myc drives the expression of ubiquitin fusion degradation 1 (UFD1), which is part of the ERAD process. This allows myc to transform the cell, which causes ER stress, but also provides a mechanism for alleviating the proteotoxic stress that this causes (Huiting et al. 2018). In contrast to other cancer models, in prostate cancer IRE1-XBP1s signalling is upstream of myc signalling, as it is required for the activation of c-myc, indicating the possible presence of a myc-UPR feedback loop, but this has only been explored in this single cancer model (Sheng et al. 2019). Other oncogenes also affect UPR signalling, such as HRas, which co-opts both aspects of the IRE1 RNase activity to direct the cell towards proliferation or senescence. HRas, like other oncogenes, cause ER stress during cellular reprogramming. $HRas^{G12V}$ was identified as an activator of IRE1-XBP1s activity, promoting proliferation through XBP1s, but was also linked to senescence through degradation of ID1 mRNA by RIDD, which directs keratinocytes towards a senescent state (Blazanin et al. 2017).

The protein quality control machineries can be influenced by the actions of mutated oncogenes, but are also impacted by the effects of mutated tumour suppressors. The classical tumour suppressor protein, p53 has many links with the UPR and other protein quality control mechanism, such as an overexpression of IRE1 in mutant p53 cell lines and the identification of p53 targets that modulate protein folding in the ER (D'Orazi and Cirone 2019). In p53 null MEFs and cancer cell lines, the IRE1-XBP1s pathway is robustly induced in response to chemically-induced ER stress. However, in cells with wild-type p53, IRE1 is degraded in a p53-dependent manner, through the E3 ligase synoviolin (SYVN1). The same study also identified that p53 can modulate the secretory capacity of the ER, again through the IRE1-XBP1s arm of the UPR (Namba et al. 2015). In triple-negative breast cancer, the UPR is affected by mutant p53. The IRE1 and PERK arms of the UPR are not activated, possibly in an attempt to stop the pro-apoptotic nature of the UPR, whereas ATF6 activation is sustained promoting cell survival (Sicari et al. 2019).

10.6.2 Solid Cancers

In breast cancer, the UPR is co-opted for cell survival, as basal XBP1s expression was observed in triple-negative breast cancer (Chen et al. 2014; Logue et al. 2018) and helped to maintain stemness after treatment with chemotherapeutic agents. UPR activation also occurs in other solid tumours such as the aggressive brain cancer, glioblastoma. IRE1 was linked to angiogenesis and invasiveness in glioblastoma by examining a genetic approach to invalidating IRE1 signalling, which revealed a balance of IRE1 regulation between upregulation of invasive and antiangiogenic factors and downregulation of proangiogenic pathways. This highlights the need for careful consideration of treatment approaches, as targeting a particular pathway may result in compensation by another (Auf et al. 2010). Follow up studies to this work revealed the mRNAs SPARC and PER1 as RIDD targets that manipulate the extracellular matrix and cytokine expression respectively (Dejeans et al. 2012; Pluquet et al. 2013). IRE1 kinase activity has also been shown to affect glioblastoma proliferation by inducing epiregulin through JNK activation, which in turn activates the epidermal growth factor receptor and initiates an autocrine loop of cell proliferation and migration (Auf et al. 2013). A comprehensive study of IRE1 in glioblastoma classified tumours based on XBP1s and RIDD expression, with high XBP1s/low RIDD tumours resembling the mesenchymal glioblastoma subtype and low XBP1s/high RIDD tumours resembling the neural phenotype, and differences in survival between the groups. XBP1s was found to promote tumour aggressiveness through angiogenesis and immune cell recruitment whereas RIDD works in an opposing manner, reducing angiogenesis and invasiveness (Lhomond et al. 2018). Again, this is something to bear in mind for potential pleiotropic effects when targeting the UPR as a cancer therapy. IRE1 is not the only branch of the UPR involved in glioblastoma as PERK was found to negatively regulate SOX2 expression independent of ATF4/eIF2α in patient-derived neurospheres thus controlling differentiation (Peñaranda-Fajardo et al. 2019). The majority of studies characterise the UPR as pro-survival in cancer, although this may vary depending on the cancer type, but clearly demonstrates that cells can take over this signalling pathway to sustain the proliferation and other oncogenic processes required for development and maintenance of cancer.

Hypoxia is an important aspect of the influence of the tumour microenvironment on proteostasis. This has been reported in breast cancer, where it was shown that the activity of the HIF1α transcription factor, the master regulator of the cell response to hypoxia, was dependent on XBP1s and that XBP1s and HIF1α form a complex in the nucleus to regulate gene expression and promote cell survival (Chen et al. 2014). PERK and eIF2α phosphorylation are also activated under hypoxic conditions as the cells seek to down regulate protein synthesis in response to inhospitable conditions, however this is independent of HIF1α (Koumenis et al. 2002). The PERK arm of the UPR was also shown to be involved in breast cancer migration under moderate hypoxic conditions through the ATF4-LAMP3 pathway (Nagelkerke et al. 2013).

10.6.3 Blood Cancers

UPR involvement has been described in several cancers of the blood and lymphatic system. Multiple myeloma originates in plasma cells, which are characterised by a reliance on the ER to facilitate the large amount of antibody production and export required to carry out its functions. High XBP1s expression has been reported in multiple myeloma (Nakamura et al. 2006; Carrasco et al. 2007) and when XBP1s was overexpressed in a mouse model this led to a multiple myeloma like phenotype, with evidence of monoclonal gammopathy of undetermined significance (MGUS), the precursor to multiple myeloma. These mice also displayed high levels of the cytokine IL-6, which could be a driver of both the cancer and transformation of the plasma cells (Carrasco et al. 2007). Another group found that multiple myeloma also has a subpopulation of XBP1 negative cells that can mediate resistance to the proteasome inhibitor bortezomib, possibly because the cells are not fully committed to secretion, and are less susceptible to proteasomal inhibition (Leung-Hagesteijn et al. 2013). The IRE1 pathway has also been described in B-cell lymphoma, specifically the germinal centre B-cell-like subtype. Here, IRE1 and XBP1s expression were low due to transcriptional repression of IRE1 by the transcriptional repressor EZH2 which methylates histones at the IRE1 promoter. The authors also demonstrated that forced expression of XBP1s could significantly decrease tumour volume in immunodeficient mice, indicating that EZH2 inhibitors could be explored as a treatment option in this cancer (Bujisic et al. 2017). Again, these cancers show a reliance on the UPR to sustain cancer development and proliferation. In these examples, cancers depend on the IRE1-XBP1 pathway in particular, which is classically involved with expansion of the ER to accommodate the increase in protein demand.

10.7 Drugging ER Proteostasis in Cancer

The protein quality control mechanisms of the cell represent attractive targets for pharmacological intervention because cancer cells rely on them to facilitate growth and maintenance of tumours. Another important aspect of drug design is that there must be targetable proteins available to manipulate, which are available in the UPR and other processes. The UPR acts as an important node in the balance of the cell between life and death, as well as being a master regulator of other processes such as invasion, migration and proliferation. Therefore, it is important to understand how pharmacological intervention affects the UPR and how this response to drugs will go on to influence the behaviour of the cell. Many inhibitors have been developed that target the UPR, both at the level of the transducers and their downstream effectors, with PERK and IRE1 are regarded as easier to target because they contain kinase and ribonuclease domains that are amenable to targeting with small molecules. There is also interest in developing molecules that could act in co-ordination with existing therapies to improve patient outcomes.

10.7.1 IRE1-Targeting Drugs

A number of IRE1 modulators have been described to be effective in cancer models (Pelizzari-Raymundo et al. 2020). Among those molecules are MKC-3946 and its recent derivatives. These IRE1 inhibitors have been examined in many different cancer models such as triple-negative breast cancer, glioblastoma, colon cancer and multiple myeloma, where they were first described. MKC-3946 is an IRE1 RNase inhibitor that inhibits the formation of XBP1s without a negative impact on IRE1 phosphorylation. It does not target the other branches of the UPR and was shown in multiple myeloma to act synergistically with the proteasome inhibitor bortezomib or the HSP-90 inhibitor 17-AAG through upregulation of CHOP in a possible compensatory mechanism by the PERK arm of the UPR (Mimura et al. 2012). A newer-generation version of MKC-3946, MKC-8866 has also been tested in a number of cancers, including breast cancer and prostate cancer. In prostate cancer, MKC8866 reduced expression of XBP1s in both *in vitro* and *in vivo* models, and when combined with taxanes or antiandrogens, both standard treatments for prostate cancer, inhibited tumour growth, suggesting a possible role for targeting the UPR in cancer to enhance already existing therapies. The reduced growth was accounted for by a reduction in c-MYC signalling when IRE1 was inhibited, as XBP1s regulated myc expression in this model. As c-MYC has been notoriously difficult to target pharmacologically, this strategy could open up a way to target this central oncogenic signalling pathway at an upstream level (Sheng et al. 2019). Another study investigating the role of MYC and MKC-8866, performed in triple-negative breast cancer showed a different role for MYC in regulation of the IRE1 pathway, as it was shown that MYC up-regulated IRE1 and then worked in co-ordination with XBP1s in the nucleus. Regardless of the exact mechanisms of the pathway, inhibition of IRE1 RNase activity by MKC-8866 in a patient-derived xenograft model in combination with the chemotherapeutic agent docetaxel suppressed the growth of these tumours, indicating a possible treatment combination for patients with high-MYC expressing breast cancer (Zhao et al. 2018). A second report of MKC-8866 in triple-negative breast cancer showed the effect of IRE1 inhibition on the tumour secretome, as levels of pro-inflammatory cytokine production was reduced following treatment with the inhibitor. This study also used MKC-8866 in combination with a chemotherapeutic agent, paclitaxel *in vivo* where again, a synergistic effect of the two therapies was observed. A role for IRE1 in tumour recurrence was also observed, as there was a reduction in tumour volume even after the withdrawal of paclitaxel treatment (Logue et al. 2018). Another IRE1 RNase inhibitor, 4μ8C impeded the proliferation of colon cancer cells through an indirect mechanism as treatment with 4μ8C increased PERK and eIF2α activity, halting the translation and the expression of β-catenin, an important regulator of colon cancer tumorigenesis (Li et al. 2017). A derivative of 4μ8C, termed B-I09, was tested in a number of haematological cancers in vitro, including mantle cell lymphoma, multiple myeloma and leukaemia cell lines. Co-treatment with the tyrosine kinase inhibitor ibrutinib showed cytotoxicity through an increase in apoptosis (Tang and Ranatunga 2014). Further studies with this inhibitor in a c-Myc driven cancer, Burkitt's lymphoma, again in conjunction with traditional chemotherapies enhanced the apoptotic effect of these drugs (Xie et al. 2018). The IRE1 RNase activity is not the only targetable domain of the protein. Kinase inhibitors of IRE1 have also been developed and tested in models such as multiple myeloma, where it was recently shown that a compound targeting the ATP-docking site of IRE1 reduced XBP1 splicing *in vivo* and decreased the growth of multiple myeloma xenografts in mice and in patient-derived samples (Harnoss et al. 2019).

10.7.2 PERK-Targeting Drugs

IRE1 is not the only targetable pathway in the UPR, there are also inhibitors of PERK and some of its downstream effectors, though not all have been considered in cancer models. The PERK inhibitors GSK2606414 and GSK2656157 were first described in 2012 and 2013 respectively (Axten et al. 2012, 2013). GSK2656157 was tested in a pancreatic cancer mouse model and showed a decrease in tumour growth, apparently because of altered amino acid metabolism and decreased vascularisation (Atkins et al. 2013). When PERK was inhibited using GSK2656157 in hepatocellular carcinoma using an *in vivo* mouse model, tumour size was reduced, most likely by overwhelming proteotoxic stress, as eIF2α levels were not significantly reduced, but gene expression of chaperone proteins that normally alleviate ER stress were decreased (Vandewynckel et al. 2015). However, these inhibitors must be used with caution, as GSK2606414 showed pancreatic toxicity in a mouse model of prion disease (Halliday et al. 2015) and there have been reported off-targets of these drugs, as they have been demonstrated to also target the kinase RIPK1 in the nanomolar range (Rojas-Rivera et al. 2017). There are other opportunities to target the PERK pathway in cancer, for example with the small molecule ISRIB (integrated stress response inhibitor), which reverses the effects of eIF2α phosphorylation. In an oncogene driven prostate cancer model which relies on the PERK pathway for survival, ISRIB treatment resulted in an increase in apoptosis and demonstrated the efficacy of ISRIB as a treatment with a metastatic PDX model (Nguyen et al. 2018).

10.7.3 Other Drug Targets in ER Proteostasis

In addition to the targeting of the PERK and IRE1 arms of the UPR described above, the inhibition of the ATF6 arm is triggered by Ceapins (Gallagher et al. 2016) through the formation of inter-organelle tethers (Torres et al. 2019). In addition, other aspects of ER proteostasis can be affected through the use of alternative molecules. As such proteasome inhibitors are of interest in cancer, because they can block the ERAD process and so remove a process that cancer cells rely on to survive. Bortezomib was the first proteasome inhibitor to be approved and is now used in the treatment of multiple myeloma along with other lymphomas and leukaemia, but other newer-generation proteasome inhibitors such as carfilzomib and ixazomib are now also used (Robak and Robak 2019). The drug eeyarestatin and its derivatives block ERAD and glycoprotein synthesis through inhibition of the p97

ATPase and the Sec61 complex, and display potent cytotoxicity when combined with proteasome inhibitor treatment. The calcium storage function of the ER is also affected by these drugs, changing the folding capacity of the ER and causing cell death (Gamayun et al. 2019). There are also new approaches to manipulating the ubiquitin proteasome system by re-engineering E3 ligases to have new and different affinities for specific targets, which are useful tools in vitro and could be extended as treatment options in the future (Guang et al. 2019).

10.8 Summary Points

- ER protein quality control mechanisms are diverse and numerous. They can be directly ER-related i.e. ERAD or protein reflux, or can serve as a signal to the rest of the cell i.e. UPR.
- Cells exploit these quality control mechanisms to survive, and cancer cells in particular are reliant on quality control mechanisms.
- The reliance of cancer cells on quality control mechanisms and the fact that these mechanisms can affect the balance between life and death of the cell offer opportunities for drug targeting.
- Numerous drugs have been developed that target ER protein quality control mechanisms, but very few are in use in the clinic.

10.9 Future Issues

- Although much is known about ER protein quality control mechanisms, we still need to fill in the gaps in our knowledge.
- A better definition of ER stress, markers of such and a way to measure them is required. Recent approaches include the development of bioinformatics-based gene signatures, which could be expanded in future to aid in this goal.
- As the ER is on contact with practically every other cellular organelle, the influence of these other signalling mechanisms could be investigated to examine their combined impact on proteostasis.
- The drugs that have been developed to target ER protein quality control mechanisms, either in cancer models or other diseases can be developed further, newer derivatives can be examined in some cases or entirely new drugs can be investigated.
- The drugs that have been developed could be brought into clinical trials, either as monotherapies or in combination with existing treatments
- Care must be given to the type of cancer that will be treated with ER protein quality control targeting drugs, as different cancers and diseases may respond differently depending on the molecular background.

Acknowledgments This work was funded by grants from the French National Cancer Institute (INCa, PLBIO), the Fondation pour la recherche Médicale (FRM; équipe labellisée 2018) to EC and EU H2020 MSCA ITN-675448 (TRAINERS) and MSCA RISE-734749 (INSPIRED) grants to AS and EC. MMM was funded by scholarships from the Irish Research Council and the "Région Bretagne" (ARED international PhD) and by a travel grant from the French embassy in Dublin, Ireland. Figures were composed with BioRender (https://biorender.com/).

References

Adams CJ, Kopp MC, Larburu N, Nowak PR, Ali MMU (2019) Structure and molecular mechanism of er stress signaling by the unfolded protein response signal activator IRE1. Front Mol Biosci 6:11. https://doi.org/10.3389/fmolb.2019.00011

Almanza A, Carlesso A, Chintha C, Creedican S, Doultsinos D, Leuzzi B, Luís A et al (2019) Endoplasmic reticulum stress signalling - from basic mechanisms to clinical applications. FEBS J 286(2):241–278. https://doi.org/10.1111/febs.14608

Atkins C, Qi L, Minthorn E, Zhang S-Y, Figueroa DJ, Moss K, Stanley TB et al (2013) Characterization of a novel PERK kinase inhibitor with antitumor and antiangiogenic activity. Cancer Res 73(6):1993 LP–1992002. https://doi.org/10.1158/0008-5472.CAN-12-3109

Auf G, Jabouille A, Guérit S, Pineau R, Delugin M, Bouchecareilh M, Magnin N et al (2010) Inositol-requiring enzyme 1alpha is a key regulator of angiogenesis and invasion in malignant glioma. Proc Natl Acad Sci U S A 107(35):15553–15558. https://doi.org/10.1073/pnas.0914072107

Auf G, Jabouille A, Delugin M, Guérit S, Pineau R, North S, Platonova N et al (2013) High epiregulin expression in human u87 glioma cells relies on IRE1α and promotes autocrine growth through EGF receptor. BMC Cancer 13(1):597. https://doi.org/10.1186/1471-2407-13-597

Axten JM, Medina JR, Feng Y, Shu A, Romeril SP, Grant SW, Li WHH et al (2012) Discovery of 7-methyl-5-(1-{[3-(trifluoromethyl)phenyl]acetyl}-2,3-dihydro-1h-indol-5-yl)-7H-pyrrolo [2,3-d]pyrimidin-4-amine (GSK2606414), a potent and selective first-in-class inhibitor of protein kinase R (PKR)-like endoplasmic reticulum kinase (PERK). J Med Chem 55 (16):7193–7207. https://doi.org/10.1021/jm300713s

Axten JM, Romeril SP, Shu A, Ralph J, Medina JR, Feng Y, Li WHH et al (2013) Discovery of GSK2656157: an optimized PERK inhibitor selected for preclinical development. ACS Med Chem Lett 4(10):964–968. https://doi.org/10.1021/ml400228e

Blazanin N, Son J, Craig-Lucas AB, John CL, Breech KJ, Podolsky MA, Glick AB (2017) ER stress and distinct outputs of the IRE1α RNase control proliferation and senescence in response to oncogenic ras. Proc Natl Acad Sci U S A 114(37):9900 LP–9909905. https://doi.org/10.1073/pnas.1701757114

Bu Y, Diehl JA (2016) PERK integrates oncogenic signaling and cell survival during cancer development. J Cell Physiol 231(10):2088–2096. https://doi.org/10.1002/jcp.25336

Bujisic B, De Gassart A, Tallant R, Demaria O, Zaffalon L, Chelbi S, Gilliet M, Bertoni F, Martinon F (2017) Impairment of both IRE1 expression and XBP1 activation is a hallmark of GCB DLBCL and contributes to tumor growth. Blood 129(17):2420–2428. https://doi.org/10.1182/blood-2016-09-741348

Carrasco DR, Sukhdeo K, Protopopova M, Sinha R, Enos M, Carrasco DE, Zheng M et al (2007) The differentiation and stress response factor XBP-1 drives multiple myeloma pathogenesis. Cancer Cell 11(4):349–360. https://doi.org/10.1016/j.ccr.2007.02.015

Chen X, Iliopoulos D, Zhang Q, Tang Q, Greenblatt MB, Hatziapostolou M, Lim E et al (2014) XBP1 promotes triple negative breast cancer by controlling the HIF1 α pathway. Nature 508 (7494):103–107. https://doi.org/10.1038/nature13119

Christianson JC, Ye Y (2014) Cleaning up in the endoplasmic reticulum: ubiquitin in charge. Nat Struct Mol Biol 21(4):325–335. https://doi.org/10.1038/nsmb.2793

D'Orazi G, Cirone M (2019) Mutant P53 and cellular stress pathways: a criminal alliance that promotes cancer progression. Cancers 11(5):614. https://doi.org/10.3390/cancers11050614

Dejeans N, Pluquet O, Lhomond S, Grise F, Bouchecareilh M, Juin A, Meynard-Cadars M et al (2012) Autocrine control of glioma cells adhesion and migration through IRE1α-mediated cleavage of SPARC MRNA. J Cell Sci 125(18):4278–4287. http://jcs.biologists.org/content/125/18/4278.abstract

Gallagher CM, Garri C, Cain EL, Ang KK-H, Wilson CG, Chen S, Hearn BR et al (2016) Ceapins are a new class of unfolded protein response inhibitors, selectively targeting the ATF6α branch. Elife 5:e11878. https://doi.org/10.7554/eLife.11878

Gamayun I, O'Keefe S, Pick T, Klein M-C, Nguyen D, McKibbin C, Piacenti M et al (2019) Eeyarestatin compounds selectively enhance sec61-mediated Ca2+ leakage from the endoplasmic reticulum. Cell Chem Biol 26(4):571–583.e6. https://doi.org/10.1016/j.chembiol.2019.01.010

Guang MHZ, Kavanagh EL, Dunne LP, Dowling P, Li Z, Lindsay S, Bazou D et al (2019) Targeting proteotoxic stress in cancer: a review of the role that protein quality control pathways play in oncogenesis. Cancers 11(1):66. https://doi.org/10.3390/cancers11010066

Halliday M, Radford H, Sekine Y, Moreno J, Verity N, le Quesne J, Ortori CA et al (2015) Partial restoration of protein synthesis rates by the small molecule ISRIB prevents neurodegeneration without pancreatic toxicity. Cell Death Dis 6(3):e1672–e1672. https://doi.org/10.1038/cddis.2015.49

Harnoss JM, Le Thomas A, Shemorry A, Marsters SA, Lawrence DA, Min L, Chen Y-CA et al (2019) Disruption of IRE1α through its kinase domain attenuates multiple myeloma. Proc Natl Acad Sci U S A 116(33):16420–16429. https://doi.org/10.1073/pnas.1906999116

Hart LS, Cunningham JT, Datta T, Dey S, Tameire F, Lehman SL, Qiu B et al (2012) ER stress–mediated autophagy promotes myc-dependent transformation and tumor growth. J Clin Invest 122(12):4621–4634. https://doi.org/10.1172/JCI62973

Hetz C, Chevet E, Oakes SA (2015) Proteostasis control by the unfolded protein response. Nat Cell Biol 17(7):829–838. https://doi.org/10.1038/ncb3184

Higa A, Taouji S, Lhomond S, Jensen D, Fernandez-Zapico ME, Simpson JC, Pasquet J-M, Schekman R, Chevet E (2014) Endoplasmic reticulum stress-activated transcription factor ATF6α requires the disulfide isomerase PDIA5 to modulate chemoresistance. Mol Cell Biol 34(10):1839–1849. https://doi.org/10.1128/MCB.01484-13

Hillary RF, FitzGerald U (2018) A lifetime of stress: ATF6 in development and homeostasis. J Biomed Sci 25(1):48. https://doi.org/10.1186/s12929-018-0453-1

Huiting LN, Samaha Y, Zhang GL, Roderick JE, Li B, Anderson NM, Wang YW et al (2018) UFD1 contributes to MYC-mediated leukemia aggressiveness through suppression of the proapoptotic unfolded protein response. Leukemia 32(11):2339–2351. https://doi.org/10.1038/s41375-018-0141-x

Igbaria A, Merksamer PI, Trusina A, Tilahun F, Johnson JR, Brandman O, Krogan NJ, Weissman JS, Papa FR (2019) Chaperone-mediated reflux of secretory proteins to the cytosol during endoplasmic reticulum stress. Proc Natl Acad Sci 116(23):11291–11298. https://doi.org/10.1073/pnas.1904516116

Kadowaki H, Nishitoh H (2019) Endoplasmic reticulum quality control by garbage disposal. FEBS J 286(2):232–240. https://doi.org/10.1111/febs.14589

Koumenis C, Naczki C, Koritzinsky M, Rastani S, Diehl A, Sonenberg N, Koromilas A, Wouters BG (2002) Regulation of protein synthesis by hypoxia via activation of the endoplasmic reticulum kinase PERK and phosphorylation of the translation initiation factor EIF2α. Mol Cell Biol 22(21):7405–7416. https://doi.org/10.1128/MCB.22.21.7405-7416.2002

Lajoie P, Snapp EL (2019) Size-dependent secretory protein reflux into the cytosol in association with acute endoplasmic reticulum stress. BioRxiv. https://doi.org/10.1101/573428

Leung-Hagesteijn C, Erdmann N, Cheung G, Keats JJ, Stewart AK, Reece DE, Chung KC, Tiedemann RE (2013) Xbp1s-negative tumor b cells and pre-plasmablasts mediate therapeutic proteasome inhibitor resistance in multiple myeloma. Cancer Cell 24(3):289–304. https://doi.org/10.1016/j.ccr.2013.08.009

Lhomond S, Avril T, Dejeans N, Voutetakis K, Doultsinos D, McMahon M, Pineau R et al (2018) Dual IRE1 RNase functions dictate glioblastoma development. EMBO Mol Med 10(3):e7929. http://embomolmed.embopress.org/content/10/3/e7929.abstract

Li X-X, Zhang H-S, Xu Y-M, Zhang R-J, Chen Y, Fan L, Qin Y-Q, Liu Y, Li M, Fang J (2017) Knockdown of IRE1α inhibits colonic tumorigenesis through decreasing β-catenin and IRE1α targeting suppresses colon cancer cells. Oncogene 36(48):6738–6746. https://doi.org/10.1038/onc.2017.284

Logue SE, McGrath EP, Cleary P, Greene S, Mnich K, Almanza A, Chevet E et al (2018) Inhibition of IRE1 RNase Activity Modulates the Tumor Cell Secretome and Enhances Response to Chemotherapy. Nat Commun 9(1):3267. https://doi.org/10.1038/s41467-018-05763-8

Madden E, Logue SE, Healy SJ, Manie S, Samali A (2019) The role of the unfolded protein response in cancer progression: from oncogenesis to chemoresistance. Biol Cell 111(1):1–17. https://doi.org/10.1111/boc.201800050

Maurel M, Chevet E, Tavernier J, Gerlo S (2014) Getting RIDD of RNA: IRE1 in cell fate regulation. Trends Biochem Sci 39(5):245–254. https://doi.org/10.1016/j.tibs.2014.02.008

McCaffrey K, Braakman I (2016) Protein quality control at the endoplasmic reticulum. Essays Biochem 60(2):227 LP–227235. https://doi.org/10.1042/EBC20160003

Mimura N, Fulciniti M, Gorgun G, Tai Y-T, Cirstea D, Santo L, Yiguo H et al (2012) Blockade of XBP1 splicing by inhibition of IRE1α is a promising therapeutic option in multiple myeloma. Blood 119(24):5772–5781. https://doi.org/10.1182/blood-2011-07-366633

Nagelkerke A, Bussink J, Mujcic H, Wouters BG, Lehmann S, Sweep FCGJ, Span PN (2013) Hypoxia stimulates migration of breast cancer cells via the PERK/ATF4/LAMP3-arm of the unfolded protein response. Breast Cancer Res 15(1):R2. https://doi.org/10.1186/bcr3373

Nakamura M, Gotoh T, Okuno Y, Tatetsu H, Sonoki T, Uneda S, Mori M, Mitsuya H, Hata H (2006) Activation of the endoplasmic reticulum stress pathway is associated with survival of myeloma cells. Leuk Lymphoma 47(3):531–539. https://doi.org/10.1080/10428190500312196

Namba T, Chu K, Kodama R, Byun S, Yoon KW, Hiraki M, Mandinova A, Lee SW (2015) Loss of P53 enhances the function of the endoplasmic reticulum through activation of the IRE1α/XBP1 pathway. Oncotarget 6(24):19990–20001. http://legacy.oncotarget.com/index.php?journal=oncotarget&amp

Needham PG, Guerriero CJ, Brodsky JL (2019) Chaperoning endoplasmic reticulum–associated degradation (ERAD) and protein conformational diseases. Cold Spring Harb Perspect Biol 11 (8):a033928. https://doi.org/10.1101/cshperspect.a033928

Nguyen HG, Conn CS, Kye Y, Xue L, Forester CM, Cowan JE, Hsieh AC et al (2018) Development of a stress response therapy targeting aggressive prostate cancer. Sci Transl Med 10(439): eaar2036. https://doi.org/10.1126/scitranslmed.aar2036

Pelizzari-Raymundo D, Doultsinos D, Guillory X, Carlesso A, Eriksson LA, Chevet E (2020) Pharmacological targeting of IRE1 in cancer. Trends Cancer 6(12):1018–1030. https://doi.org/10.1016/j.trecan.2020.07.006

Peñaranda-Fajardo NM, Meijer C, Liang Y, Dijkstra BM, Aguirre-Gamboa R, den Dunnen WFA, Kruyt FAE (2019) ER stress and UPR activation in glioblastoma: identification of a noncanonical PERK mechanism regulating GBM stem cells through SOX2 modulation. Cell Death Dis 10(10):690. https://doi.org/10.1038/s41419-019-1934-1

Pluquet O, Dejeans N, Bouchecareilh M, Lhomond S, Pineau R, Higa A, Delugin M et al (2013) Posttranscriptional regulation of PER1 underlies the oncogenic function of IREα. Cancer Res 73 (15):4732–4743. https://doi.org/10.1158/0008-5472.CAN-12-3989

Robak P, Robak T (2019) Bortezomib for the treatment of hematologic malignancies: 15 years later. Drugs R&D 19(2):73–92. https://doi.org/10.1007/s40268-019-0269-9

Rojas-Rivera D, Delvaeye T, Roelandt R, Nerinckx W, Augustyns K, Vandenabeele P, Bertrand MJM (2017) When PERK inhibitors turn out to be new potent RIPK1 inhibitors: critical issues on the specificity and use of GSK2606414 and GSK2656157. Cell Death Differ 24 (6):1100–1110. https://doi.org/10.1038/cdd.2017.58

Rubio-Patiño C, Bossowski JP, Chevet E, Ricci J-E (2018) Reshaping the immune tumor microenvironment through IRE1 signaling. Trends Mol Med 24(7):607–614. https://doi.org/10.1016/j.molmed.2018.05.005

Sheng X, Nenseth HZ, Su Q, Kuzu OF, Frahnow T, Simon L, Greene S et al (2019) IRE1α-XBP1s pathway promotes prostate cancer by activating c-MYC signaling. Nat Commun 10(1):323. https://doi.org/10.1038/s41467-018-08152-3

Sicari D, Fantuz M, Bellazzo A, Valentino E, Apollonio M, Pontisso I, Di Cristino F et al (2019) Mutant P53 improves cancer cells' resistance to endoplasmic reticulum stress by sustaining activation of the UPR regulator ATF6. Oncogene 38(34):6184–6195. https://doi.org/10.1038/s41388-019-0878-3

Spang A (2018) The endoplasmic reticulum-the caring mother of the cell. Curr Opin Cell Biol 53:92–96. https://doi.org/10.1016/j.ceb.2018.06.004

Sun Z, Brodsky JL (2019) Protein quality control in the secretory pathway. J Cell Biol 218 (10):3171–3187. https://doi.org/10.1083/jcb.201906047

Tang C-HA, Ranatunga S, Kriss CL, Cubitt CL, Tao J, Pinilla-Ibarz JA, Del Valle JR, Hu C-CA (2014) Inhibition of ER stress–associated IRE-1/XBP-1 pathway reduces leukemic cell survival. J Clin Invest 124(6):2585–2598. https://doi.org/10.1172/JCI73448

Torres SE, Gallagher CM, Plate L, Gupta M, Liem CR, Guo X, Tian R et al (2019) Ceapins block the unfolded protein response sensor ATF6α by inducing a neomorphic inter-organelle tether. elife 8:e46595. https://doi.org/10.7554/eLife.46595

Urra H, Dufey E, Avril T, Chevet E, Hetz C (2017) Endoplasmic reticulum stress and the hallmarks of cancer. Trends Cancer 2(5):252–262. https://doi.org/10.1016/j.trecan.2016.03.007

Urra H, Daniel R Henriquez, José Cánovas, David Villarroel-Campos, Amado Carreras-Sureda, Eduardo Pulgar, Emiliano Molina, et al. (2018) IRE1α governs cytoskeleton remodelling and cell migration through a direct interaction with filamin A. Nat Cell Biol 20(8):942–953. https://doi.org/10.1038/s41556-018-0141-0

Vandewynckel Y-P, Laukens D, Bogaerts E, Paridaens A, Van den Bussche A, Verhelst X, Van Steenkiste C et al (2015) Modulation of the unfolded protein response impedes tumor cell adaptation to proteotoxic stress: a PERK for hepatocellular carcinoma therapy. Hepatol Int 9 (1):93–104. https://doi.org/10.1007/s12072-014-9582-0

Vliet V, Alexander FG, Gerlo S, Segura I, Van Eygen S, Molenberghs G, Rocha S et al (2017) The ER stress sensor PERK coordinates ER-plasma membrane contact site formation through interaction with Filamin-A and F-Actin remodeling. Mol Cell 65(5):885–899.e6. https://doi.org/10.1016/j.molcel.2017.01.020

Xie H, Tang C-HA, Song JH, Mancuso A, Del Valle JR, Cao J, Xiang Y et al (2018) IRE1α RNase–dependent lipid homeostasis promotes survival in Myc-transformed cancers. J Clin Invest 128 (4):1300–1316. https://doi.org/10.1172/JCI95864

Zhao N, Cao J, Xu L, Tang Q, Dobrolecki LE, Lv X, Talukdar M et al (2018) Pharmacological targeting of MYC-regulated IRE1/XBP1 pathway suppresses MYC-driven breast cancer. J Clin Invest 128(4):1283–1299. https://doi.org/10.1172/JCI95873

Chapter 11
IP_3 Receptor Biology and Endoplasmic Reticulum Calcium Dynamics in Cancer

Jan B. Parys, Geert Bultynck, and Tim Vervliet

Abstract Intracellular Ca^{2+} signaling regulates a plethora of cellular functions. A central role in these processes is reserved for the inositol 1,4,5-trisphosphate receptor (IP_3R), a ubiquitously expressed Ca^{2+}-release channel, mainly located in the endoplasmic reticulum (ER). Three IP_3R isoforms (IP_3R1, IP_3R2 and IP_3R3) exist, encoded respectively by *ITPR1*, *ITPR2* and *ITPR3*. The proteins encoded by these genes are each about 2700 amino acids long and assemble into large tetrameric channels, which form the target of many regulatory proteins, including several tumor suppressors and oncogenes. Due to the important role of the IP_3Rs in cell function, their dysregulation is linked to multiple pathologies. In this review, we highlight the complex role of the IP_3R in cancer, as it participates in most of the so-called "hallmarks of cancer". In particular, the IP_3R directly controls cell death and cell survival decisions *via* regulation of autophagy and apoptosis. Moreover, the IP_3R impacts cellular proliferation, migration and invasion. Typical examples of the role of the IP_3Rs in these various processes are discussed. The relative levels of the IP_3R isoforms expressed and their subcellular localization, *e.g.* at the ER-mitochondrial interface, is hereby important. Finally, evidence is provided about how the knowledge of the regulation of the IP_3R by tumor suppressors and oncogenes can be exploited to develop novel therapeutic approaches to fight cancer.

Keywords IP_3 receptor · Ca^{2+} signaling · Cancer · Migration · Invasion · Apoptosis · Autophagy

Abbreviations

AMPK	Adenosine monophosphate-activated protein kinase
APAF1	Apoptotic protease-activating factor 1
ARM	Armadillo solenoid fold

J. B. Parys (✉) · G. Bultynck · T. Vervliet
Laboratory for Molecular and Cellular Signaling, Department of Cellular and Molecular Medicine & Leuven Cancer Institute, KU Leuven, Leuven, Belgium
e-mail: jan.parys@kuleuven.be; geert.bultynck@kuleuven.be; tim.vervliet@kuleuven.be

L. B. Agellon, M. Michalak (eds.), *Cellular Biology of the Endoplasmic Reticulum*, Progress in Molecular and Subcellular Biology 59,
https://doi.org/10.1007/978-3-030-67696-4_11

ATG	Autophagy-related
BK_{Ca} Channel	large conductance Ca^{2+}-activated K^+ channel
ER	Endoplasmic reticulum
IICR	IP_3-induced Ca^{2+} release
IP_3	Inositol 1,4,5-trisphosphate
IP_3R	Inositol 1,4,5-trisphosphate receptor
ITPR	Inositol 1,4,5-trisphosphate receptor gene
MOMP	Mitochondrial outer membrane permeabilization
mPTP	Mitochondrial permeability transition pore
mTORC1	Mechanistic target of rapamycin complex 1
ROS	Reactive oxygen species
SMAC	Second mitochondria-derived activator of caspases
SOCE	Store-operated Ca^{2+} entry
XIAP	X-linked inhibitor of apoptosis protein

11.1 Introduction

Ca^{2+} is in all cell types a crucial intracellular messenger, controlling a plethora of cellular functions, including fertilization, differentiation, secretion, muscle contraction, synaptic plasticity and memory formation, gene transcription and metabolism (Berridge et al. 2000). Meanwhile it became clear that Ca^{2+} is also controlling cell proliferation (Humeau et al. 2018; Roderick and Cook 2008) and processes as macroautophagy (further named autophagy) and apoptosis (Dubois et al. 2016; Hu et al. 2019; La Rovere et al. 2016; Smaili et al. 2013) thereby directly affecting cell death and cell survival decisions. To perform this wide variety of functions, Ca^{2+} is encoded as complex spatio-temporal signals varying in frequency, amplitude, kinetics and subcellular localization which depend on the interplay between the members of the so-called "Ca^{2+}-signaling toolkit", a whole set of Ca^{2+}-binding proteins, Ca^{2+} transporters and modulators of these proteins (Berridge et al. 2000; Bootman and Bultynck 2020).

A central tenant hereby is the inositol 1,4,5-trisphosphate (IP_3) receptor (IP_3R), a ubiquitously expressed, intracellular Ca^{2+}-release channel mainly localized on the endoplasmic reticulum (ER). This ER Ca^{2+} channel is activated by IP_3, which is produced from phosphatidylinositol 1,4-bisphosphate by various phospholipase C isoforms after cellular stimulation by *e.g.* hormones, growth factors or neurotransmitters (Foskett et al. 2007; Parys and De Smedt 2012; Parys and Vervliet 2020; Prole and Taylor 2019).

Activation of the IP_3R leads to Ca^{2+} release from the ER into the cytoplasm, setting in motion various cellular signaling cascades. Depending on the cellular situation, the released Ca^{2+} can be funneled to other organelles, transported back into the ER or out of the cell. Meanwhile, the decreased Ca^{2+} load in the ER may lead to the dissociation of Ca^{2+} from the ER Ca^{2+} sensors Stim1/Stim2, leading to their oligomerisation and a rearrangement of their cytosolic domains. This enables the interaction of Stim1 and/or Stim2 with the plasma membrane Orai channels resulting in their opening (Lewis 2020; Zhou et al. 2019). The subsequent Ca^{2+} entry in the cell has a dual role, allowing a more prolonged Ca^{2+} signaling in the cell as well as a replenishment of the ER Ca^{2+} stores. As this form of Ca^{2+} entry primarily depends on the Ca^{2+}-store content, this process is known as store-operated Ca^{2+} entry (SOCE).

11.2 IP_3R Structure-Function Relationship

In higher organisms, there are three IP_3R isoforms (IP_3R1, IP_3R2, and IP_3R3), encoded by three different genes (*ITPR1*, *ITPR2* and *ITPR3*) and further diversity is provided at the protein level by splicing events. The homology among the three isoforms is rather high (around 75% at the amino acid level) and the IP_3Rs form homo- or heterotetrameric complexes with each a total molecular mass of about 1.2 MDa (Foskett et al. 2007; Parys and De Smedt 2012).

Each monomer has a length of about 2700 amino acids and can be divided in three large regions: at its N-terminal extremity the cytosolic ligand-binding region (for IP_3R1: amino acids 1–578), subsequently the large regulatory and coupling region (amino acids 579–2275), and finally the C-terminal part (amino acids 2276–2749) consisting of 6 transmembrane helices forming the ion channel proper, followed by a short cytosolic tail (Südhof et al. 1991) (Fig. 11.1).

The ligand-binding region can functionally be divided in a suppressor (amino acids 1–225) and an IP_3-binding core (amino acids 226–578). The IP_3-binding core of each of the three IP_3R isoforms forms a cleft with very high affinity for IP_3 (K_d 2 nM), which however is to a variable extent affected by its respective upstream suppressor sequence. The rank order of eventual affinities of the IP_3Rs for IP_3 is therefore IP_3R2 (K_d 14 nM) > IP_3R1 (K_d 50 nM) > IP_3R3 (K_d 163 nM) (Iwai et al. 2007).

IP_3R channel opening needs the binding of IP_3 to each of the four IP_3R monomers (Alzayady et al. 2016). This activation of the channel must be fine-tuned and occurs via a multitude of mechanisms. The regulators can be divided in 3 broad classes, small soluble cytosolic factors (*e.g.* Ca^{2+}, ATP, . . .) (Foskett et al. 2007; Taylor and Tovey 2010), regulatory proteins (Parys and Vervliet 2020; Prole and Taylor 2016) and post-translational modifications such as phosphorylation/dephosphorylation events (Vanderheyden et al. 2009).

The original analysis of the crystal structure of the N-terminal ligand-binding region (Bosanac et al. 2002, 2005), only quite recently extended to the full cytosolic part of the IP_3R (Hamada et al. 2017), has been complemented by high-quality cryo-electron microscopy (Taylor et al. 2004). The latter studies have tremendously increased our knowledge of the structure of the IP_3R.

Presently, the view of the general structure of the IP_3R1 can be summarized as follows. The IP_3R1 has a mushroom-like structure with a height of 19 nm.

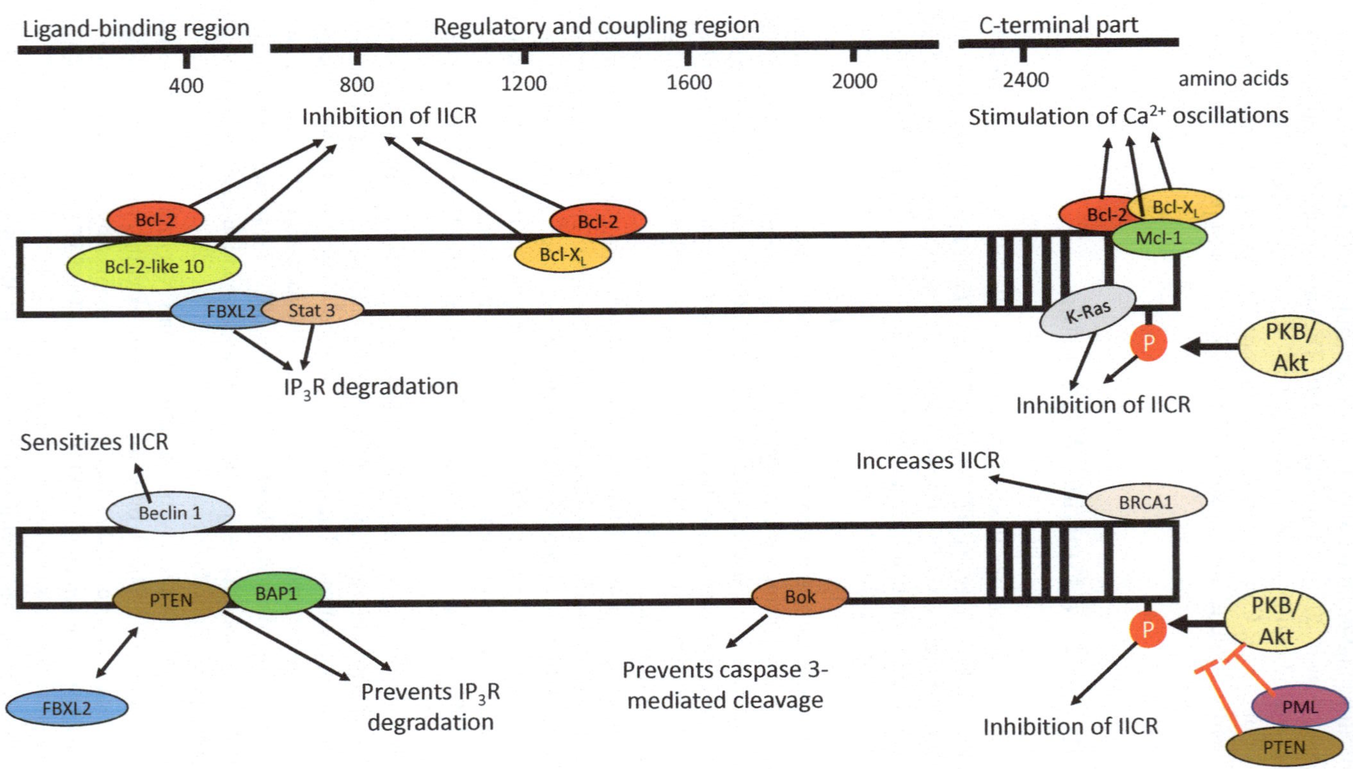

Fig. 11.1 Oncogenes and tumor suppressors interact with and influence IP_3R function. Linear representation of a generic IP_3R showing the location of the ligand-binding region, the regulatory and coupling region and the C-terminal part. A scale in amino acids is also provided. The oncogenes (top panel) and tumor suppressors (bottom panel) listed in the Tables are depicted at the proposed site of their interaction with the IP_3R. Some of them have multiple interaction sites on the IP_3R, and are therefore depicted multiple times, the same color always denoting the same protein. From only some of those proteins the exact interaction sites are known at the amino acid level. These are (in alphabetical order): Bcl-2 in the regulatory and coupling region (amino acids 1389–1408 of IP_3R1), Bcl-X_L in the C-terminal part (amino acids 2571–2606 and 2690–2732), Bok (amino acids 1895–1903 of IP_3R1), FBXL2 (amino acids 545–566 of IP_3R3), and the

PKB/Akt phosphorylation site (S^{2681} of IP_3R1). In general, oncogenes stimulate pro-survival Ca^{2+} oscillations while inhibiting pro-apoptotic Ca^{2+} signaling. The latter can occur by direct inhibition of the IP_3R, by inducing post-translational modification of the IP_3R or by stimulating its degradation. Tumor suppressors in contrast will stimulate pro-apoptotic Ca^{2+} signaling by either sensitizing the IP_3R or by preventing its degradation. Of note, there are special situations in which a tumor suppressor may compete with an oncogene for binding to the IP_3R, *e.g.* PTEN and FBXL2, or can reverse post-translational modifications thereby negating the pro-survival effects of the oncogenes. IICR: IP_3-induced Ca^{2+} release

High-resolution analysis of its closed form (no IP_3, no Ca^{2+} present) allowed the identification of ten domains: β-trefoil domains 1 and 2 (amino acids 1–436), followed by armadillo solenoid folds (ARM1–ARM3, amino acids 437–2192) with an α-helical domain between ARM1 and 2, the intervening lateral domain (amino acids 2193–2272), the transmembrane region with six transmembrane α-helices (amino acids 2273–2600), a linker domain (amino acids 2601–2680) and the C-terminal domain (amino acids 2681–2731) (Serysheva et al. 2017). The activation mechanism responsible for the channel opening in the presence of IP_3 however remains to be elucidated. Recent studies have provided additional information on the structure of IP_3R3 both in the presence and absence of IP_3 and/or Ca^{2+} (Paknejad and Hite 2018) and indicated in the regulatory and coupling region the presence of an auto-inhibitory peptide that can modulate the IP_3-binding site (Azumaya et al. 2020).

Future analysis of the structure of the various IP_3R isoforms under different experimental conditions will allow to better understand the activation and regulation mechanisms of this important class of channels as well as the physiological function of the various isoforms.

11.3 Cellular and Subcellular Localization of the IP_3R

The expression pattern of the various IP_3R isoforms as well as their subcellular localization is strongly tissue-, cell type-, and differentiation stage-dependent (Vermassen et al. 2004). Although most cells express more than one IP_3R isoform, IP_3R1 is the most ubiquitous isoform and can be expressed at very high levels in some tissues as *e.g.* Purkinje neurons and oocytes. IP_3R2 has a quite restricted expression pattern and is especially present in hepatocytes, glial cells, cardiomyocytes, various types of secretory cells as well as in some hematological cells (Vervloessem et al. 2015). IP_3R3 levels are generally elevated in rapidly proliferating cells, as well as, together with IP_3R2, in secretory cells (Ivanova et al. 2014).

The fact that IP_3Rs can be expressed at well-determined subcellular localizations strongly affects their function (Alonso and Garcia-Sancho 2011; Hohendanner et al. 2014; La Rovere et al. 2016; Lock et al. 2019). Examples of very specific subcellular localization of IP_3Rs are at the ER-mitochondrial contact sites (Csordas et al. 1999; Rizzuto et al. 1993; Szabadkai et al. 2006), in the junctional ER close to the plasma membrane (Okeke et al. 2016; Thillaiappan et al. 2017) as well as at contact sites between ER and lysosomes (Atakpa et al. 2018; Kilpatrick et al. 2013; Lopez-Sanjurjo et al. 2013; Morgan et al. 2013).

11.4 Intracellular Ca^{2+} Signaling and Cancer

As IP_3Rs control a plethora of cellular and physiological functions and processes, IP_3R dysregulation (Berridge 2016) and *ITPR* mutations (Hisatsune and Mikoshiba 2017; Kerkhofs et al. 2018; Terry et al. 2018; Gambardella et al. 2020) have patho (physio) logical consequences. While a number of those defects correlate with one specific disease situation (*e.g.* the occurrence of spinocerebellar ataxia or anhidrosis), this review will demonstrate that the role of the IP_3R in cancer is more complex and so illustrate the multifaceted aspects of IP_3R-mediated Ca^{2+} signaling.

Hanahan and Weinberg (Hanahan and Weinberg 2011) have put forward the concept that tumor formation depends on the acquisition of the so-called "hallmarks of cancer". These are the sum of the capabilities of the cells to sustain proliferation, evade growth suppressors, resist to cell death, possess replicative immortality, stimulate angiogenesis, invasion and metastasis, reprogram energy metabolism and evade the immune response. Interestingly, in most of these processes, Ca^{2+} can play a role (Monteith et al. 2017; Prevarskaya et al. 2018; Roberts-Thomson et al. 2019; Roderick and Cook 2008), and in various cancers a changed expression pattern was found for members of the Ca^{2+}-signaling toolkit, including the IP_3R, as was recently reviewed (Pierro et al. 2019). Modulation of IP_3R function in cancer cells can be attributed to one or several of the following factors: mutations, change in expression levels, change in subcellular localization or changes in Ca^{2+}-release activity (due to post-translational modifications, interaction with regulatory proteins or protein degradation). Especially noteworthy is that various recognized oncogenes and tumor suppressors directly interact with the IP_3Rs, thereby regulating IP_3R activity or stability (Bittremieux et al. 2016). For all these reasons, members of the Ca^{2+}-signaling toolkit have been proposed as novel therapeutic targets (Bong and Monteith 2018; Cui et al. 2017).

In this review, we will start by explaining the role of IP_3R signaling in autophagy and apoptosis as these processes are critical for cancer development. Next, the complex role that IP_3Rs can play during the various phases of cancer development will be discussed. Finally, we will highlight the interrelation that exist between the IP_3R and a number of tumor suppressors and oncogenes and how these interrelations can be exploited for therapeutic purposes.

11.5 The IP_3R as Regulator of Autophagy

To support cellular homeostasis and to protect themselves from cellular stress, cells make use of a process called autophagy. In this process, long-lived proteins as well as dysfunctional organelles and intracellular pathogens are engulfed by newly formed double-membrane vesicles named autophagosomes. The latter will eventually fuse with lysosomes, forming autolysosomes in which both the autophagosomes

Table 11.1 List of tumor suppressors interacting with the IP_3R and their effects

Tumor suppressor	Interaction sites	Functional effects
BAP1	N-terminus (IP_3R3)	Prevents IP_3R3 degradation
Beclin 1	Ligand-binding region (IP_3R1 and IP_3R3)	Sensitizes IICR
Bok	Regulatory and coupling region (IP_3R1 and IP_3R2)	Protects IP_3R1 and IP_3R2 from caspase 3-mediated cleavage
BRCA1	C-terminal part	Increases IICR and sensitivity to apoptosis
PML	Reduces PKB/Akt-mediated phosphorylation (IP_3R3)	Stimulates Ca^{2+} transfer to the mitochondria via IP_3R3
PTEN	Ligand-binding region (IP_3R3)	Displaces FBXL2 thereby inhibiting IP_3R3 degradation
	Inhibits PKB/Akt-mediated phosphorlyation	Stimulates IICR

Tumor suppressors are alphabetically listed. Interaction sites mention the IP_3R isoform involved when it was specifically described. When more than one interaction site was reported, the functional effects are separately indicated for each one. For the location of known interaction sites, please see Fig. 11.1. *IICR*: IP_3-induced Ca^{2+} release

and their content are fully degraded, after which the materials can be recycled. Autophagy forms a multi-step process that can be activated by adenosine monophosphate-activated protein kinase (AMPK), inhibited by the mechanistic target of rapamycin complex 1 (mTORC1) and in which over 30 autophagy-related (ATG) proteins play crucial roles (Ravikumar et al. 2010).

As autophagy has essentially a protective function, it can play a dual role in cancer. Initially, by supporting cellular homeostasis and thus antagonizing cellular stress, autophagy will protect cells from tumorigenicity, but once tumorigenesis has started it similarly will protect the tumoral cells from the harsh conditions in the tumor microenvironment (Eskelinen 2011; Liu and Ryan 2012). Due to the oncosuppressive role of several ATG proteins, including Beclin 1 (Table 11.1), these are considered as tumor suppressors (Yue et al. 2003).

Although the first effects of Ca^{2+} on autophagy have been described more than 2 decades ago (Gordon et al. 1993), the exact role(s) of Ca^{2+} signaling in the regulation of autophagy is only now slowly emerging. The main reason for this are the various, apparently contradictory results that have been published about the role of Ca^{2+}, as has been reviewed (Decuypere et al. 2011a; Sun et al. 2016). The reasons for these controversies are multiple, and can be due to cell-type dependent context, the occurrence of various Ca^{2+}-sensitive steps and the use of non-physiological conditions or of drugs with pleiotropic effects in a number of studies (Bootman et al. 2018).

Also the role of the IP_3R itself in autophagy is complex (Parys et al. 2012). On the one hand, since the IP_3R controls mitochondrial $[Ca^{2+}]$ and therefore mitochondrial metabolism (*vide infra*), it is clear that when Ca^{2+} transfer from the ER to the mitochondria is suppressed, ATP production drops, the AMP/ATP ratio increases and AMPK is activated. This increase in AMPK activity will lead to autophagy induction (Cardenas et al. 2010). Therefore, in basal conditions, the IP_3R acts as an inhibitor of autophagy. On the other hand, intracellular Ca^{2+} signals occurring in the cytosol can affect autophagy in a different way, whereby IP_3R-mediated Ca^{2+} release stimulates the autophagy process (Høyer-Hansen et al. 2007; Lam et al. 2008; Liu et al. 2013; Luyten et al. 2017; Wang et al. 2008). At the molecular level, IP_3Rs are directly sensitized by the pro-autophagic protein Beclin 1, driving starvation-induced autophagic flux (Decuypere et al. 2011b). Whether Beclin 1 recruitment by IP_3Rs and subsequent IP_3R sensitization also occurs in other authophagy-inducing conditions remains currently not known.

Therefore, the IP_3R, via its actions on autophagy, can also affect cancer cells (Kania et al. 2017). Interestingly, in contrast to other cells, cancer cells appear extremely sensitive to conditions in which the Ca^{2+} transfer to the mitochondria is inhibited. Autophagy induction does not guarantee their survival as the cancer cells, in spite of mitochondrial function decline, do not slow down their proliferation rate and thus undergo a necrotic collapse during cytokinesis (Cardenas et al. 2016).

A specific increase in both IP_3R2 and IP_3R3 expression levels in breast tumor tissues relative to the adjacent non-tumoral tissue was observed in patients (Singh et al. 2017b). These increased IP_3R expression levels correlated with metabolic changes as a higher rate of glycolysis and an increased glutaminolysis. IP_3R inhibition or siRNA-mediated silencing of IP_3R2 and -3 in breast cancer cell models affected exactly those same pathways, showing an association between IP_3R levels and the dysregulated metabolism of cancer cells. Interestingly, and similar to the study by Cardenas et al. mentioned above, the inhibition of IP_3Rs in breast cancer cells induced autophagy but the compromised energetics together with reactive oxygen species (ROS) production led to their demise (Singh et al. 2017a). Conversely, in clear cell renal cell carcinomas the increased hypoxia-inducible factor 2α level leads to IP_3R1 overexpression and autophagy induction (Messai et al. 2014). This autophagy induction protects the cancer cells against lysis mediated by natural killer cells.

11.6 The IP_3R in Apoptosis

An important hallmark of cancer is the avoidance of cell death and cancer cells therefore often have upregulated anti-apoptotic mechanisms. Apoptosis can originate as the consequence of perturbations in the intracellular milieu (intrinsic apoptosis) or due to the presence of extracellular factors (extrinsic apoptosis) (Galluzzi and Vanpouille-Box 2018).

In intrinsic apoptosis, the crucial step is the large structural changes occurring at the level of the mitochondria, called mitochondrial outer membrane permeabilization (MOMP), a process that occurs upon Bax/Bak activation and that is mediated by proteinaceous channel complexes formed by Bax/Bak oligomers (Kale et al. 2018; Singh et al. 2019). MOMP results in the release of pro-apoptotic factors as cytochrome c and second mitochondria-derived activator of caspases (SMAC) from the mitochondrial intermembrane space into the cytosol. Here, cytochrome c forms a complex with apoptotic protease-activating factor 1 (APAF1) forming the apoptosome, leading to subsequent activation of caspase 9 and of the executioner caspases 3 and 7 and thus apoptosis (Galluzzi and Vanpouille-Box 2018). SMAC supports these processes by inhibiting X-linked inhibitor of apoptosis protein (XIAP), a factor that blocks caspase activation.

It is known that IP_3Rs play a crucial role in the various processes linked to cellular survival or cell death. These properties are linked to the fact that (a fraction of the) IP_3Rs are localized at ER-mitochondrial contact sites and control therefore the transfer of Ca^{2+} from the ER to the mitochondria (Csordas et al. 1999; Giacomello et al. 2010; Rizzuto et al. 1993). If a too high level of Ca^{2+} transfer occurs from the ER to the mitochondria, mitochondrial Ca^{2+} overload follows, which leads to the opening of the mitochondrial permeability transition pore (mPTP) in the inner mitochondrial membrane. The molecular nature of the mPTP is still controversial (Carroll et al. 2019; Urbani et al. 2019) and the link between mitochondrial Ca^{2+} overload and mPTP opening is also not completely understood. The latter can involve Ca^{2+} binding to cardiolipin leading to the disintegration of mitochondrial complex II and a massive production of ROS that triggers mPTP opening (Hwang et al. 2014) and/or a Ca^{2+}-mediated conformational change of the F-ATP synthase which was proposed to form an essential component of the mPTP (Giorgio et al. 2017). Independently of the exact mechanism, the opening of the mPTP corresponds to the opening of a large, non-specific channel in the inner mitochondrial membrane, and consequently depolarization of the mitochondria, ATP hydrolysis, matrix swelling due to osmotic forces, and ultimately rupture of the outer mitochondrial membrane and the release of the above-mentioned pro-apoptotic factors.

Although originally the idea prevailed that IP_3R3 was particularly expressed at the ER-mitochondria contact sites and therefore most important for triggering Ca^{2+}-dependent apoptosis (Blackshaw et al. 2000; Mendes et al. 2005), other evidence also pointed to an important role in these processes for IP_3R1 (Feriod et al. 2017; Szabadkai et al. 2006) or IP_3R2 (Akl et al. 2013), suggesting a cell-type rather than a function-specific isoform distribution. Recent work extended these findings by conclusively demonstrating in cell line models that all three IP_3R isoforms can structurally participate in the ER-mitochondria contacts and that all three can flux Ca^{2+} to the mitochondria (Bartok et al. 2019). IP_3R2 appeared in fact the most effective for this process followed by IP_3R3 and IP_3R1. The latter was however the most effective inducer of SOCE leading to a sustained mitochondrial Ca^{2+} uptake.

The most important factor determining apoptosis sensitivity will therefore rather be the amplitude of the IP_3R-mediated Ca^{2+} transfer between ER and mitochondria than which IP_3R isoform is present at the ER-mitochondria contact sites. Interestingly, it was shown that cytochrome c, which is released from the mitochondria after MOMP, not only induces apoptosome formation, but also interacts with the IP_3R. This interaction antagonizes Ca^{2+}-mediated inhibition of the IP_3R and thus amplifies pro-apoptotic Ca^{2+} signaling (Boehning et al. 2003).

Extrinsic apoptosis is initiated by extracellular factors interacting with the so-called death receptors at the plasma membrane as the FAS and TRAIL receptors. Binding of respectively FAS ligand and TRAIL leads to a conformational change that allows for the formation of an intracellular "death-inducing signaling complex" leading eventually to activation of the initiator caspases 8 and 10. These caspases can either directly activate the effector caspases and/or activate them via cleaving the pro-apoptotic protein Bid thus triggering MOMP and apoptosome formation. This shows a convergence of the extrinsic and the intrinsic pathways (Barnhart et al. 2003; Scaffidi et al. 1998). Interestingly, it was shown that in some cells, including Jurkat T cells and hepatocytes, both phospholipase C γ1 and IP_3R1 were needed for Fas-mediated apoptosis (Wozniak et al. 2006). Moreover, Fas-dependent killing of Jurkat T cells by colon cancer cells depends on IP_3R-mediated Ca^{2+} release, which in turn is amplified by cytochrome c release (Steinmann et al. 2008). Apart from the mitochondrial Ca^{2+} overload that leads to a generalized MOMP and apoptosis, it has been proposed that IP_3R-mediated Ca^{2+} release also stimulates apoptosis by supporting lipidation of various proteins including the IP_3R itself (Fredericks et al. 2014), various components of the T cell receptor complex and proteins of the Wnt signaling pathway (Chen and Boehning 2017).

11.7 The Complex Role of the IP_3R in Tumorigenesis: An Integrated View

Tumorigenesis is typically a multi-step process. After oncogenic transformation, a first characteristic of tumor growth is the acquired capacity for quasi-unlimited cellular proliferation. The metastasis of the tumor subsequently depends on the migration capabilities of the cells followed by the invasion of other tissues. As will be discussed below, IP_3Rs actively participate in these key processes of tumorigenesis in various cancer types. The role of the IP_3R has particularly been investigated in breast and colorectal cancer and these cancer types will be discussed in more detail.

11.7.1 Breast Cancer

The growth stimulation of the MCF-7 breast cancer cell line by 17β-estradiol is coupled to an increased expression of IP_3R3 and a corresponding change in intracellular Ca^{2+} signaling (Szatkowski et al. 2010). siRNA-mediated IP_3R3 down-regulation modified ATP-evoked Ca^{2+} signals from sustained to oscillatory and antagonized the growth-promoting effect of 17β-estradiol, supporting a role for IP_3R3 in proliferation. This latter effect appeared to depend on a direct interaction between ER-localized IP_3R3 and the plasma membrane large conductance

Ca^{2+}-activated K^+ (BK_{Ca}) channels (Mound et al. 2013). This coupling allows IP_3R3 to activate the BK_{Ca} channels leading to cellular hyperpolarization. Down-regulation in MCF-7 cells of either IP_3R3 or BK_{Ca} (or of both) led to a decrease in cyclin D1 and in cyclin-dependent kinase 4 levels and to cell cycle arrest in the G0/G1 phase. Furthermore, when comparing the low-migrating MCF-7 and the highly migrating and invasive MDA-MB-231 and MDA-MB-435S cell lines, it appears that specifically a higher expression level of IP_3R3 correlates with a stronger migration capacity that can be reversed by IP_3R3 down-regulation or amplified by IP_3R3 overexpression (Mound et al. 2017). This property could be linked to the fact that the Ca^{2+} signals evoked by IP_3R3 affect the cellular morphology by impacting profilin levels and therefore the actin cytoskeleton (Vautrin-Glabik et al. 2018).

11.7.2 Colorectal Cancer

A study performed in 116 patients indicated that while in normal colorectal mucosa only IP_3R1 and IP_3R2 are expressed, in colorectal carcinomas all three IP_3R isoforms are expressed (Shibao et al. 2010). Moreover, not unlike the case of breast cancer, the IP_3R3 expression was especially high in actively invading cancer cells. IP_3R3 levels thus correlated with cancer aggressiveness and poorer patient survival (Shibao et al. 2010). Noteworthy is the fact that also in Caco-2 colon adenocarcinoma cells IP_3R3 levels inversely correlated with their sensitivity to apoptosis.

The role of IP_3R3 in tumor progression was further highlighted in a colorectal adenocarcinoma cell line DLD1 expressing a single oncogenic K-Ras mutation (G13D) making the protein constitutively active. K-Ras is a small monomeric GTPase and a well-known proto-oncogene. Mutations in the K-Ras isoform occur in 30–40% of colorectal cancers (Arrington et al. 2012). In accordance with the results discussed above (Shibao et al. 2010), IP_3R3 silencing made DLD1 colorectal adenocarcinoma cells more sensitive to apoptosis and decreased their migration capacity (Rezuchova et al. 2019). Moreover, inoculation in nude mice of DLD1 cells with complete IP_3R3 knockout led to smaller tumors than when parental DLD1 cells were inoculated.

Furthermore, comparison of colorectal cancer cell lines (expressing $K\text{-}Ras^{G13D}$) with their isogenic derivatives in which the mutated K-Ras allele was deleted (HKH2 and DKO4 cells), allowed a detailed investigation of the relation between oncogenic Ras and Ca^{2+} handling by the IP_3R (Pierro et al. 2014). In absence of oncogenic Ras, a remodeling occurred towards an increased expression of SERCA2B and of IP_3R3 while IP_3R1 level decreased. This remodeling correlated with an increased ER Ca^{2+} content, a higher level of IP_3-induced Ca^{2+} release (IICR), and subsequently more mitochondrial Ca^{2+} uptake and an increased sensitivity to apoptosis. A follow-up study furthermore indicated that the suppression of the oncogenic K-Ras also led to an increased expression of Stim1 (in the HKH2 cell line) or of Stim2 (in the DKO4 cell line) and a larger SOCE in both cell lines, which can contribute to the larger filling state of the ER (Pierro et al. 2018). Although these results fit in the context that increased IP_3R activity is linked to an increased propensity to apoptosis (*vide supra*) they appear to be in contradiction with the finding mentioned above that higher IP_3R3 levels correlate with increased cancer aggressiveness. Interestingly, the decreased IP_3R activity when oncogenic K-Ras is expressed, is potentially related to the finding that GTP-bound, phosphorylated K-Ras4B forms a trimolecular complex with the IP_3R and the anti-apoptotic protein Bcl-Xl (Sung et al. 2013). Under these conditions, the observed pro-survival stimulatory effect of Bcl-Xl (White et al. 2005) on the IP_3R is counteracted, leading to a less efficient respiration process, induction of autophagy, and ultimately cell death (Sung et al. 2013).

11.7.3 Other Cancers

Also in other cancer types, a relation between IP_3R levels and/or activity has been found with oncogenic progression. Interestingly, in several types of cancer a specific role was found for IP_3R3, though its role appears to differ between the various cancers, as will be further explained below.

In contrast to healthy hepatocytes, IP_3R3 is for example expressed in patients with hepatocellular carcinoma and this independently of the underlying liver disease (Guerra et al. 2019). This *de novo* expression of IP_3R3 appears to be due to a specific demethylation of its promoter. Although further work is needed to unravel the mechanisms, the high expression level of IP_3R3 leads not only to enhanced Ca^{2+} signaling but also to changes in expression of various proteins, including upregulation of the anti-apoptotic, pro-survival transcription factor POUAF1. Such remodeling might explain the lower sensitivity to apoptosis and the poorer patient survival.

In cholangiocarcinoma, the second most important liver malignancy, IP_3R3 levels are increased above normal level in both patient biopsies and in biliary adenocarcinoma cell lines as MzCha1 and HuCCA1 (Ueasilamongkol et al. 2020). Moreover, in MzCha1 cells, the IP_3R3, instead of being concentrated in the subapical ER, was expressed in a more diffuse way, including in close association with the mitochondria. Knockout of IP_3R3 in the cell models led to suppression of mitochondrial Ca^{2+} signaling, arrest of the cells in S-phase, increased cell death, and impaired proliferation and migration.

In clear cell renal cell carcinoma tumors, IP_3R3 mRNA levels were generally increased, while the mRNA levels of IP_3R1 and IP_3R2 appeared unaffected or even downregulated (Rezuchova et al. 2019). In the corresponding RCC4 cell line model, the silencing of either IP_3R1 or IP_3R3 affected cytosolic Ca^{2+} levels, though interestingly IP_3R1 silencing abolished apoptosis while IP_3R3 silencing strongly upregulated it.

In contrast with these various cancer cell types, in the lung adenocarcinoma cell line A549, IP_3R2 appeared the most important for cell migration (Huang et al. 2016). Interestingly, overexpression of the intraluminal ER protein PDIA10 (a.k.a. ERp44) inhibited IICR and decreased migration. It however remains to be elucidated how

PDIA10 is affecting in those cells IICR, as it was originally described to uniquely interact with IP_3R1 (Higo et al. 2005).

From the data presented here, it thus clearly appears that IP_3Rs and IICR are involved in various stages of tumorigenesis, but that their role is not fixed. Depending on the IP_3R isoform expressed, their intracellular localization, and the general cellular context, including cell type-specific expression of various other proteins, IICR can drive or regulate different intracellular pathways and consequently exert another physiological function.

11.8 IP_3Rs Are Targeted by Oncogenes and Tumor Suppressors: A Therapeutic Opportunity?

Several oncogenes and tumor suppressors were found to regulate IP_3R activity or stability and so to modulate their propensity for apoptosis (Bittremieux et al. 2016; Rosa et al. 2020). Generally speaking, oncogenes promote cell survival and thus tumor formation by suppressing pro-apoptotic Ca^{2+} signaling leading to mitochondrial Ca^{2+} overload, while tumor suppressors will rather stimulate these processes in order to induce apoptosis. Various regulatory proteins derive (part of) their oncogenic or tumor suppressive ability from interfering with the IP_3R-mediated Ca^{2+} signals. As we have reviewed this topic extensively in recent publications (Bittremieux et al. 2016; Parys and Vervliet 2020; Rosa et al. 2020), we have summarized the major properties of these proteins with respect to IP_3R-mediated Ca^{2+} signaling in Table 11.1 (for the tumor suppressors) and Table 11.2 (for the oncogenes) as well as their approximate binding sites on the IP_3R in Fig. 11.1. It should be emphasized that those proteins are not necessarily expressed in all cell types, and that their expression patterns and their functional importance can thus be cancer specific (Morciano et al. 2018). A special case concerns the interaction of various proteins of the Bcl-2 family with the IP_3R. With multiple binding sites and with both inhibitory and stimulatory effects described, the situation is quite complex. As a full description of the effects of those proteins on IICR fall outside the scope of this review, we refer to a recent review for that matter (Ivanova et al. 2020).

To illustrate the potential of the IP_3R in complex with its regulatory proteins to be used as a target for drug development in cancer treatment we would like to highlight some studies in which interfering with IP_3R/Bcl-2 complexes can provoke the death of cancer cells. These translational applications were driven by studies unraveling the molecular determinants underlying IP_3R/Bcl-2 complex formation. In particular, disruption of IP_3R/Bcl-2 complexes can be achieved using peptides containing the interaction site of Bcl-2 in the regulatory and coupling region of the IP_3R1 (amino acids 1389–1408). The binding of Bcl-2 to this site was demonstrated to inhibit IICR and to have an anti-apoptotic action (Rong et al. 2008). Various cell-permeable forms of this disrupting peptide have been developed, which differ in their stability/efficacy, and these have successfully been used to induce apoptosis via enhanced

Table 11.2 List of oncogenes interacting with the IP_3R and their effects

Oncogene	Interaction sites	Functional effects
Bcl-2	Ligand-binding region	Inhibits IICR
	Regulatory and coupling region (all IP_3R isoforms)	Inhibits IICR
	C-terminal part (all IP_3R isoforms)	Stimulates Ca^{2+} oscillations
Bcl-2 like 10	Ligand-binding region (IP_3R1)	Inhibits IICR
Bcl-X_L	Regulatory and coupling region	Inhibits IICR
	C-terminal part (all IP_3R isoforms)	Stimulates Ca^{2+} oscillations
FBXL2	Ligand-binding region (IP_3R3)	Promotes IP_3R3 degradation
K-Ras	C-terminal part	Suppresses IICR
Mcl-1	C-terminal part (all IP_3R isoforms)	Stimulates Ca^{2+} oscillations
PKB/Akt	C-terminal part (all IP_3R isoforms)	Phosphorylation of the IP_3R, leading to IICR inhibition
Stat3	Ligand-binding core/proximal part of the regulatory and coupling region (IP_3R3)	Induces IP_3R3 degradation

Oncogenes are alphabetically listed. Interaction sites mention the IP_3R isoform involved when it was specifically described. When more than one interaction site was reported, the functional effects are separately indicated for each one. For the location of known interaction sites, please see Fig. 11.1. *IICR*: IP_3-induced Ca^{2+} release

Ca^{2+} release from the ER in chronic lymphocytic leukemia (Zhong et al. 2011), diffuse large B cell lymphoma (Akl et al. 2013; Bittremieux et al. 2019), multiple myeloma and follicular lymphoma (Lavik et al. 2015) and small cell lung cancer (Greenberg et al. 2015) as well as to enhance the sensitivity of ovarian cancer cells towards cisplatin (Xie et al. 2018). Such peptides can also cooperate with BH3-mimetic drugs currently used in the clinic, including venetoclax (Akl et al. 2015; Greenberg et al. 2015; Lavik et al. 2015; Vervloessem et al. 2017). This clearly demonstrates that insights in the molecular underpinnings of IP_3R regulation by regulatory proteins, including oncogenes and tumor suppressors, can successfully be exploited therapeutically.

11.9 Conclusions

The IP_3R is a central player controlling many cellular functions. To that effect, IP_3R activity and stability are controlled by a plethora of mechanisms, often involving interacting proteins. Several of these interacting proteins are known oncogenes and tumor suppressors. As such, it comes as no surprise that altered IP_3R signaling is critically altered and contributes to several hallmarks of cancer. This is in large part due to IP_3R localization at the ER-mitochondrial contact sites allowing it to regulate mitochondrial Ca^{2+} levels and subsequently mitochondrial processes including energy production and apoptosis. Interestingly, depending on the cell type under investigation, the various IP_3R isoforms can have different properties and consequently different effects. This can be due to a multitude of reasons, including the presence or absence of regulatory proteins, their subcellular localization and the amplitude of their subsequent Ca^{2+} signals that together will determine which intracellular pathways will be activated. Given the central role of IP_3Rs in most if not all hallmarks of cancer, the IP_3R may have crucial roles in cancer prevention, neoplasia and cancer progression. These insights can be employed to develop novel anti-cancer strategies, as well as providing interesting opportunities for future research and the development of translational applications.

11.10 Summary Points

- IP_3R receptors control cell death and cell survival decisions critically involved in cancer development
- IP_3R receptors control cellular proliferation, migration and invasion
- IP_3R are targets of oncogenes and tumor suppressors
- Expression of specific IP_3R isoforms may differentially impact cancer progression
- Targeting IP_3R-mediated signaling could provide novel therapeutic avenues to fight cancer

11.11 Future Issues

- Obtaining detailed structural information about the open and closed states of the various IP_3R isoforms will provide important information about the mode of action of regulatory oncogenes and tumor suppressors
- It is anticipated that not all regulatory proteins acting on the IP_3R in cancer have been identified. Moreover, such proteins likely are highly cancer specific, underlying unique opportunities to target such complexes in different cancers. Further research on this topic is urgently needed
- The possibility for structural and/or functional interactions between various oncogenes and/or tumor suppressors in the regulation of the IP_3R is presently very insufficiently appreciated. It is however crucial to understand how these proteins affect each other in the regulation of the IP_3R
- Development of (isoform) specific IP_3R inhibitors will further increase the understanding of their exact role in cancer
- Targeting the cell-type specific interaction of tumor suppressors and oncogenes with the IP_3R as therapeutic avenue must be further explored and stable peptides and small molecules developed to that effect

Acknowledgements Research work performed in our laboratory on this topic was supported by grants from the Research Foundation—Flanders (FWO) (G073109N, G057112N, G081913N, G063413N, G0C9114N, G092715N, G0A3416N, G090118N and G0A6919N), the Inter University Attraction Poles program P7/13 and by the Research Council of the KU Leuven (Concerted Actions G.O.A. 09/012 and grants 14/101 and C14/19/099). T.V. obtained a postdoctoral fellowship of the FWO. G.B. and J.B.P. are part of the FWO Scientific Research Network CaSign (W0.019.17N).

Conflict-of-Interest Disclosure Statement The authors declare no conflict of interest.

References

Akl H, Monaco G, La Rovere R, Welkenhuyzen K, Kiviluoto S et al (2013) IP_3R2 levels dictate the apoptotic sensitivity of diffuse large B-cell lymphoma cells to an IP_3R-derived peptide targeting the BH4 domain of Bcl-2. Cell Death Dis 4:e632. https://doi.org/10.1038/cddis.2013.140

Akl H, La Rovere RM, Janssens A, Vandenberghe P, Parys JB et al (2015) HA14-1 potentiates apoptosis in B-cell cancer cells sensitive to a peptide disrupting IP_3 receptor/Bcl-2 complexes. Int J Dev Biol 59(7–9):391–398. https://doi.org/10.1387/ijdb.150213gb

Alonso MT, Garcia-Sancho J (2011) Nuclear Ca^{2+} signalling. Cell Calcium 49(5):280–289. https://doi.org/10.1016/j.ceca.2010.11.004

Alzayady KJ, Wang L, Chandrasekhar R, Wagner LE II, Van Petegem F et al (2016) Defining the stoichiometry of inositol 1,4,5-trisphosphate binding required to initiate Ca^{2+} release. Sci Signal 9(422):ra35. https://doi.org/10.1126/scisignal.aad6281

Arrington AK, Heinrich EL, Lee W, Duldulao M, Patel S et al (2012) Prognostic and predictive roles of KRAS mutation in colorectal cancer. Int J Mol Sci 13(10):12153–12168. https://doi.org/10.3390/ijms131012153

Atakpa P, Thillaiappan NB, Mataragka S, Prole DL, Taylor CW (2018) IP_3 receptors preferentially associate with ER-lysosome contact sites and selectively deliver Ca^{2+} to lysosomes. Cell Rep 25(11):3180–3193 e7. https://doi.org/10.1016/j.celrep.2018.11.064

Azumaya CM, Linton EA, Risener CJ, Nakagawa T, Karakas E (2020) Cryo-EM structure of human type-3 inositol triphosphate receptor reveals the presence of a self-binding peptide that acts as an antagonist. J Biol Chem 295(6):1743–1753. https://doi.org/10.1074/jbc.RA119.011570

Barnhart BC, Alappat EC, Peter ME (2003) The CD95 type I/type II model. Semin Immunol 15(3):185–193. https://doi.org/10.1016/s1044-5323(03)00031-9

Bartok A, Weaver D, Golenar T, Nichtova Z, Katona M et al (2019) IP_3 receptor isoforms differently regulate ER-mitochondrial contacts and local calcium transfer. Nat Commun 10(1):3726. https://doi.org/10.1038/s41467-019-11646-3

Berridge MJ (2016) The inositol trisphosphate/calcium signaling pathway in health and disease. Physiol Rev 96(4):1261–1296. https://doi.org/10.1152/physrev.00006.2016

Berridge MJ, Lipp P, Bootman MD (2000) The versatility and universality of calcium signalling. Nat Rev Mol Cell Biol 1(1):11–21. https://doi.org/10.1038/35036035

Bittremieux M, Parys JB, Pinton P, Bultynck G (2016) ER functions of oncogenes and tumor suppressors: Modulators of intracellular Ca^{2+} signaling. Biochim Biophys Acta 1863(6 Pt B):1364–1378. https://doi.org/10.1016/j.bbamcr.2016.01.002

Bittremieux M, La Rovere RM, Akl H, Martines C, Welkenhuyzen K et al (2019) Constitutive IP_3 signaling underlies the sensitivity of B-cell cancers to the Bcl-2/IP_3 receptor disruptor BIRD-2. Cell Death Differ 26(3):531–547. https://doi.org/10.1038/s41418-018-0142-3

Blackshaw S, Sawa A, Sharp AH, Ross CA, Snyder SH et al (2000) Type 3 inositol 1,4,5-trisphosphate receptor modulates cell death. FASEB J 14(10):1375–1379. https://doi.org/10.1096/fj.14.10.1375

Boehning D, Patterson RL, Sedaghat L, Glebova NO, Kurosaki T et al (2003) Cytochrome c binds to inositol (1,4,5) trisphosphate receptors, amplifying calcium-dependent apoptosis. Nat Cell Biol 5(12):1051–1061. https://doi.org/10.1038/ncb1063

Bong AHL, Monteith GR (2018) Calcium signaling and the therapeutic targeting of cancer cells. Biochim Biophys Acta Mol Cell Res 1865(11 Pt B):1786–1794. https://doi.org/10.1016/j.bbamcr.2018.05.015

Bootman MD, Bultynck G (2020) Fundamentals of cellular calcium signaling: a primer. Cold Spring Harb Perspect Biol 12(1):a038802. https://doi.org/10.1101/cshperspect.a038802

Bootman MD, Chehab T, Bultynck G, Parys JB, Rietdorf K (2018) The regulation of autophagy by calcium signals: Do we have a consensus? Cell Calcium 70:32–46. https://doi.org/10.1016/j.ceca.2017.08.005

Bosanac I, Alattia JR, Mal TK, Chan J, Talarico S et al (2002) Structure of the inositol 1,4,5-trisphosphate receptor binding core in complex with its ligand. Nature 420(6916):696–700. https://doi.org/10.1038/nature01268

Bosanac I, Yamazaki H, Matsu-Ura T, Michikawa T, Mikoshiba K et al (2005) Crystal structure of the ligand binding suppressor domain of type 1 inositol 1,4,5-trisphosphate receptor. Mol Cell 17(2):193–203. https://doi.org/10.1016/j.molcel.2004.11.047

Cardenas C, Miller RA, Smith I, Bui T, Molgó J et al (2010) Essential regulation of cell bioenergetics by constitutive $InsP_3$ receptor Ca^{2+} transfer to mitochondria. Cell 142 (2):270–283. https://doi.org/10.1016/j.cell.2010.06.007

Cardenas C, Muller M, McNeal A, Lovy A, Jana F et al (2016) Selective vulnerability of cancer cells by inhibition of Ca^{2+} transfer from endoplasmic reticulum to mitochondria. Cell Rep 14 (10):2313–2324. https://doi.org/10.1016/j.celrep.2016.02.030

Carroll J, He J, Ding S, Fearnley IM, Walker JE (2019) Persistence of the permeability transition pore in human mitochondria devoid of an assembled ATP synthase. Proc Natl Acad Sci U S A 116(26):12816–12821. https://doi.org/10.1073/pnas.1904005116

Chen JJ, Boehning D (2017) Protein lipidation as a regulator of apoptotic calcium release: relevance to cancer. Front Oncol 7:138. https://doi.org/10.3389/fonc.2017.00138

Csordas G, Thomas AP, Hajnoczky G (1999) Quasi-synaptic calcium signal transmission between endoplasmic reticulum and mitochondria. EMBO J 18(1):96–108. https://doi.org/10.1093/emboj/18.1.96

Cui C, Merritt R, Fu L, Pan Z (2017) Targeting calcium signaling in cancer therapy. Acta Pharm Sin B 7(1):3–17. https://doi.org/10.1016/j.apsb.2016.11.001

Decuypere JP, Bultynck G, Parys JB (2011a) A dual role for Ca^{2+} in autophagy regulation. Cell Calcium 50(3):242–250. https://doi.org/10.1016/j.ceca.2011.04.001

Decuypere JP, Welkenhuyzen K, Luyten T, Ponsaerts R, Dewaele M et al (2011b) Ins(1,4,5)P3 receptor-mediated Ca^{2+} signaling and autophagy induction are interrelated. Autophagy 7 (12):1472–1489. https://doi.org/10.4161/auto.7.12.17909

Dubois C, Prevarskaya N, Vanden Abeele F (2016) The calcium-signaling toolkit: updates needed. Biochim Biophys Acta 1863(6 Pt B):1337–1343. https://doi.org/10.1016/j.bbamcr.2015.11.033

Eskelinen EL (2011) The dual role of autophagy in cancer. Curr Opin Pharmacol 11(4):294–300. https://doi.org/10.1016/j.coph.2011.03.009

Feriod CN, Oliveira AG, Guerra MT, Nguyen L, Richards KM et al (2017) Hepatic inositol 1,4,5 trisphosphate receptor type 1 mediates fatty liver. Hepatol Commun 1(1):23–35. https://doi.org/10.1002/hep4.1012

Foskett JK, White C, Cheung KH, Mak DO (2007) Inositol trisphosphate receptor Ca^{2+} release channels. Physiol Rev 87(2):593–658. https://doi.org/10.1152/physrev.00035.2006

Fredericks GJ, Hoffmann FW, Rose AH, Osterheld HJ, Hess FM et al (2014) Stable expression and function of the inositol 1,4,5-triphosphate receptor requires palmitoylation by a DHHC6/

selenoprotein K complex. Proc Natl Acad Sci U S A 111(46):16478–16483. https://doi.org/10.1073/pnas.1417176111

Galluzzi L, Vanpouille-Box C (2018) BAX and BAK at the gates of innate immunity. Trends Cell Biol 28(5):343–345. https://doi.org/10.1016/j.tcb.2018.02.010

Gambardella J, Lombardi A, Morelli MB, Ferrara J, Santulli G (2020) Inositol 1,4,5-trisphosphate receptors in human disease: a comprehensive update. J Clin Med 9(4):1096. https://doi.org/10.3390/jcm9041096

Giacomello M, Drago I, Bortolozzi M, Scorzeto M, Gianelle A et al (2010) Ca^{2+} hot spots on the mitochondrial surface are generated by Ca^{2+} mobilization from stores, but not by activation of store-operated Ca^{2+} channels. Mol Cell 38(2):280–290. https://doi.org/10.1016/j.molcel.2010.04.003

Giorgio V, Burchell V, Schiavone M, Bassot C, Minervini G et al (2017) Ca^{2+} binding to F-ATP synthase β subunit triggers the mitochondrial permeability transition. EMBO Rep 18 (7):1065–1076. https://doi.org/10.15252/embr.201643354

Gordon PB, Holen I, Fosse M, Rotnes JS, Seglen PO (1993) Dependence of hepatocytic autophagy on intracellularly sequestered calcium. J Biol Chem 268(35):26107–26112

Greenberg EF, McColl KS, Zhong F, Wildey G, Dowlati A et al (2015) Synergistic killing of human small cell lung cancer cells by the Bcl-2-inositol 1,4,5-trisphosphate receptor disruptor BIRD-2 and the BH3-mimetic ABT-263. Cell Death Dis 6:e2034. https://doi.org/10.1038/cddis.2015.355

Guerra MT, Florentino RM, Franca A, Lima Filho AC, Dos Santos ML et al (2019) Expression of the type 3 $InsP_3$ receptor is a final common event in the development of hepatocellular carcinoma. Gut 68(9):1676–1687. https://doi.org/10.1136/gutjnl-2018-317811

Hamada K, Miyatake H, Terauchi A, Mikoshiba K (2017) IP_3-mediated gating mechanism of the IP_3 receptor revealed by mutagenesis and X-ray crystallography. Proc Natl Acad Sci U S A 114 (18):4661–4666. https://doi.org/10.1073/pnas.1701420114

Hanahan D, Weinberg RA (2011) Hallmarks of cancer: the next generation. Cell 144(5):646–674. https://doi.org/10.1016/j.cell.2011.02.013

Higo T, Hattori M, Nakamura T, Natsume T, Michikawa T et al (2005) Subtype-specific and ER lumenal environment-dependent regulation of inositol 1,4,5-trisphosphate receptor type 1 by ERp44. Cell 120(1):85–98. https://doi.org/10.1016/j.cell.2004.11.048

Hisatsune C, Mikoshiba K (2017) IP_3 receptor mutations and brain diseases in human and rodents. J Neurochem 141(6):790–807. https://doi.org/10.1111/jnc.13991

Hohendanner F, McCulloch AD, Blatter LA, Michailova AP (2014) Calcium and IP_3 dynamics in cardiac myocytes: experimental and computational perspectives and approaches. Front Pharmacol 5:35. https://doi.org/10.3389/fphar.2014.00035

Høyer-Hansen M, Bastholm L, Szyniarowski P, Campanella M, Szabadkai G et al (2007) Control of macroautophagy by calcium, calmodulin-dependent kinase kinase-β, and Bcl-2. Mol Cell 25 (2):193–205. https://doi.org/10.1016/j.molcel.2006.12.009

Hu YX, Han XS, Jing Q (2019) Ca^{2+} ion and autophagy. Adv Exp Med Biol 1206:151–166. https://doi.org/10.1007/978-981-15-0602-4_7

Huang X, Jin M, Chen YX, Wang J, Zhai K et al (2016) ERP44 inhibits human lung cancer cell migration mainly via IP_3R2. Aging (Albany NY) 8(6):1276–1286. https://doi.org/10.18632/aging.100984

Humeau J, Bravo-San Pedro JM, Vitale I, Nunez L, Villalobos C et al (2018) Calcium signaling and cell cycle: progression or death. Cell Calcium 70:3–15. https://doi.org/10.1016/j.ceca.2017.07.006

Hwang MS, Schwall CT, Pazarentzos E, Datler C, Alder NN et al (2014) Mitochondrial Ca^{2+} influx targets cardiolipin to disintegrate respiratory chain complex II for cell death induction. Cell Death Differ 21(11):1733–1745. https://doi.org/10.1038/cdd.2014.84

Ivanova H, Vervliet T, Missiaen L, Parys JB, De Smedt H et al (2014) Inositol 1,4,5-trisphosphate receptor-isoform diversity in cell death and survival. Biochim Biophys Acta 1843 (10):2164–2183. https://doi.org/10.1016/j.bbamcr.2014.03.007

Ivanova H, Vervliet T, Monaco G, Terry LE, Rosa N et al (2020) Bcl-2-protein family as modulators of IP_3 receptors and other organellar Ca^{2+} channels. Cold Spring Harb Perspect Biol 12(4):a035089. https://doi.org/10.1101/cshperspect.a035089

Iwai M, Michikawa T, Bosanac I, Ikura M, Mikoshiba K (2007) Molecular basis of the isoform-specific ligand-binding affinity of inositol 1,4,5-trisphosphate receptors. J Biol Chem 282 (17):12755–12764. https://doi.org/10.1074/jbc.M609833200

Kale J, Osterlund EJ, Andrews DW (2018) BCL-2 family proteins: changing partners in the dance towards death. Cell Death Differ 25(1):65–80. https://doi.org/10.1038/cdd.2017.186

Kania E, Roest G, Vervliet T, Parys JB, Bultynck G (2017) IP_3 receptor-mediated calcium signaling and its role in autophagy in cancer. Front Oncol 7:140. https://doi.org/10.3389/fonc.2017.00140

Kerkhofs M, Seitaj B, Ivanova H, Monaco G, Bultynck G et al (2018) Pathophysiological consequences of isoform-specific IP_3 receptor mutations. *Biochim Biophys Acta Mol Cell Res* 1865(11 Pt B):1707–1717. https://doi.org/10.1016/j.bbamcr.2018.06.004

Kilpatrick BS, Eden ER, Schapira AH, Futter CE, Patel S (2013) Direct mobilisation of lysosomal Ca^{2+} triggers complex Ca^{2+} signals. J Cell Sci 126(Pt 1):60–66. https://doi.org/10.1242/jcs.118836

La Rovere RM, Roest G, Bultynck G, Parys JB (2016) Intracellular Ca^{2+} signaling and Ca^{2+} microdomains in the control of cell survival, apoptosis and autophagy. Cell Calcium 60 (2):74–87. https://doi.org/10.1016/j.ceca.2016.04.005

Lam D, Kosta A, Luciani MF, Golstein P (2008) The inositol 1,4,5-trisphosphate receptor is required to signal autophagic cell death. Mol Biol Cell 19(2):691–700. https://doi.org/10.1091/mbc.e07-08-0823

Lavik AR, Zhong F, Chang MJ, Greenberg E, Choudhary Y et al (2015) A synthetic peptide targeting the BH4 domain of Bcl-2 induces apoptosis in multiple myeloma and follicular lymphoma cells alone or in combination with agents targeting the BH3-binding pocket of Bcl-2. Oncotarget 6(29):27388–27402. https://doi.org/10.18632/oncotarget.4489

Lewis RS (2020) Store-operated calcium channels: from function to structure and back again. Cold Spring Harb Perspect Biol 12(5):a035055. https://doi.org/10.1101/cshperspect.a035055

Liu EY, Ryan KM (2012) Autophagy and cancer--issues we need to digest. J Cell Sci 125 (Pt 10):2349–2358. https://doi.org/10.1242/jcs.093708

Liu AJ, Wang SH, Chen KC, Kuei HP, Shih YL et al (2013) Evodiamine, a plant alkaloid, induces calcium/JNK-mediated autophagy and calcium/mitochondria-mediated apoptosis in human glioblastoma cells. Chem Biol Interact 205(1):20–28. https://doi.org/10.1016/j.cbi.2013.06.004

Lock JT, Smith IF, Parker I (2019) Spatial-temporal patterning of Ca^{2+} signals by the subcellular distribution of IP_3 and IP_3 receptors. Semin Cell Dev Biol 94:3–10. https://doi.org/10.1016/j.semcdb.2019.01.012

Lopez-Sanjurjo CI, Tovey SC, Prole DL, Taylor CW (2013) Lysosomes shape Ins(1,4,5)P_3-evoked Ca^{2+} signals by selectively sequestering Ca^{2+} released from the endoplasmic reticulum. J Cell Sci 126(Pt 1):289–300. https://doi.org/10.1242/jcs.116103

Luyten T, Welkenhuyzen K, Roest G, Kania E, Wang L et al (2017) Resveratrol-induced autophagy is dependent on IP_3Rs and on cytosolic Ca^{2+}. Biochim Biophys Acta, Mol Cell Res 1864 (6):947–956. https://doi.org/10.1016/j.bbamcr.2017.02.013

Mendes CC, Gomes DA, Thompson M, Souto NC, Goes TS et al (2005) The type III inositol 1,4,5-trisphosphate receptor preferentially transmits apoptotic Ca^{2+} signals into mitochondria. J Biol Chem 280(49):40892–40900. https://doi.org/10.1074/jbc.M506623200

Messai Y, Noman MZ, Hasmim M, Janji B, Tittarelli A et al (2014) ITPR1 protects renal cancer cells against natural killer cells by inducing autophagy. Cancer Res 74(23):6820–6832. https://doi.org/10.1158/0008-5472.CAN-14-0303

Monteith GR, Prevarskaya N, Roberts-Thomson SJ (2017) The calcium-cancer signalling nexus. Nat Rev Cancer 17(6):367–380. https://doi.org/10.1038/nrc.2017.18

Morciano G, Marchi S, Morganti C, Sbano L, Bittremieux M et al (2018) Role of mitochondria-associated ER membranes in calcium regulation in cancer-specific settings. Neoplasia 20 (5):510–523. https://doi.org/10.1016/j.neo.2018.03.005

Morgan AJ, Davis LC, Wagner SK, Lewis AM, Parrington J et al (2013) Bidirectional Ca^{2+} signaling occurs between the endoplasmic reticulum and acidic organelles. J Cell Biol 200 (6):789–805. https://doi.org/10.1083/jcb.201204078

Mound A, Rodat-Despoix L, Bougarn S, Ouadid-Ahidouch H, Matifat F (2013) Molecular interaction and functional coupling between type 3 inositol 1,4,5-trisphosphate receptor and BK_{Ca} channel stimulate breast cancer cell proliferation. Eur J Cancer 49(17):3738–3751. https://doi.org/10.1016/j.ejca.2013.07.013

Mound A, Vautrin-Glabik A, Foulon A, Botia B, Hague F et al (2017) Downregulation of type 3 inositol (1,4,5)-trisphosphate receptor decreases breast cancer cell migration through an oscillatory Ca^{2+} signal. Oncotarget 8(42):72324–72341. https://doi.org/10.18632/oncotarget.20327

Okeke E, Parker T, Dingsdale H, Concannon M, Awais M et al (2016) Epithelial-mesenchymal transition, IP_3 receptors and ER-PM junctions: translocation of Ca^{2+} signalling complexes and regulation of migration. Biochem J 473(6):757–767. https://doi.org/10.1042/BJ20150364

Paknejad N, Hite RK (2018) Structural basis for the regulation of inositol trisphosphate receptors by Ca^{2+} and IP_3. Nat Struct Mol Biol 25(8):660–668. https://doi.org/10.1038/s41594-018-0089-6

Parys JB, De Smedt H (2012) Inositol 1,4,5-trisphosphate and its receptors. Adv Exp Med Biol 740:255–279. https://doi.org/10.1007/978-94-007-2888-2_11

Parys JB, Vervliet T (2020) New insights in the IP_3 receptor and its regulation. Adv Exp Med Biol 1131:243–270. https://doi.org/10.1007/978-3-030-12457-1_10

Parys JB, Decuypere JP, Bultynck G (2012) Role of the inositol 1,4,5-trisphosphate receptor/Ca^{2+}-release channel in autophagy. Cell Commun Signal 10(1):17. https://doi.org/10.1186/1478-811X-10-17

Pierro C, Cook SJ, Foets TC, Bootman MD, Roderick HL (2014) Oncogenic K-Ras suppresses IP_3-dependent Ca^{2+} release through remodelling of the isoform composition of IP_3Rs and ER luminal Ca^{2+} levels in colorectal cancer cell lines. J Cell Sci 127(Pt 7):1607–1619. https://doi.org/10.1242/jcs.141408

Pierro C, Zhang X, Kankeu C, Trebak M, Bootman MD et al (2018) Oncogenic KRAS suppresses store-operated Ca^{2+} entry and I_{CRAC} through ERK pathway-dependent remodelling of STIM expression in colorectal cancer cell lines. Cell Calcium 72:70–80. https://doi.org/10.1016/j.ceca.2018.03.002

Pierro C, Sneyers F, Bultynck G, Roderick HL (2019) ER Ca^{2+} release and store-operated Ca^{2+} entry - partners in crime or independent actors in oncogenic transformation? Cell Calcium 82:102061. https://doi.org/10.1016/j.ceca.2019.102061

Prevarskaya N, Skryma R, Shuba Y (2018) Ion channels in cancer: are cancer hallmarks oncochannelopathies? Physiol Rev 98(2):559–621. https://doi.org/10.1152/physrev.00044.2016

Prole DL, Taylor CW (2016) Inositol 1,4,5-trisphosphate receptors and their protein partners as signalling hubs. J Physiol 594(11):2849–2866. https://doi.org/10.1113/JP271139

Prole DL, Taylor CW (2019) Structure and function of IP_3 receptors. Cold Spring Harb Perspect Biol 11(4):a035063. https://doi.org/10.1101/cshperspect.a035063

Ravikumar B, Sarkar S, Davies JE, Futter M, Garcia-Arencibia M et al (2010) Regulation of mammalian autophagy in physiology and pathophysiology. Physiol Rev 90(4):1383–1435. https://doi.org/10.1152/physrev.00030.2009

Rezuchova I, Hudecova S, Soltysova A, Matuskova M, Durinikova E et al (2019) Type 3 inositol 1,4,5-trisphosphate receptor has antiapoptotic and proliferative role in cancer cells. Cell Death Dis 10(3):186. https://doi.org/10.1038/s41419-019-1433-4

Rizzuto R, Brini M, Murgia M, Pozzan T (1993) Microdomains with high Ca^{2+} close to IP_3-sensitive channels that are sensed by neighboring mitochondria. Science 262(5134):744–747. https://doi.org/10.1126/science.8235595

Roberts-Thomson SJ, Chalmers SB, Monteith GR (2019) The calcium-signaling toolkit in cancer: remodeling and targeting. Cold Spring Harb Perspect Biol 11(8):a035204. https://doi.org/10.1101/cshperspect.a035204

Roderick HL, Cook SJ (2008) Ca^{2+} signalling checkpoints in cancer: remodelling Ca^{2+} for cancer cell proliferation and survival. Nat Rev Cancer 8(5):361–375. https://doi.org/10.1038/nrc2374

Rong YP, Aromolaran AS, Bultynck G, Zhong F, Li X et al (2008) Targeting Bcl-2-IP_3 receptor interaction to reverse Bcl-2's inhibition of apoptotic calcium signals. Mol Cell 31(2):255–265. https://doi.org/10.1016/j.molcel.2008.06.014

Rosa N, Sneyers F, Parys JB, Bultynck G (2020) Type 3 IP_3 receptors: the chameleon in cancer. Int Rev Cell Mol Biol 351:101–148. https://doi.org/10.1016/bs.ircmb.2020.02.003

Scaffidi C, Fulda S, Srinivasan A, Friesen C, Li F et al (1998) Two CD95 (APO-1/Fas) signaling pathways. EMBO J 17(6):1675–1687. https://doi.org/10.1093/emboj/17.6.1675

Serysheva II, Baker MR, Fan G (2017) Structural insights into IP_3R function. Adv Exp Med Biol 981:121–147. https://doi.org/10.1007/978-3-319-55858-5_6

Shibao K, Fiedler MJ, Nagata J, Minagawa N, Hirata K et al (2010) The type III inositol 1,4,5-trisphosphate receptor is associated with aggressiveness of colorectal carcinoma. Cell Calcium 48(6):315–323. https://doi.org/10.1016/j.ceca.2010.09.005

Singh A, Chagtoo M, Tiwari S, George N, Chakravarti B et al (2017a) Inhibition of inositol 1, 4, 5-trisphosphate receptor induce breast cancer cell death through deregulated autophagy and cellular bioenergetics. J Cell Biochem 118(8):2333–2346. https://doi.org/10.1002/jcb.25891

Singh A, Sharma RK, Chagtoo M, Agarwal G, George N et al (2017b) ^{1}H NMR metabolomics reveals association of high expression of inositol 1, 4, 5 trisphosphate receptor and metabolites in breast cancer patients. PLoS One 12(1):e0169330. https://doi.org/10.1371/journal.pone.0169330

Singh R, Letai A, Sarosiek K (2019) Regulation of apoptosis in health and disease: the balancing act of BCL-2 family proteins. Nat Rev Mol Cell Biol 20(3):175–193. https://doi.org/10.1038/s41580-018-0089-8

Smaili SS, Pereira GJ, Costa MM, Rocha KK, Rodrigues L et al (2013) The role of calcium stores in apoptosis and autophagy. Curr Mol Med 13(2):252–265. https://doi.org/10.2174/156652413804810772

Steinmann C, Landsverk ML, Barral JM, Boehning D (2008) Requirement of inositol 1,4,5-trisphosphate receptors for tumor-mediated lymphocyte apoptosis. J Biol Chem 283(20):13506–13509. https://doi.org/10.1074/jbc.C800029200

Südhof TC, Newton CL, Archer BT 3rd, Ushkaryov YA, Mignery GA (1991) Structure of a novel $InsP_3$ receptor. EMBO J 10(11):3199–3206

Sun F, Xu X, Wang X, Zhang B (2016) Regulation of autophagy by Ca^{2+}. Tumour Biol 37:15467–15476. https://doi.org/10.1007/s13277-016-5353-y

Sung PJ, Tsai FD, Vais H, Court H, Yang J et al (2013) Phosphorylated K-Ras limits cell survival by blocking Bcl-xL sensitization of inositol trisphosphate receptors. Proc Natl Acad Sci U S A 110(51):20593–20598. https://doi.org/10.1073/pnas.1306431110

Szabadkai G, Bianchi K, Varnai P, De Stefani D, Wieckowski MR et al (2006) Chaperone-mediated coupling of endoplasmic reticulum and mitochondrial Ca^{2+} channels. J Cell Biol 175(6):901–911. https://doi.org/10.1083/jcb.200608073

Szatkowski C, Parys JB, Ouadid-Ahidouch H, Matifat F (2010) Inositol 1,4,5-trisphosphate-induced Ca^{2+} signalling is involved in estradiol-induced breast cancer epithelial cell growth. Mol Cancer 9:156. https://doi.org/10.1186/1476-4598-9-156

Taylor CW, Tovey SC (2010) IP_3 receptors: toward understanding their activation. Cold Spring Harb Perspect Biol 2(12):a004010. https://doi.org/10.1101/cshperspect.a004010

Taylor CW, da Fonseca PC, Morris EP (2004) IP_3 receptors: the search for structure. Trends Biochem Sci 29(4):210–219. https://doi.org/10.1016/j.tibs.2004.02.010

Terry LE, Alzayady KJ, Furati E, Yule DI (2018) Inositol 1,4,5-trisphosphate receptor mutations associated with human disease. Messenger (Los Angel) 6(1–2):29–44

Thillaiappan NB, Chavda AP, Tovey SC, Prole DL, Taylor CW (2017) Ca^{2+} signals initiate at immobile IP_3 receptors adjacent to ER-plasma membrane junctions. Nat Commun 8(1):1505. https://doi.org/10.1038/s41467-017-01644-8

Ueasilamongkol P, Khamphaya T, Guerra MT, Rodrigues MA, Gomes DA et al (2020) Type 3 inositol 1,4,5-trisphosphate receptor is increased and enhances malignant properties in cholangiocarcinoma. Hepatology 71(2):583–599. https://doi.org/10.1002/hep.30839

Urbani A, Giorgio V, Carrer A, Franchin C, Arrigoni G et al (2019) Purified F-ATP synthase forms a Ca^{2+}-dependent high-conductance channel matching the mitochondrial permeability transition pore. Nat Commun 10(1):4341. https://doi.org/10.1038/s41467-019-12331-1

Vanderheyden V, Devogelaere B, Missiaen L, De Smedt H, Bultynck G et al (2009) Regulation of inositol 1,4,5-trisphosphate-induced Ca^{2+} release by reversible phosphorylation and dephosphorylation. Biochim Biophys Acta 1793(6):959–970. https://doi.org/10.1016/j.bbamcr.2008.12.003

Vautrin-Glabik A, Botia B, Kischel P, Ouadid-Ahidouch H, Rodat-Despoix L (2018) IP_3R3 silencing induced actin cytoskeletal reorganization through ARHGAP18/RhoA/mDia1/FAK pathway in breast cancer cell lines. Biochim Biophys Acta, Mol Cell Res 1865(7):945–958. https://doi.org/10.1016/j.bbamcr.2018.04.002

Vermassen E, Parys JB, Mauger JP (2004) Subcellular distribution of the inositol 1,4,5-trisphosphate receptors: functional relevance and molecular determinants. Biol Cell 96 (1):3–17. https://doi.org/10.1016/j.biolcel.2003.11.004

Vervloessem T, Yule DI, Bultynck G, Parys JB (2015) The type 2 inositol 1,4,5-trisphosphate receptor, emerging functions for an intriguing Ca^{2+}-release channel. Biochim Biophys Acta 1853(9):1992–2005. https://doi.org/10.1016/j.bbamcr.2014.12.006

Vervloessem T, Akl H, Tousseyn T, De Smedt H, Parys JB et al (2017) Reciprocal sensitivity of diffuse large B-cell lymphoma cells to Bcl-2 inhibitors BIRD-2 versus venetoclax. Oncotarget 8 (67):111656–111671. https://doi.org/10.18632/oncotarget.22898

Wang SH, Shih YL, Ko WC, Wei YH, Shih CM (2008) Cadmium-induced autophagy and apoptosis are mediated by a calcium signaling pathway. Cell Mol Life Sci 65(22):3640–3652. https://doi.org/10.1007/s00018-008-8383-9

White C, Li C, Yang J, Petrenko NB, Madesh M et al (2005) The endoplasmic reticulum gateway to apoptosis by Bcl-X_L modulation of the $InsP_3R$. Nat Cell Biol 7(10):1021–1028. https://doi.org/10.1038/ncb1302

Wozniak AL, Wang X, Stieren ES, Scarbrough SG, Elferink CJ et al (2006) Requirement of biphasic calcium release from the endoplasmic reticulum for Fas-mediated apoptosis. J Cell Biol 175(5):709–714. https://doi.org/10.1083/jcb.200608035

Xie Q, Xu Y, Gao W, Zhang Y, Su J et al (2018) TAT-fused IP_3R-derived peptide enhances cisplatin sensitivity of ovarian cancer cells by increasing ER Ca^{2+} release. Int J Mol Med 41 (2):809–817. https://doi.org/10.3892/ijmm.2017.3260

Yue Z, Jin S, Yang C, Levine AJ, Heintz N (2003) Beclin 1, an autophagy gene essential for early embryonic development, is a haploinsufficient tumor suppressor. Proc Natl Acad Sci U S A 100 (25):15077–15082. https://doi.org/10.1073/pnas.2436255100

Zhong F, Harr MW, Bultynck G, Monaco G, Parys JB et al (2011) Induction of Ca^{2+}-driven apoptosis in chronic lymphocytic leukemia cells by peptide-mediated disruption of Bcl-2-IP_3 receptor interaction. Blood 117(10):2924–2934. https://doi.org/10.1182/blood-2010-09-307405

Zhou Y, Nwokonko RM, Baraniak JH Jr, Trebak M, Lee KPK et al (2019) The remote allosteric control of Orai channel gating. PLoS Biol 17(8):e3000413. https://doi.org/10.1371/journal.pbio.3000413

Chapter 12
Disruption of Endoplasmic Reticulum Proteostasis in Age-Related Nervous System Disorders

Danilo B. Medinas, Younis Hazari, and Claudio Hetz

Abstract Endoplasmic reticulum (ER) stress is a prominent cellular alteration of diseases impacting the nervous system that are associated to the accumulation of misfolded and aggregated protein species during aging. The unfolded protein response (UPR) is the main pathway mediating adaptation to ER stress, but it can also trigger deleterious cascades of inflammation and cell death leading to cell dysfunction and neurodegeneration. Genetic and pharmacological studies in experimental models shed light into molecular pathways possibly contributing to ER stress and the UPR activation in human neuropathies. Most of experimental models are, however, based on the overexpression of mutant proteins causing familial forms of these diseases or the administration of neurotoxins that induce pathology in young animals. Whether the mechanisms uncovered in these models are relevant for the etiology of the vast majority of age-related sporadic forms of neurodegenerative diseases is an open question. Here, we provide a systematic analysis of the current evidence linking ER stress to human pathology and the main mechanisms elucidated in experimental models. Furthermore, we highlight the recent association of

---

"Danilo B. Medinas" and "Younis Hazari" contributed equally

D. B. Medinas (✉) · Y. Hazari
Biomedical Neuroscience Institute, Faculty of Medicine, University of Chile, Santiago, Chile

Program of Cellular and Molecular Biology, Center for Molecular Studies of the Cell, Institute of Biomedical Sciences, University of Chile, Santiago, Chile

Center for Geroscience, Brain Health and Metabolism, Santiago, Chile
e-mail: dmedinas@med.uchile.cl

C. Hetz (✉)
Biomedical Neuroscience Institute, Faculty of Medicine, University of Chile, Santiago, Chile

Program of Cellular and Molecular Biology, Center for Molecular Studies of the Cell, Institute of Biomedical Sciences, University of Chile, Santiago, Chile

Center for Geroscience, Brain Health and Metabolism, Santiago, Chile

Buck Institute for Research on Aging, Novato, CA, USA
e-mail: chetz@uchile.cl

L. B. Agellon, M. Michalak (eds.), *Cellular Biology of the Endoplasmic Reticulum*, Progress in Molecular and Subcellular Biology 59,
https://doi.org/10.1007/978-3-030-67696-4_12

metabolic syndrome to increased risk to undergo neurodegeneration, where ER stress arises as a common denominator in the pathogenic crosstalk between peripheral organs and the nervous system.

Keywords ER stress · Protein misfolding · Neurodegenerative diseases · Metabolic syndrome · Aging

Abbreviations

AD	Alzheimer's disease
AgRP	agouti-related protein
ALS	amyotrophic lateral sclerosis
APP	amyloid precursor protein
ASK1	apoptosis signal-regulating kinase 1
ATF4	activating transcription factor 4
ATF6	activating transcription factor 6
ATF6f	ATF6 fragment
BDNF	brain-derived neurotrophic factor
BiP	binding-immunoglobulin protein
CHOP	C/EBP homologous protein
CNS	central nervous system
Canx	calnexin
CREB	c-AMP response element-binding protein
Calr	calreticulin
DBM	dibenzoylmethane
EDEM	endoplasmic reticulum degradation-enhancing α-mannosidase-like protein
eIF2α	eukaryotic initiation factor 2α
ER	endoplasmic reticulum
ERAD	endoplasmic reticulum-associated protein degradation
ERp57	endoplasmic reticulum-related protein 57
FoxO1	Forkhead box protein O1
FUS	fused in sarcoma
GADD34	growth arrest and DNA damage-inducible protein 34
GalE	UDP-galactose-4-epimerase
GSK3β	glycogen synthase kinase 3β
GT1	UDP-glucose:glycoprotein glucosyltransferase
HD	Huntington's disease
HERPUD1	homocysteine-responsive ER-resident ubiquitin-like domain member 1 protein
HFD	high-fat-diet
HIPK2	homeodomain-interacting protein kinase 2
IκB	inhibitor of NF-κB
IKK	inhibitor of NF-κB kinase

IL-1β	interleukin 1β
iNOS	inducible nitric oxide synthase
IRE1α	Inositol-requiring enzyme 1α
ISR	integrated stress response
ISRIB	integrated stress response inhibitor
JNK	c-Jun N-terminal kinase
KIRA	kinase inhibitor RNase attenuator
MSA	multiple system atrophy
α-MSH	α-melanocyte-stimulating hormone
NF-κB	nuclear factor-κB
NLRP3	nucleotide-binding oligomerization domain (NOD)-like receptor family pyrin domain-containing 3
4-PBA	4-phenybutyrate
PD	Parkinson's disease
PDI	protein disulfide isomerase
PERK	PKR-like endoplasmic reticulum kinase
PI3K	phosphoinositide 3-kinases
PMD	protein misfolding disorder
POMC	pro-opiomelanocortin
PPI	protein phosphatase 1
PRD	prion-related disorder
RIDD	regulated IRE1-dependent decay
RTCB	RNA 2′,3′-cyclic phosphate and 5'-OH ligase
S1P	site-1 protease
S2P	site-2 protease
SIL1	nucleotide exchanging factor for BiP
SOCS3	suppressor of cytokine signaling 3
SOD1	superoxide dismutase 1
TARDBP	TAR DNA binding protein
TNF	tumor necrosis factor
TRAF2	TNF receptor-associated factor 2
TXNIP	thioredoxin-interacting protein
UPR	unfolded protein response
UPS	ubiquitin proteasome system
VCP	valosin-containing protein
XBP1	X-box binding protein 1

12.1 Introduction

Alterations to the mechanisms that control protein homeostasis (also termed proteostasis) are emerging as salient features of age-related diseases. Among these, the endoplasmic reticulum (ER) and the secretory pathway compose a node of the proteostasis network where distinct pathological inputs converge to cause neuropathies. ER stress, condition marked by the accumulation of misfolded proteins in the

ER lumen, has been proposed to exert a fundamental role in pathogenic mechanisms leading to cell demise of specific neuronal populations in neurodegenerative diseases including Alzheimer's disease (AD), Parkinson's disease (PD), Huntington's disease (HD), prion-related disorders (PRDs), amyotrophic lateral sclerosis (ALS), among other pathological conditions of the nervous system (Hetz and Saxena 2017). The accumulation of misfolded and aggregated proteins constitutes a hallmark of many neurological pathologies, which are collective referred to as protein misfolding disorders (PMDs) (Soto and Pritzkow 2018). Aging is the common denominator of PMDs, possibly due to a decline of mechanisms supporting proteostasis (Hipp et al. 2014). The folding and maturation of proteins in the ER involves steps of glycosylation and oxidative modification that are tightly coupled to the metabolic status of the cell (Tu and Weissman 2002; Harada et al. 2013). This intricate posttranslational processing of proteins trafficking through the secretory pathway, which comprise around one third of the proteome (Ghaemmaghami et al. 2003), makes the ER highly vulnerable to disturbance of proteostasis networks.

ER stress activates a signal transduction pathway termed unfolded protein response (UPR) (Hetz et al. 2020). The UPR communicates disturbances of protein folding in the ER lumen to the cytosol and nucleus, causing attenuation of translation and reprogramming gene expression through the action of UPR-related transcription factors (Hetz et al. 2020). The UPR can cope with transient ER stress by adjusting translation rates to protein folding capacity, in addition to inducing expression of chaperones, folding catalysts and degradation machinery. In this manner, proteostasis is restored and the cell resumes its physiological function (Hetz et al. 2020). On the other hand, chronic ER stress causes sustained activation of the UPR, which shifts its output leading to cell dysfunction and eventually cell death (Hetz et al. 2020). An important question pertains to the origin of ER stress observed in neurological conditions. Inherited or somatic mutations can lead to expression of unstable, aggregation-prone proteins that will impose a challenge to the maintenance of functional proteostasis networks in specific neuronal populations during aging, with some disease genes encoding for known cargoes of the secretory pathway (Hipp et al. 2014; Fu et al. 2018). Furthermore, genetic mutations can also impair components of the proteostasis networks, causing age-related alterations in the proteome that render subsets of neurons dysfunctional and vulnerable to degeneration (Labbadia and Morimoto 2015).

There is clear evidence in the literature indicating that genetic mutations linked to familial cases of neurodegenerative diseases impair ER proteostasis (Hetz and Saxena 2017). Studies in cellular and animal models of PMDs expressing mutant proteins have established the dysfunction of the ER folding machinery and the activation of the UPR in pathogenic mechanisms (Hetz and Saxena 2017). However, the vast majority of PMDs are sporadic (*i.e.* have no familial history), which share common clinical and histopathological features of familial cases but have genetic factors largely undefined (Turner et al. 2013; Masters et al. 2015; Langston et al. 2015). The occurrence of ER stress reported in these cases may be rooted, at least partially, on metabolic dysfunction, a transversal risk factor for the emergence of myriad age-related comorbidities that establish a deleterious crosstalk with the nervous system (Hotamisligil 2017; Mattson and Arumugam 2018).

In this chapter, we will provide an integrative view of the relationship between ER stress and neurodegenerative diseases, highlighting the interplay existing among different nodes of the ER proteostasis, PMDs-linked protein species and metabolic disorders.

12.2 ER Machinery and UPR Signaling

The ER consists of a network of membranous sacs and tubules dedicated to the synthesis of lipids and proteins, in addition to storage of calcium ions, that is highly interconnected to other organelles (Hetz et al. 2020). The ER responds to the nutrient status of the cell by fine tuning synthetic pathways and bioenergetic demands (Hetz et al. 2020), employing a vast array of enzymatic reactions and signaling platforms (Hetz et al. 2020). The synthesis of membrane and secreted proteins occurs in ribosomes docked on the cytoplasmic leaflet of the ER membrane, enabling the co-translational translocation and folding of the nascent polypeptide in the ER lumen. The folding and quality control of proteins in the ER is extremely complex and a detailed description is beyond the scope of this chapter (for specialized reviews see (Calì et al. 2008; Tannous et al. 2015)). Here we will describe some fundamental mechanisms and their components, together with the available tools for their pharmacological manipulation, so the reader can follow the evidence discussed in the next sections linking ER stress to the etiology of human disorders (see Figs. 12.1 and 12.2).

The translocation of a polypeptide across the ER membrane starts with the passage of the N-terminal signal sequence through a translocon complex containing Sec61 as a core component (Fig. 12.1). The polypeptide that emerges in the ER lumen is bound by chaperones of the heat shock family like binding-immunoglobulin protein (BiP) that prevent protein aggregation by favoring folding-competent conformations (Calì et al. 2008). The polypeptide also forms intermolecular disulfide bonds with protein disulfide isomerase (PDI), precluding formation of aberrant disulfide crosslinks that could result in protein misfolding and aggregation (Calì et al. 2008). Most of the proteins trafficking through the ER suffer N-glycosylation in asparagine residues present in the consensus sequence Axn-Xxx-Ser/Thr. The glycosylation increases polypeptide solubility and determines the folding and quality control mechanisms (Fig. 12.1). Mature proteins are then exported to the Golgi apparatus through vesicular transport while misfolded proteins can go through additional folding cycles or be directed to degradation by the ER-associated protein degradation (ERAD) pathway that retro-translocates the polypeptide to the cytosol for subsequent proteolysis by the ubiquitin-proteasome system (UPS) or by autophagic and lysosomal routes (Fig. 12.1) (Fregno and Molinari 2019).

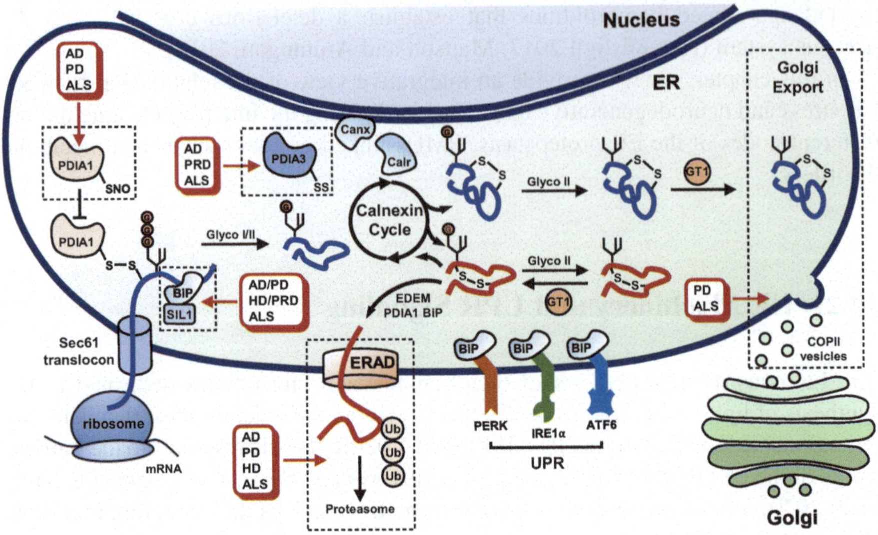

Fig. 12.1 ER proteostasis network is impaired in neurodegenerative diseases. Proteins trafficking through the secretory pathway are co-translationally translocated into the ER lumen through the Sec61 translocon and readily N-glycosylated and associated to BiP and PDIA1 through disulfide bonds. Glycosylated proteins are processed by glycosidases, undergo protein folding interactions in the calnexin cycle (Canx/Calr/PDIA3) and quality control by GT1, being directed to Golgi export or degradation by ERAD. Under normal conditions, the UPR transducers PERK, IRE1α and ATF6 are maintained inactive by association with BiP. PDIA1 nitrosylation (SNO), deficiency of the BiP co-chaperone SIL1, altered calnexin cycle, inhibited ERAD and impaired Golgi export are common disturbances detected in neurodegenerative diseases as indicated and arise from aberrant interactions of misfolded proteins with different components of the ER proteostasis network, production of reactive oxygen species, among others. *ATF6* activating transcription factor 6; *BiP* binding-immunoglobulin protein; *Canx* calnexin; *Calr* calreticulin; *COPII* coat complex protein II; *EDEM* ER degradation-enhancing α-mannosidase-like protein; *ERAD* ER-associated degradation; *GT1* UDP-glucose:glycoprotein glucosyltransferase; *IRE1α* inositol-requiring protein 1α; *PDIA1 and PDIA3* protein disulfide isomerase A1 and A3; *PERK* protein kinase RNA-like ER kinase; *SIL1* nucleotide exchanging factor for BiP; *Ub* ubiquitin; *UPR* unfolded protein response

Disulfide bond formation is a key posttranslational modification conferring structural stability as well as functional versatility to proteins (Cook and Hogg 2013; Feyertag and Alvarez-Ponce 2017). The correct formation of disulfide bonds depends on the concerted action of members of the PDI family, such as PDIA1 (also known as PDI or P4HB) and PDIA3 (also known as ERp57 or GRP58) (Hatahet and Ruddock 2009). The N-glycosylation of asparagine residues suffers trimming of glucose units by glycosidases I and II, directing the polypeptide to interact with the lectin-like chaperones calnexin (Canx) and calreticulin (Calr) (Fig. 12.1). The protein folding in the so-called calnexin cycle involves disulfide bond formation by PDIA3 (Calì et al. 2008). The protein released from the calnexin cycle suffers trimming of the last glucose unit by glycosidase II and pass through quality control by UDP-glucose:glycoprotein glucosyltransferase (GT1). Mature proteins are exported to the Golgi apparatus, while misfolded proteins receive a glucose unit transferred by GT1 and re-enter the calnexin cycle (Fig. 12.1). Terminally misfolded proteins suffer de-glycosylation by mannosidases (*e.g.* EDEM1), are recognized by PDIA1/BiP and eventually directed to ERAD for degradation (Fig. 12.1) (Calì et al. 2008).

The accumulation of misfolded proteins in the ER lumen causes activation of the UPR (Hetz et al. 2020). The adaptive UPR counteract ER stress to reestablish homeostasis, whereas chronic or irreversible ER stress may trigger the terminal UPR to eliminate the cell through apoptosis (Walter and Ron 2011). The UPR is mediated by UPR transducers located in the ER membrane, named Inositol-requiring enzyme 1α (IRE1α), PKR-like ER kinase (PERK) and activating transcription factor 6 (ATF6) (Fig. 12.1). Two main mechanisms have been proposed for ER stress sensing and activation of the UPR: a direct-recognition model where misfolded proteins act as ligands of UPR stress sensors (Karagöz et al. 2019), and an indirect model where ER chaperones operate as sensors that transmit ER stress signal to UPR transducers to engage the pathway (Preissler and Ron 2019). In this second model, the activity of the UPR transducers is intimately coupled to the folding machinery, with association of BiP to the luminal domain of IRE1, PERK and ATF6 maintaining these transducers inactive under basal conditions (Fig. 12.1). BiP dissociation from ER stress transducers to bind misfolded proteins leads to UPR activation (Fig. 12.2). Other components of the folding machinery also serve as regulators for the UPR transducers, for instance PDIA1, PDIA6 and HSP47 for IRE1, PDIA5 and ERp18/PDIA16 for ATF6 (Hetz et al. 2020).

IRE1α has kinase and RNase activities and its activation involves protein dimerization and trans-autophosphorylation (Hetz et al. 2020). Active IRE1α catalyzes the unconventional splicing of *Xbp1* mRNA in conjunction with the RNA ligase RTCB (RNA 2′,3′-cyclic phosphate and 5'-OH ligase) (Fig. 12.2). The spliced *Xbp1* mRNA is translated into a potent transcription factor termed XBP1s to promote adaptation to ER stress by inducing the expression of chaperones, ERAD components, among others (Hetz et al. 2020). In addition, IRE1α is also capable of degrading a subset of mRNA and miRNA with specific structural features, an activity termed regulated IRE1-dependent decay (RIDD) (Fig. 12.2) (Hollien and Weissman 2006; Upton et al. 2012). Finally, IRE1α has emerged as an unique signaling platform that integrates the UPR with other homeostatic mechanisms of the cell (Fig. 12.2), contributing to the control of the actin cytoskeleton, mitochondria bioenergetic, inflammatory responses, apoptotic cascades, among others (Hetz et al. 2020). The fine balance of *Xbp1* mRNA splicing, RIDD and scaffold activities of IRE1α is proposed to determine the transition between adaptive and terminal UPR (Hetz et al. 2020). Different classes of inhibitors have been developed for IRE1α, either by targeting the RNase domain such as STF-083010, MKC 3946, 4μ8c, or the kinase domain such as Compound 18 or KIRA (kinase inhibitor RNase attenuator) (Maly and Papa 2014; Hetz et al. 2019), in addition to recently discovered IRE1α activators (Grandjean et al. 2020).

PERK is a kinase that is also activated by dimerization and trans-autophosphorylation (Hetz et al. 2020). Upon activation, PERK phosphorylates

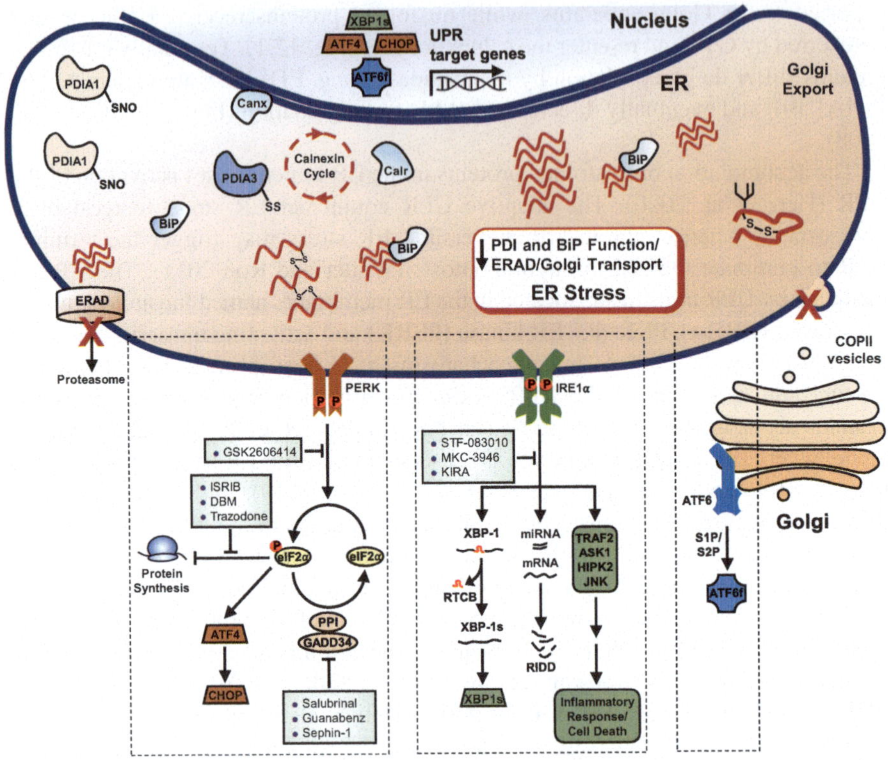

Fig. 12.2 UPR activation and its pharmacological regulation. The accumulation of misfolded and aggregated protein species resulting from defective ER machinery in neurodegenerative diseases activates the UPR transducers through dissociation of BiP. Additionally, PERK and IRE1α suffer dimerization and trans-autophosphorylation, while ATF6 translocates to Golgi for its proteolytic processing by S1P and S2P leading to the cytosolic fragment ATF6f. PERK phosphorylates eIF2α inhibiting protein synthesis while favoring translation of ATF4 and CHOP. IRE1α catalyzes *Xbp1* mRNA splicing in concert with RTCB, promoting translation of XBP1s. In addition, IRE1α catalyzes degradation of mRNA and miRNA through RIDD and serves as scaffold for a pro-inflammatory/pro-apoptotic signaling pathway involving TRAF2-ASK1-HIPK2-JNK. Phosphorylated PERK and IRE1α are proximal markers of ER stress reported in neurodegenerative disease, along with increased levels of phosphorylated eIF2α, ATF4, CHOP and XBP1s. The evidence for ATF6 activation is scarce. ATF4, CHOP, XBP1s and ATF6f are UPR transcription factors that reprogram gene expression to enhance adaptive capacity or trigger death of cells. The genetic and pharmacological manipulation of the UPR has established the functional involvement of the pathway in disease pathogenesis. The most common drugs available to inhibit each UPR branch are indicated. *ASK1* apoptosis signal-regulating kinase 1; *ATF4* activating transcription factor 4; *ATF6* activating transcription factor 6; *ATF6f* activating transcription factor 6 fragment; *BiP* binding-immunoglobulin protein; *Canx* calnexin; *Calr* calreticulin; *CHOP* C/EBP homologous protein; *COPII* coat complex protein II; *DBM* dibenzoylmethane; *eIF2α* eukaryotic initiation factor 2α; *ERAD* ER-associated degradation; *GADD34* growth arrest and DNA damage-inducible protein 34; *HIPK2* homeodomain-interacting protein kinase 2; *IRE1α* inositol-requiring protein 1α; *ISRIB* integrated stress response inhibitor; *JNK* c-Jun N-terminal kinase; *KIRA* kinase inhibitor RNase attenuator; *PDIA1 and PDIA3* protein disulfide isomerase A1 and A3; *PERK* protein kinase RNA-like ER kinase; *PPI* protein phosphatase 1; *RIDD* regulated IRE1-dependent decay; *RTCB* RNA 2′,3′-cyclic phosphate and 5'-OH ligase; *S1P* site-1 protease; *S2P* site-2 protease; *TRAF2* TNF receptor-associated factor 2; *UPR* unfolded protein response; *XBP1* X-box binding protein 1; *XBP1s* X-box binding protein 1 spliced form

the eukaryotic initiation factor 2 subunit α (eIF2α) to inhibit translation and decrease the demand for protein folding in the ER during adaptive UPR (Fig. 12.2) (Costa-Mattioli and Walter 2020). eIF2α phosphorylation also favors translation of mRNA containing upstream open reading frames (uORF) in the 5′ untranslated region such as *Atf4* (Fig. 12.2). Activation transcription factor 4 (ATF4) contributes to the adaptive UPR by controlling expression of genes involved in redox control, amino acid metabolism, among others (Hetz et al. 2020). Besides, ATF4 also induces the expression of C/EBP homologous protein (CHOP), a transcription factor mediating apoptotic cell death in the terminal UPR (Fig. 12.2) (Walter and Ron 2011; Hetz et al. 2020). Different classes of drugs have been developed to target the PERK/eIF2α pathway, including PERK inhibitors such as GSK2606414, eIF2α phosphatase inhibitors such as salubrinal, guanabenz and sephin-1, and compounds that reverse translation repression by phosphorylated eIF2α (p-eIF2α) such as ISRIB, trazodone hydrochloride and dibenzoylmethane (DBM) (Fig. 12.2) (Gonzalez-Teuber et al. 2019).

Upon ER stress, ATF6 suffers redox activation and is translocated to the Golgi apparatus where it is cleaved by site-1 protease (S1P) and site-2 protease (S2P) releasing an active cytosolic fragment (ATF6f) that migrates to the nucleus inducing transcription of genes involved in protein folding and quality control in the ER, in addition to *Xbp1* (Fig. 12.2) (Hetz et al. 2020). Small molecules that act as activators and inhibitors of ATF6 have been also recently reported (Hetz et al. 2019).

12.3 ER Stress in Neurodegenerative Diseases

The involvement of ER stress in the pathogenesis of different neurodegenerative diseases can be inferred from studies ranging from histopathological and biochemical examination of patient postmortem tissue to genetic and pharmacological manipulations of experimental models. Some molecular alterations are consistently reported across diseases, such as activation of the UPR transducers and induction of downstream transcription factors, in addition to up-regulation of specific chaperones and foldases. In this section, we will highlight major alterations to the ER proteostasis network and their functional connection to pathogenic mechanisms.

12.3.1 IRE1α Pathway

The accumulation of amyloid plaques, extracellular protein aggregates enriched in β-sheet structure assuming fibrillar morphology, and neurofibrillary tangles, intracellular aggregates of hyperphosphorylated tau protein, are hallmarks of AD (Masters et al. 2015). The proteolytic processing of amyloid precursor protein (APP) by β-secretase (β-APP cleaving enzyme 1; BACE1) and γ-secretase yields amyloid beta (Aβ) peptides in the extracellular space, which initially combine into

Aβ oligomers, considered the actual neurotoxic species in AD, and then grow to form amyloid plaques (Selkoe and Hardy 2016). The activation of IRE1α-XBP1 axis may be closely related to this amyloidogenic process. In a genome-wide screening, genes encoding for proteins of the γ-secretase complex, named presenilin 1 and 2, nicastrin and the presenilin enhancer 2, were found to be possibly regulated by XBP1s (Acosta-Alvear et al. 2007). Furthermore, the tau kinase cyclin-dependent kinase 5 (Cdk5) also emerged in the universe of putative XBP1s targets, potentially linking IRE1α-XBP1 axis to formation of neurofibrillary tangles (Acosta-Alvear et al. 2007; Kimura et al. 2014). The regulation of APP and tau protein by XBP1s in AD, however, remains to be demonstrated. Supporting this notion, the detection of phosphorylated IRE1α (p-IRE1α) in patient *postmortem* tissue correlated with AD progression (Braak stage), with most neurons containing neurofibrillary tangles also positive for p-IRE1α (Duran-Aniotz et al. 2017). In an AD mouse model overexpressing mutant APP, the genetic ablation of IRE1α RNAse activity afforded neuroprotection by reducing deposition of amyloid plaques (Duran-Aniotz et al. 2017). On the contrary, a polymorphism in the *XBP1* promoter associated with reduced *XBP1* expression was identified as a risk factor to develop sporadic AD in the Chinese population (Liu et al. 2013). In line with this genetic trait, the forced expression of XBP1s restored synaptic plasticity and memory in experimental AD independently of accumulation of amyloid plaques (Cissé et al. 2017). Thus, the IRE1α-XBP1 axis may exert multiple functions in different stages of AD pathology, either by modulating proteostasis or synaptic function. Indeed, XBP1s has been recently shown to contribute to learning and memory through induction of brain-derived neurotrophic factor (BDNF) leading to enhanced synaptic plasticity (Martínez et al. 2016).

Contrary to AD, activation of the IRE1α pathway is not detected in *postmortem* tissue of PRDs patients (Wiersma et al. 2016). Accordingly, the conditional knock-out of *Xbp1* in the nervous system had no effect on disease progression or prion replication in mice (Hetz et al. 2008). PD is marked by the accumulation of modified and aggregated species of α-synuclein forming Lewy bodies (Spillantini et al. 1997; Henderson et al. 2019). Activation of IRE1α has been reported to promote neurodegeneration independently of XBP1 in a *Drosophila* model of PD overexpressing α-synuclein (Yan et al. 2019). In addition, detection of p-IRE1α has been reported in patient *postmortem* samples of multiple system atrophy (MSA), another synucleinopathy (Makioka et al. 2010). On the other hand, the overexpression of XBP1s has been show to afford neuroprotection in pharmacological mouse models of PD (Sado et al. 2009; Valdés et al. 2014). Furthermore, *Xbp1* knockdown renders dopaminergic neurons in the *substantia nigra pars compacta* (*SNpc*), which are selectively affected in PD, vulnerable to degeneration (Valdés et al. 2014). Regarding pathological insults to the motor system, the activation of IRE1α pathway has been correlated with axonal regeneration after sciatic nerve crush in mice (Ohtake et al. 2018). In addition, the genetic ablation of *Xbp1* or overexpression of XBP1s using gene therapy deteriorate or improve, respectively, locomotor recovery in both spinal cord injury and sciatic nerve crush mouse models (Valenzuela et al. 2012; Oñate et al. 2016).

ALS is a fatal degenerative disease affecting motor neurons, leading to muscle weakness, paralysis and death (Turner et al. 2013). About 10% of ALS cases are familial, being linked to mutations in several genes including *C9ORF72*, *SOD1*, *TARDBP* (encoding the protein TDP-43), *FUS*, among others (Turner et al. 2013). Despite the linear association of XBP1 levels and locomotor recovery after mechanical injury to peripheral nerve (Valenzuela et al. 2012; Oñate et al. 2016), a complex scenario has emerged regarding the involvement of IRE1α and XBP1 in ALS pathogenesis. Increased levels of IRE1α and XBP1s have been documented in sporadic cases (Atkin et al. 2008; Hetz et al. 2009). In a mutant SOD1 mouse model of familial ALS, chronic ER stress has been suggested to trigger apoptosis of motoneurons through the IRE1α-TRAF2-ASK1 pathway (Nishitoh et al. 2008). Relevantly, the IRE1α-ASK1 apoptotic signaling has also been extended to sporadic and *C9ORF72*-linked ALS, which account for most of cases (Taylor et al. 2016; Lee et al. 2016). Using induced pluripotent stem cell technology, studies in motoneurons of SOD1-linked ALS patients showed that ER stress is an intrinsic property of these cells linked to their hyperexcitability, with knockdown of *XBP1* affording protection (Kiskinis et al. 2014; Wainger et al. 2014). The conditional knockout of *Xbp1* in the nervous system also led to neuroprotection in a SOD1 mouse model of ALS (Hetz et al. 2009). The observed neuroprotection through genetic ablation of a master regulator of the ER proteostasis seemed contradictory. However, virtually the same outcome was obtained using this genetic strategy in a HD mouse model that overexpress mutant huntingtin with polyglutamine expansion (Vidal et al. 2012). It was discovered that under pathological conditions associated to PMDs, *Xbp1* knockout during brain development triggers a homeostatic response to up-regulate autophagic degradation of protein aggregates leading to functional improvement (Hetz et al. 2009; Vidal et al. 2012). In fact, whereas knockdown of *Xbp1* in dopaminergic neurons of the *SNpc* in adult brain triggers spontaneous neurodegeneration as mentioned, *Xbp1* genetic ablation during development using conditional knockout mice also provides protection against PD-inducing neurotoxins involving a switch in the proteostasis network to up-regulate ER chaperones and autophagy components (Valdés et al. 2014). These results support that IRE1α signaling is highly interconnected to proteostasis networks in different cellular compartments. Increased levels of p-IRE1α and XBP1s have been reported in HD mouse models and patients *postmortem* tissue (Vidal et al. 2012; Lee et al. 2012). While administration of a gene therapy to overexpress XBP1s is protective by reducing mutant huntingtin aggregation in a mouse model of HD (Zuleta et al. 2012), the activation of IRE1α-TRAF2 pathway has been linked to accumulation of mutant huntingtin through inhibition of autophagic flux in *Drosophila* (Lee et al. 2012). Moreover, the activation of IRE1α has also been linked to protein aggregation in mice overexpressing mutant huntingtin (Hyrskyluoto et al. 2014).

12.3.2 PERK Pathway

Most of the literature associating ER stress to neurodegenerative diseases is based on activation of the PERK pathway (Scheper and Hoozemans 2015). However, the conclusions of such studies should be examined with great caution as much of the experimental evidence relies on the detection of p-eIF2α, a downstream event of the integrated stress response (ISR) that can reflect activation of other kinases independently of ER stress (Costa-Mattioli and Walter 2020).

In AD, there are studies reporting detection of both phosphorylated PERK (p-PERK) and p-eIF2α in association with abnormally phosphorylated tau in patient *postmortem* tissue (Unterberger et al. 2006; Hoozemans et al. 2012). The translational repression caused by sustained eIF2α phosphorylation can lead to synaptic dysfunction and memory impairment (Costa-Mattioli et al. 2009). In the pathological context of AD mouse models, the genetic deletion of PERK prevented enhanced eIF2α phosphorylation and deficits in protein synthesis, synaptic plasticity and memory (Ma et al. 2013). Treatment with a PERK inhibitor and compounds that reverse translational repression by p-eIF2α presented therapeutic effects in a mouse model of tau pathology (Radford et al. 2015; Halliday et al. 2017). Besides translational repression, eIF2α phosphorylation may contribute to AD pathology due to induction of the transcription factor ATF4 (Baleriola et al. 2014). It has been reported that Aβ peptides induce phosphorylation of eIF2α in the axon, inducing the local synthesis of ATF4, which can retro-translocate to the soma propagating degeneration in the brain via CHOP (Baleriola et al. 2014). The intermediacy of PERK in this mechanism, however, was not demonstrated (Baleriola et al. 2014). Of note, in progressive supranuclear palsy, a neurodegenerative disease presenting neurofibrillary tangles, PERK activation dissociated from eIF2α phosphorylation is proposed to exert beneficial effects on memory and locomotor function with preservation of synapses (Bruch et al. 2017). In addition, loss-of-function mutations *EIF2AK3*, encoding PERK, causes Wolcott–Rallison syndrome, a rare autosomal recessive disorder of infants marked by diabetes and neurological problems associated to neurofibrillary tangles (Bruch et al. 2015).

The detection of p-PERK and p-eIF2α has also been reported in *postmortem* tissue of PD patients in association with α-synuclein, being regarded as an early pathological event (Hoozemans et al. 2007, 2012; Mercado et al. 2018). As for p-IRE1α, detection of p-PERK and p-eIF2α has been reported in MSA (Makioka et al. 2010). The pharmacological inhibition of PERK afforded protection in a drug-induced model of PD, attenuating loss of dopaminergic neurons in *SNpc*, with sustained levels of synaptic proteins and improved motor performance (Mercado et al. 2018). The inhibitor of eIF2α phosphatase, Salubrinal, has been reported beneficial against α-synucleinopathy in rodent models despite the fact that phosphorylation of eIF2α could not be consistently determined (Colla et al. 2012). The knockout of *Ddit3*, encoding CHOP, reduced apoptosis in a drug-induced model of PD (Silva et al. 2005). In PRDs, the current evidence in patient *postmortem* tissue does not support the involvement of p-PERK/p-eIF2α in neurodegeneration

(Unterberger et al. 2006; Wiersma et al. 2016), despite a wealth of studies in mouse models (Moreno et al. 2012, 2013; Halliday et al. 2015, 2017). The phosphorylation of PERK and eIF2α has been documented in mice infected with prions, leading to decreased synthesis of synaptic proteins, reduction of synapses and neuronal loss (Moreno et al. 2012). The overexpression of eIF2α phosphatase or its pharmacological inhibition with Salubrinal protected or enhanced pathology, respectively, without affecting prion levels (Moreno et al. 2012). Similarly, administration of a PERK inhibitor or compounds that reverse translational repression by p-eIF2α also presented therapeutic effects against prion-induced pathology (Moreno et al. 2013; Halliday et al. 2015, 2017).

The activation of the PERK pathway has been correlated with axonal regeneration after sciatic nerve crush in mice (Ohtake et al. 2018). The downstream targets of PERK exerting protective effects, however, remained unidentified (Ohtake et al. 2018). While knockout of *Atf4* deteriorate locomotor recovery upon spinal cord injury (Valenzuela et al. 2012), it had no effect on sciatic nerve crush model (Oñate et al. 2016). There are several studies reporting the participation of the PERK pathway in the etiology of ALS, with contrasting conclusions. Increased levels of components of the PERK pathway, including PERK, p-eIF2α, ATF4 and CHOP have been reported in patient *postmortem* tissue of sporadic ALS cases (Ilieva et al. 2007; Atkin et al. 2008; Hetz et al. 2009; Ito et al. 2009). Consistent with an adaptive advantage conferred by the PERK pathway in ALS, it has been shown that *Eif2ak3* haploinsufficiency accelerates disease onset and progression in a mutant SOD1 mouse model of ALS (Wang et al. 2011). Further in mutant SOD1 mice, engagement of the PERK pathway has been detected in vulnerable motor neurons and treatment with Salubrinal improved muscle strength and survival (Saxena et al. 2009). Other studies corroborated this approach, showing that either pharmacological inhibition or genetic ablation of the eIF2α phosphatase are protective in mutant SOD1 mouse models of ALS (Wang et al. 2014c, b; Jiang et al. 2014; Das et al. 2015). One study using Guanabenz, however, reported acceleration of disease progression with pharmacological inhibition of the eIF2α phosphatase (Vieira et al. 2015). Furthermore, the genetic ablation of *Atf4* showed protective effects in a mouse model of ALS (Matus et al. 2013), possibly due to reduced levels of apoptosis mediators. Challenging these previous reports, a systematic study performing genetic manipulation of *Eif2ak3*, *Ddit3* and the eIF2α phosphatase on several mutant SOD1 mouse lines suggested that the PERK pathway is not a relevant target for therapeutic intervention (Dzhashiashvili et al. 2019).

In ALS linked to *C9ORF72* G_4C_2 hexanucleotide repeat expansion, up-regulation of ATF4 and CHOP has been reported in patient *postmortem* tissue (Zhang et al. 2014). The phosphorylation of eIF2α may play a role amplifying the so-called 'C9-ALS' pathology (Green et al. 2017; Cheng et al. 2018). The G_4C_2 hexanucleotide repeat expansion can suffer repeat-associated non-AUG (RAN) translation to produce neurotoxic dipeptide repeat (DPR) proteins in 'C9-ALS' (Ash et al. 2013; Gendron et al. 2013). The DPR proteins have been described to induce ER stress marked by PERK phosphorylation (Zhang et al. 2014), which in turn may close a toxic feed-forward cycle of RAN translation through phosphorylation of eIF2α

(Green et al. 2017; Cheng et al. 2018). Furthermore, evoking ionic currents in motor neurons also induce p-PERK and p-eIF2α and facilitate RAN translation (Westergard et al. 2019). On the other hand, sustaining p-eIF2α with Salubrinal treatment was shown to protect against toxicity of DPR proteins associated to ER stress (Zhang et al. 2014). In the context of TDP-43 pathology, a genetic screening identified components of stress granules converging on p-eIF2α as enhancers of TDP-43 toxicity (Kim et al. 2014). The knockdown of *PEK* or *Gadd34*, which encode for *Drosophila* homologous of PERK and the eIF2α phosphatase GADD34, protected against or aggravated TDP-43 toxicity *in vivo*, respectively, corroborating that sustained phosphorylation of eIF2α is detrimental (Kim et al. 2014). In addition, treatment with the PERK inhibitor GSK2606414 afforded neuroprotection in the *Drosophila* model of TDP-43 neurodegeneration (Kim et al. 2014). In contradiction, a study of TDP-43 toxicity in *C. elegans* and zebrafish showed that Salubrinal and Guanabenz recovered locomotor function in both models (Vaccaro et al. 2013).

12.3.3 ATF6 Pathway

The ATF6 pathway is the least UPR branch characterized in neurodegenerative diseases. This may be, at least in part, due to scarce evidence of ATF6 activation in patient *postmortem* samples (Scheper and Hoozemans 2015). ATF6 levels have been reported up-regulated in *postmortem* tissue of sporadic ALS patients (Atkin et al. 2008), while ATF6 processing appeared compromised in HD (Fernandez-Fernandez et al. 2011). Despite higher levels of full-length ATF6 in patient *post-mortem* tissue of HD and mouse models overexpressing mutant huntingtin, there was lower content of the active fragment of ATF6 with no nuclear accumulation (Fernandez-Fernandez et al. 2011). The rescue of ATF6 processing is neuroprotective in HD (Naranjo et al. 2016). It has been reported that α-synuclein can interfere with ATF6 transport from the ER to Golgi disrupting the UPR (Credle et al. 2015). Furthermore, the knockout of *Atf6* enhances susceptibility to degeneration of dopaminergic neurons associated with increased accumulation of ubiquitinated protein aggregates in a drug-induced mouse model of PD (Egawa et al. 2011; Hashida et al. 2012). Recent studies using ATF6 activator compounds indicate that the pathway has neuroprotective effects in models of brain ischemia (Blackwood et al. 2019).

12.3.4 The Folding Machinery and Protein Quality Control

Besides the UPR activation, the up-regulation and post-translational modification of chaperones and enzymes in patient *postmortem* samples and animal models also support that protein folding stress in the ER contributes to neurodegeneration. The accumulation of misfolded proteins in the ER lumen may arise from disruption of different aspects of the secretory pathway, from protein export to quality control (Fig. 12.1). Here we discuss major evidence showing how proteins linked to neurodegenerative diseases can disturb ER proteostasis (Fig. 12.1) and the functional relationship of components of the ER machinery to pathological outcomes.

The specific up-regulation of the chaperone BiP (also known as GRP78) only in normal-appearing neurons from AD cases indicated an adaptive advantage of the ER folding machinery against proteotoxic stress (Hamos et al. 1991; Hoozemans et al. 2005). In addition to AD, BiP has been found up-regulated in PRDs, ALS and HD (Hetz et al. 2003; Carnemolla et al. 2009; Sasaki 2010; Lee et al. 2012). In PD, a study reported decreased BiP protein levels despite higher mRNA content (Baek et al. 2019). Studies in mouse models suggest that BiP dysfunction could contribute to pathological mechanisms. For instance, heterozygous knock-in mice for mutant *Hspa5*, encoding BiP, developed serious motor problems during aging accompanied by selective degeneration of motoneurons in the spinal cord and accumulation of wild-type SOD1 aggregates reminiscent of species associated to sporadic ALS (Jin et al. 2014). In addition, BiP-deficient mice show impaired sleep behavior during aging (Naidoo et al. 2018). Loss-of-function mutations in SIL1, a nucleotide exchange factor for BiP, cause Marinesco–Sjögren syndrome, a condition marked by cerebellar ataxia and intellectual disability (Anttonen et al. 2005; Senderek et al. 2005). SIL1 deficiency has been found to confer susceptibility of vulnerable motoneurons in ALS, with SIL1 overexpression using a gene therapy strategy affording protection in a mutant SOD1 mouse model of the disease (Filézac de L'Etang et al. 2015). Furthermore, mutant SOD1 aggregates have been detected interacting with BiP in microsomal fractions of spinal cord of diseased mice (Kikuchi et al. 2006). Increased levels of BiP were found at pre-symptomatic stages in AD mouse models (Soejima et al. 2013; Liu et al. 2016a), suggesting that BiP could mediate early protective mechanisms in affected neurons. However, one study has shown that BiP up-regulation occurred concomitantly with a decrease in SIL1 levels, which was linked to stimulation of tau phosphorylation through activation of glycogen synthase kinase 3β (GSK3β) (Liu et al. 2016a). The overexpression of SIL1 in cell culture system was able to prevent tau phosphorylation triggered by increased levels of BiP (Liu et al. 2016a). Thus, the expected neuroprotective effects of BiP likely depends on its proper association with co-chaperones (Casas 2017). In PD mouse models, BiP levels were augmented and physical interaction detected between the chaperone and α-synuclein aggregates recovered from microsomes (Bellucci et al. 2011; Colla et al. 2012). Corroborating this observation, gene therapy to overexpress BiP in a rat model of PD rescued dopaminergic neuron loss and behavior deficit possibly by sequestering α-synuclein (Gorbatyuk et al. 2012). In a knock-in mouse model of HD, BiP has been found up-regulated in pre-symptomatic stages before appearance of mutant huntingtin inclusions (Carnemolla et al. 2009). In prion infected mice, reduction of BiP expression accelerates the pathology (Park et al. 2017).

Members of the PDI family have also been consistently implicated in neurodegenerative diseases (Hetz et al. 2003, 2009; Uehara et al. 2006; Atkin et al. 2008; Walker et al. 2010; Nardo et al. 2011; Kwok et al. 2013; Gonzalez-Perez et al. 2015; Medinas et al. 2018; Montibeller and de Belleroche 2018; Montibeller et al. 2020). A common pathological signature emerged in AD, PD and sporadic ALS, with detection of S-nitrosylated PDIA1 possibly reflecting inactivation of the enzyme and, as a consequence, dysfunction of redox folding in the ER contributing to proteostasis decay in these neurodegenerative diseases (Uehara et al. 2006; Walker et al. 2010). Accordingly, increased levels of PDIA3, PDIA4 and PDIA6 have been reported in AD (Montibeller and de Belleroche 2018) and PDIA1, PDIA3 and PDIA4 in ALS (Atkin et al. 2008; Hetz et al. 2009; Medinas et al. 2018; Montibeller et al. 2020). Intronic polymorphisms in *PDIA1* have been associated to decreased survival in sporadic ALS (Kwok et al. 2013; Yang and Guo 2015), while missense mutations in *PDIA1* and *PDIA3* have been described as risk factors for ALS (Gonzalez-Perez et al. 2015). Furthermore, PDIA1 and PDIA3 have been proposed as possible biomarkers for ALS, with increased levels in cerebrospinal fluid and blood of patients, respectively (Atkin et al. 2008; Nardo et al. 2011). In PRDs, increased levels of PDIA3 have been reported (Hetz et al. 2003).

The exact participation of PDIs in pathological mechanisms remain debatable. In prion infected animals, induction of PDIA3 in different brain areas was the earliest alteration to ER proteostasis accompanying prion spreading, with its overexpression mitigating prion toxicity in cell culture (Hetz et al. 2005). In line with this observation, PDIA3 has been shown to participate in the maturation of the prion protein (Torres et al. 2015), and also could reduce prion propagation and ER stress in prion-infected cells (Thapa et al. 2018). Relevantly, gene therapy to deliver PDIA3 into the brain protected prion-infected mice against neurodegeneration (Thapa et al. 2018). In ALS, induction of PDIs is proposed to constitute an adaptive response to counteract protein aggregation (Walker et al. 2010; Jeon et al. 2014; Parakh et al. 2018; Medinas et al. 2018). The ALS-linked PDIA1 lose the capacity to reduce mutant SOD1 and TDP-43 aggregation (Parakh et al. 2020). Despite PDIs capacity to reduce aggregation of ALS proteins, experimental evidence for this function is lacking *in vivo*. On the other hand, both PDIA1 and PDIA3 have been shown to promote neuritogenesis in cell culture and their ALS-linked mutants to impair neuromuscular junction (NMJ) in zebrafish (Woehlbier et al. 2016). Moreover, the conditional knockout of *Pdia3* in the nervous system led to disruption of NMJ in mice, which was correlated to altered maturation of synaptic proteins and motor impairment (Woehlbier et al. 2016). Supporting this notion, calreticulin deficiency accelerated muscle denervation in a mutant SOD1 mouse model (Bernard-Marissal et al. 2015). PDIA3 overexpression in mice protected against muscle denervation in experimental ALS and improved locomotor recovery in a model of sciatic nerve crush, but did not exert any protection in a pharmacological model of PD (Rozas et al. 2021; Castillo et al. 2015).

ERAD is another critical arm of the ER proteostasis network possibly affected in neurodegenerative diseases. In AD, expression of the E3 ubiquitin-protein ligase hHRD1 involved in ERAD was observed in neurons and reactive astrocytes in the hippocampus and possibly associated with degradation of hyperphosphorylated tau (Hou et al. 2006). Indeed, interaction of tau with ERAD components HRD1 and valosin-containing protein (VCP) has been detected in transgenic tau mice exhibiting accumulation of an ERAD substrate, which also increase in the brain of AD patients

(Abisambra et al. 2013). In HD, increased expression of the ERAD component homocysteine-responsive ER-resident ubiquitin-like domain member 1 protein (HERPUD1) has been documented (Carnemolla et al. 2009). In fact, a polyglutamine tract of mutant huntingtin has been shown to impair ERAD by sequestering specific components (Duennwald and Lindquist 2008). Mutant SOD1 linked to ALS has been demonstrated to disturb ERAD through a specific interaction with Derlin-1, triggering ER stress and motoneuron demise (Nishitoh et al. 2008). The binding of mutant SOD1 to Derlin-1 was mapped and shown to be a common property of the majority of ALS-linked mutants and Zn-deficient wild-type SOD1 (Fujisawa et al. 2012). Furthermore, a conformation-specific antibody has been developed to the epitope corresponding to the Derlin-1 binding region and used to detect misfolded SOD1 in blood cells of SOD1-positive patients (Fujisawa et al. 2012). In iPSC-derived neurons from patients carrying mutant α-synuclein, ERAD dysfunction has been detected as an early and progressive impairment yielding molecular fingerprints of substrate trafficking and accumulation that was also observed in *postmortem* brain samples of familial and sporadic PD cases (Chung et al. 2013).

The ER-to-Golgi transport can also be affected by misfolded proteins related to neurodegenerative diseases, impacting ER exit and resulting in ER stress. The clearest evidence for this pathological mechanism has been provided for ALS, with mutant SOD1, TDP-43 and FUS inhibiting Rab1-dependent vesicular traffic between the ER and Golgi (Soo et al. 2015). Rab1 co-localized with mutant SOD1, TDP-43 and FUS in cell culture and formed inclusions with TDP-43 in motoneurons of sporadic ALS cases, evidencing the broad relevance of this pathway to disease pathogenesis (Soo et al. 2015). Furthermore, the overexpression of Rab1 was enough to restore ER-to-Golgi transport and reduce ER stress in the ALS cellular models (Soo et al. 2015). A genetic screening for modifiers of α-synuclein toxicity in yeast found components of the ER-to-Golgi transport as the most effective suppressors (Cooper 2006). Functional testing *in vivo* using *Drosophila* and *C. elegans* models of PD demonstrated that overexpression of Rab1 prevented loss of dopaminergic neurons (Cooper 2006).

12.4 Diet-Induced Comorbidities: The Road to Deleterious Cascades of ER Stress

Aging involves an orchestrated decline of organism functions (López-Otín et al. 2013), comprising the main risk factor for the development of neurodegenerative diseases (Kennedy et al. 2014). Common cellular and molecular pathways, collectively termed the hallmarks of aging, are proposed to underlie normal aging in different parts of the body (López-Otín et al. 2013). The acceleration or exacerbation of the hallmarks of aging, either due to individual genetics or the environment, could trigger the switch from normal to pathological aging (Kennedy et al. 2014).

Although each organ exhibits particular trajectories of gene expression and tissue renewal, remodeling, scarring and inflammation during lifespan (Schaum et al. 2020), age-related pathological processes can be intercommunicable between organs through systemic mediators as shown for blood components modulating the risk of neurodegeneration (Pluvinage and Wyss-Coray 2020). Dietary restriction has been proven extraordinarily efficient at improving most aspects of aging, extending healthspan and lifespan (Fontana and Partridge 2015). Of note, the brain alone consumes about one quarter of total oxygen and glucose ingested by the organism, despite accounting for only 2–3% of the body weight (Wahl et al. 2019). The high demand for energy resources placed evolutionary constraints for optimal brain functioning, which adapted for intermittent feeding (Mattson and Arumugam 2018). The pace of brain aging is tightly coupled to metabolic states, with calorie restriction and exercise contributing to structural integrity and preserved function of the nervous system whereas excessive energy intake and obesity accelerate brain atrophy and cognitive impairment (Mattson 2012; Mattson and Arumugam 2018). Indeed, the relationship between metabolic disorders and the risk of neurological disease marked by cognitive problems has been established (Mattson 2012; Mattson and Arumugam 2018; Wahl et al. 2019), and the underlying mechanisms constitute promising research areas for therapeutic development.

Metabolic syndrome is an intertwined association of various complex conditions like insulin resistance, abdominal obesity, dyslipidemia, hypertension, and hyperglycemia that is reaching epidemic proportions due to the increased obesity, with around 30% of the adult US population affected (Aguilar et al. 2015). A combination of genetic background, high-calorie diet, lack of physical exercise, aging and environmental factors precipitates the development of metabolic syndrome, predisposing an individual for myriad diseases including cardiovascular disease, chronic kidney disease, cancers, type-2 diabetes (T2D) and dementia (Alberti et al. 2009). The ER is a highly sensitive organelle to nutrient status, where the UPR coordinates the metabolic function of different organs such as the liver, the adipose tissue, the pancreas and the brain (Fig. 12.3). The integration of energy metabolism with immune responses, which presumably has evolved to support organismal adaptation to environmental threats including starvation and pathogens (Hotamisligil 2017), is burdened by the excess of nutrient consumption of modern lifestyle, featuring maladaptive UPR and 'low-grade' inflammation driving metabolic collapse.

In this section, we overview current evidence linking ER stress and the UPR to pathophysiology of key metabolic organs to set the context for interpretation of the crosstalk between metabolic and nervous system disorders.

12.4.1 Liver

The liver is the central organ orchestrating glucose and lipid metabolism at the organismal level by controlling glycemia and secretion of lipoproteins. XBP1s,

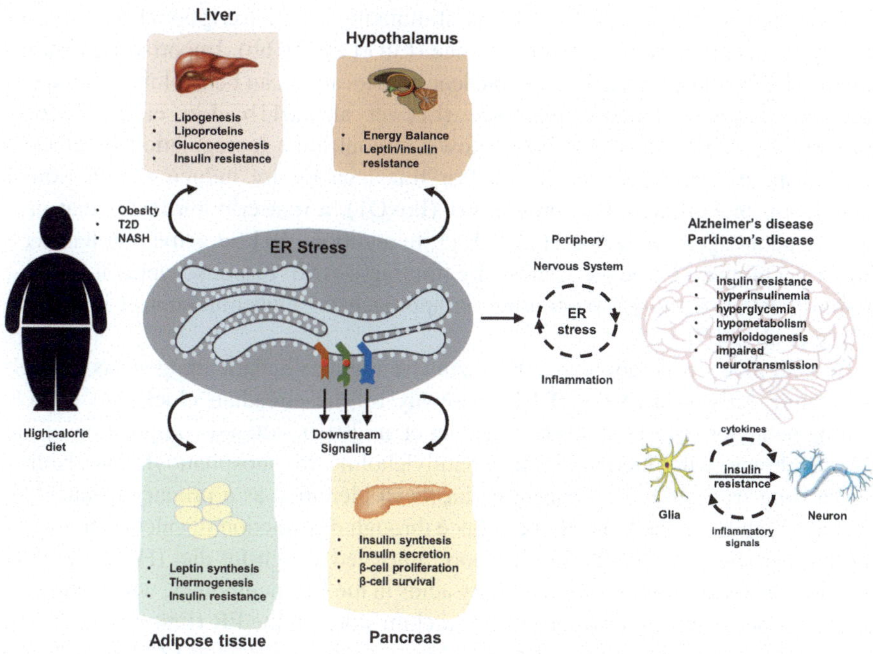

Fig. 12.3 ER stress fuels a deleterious crosstalk between metabolic syndrome and neurodegenerative diseases. The ER and the UPR sustain function of different metabolic organs, such as lipogenesis and lipoprotein synthesis in liver, leptin and adipokines synthesis in adipose tissue, insulin synthesis and glucose-stimulated secretion in pancreas and neuropeptides synthesis in response to insulin and leptin in the hypothalamus. The excessive nutrient intake in high-calorie diet disturbs ER proteostasis in these different organs, leading to sustained UPR activation and 'low-grade' inflammation that perpetuates deleterious vicious cycle of ER stress in the periphery and the central nervous system, resulting in insulin and leptin resistance, hyperinsulinemia, hyperglycemia, among other metabolic disarrangements that contribute to development of metabolic and neurodegenerative diseases. *T2D* type-2 diabetes; *NASH* non-alcoholic steatohepatitis

through UDP-galactose-4-epimerase (GalE), promotes postprandial metabolic remodeling by coupling glucose uptake to increased synthesis of glycosylated proteins (Deng et al. 2013). Further underscoring XBP1s function in glucose metabolism, insulin receptor signaling after refeeding of fasted mice induces nuclear translocation of XBP1s in the liver upon association with p85α and p85β, regulatory subunits of the phosphoinositide 3-kinases (PI3K) that were identified as XBP1s interactors (Park et al. 2010). Importantly, this interaction is disrupted in the liver of obese and insulin-resistant *ob/ob* mice and can be restored upon overexpression of p85α or p85β, with normalization of glycemia and reduction of ER stress in the liver after refeeding (Park et al. 2010). Using the same fasting-refeeding paradigm, the stress-activate protein kinase p38 MAPK has been shown to promote nuclear translocation of XBP1s through phosphorylation, a pathway impaired in *ob/ob* mice (Lee et al. 2011b). Along the same line, IkB kinase beta (IKKβ), a major inflammatory mediator downstream of tumor necrosis factor (TNF), has been demonstrated to phosphorylate XBP1s stimulating its transcriptional activity, an activity that is compromised in obese mice (Liu et al. 2016b). Importantly, restoration of XBP1s phosphorylation and nuclear translocation can be exploited therapeutically to improve glucose tolerance (Lee et al. 2011b; Liu et al. 2016b). Surprisingly, XBP1s has also been shown to contribute to regulation of glucose metabolism independently of its nuclear translocation by inducing proteasomal degradation of Forkhead box protein O1 (FoxO1), a transcription factor that promotes gluconeogenesis (Zhou et al. 2011). In addition, ATF6α activation has been shown to counteract gluconeogenesis by limiting c-AMP response element-binding protein (CREB) activity, constituting an approach for glycemia control in obesity (Wang et al. 2009).

Beyond glucose metabolism, XBP1 can also regulate hepatic lipogenesis through a rhythmic activation of the IRE1α pathway by the circadian clock that reflects feeding behavior (Lee et al. 2008; Cretenet et al. 2010). Altered composition of the ER membrane with increased phosphatidylcholine to phosphatidylethanolamine ratio arising from *de novo* lipogenesis has been identified as a primary mechanism driving ER stress in the liver of obese mice through dysfunction of calcium transport into the ER (Fu et al. 2011). Furthermore, mice fed a high-fat-diet (HFD) exhibits increased incorporation of saturated fatty acids in the ER membranes, which can also lead to ER stress through compromised calcium stores in the ER (Ersoy et al. 2017). On the other hand, chronic ER stress in obesity leads to sustained IRE1α activation and development of insulin resistance through c-Jun N-terminal kinases (JNK)-mediated serine phosphorylation of insulin receptor substrate (IRS), a mechanism exacerbated upon deficiency of XBP1 (Ozcan 2004). Moreover, ER stress and the UPR are also intimately connected to multiple layers of lipid metabolism that can result in hepatic steatosis, non-alcoholic fat liver disease (NAFLD) and non-alcoholic steatohepatitis (NASH) (Lebeaupin et al. 2018; Henkel 2018).

Indeed, sustained UPR can lead to induction of inflammatory pathways, with IRE1α activating JNK and IKK (inhibitor of nuclear factor-κB (NF-κB) kinase), PERK releasing inhibition of NF-κB through translational suppression of IκB (inhibitor of NF-κB), and ATF6 activating NF-κB (Hotamisligil 2010). Metabolic inflammation, or metaflammation, emerged as a critical mechanism disrupting energy metabolism in obesity and related disorders, featuring infiltration of immune cells in tissue stroma and execution of deleterious inflammatory cascades by immune and non-immune cells (Hotamisligil 2017). For instance, IRE1α nitrosylation by inducible nitric oxide synthase (iNOS) in the liver has been shown to suppress its RNase activity, compromising XBP1 splicing and control of glucose metabolism (Yang et al. 2015). Besides, compromised RIDD activity due to IRE1α nitrosylation can lead to increased levels of miRNA involved in lipid mobilization contributing to hepatic steatosis (Wang et al. 2018; Henkel 2018).

It should be noted that controversies remain about the role of the UPR in regulating different classes of metabolites and the exact implication of metabolic and inflammatory crosstalk to pathogenic mechanisms in humans. Nonetheless, the accumulated evidence supports that metabolic and inflammatory pathways are intimately connected by the UPR, which sustain vicious cycles of abnormal glucose and lipids utilization in the context of excess of nutrients and precipitate insulin resistance, hyperglycemia and dyslipidemia that burden and predispose the organism for myriad diseases.

12.4.2 Adipose Tissue

The adipose tissue is the site for energy storage in the organism in the form of triacylglycerides. As in the liver, ER stress constitutes a cellular signature of adipose tissue in obesity (Ozcan 2004). The enlargement of adipose tissue in obesity can occur by both hyperplasia and hypertrophy of adipocytes, which implicate extensive cell remodeling marked by ER stress, cell death and recruitment of macrophages (Chawla et al. 2011). A landmark study by Hotamisligil *et al* showing TNF-α induction in the adipose tissue of obese mice provided the first evidence linking inflammation to metabolic syndrome (Hotamisligil et al. 1993). In fact, obesity causes a dramatic shift in the macrophage population residing in the adipose tissue, inducing a shift from alternative M2 state supporting tissue homeostasis to classical M1 state that generates a noxious environment marked by secretion of pro-inflammatory cytokines and reactive oxygen species (Chawla et al. 2011). The exposure of adipose tissue macrophages to saturated fatty acids from obesogenic diet, and possibly from the phagocytosis of lipids released by dying adipocytes, may induce a state of 'metabolic activation' (M_{Me}) of these cells that is characterized by IRE1α activation and, as a consequence, assembly of nucleotide-binding oligomerization domain (NOD)-like receptor family, pyrin domain-containing 3 (NLRP3) inflammasome and secretion IL-1β, a pro-inflammatory cytokine associated with insulin resistance (Robblee et al. 2016). Moreover, IRE1α has been shown to suppress M2 state of macrophages, impairing energy expenditure associated to adipose tissue browning and thermogenesis (Shan et al. 2017). On the other hand, inducible XBP1s overexpression in adipocytes results in reduced adiposity in genetic and diet-induced obese mouse models through stimulation uridine biosynthesis (Deng et al. 2018), which has been linked to maintaining whole-body metabolic balance, including thermogenesis (Deng et al. 2017a). Additionally, XBP1s overexpression in adipocytes was also reported in lean and obese mouse models that exhibited improvement in insulin resistance and glucose intolerance (Sha et al. 2014), largely because of the enhancement of ER chaperone-mediated maturation of high-molecular-weight adiponectin, the multifunctional adipokine that exerts a broad range of beneficial effects upon metabolic health (Turer and Scherer 2012).

12.4.3 Pancreas

The pancreatic islets are the structures in the pancreas controlling systemic glucose metabolism through the production and secretion of glucagon and insulin. The ER of the β-cell is highly demanded for production of insulin, experiencing a flux of approximately one million molecules per minute (Scheuner and Kaufman 2008). The burden on the ER of β-cells is further aggravated under conditions marked by insulin resistance since the requirement for insulin synthesis sharply increases (Karam et al. 1963). The activity of the PERK pathway is fundamental to the function and viability of β-cells by fine-tuning translation rates and, therefore, preventing the protein folding capacity of the ER from being overwhelmed (Scheuner and Kaufman 2008). Indeed, the pancreatic toxicity of PERK inhibitors restrict their use as therapeutic intervention to neurodegenerative diseases (Hetz and Saxena 2017). Glucose levels can influence insulin synthesis and secretion at multiple layers (Scheuner and Kaufman 2008), with activation of protein phosphatase 1 (PP1) to catalyze dephosphorylation of eIF2α directly coupling the UPR to metabolic regulation (Vander Mierde et al. 2007). Furthermore, the IRE1α pathway has been shown to regulate insulin metabolism in β-cells at different molecular levels, such as insulin biosynthesis, processing, and secretion through both *Xbp1* mRNA splicing and RIDD activities. This molecular regulation must be minutely adjusted given that sustained production of XBP1s induces dysfunction and apoptosis of primary β-cells (Allagnat et al. 2010), whereas XBP1 deficiency in β-cells of conditional knockout mice results in impaired insulin maturation, reduced glucose-stimulated insulin secretion, inhibited cell proliferation, and hyperactivation of IRE1α driving degradation of mRNAs encoding proinsulin-processing enzymes (Lee et al. 2011a). In the same line, conditional deletion of IRE1α in β-cells affects insulin secretion mainly by decreasing levels of members of the PDI family involved in the oxidative folding of proinsulin (Tsuchiya et al. 2018). On the other hand, IRE1α hyperactivation in β-cells of the Akita mouse model of diabetes can induce thioredoxin-interacting protein (TXNIP), involved in NLRP3 inflammasome activation, and degradation of mRNA encoding for proinsulin, which can be blocked by inhibiting RIDD activity with KIRA (Morita et al. 2017). Thus, the balance between *Xbp1* mRNA splicing versus RIDD activities appears to be determinant for IRE1α contribution to β-cell function and vulnerability in diabetes context.

12.4.4 Hypothalamus: Central Nervous System Control of Energy Homeostasis

The hypothalamus is the brain region integrating signals from the periphery to regulate organism metabolism, influencing food intake and energy expenditure in response to the anorexic hormones insulin and leptin secreted by the pancreas and adipose tissue, respectively (Varela and Horvath 2012). The dysregulation of hypothalamic function due to ER stress and inflammation also occurs in obesity, leading to insulin and leptin resistance (Cakir and Nillni 2019). Two neuronal populations in the arcuate nucleus of hypothalamus, agouti-related protein (AgRP) and pro-opiomelanocortin (POMC) neurons, play opposing roles on feeding behavior that are intimately linked with the UPR (Varela and Horvath 2012; Cakir and Nillni 2019).

The AgRP neurons exhibit orexigenic function and induce XBP1s in fasted mice (Henry et al. 2015). On the other hand, conditional knockout of *Atf4* in AgRP neurons reduce food intake and increase energy expenditure mainly because of thermogenesis in brown fat tissue (Deng et al. 2017b). Importantly, ablation of *Atf4* in AgRP neurons protected against HFD-induced obesity, insulin resistance and liver steatosis, effects associated to decreased expression of FoxO1 in the hypothalamus (Deng et al. 2017b). Conversely, induction of ATF4 expression in the hypothalamus leads to insulin resistance in the liver via the vagus nerve (Zhang et al. 2013). In POMC neurons, conditional knockout of *Atf4* promotes increased energy expenditure and protects against HFD-induced obesity, glucose intolerance and leptin resistance, effects that were associated to enhanced autophagy and production of α-melanocyte-stimulating hormone (α-MSH) (Xiao et al. 2017). Contrarily, conditional knockout of *Xbp1* in the nervous system aggravates obesity in mice, worsening insulin resistance accompanied by marked hyperplasia of pancreatic islets, in addition to leptin resistance and augmented ER stress in the hypothalamus (Ozcan et al. 2009). Accordingly, genetic ablation of IRE1α RNase activity in POMC neurons causes similar detrimental effects (Yao et al. 2017), while overexpression of XBP1s protects against diet-induced obesity and leptin/insulin resistance (Williams et al. 2014). Remarkably, overexpression of XBP1s in POMC neurons engages XBP1s in the liver through cell-nonautonomous UPR, correlating with improved hepatic insulin signaling and repression of gluconeogenesis (Williams et al. 2014). Along the same lines, food perception stimulates POMC neurons to induce cephalic response via sympathetic nervous system, priming the liver for nutrient ingestion through activation of mTOR and *Xbp1* mRNA splicing (Brandt et al. 2018). This cell-nonautonomous UPR involving the hypothalamus-liver axis is analogous to the mechanism operating in *C. elegans* with XBP1s expression in neurons instigating mirror responses in the intestine, where XBP1s engages responses linked to lipid metabolism to extend lifespan (Taylor and Dillin 2013; Imanikia et al. 2019; Daniele et al. 2020).

Similarly to the periphery, excess nutrients leads to ER stress-mediated activation of IKK/NF-κB inflammatory pathway in the hypothalamus and induction of suppressor of cytokine signaling 3 (SOCS3), an inhibitor of leptin and insulin signaling (Zhang et al. 2008). This inflammatory response may be mediated in part by IRE1α (Hotamisligil 2010), possibly reconciling the protective effects of complete deletion of IRE1α protein in POMC neurons of obese mice (Xiao et al. 2016). Impaired processing of neuropeptides constitutes another important negative effect of ER stress in the hypothalamus, with obese animals, for instance, displaying accumulation of POMC and decreased levels of adrenocorticotropic hormone and α-MSH (Cakir and Nillni 2019). Thus, ER stress in the hypothalamus can contribute to obesity through similar mechanisms operating in the periphery, in addition to affect the intrinsic balance of protein folding and processing taking place within specific neuronal populations.

12.5 Metabolic Syndrome and Neurodegeneration: Cause and Effect Relationship?

Endogenous and exogenous risk factors may aggravate the natural decay of proteostasis networks, accelerating the burden of misfolded and aggregated proteins to neuronal physiology and triggering neurodegenerative cascades. In this section, we will present evidence supporting shared molecular pathways between metabolic syndrome and neurodegeneration emanating from ER stress and the UPR.

As discussed in the previous section, overnutrition can result in the development of chronic pathological states marked by insulin resistance. Epidemiological studies have showed that diabetic individuals are at increased risk of developing dementia (Biessels et al. 2006; Haan 2006). Analysis of patient *postmortem* tissue found brain insulin resistance with reduced expression and function of insulin receptor as a consistent feature of AD brains that correlates with declined cognitive capacity (Talbot et al. 2012). In addition, reduced glucose metabolism is an early pathological event in AD, comprising a better disease biomarker than levels of Aβ peptides (Dukart et al. 2013; Chen and Zhong 2013). The overexpression of glucose transporters in neurons has been showed to rescue neuropathology and extend lifespan in a *Drosophila* model of AD (Niccoli et al. 2016), an effect accompanied by the restoration of ER proteostasis through reduction of BiP protein levels and the UPR activation. This study also showed that treatment of *Drosophila* models with metformin, an anti-diabetogenic drug that increases glucose uptake by increasing the translocation of glucose transporters to the plasma membrane, affords neuroprotection in addition to normalizing BiP protein levels. Currently, the use of metformin to treat AD patients is under scrutiny (Koenig et al. 2017). Impaired glucose utilization in AD may result in hypometabolism (La Joie et al. 2012), a condition triggering tau phosphorylation in neurons through intermediacy of ER stress and the UPR activation (van der Harg et al. 2014). Of note, polymorphisms in *GLIS3*, a key transcription factor in glucose metabolism also associated to sporadic neonatal diabetes (Senée et al. 2006), have been identified as risk variants for AD through correlation with increased levels of tau and phosphorylated tau in cerebrospinal cord fluid (CSF) (Cruchaga et al. 2013). Moreover, hyperglycemia has been linked to hypometabolism in brain areas affected in AD (Burns et al. 2013).

Peripheral insulin is transported across the blood-brain-barrier via a carrier-facilitated process (Banks 2004). Insulin signaling in the brain is mainly involved in regulation of appetite and cognitive function, not exerting a significant control on glucose uptake (Heni et al. 2015). In mice, age-related peripheral hyperinsulinemia has been show to translate into higher insulin levels in CSF, which in turn can cause insulin resistance and cellular senescence in neurons marked by exhaustion of synaptic activity (Chow et al. 2019). Interestingly, a possible relationship between deficiency of IRE1α/XBP1 axis and neuronal senescence has been proposed (Cabral-Miranda et al. 2020). Insulin resistance may predispose middle-aged adults to amyloid deposition, despite no increase in amyloid burden being reported in diabetic patients (Biessels and Reagan 2015). In experimental AD, diet-induced insulin resistance can promote amyloidosis (Ho et al. 2004). These observations indicate that insulin resistance may facilitate initial steps of amyloid cascade. Indeed, insulin can stimulate secretion of Aβ peptides while inhibiting its extracellular degradation by competition for the insulin-degrading enzyme, a metallopeptidase acting on both insulin and Aβ peptides (Gasparini et al. 2001). On the other hand, Aβ oligomers can impair function of neuronal insulin receptors (Zhao et al. 2008). Thus, hyperinsulinemia can directly contribute to feed a vicious cycle of Aβ peptide production that could aggravate insulin resistance in AD.

Aβ oligomers cause activation of AMPK in neurons with consequent inhibition of the mTOR pathway (Yoon et al. 2012). This metabolic modulation is accompanied by ER stress and activation of JNK3 (Yoon et al. 2012). Activated JNK3 has been detected in brain tissue of AD patients and mice, being capable of phosphorylating APP favoring its amyloidogenic processing (Yoon et al. 2012). Moreover, ER stress originating from Aβ oligomers activates IRE1α (Lourenco et al. 2013), which in turn can promote further activation of JNK3 feeding a vicious cycle of Aβ peptide production and ER stress (Yoon et al. 2012). Indeed, the genetic ablation of JNK3 in an AD model prevents ER stress (Yoon et al. 2012). Furthermore, ER stress can blunt insulin signaling by inducing phosphorylation of insulin receptor substrate 1 (IRS-1), and effect blocked by 4-phenybutyrate (4-PBA) (Lourenco et al. 2013). Of note, the anti-diabetogenic drug liraglutide reduces ER stress and rescues cognitive capacity of AD mice (Lourenco et al. 2013).

AD pathology also features neuroinflammatory responses with activation of astrocytes and microglia surrounding amyloid plaques (Glass et al. 2010). Delivery of Aβ oligomers to the brain of mice results in memory impairment and insulin resistance that can be prevented by blocking the action of TNF-α (Lourenco et al. 2013). Furthermore, TNF-α also mediates inflammatory responses in the hypothalamus of mice receiving cerebral delivery of Aβ oligomers, causing peripheral metabolic dysregulation marked by glucose intolerance (Clarke et al. 2015). Thus, the development of insulin resistance in the brain likely results from a pathological interaction between glia and neurons. The pathogenic mechanisms emerging in AD are remarkably similar to those described in metabolic disorders, which could operate both locally in the central nervous system (CNS) and through deleterious crosstalk with the periphery as well (Fig. 12.3).

Obesity and accompanying conditions such as NASH comprise risk factors for cognitive impairment and neurodegeneration (de la Monte and Tong 2014), featuring ER stress, chronic inflammation and development of insulin resistance dependent on IRE1α signaling (Lebeaupin et al. 2018; Henkel 2018). In patients and mouse model of NASH, IRE1α/XBP1 pathway in the liver stimulates *de novo* synthesis and secretion of ceramides (Dasgupta et al. 2020), toxic lipids that can modulate systemic metabolism and exert detrimental effects in the CNS (Bikman and Summers 2011; de la Monte and Tong 2014). Interestingly, increased levels of ceramides normally produced in the liver were found in plasma and brain of AD patients (Kurz et al. 2019), illustrating a possible peripheral insult contributing to dementia under metabolic dysregulation.

An association between diabetes and increased risk of PD has also been described (Hu et al. 2007; Schernhammer et al. 2011). Indeed, altered glucose metabolism and insulin signaling has been described in PD patients (Aviles-Olmos et al. 2013; Sekar and Taghibiglou 2018), with insulin pathways being intimately related to the control of dopamine metabolism (Fiory et al. 2019). Furthermore, pharmacological and genetic mouse models of PD develop insulin resistance in the brain (Fiory et al. 2019). There is evidence showing ER stress in the midbrain of ob/ob and db/db diabetic mice, with increased levels of BiP and CHOP (Wang et al. 2014b). Interestingly, these mice are more susceptible to a parkinsonian toxin (Wang et al. 2014b). Further investigation on the role of ER stress in the development of insulin resistance in PD is warranted, where we highlight possible interrelationships with pathogenic mechanisms in AD driving non-motor symptoms in PD ranging from mild cognitive impairment to dementia (Athauda and Foltynie 2016).

While insulin resistance has been demonstrated in ALS patients (Perurena and Festoff 1987; Ahmed et al. 2019), the possible relationship between metabolic syndrome and ALS remains elusive. ALS patients are hypermetabolic, with long-term high-calorie diet associated to extended survival (Ahmed et al. 2016). Contrary to AD and PD, metabolic syndrome and diabetes seem to exert protective effects correlating with later onset in ALS (Ahmed et al. 2016; Jawaid et al. 2018). However, important differences may exist regarding the impact of metabolic perturbations on motor and cognitive function of the nervous system (Jawaid et al. 2018). Further studies are needed for a mechanistic comprehension of insulin pathways in ALS pathogenesis.

12.6 Conclusions

The ER is a major platform integrating proteostasis networks and metabolic pathways, promoting inter-organelle communication and coordinating cellular response to stress, including induction of inflammatory cascades and cell death. There is undeniable evidence for a transversal role of ER stress in the etiology of myriad pathological conditions of the nervous system. The development of a certain neurodegenerative disease may reflect a complex interplay of genetic mutations, lifestyle and aging, conferring selective vulnerability to neuronal populations that are developmentally, functionally and spatially unique. The molecular pathways disrupted in metabolic syndrome leading to failure of peripheral organs are mirrored in the nervous system, contributing to neurodegenerative diseases either because they intrinsically affect neuronal function and neuron-glia balance or because they cause systemic alterations that establish a deleterious crosstalk with the CNS (Fig. 12.3). Thus, the accumulation of misfolded and aggregated proteins in neurons and glia impacting ER proteostasis through multiple mechanisms can synergize with metabolic dysregulation to amplify pathogenic cascades of ER stress, inflammation, and proteostasis impairment. Furthermore, the pathophysiological effects of metabolic syndrome on vascular system not discussed here may also have a major impact on the etiology of neurodegenerative disease due to microinfarcts. In conclusion, the concomitant targeting of protein misfolding and metabolic unbalance may prove a realistic strategy for development of much needed therapeutics for nervous system disorders.

12.7 Summary Points

- ER stress is a transversal pathogenic mechanism in nervous system disorders.
- Misfolded and aggregated protein species related to neurodegenerative diseases can impair different nodes of the ER proteostasis network.
- Manipulation of the UPR has therapeutic potential to treat nervous system disorders.
- Metabolic syndrome constitutes a risk factor for neurodegeneration.
- Shared molecular pathways between metabolic syndrome and neurodegenerative diseases open novel therapeutic opportunities to treat these comorbidities.

12.8 Future Issues

- What is the cause of the ER stress observed in different neurodegenerative diseases?
- Is ER stress associated to selective vulnerability of different neuronal populations in nervous system disorders?
- Is there a cause effect relationship between metabolic syndrome and neurodegenerative diseases or are these epiphenomenal?
- Can ER stress originating from metabolic syndrome be propagated from peripheral organs to the CNS in a cell nonautonomous manner?
- What is the spatial and temporal sequence of ER stress responses in the CNS under metabolic dysregulation?
- Does metabolic syndrome contribute to the pathological threshold for clinical manifestation of neurodegenerative diseases?
- What are the key molecular pathways to therapeutically target the deleterious crosstalk between metabolic syndrome and neurodegeneration?

Conflict-of-Interest Disclosure Statement
Claudio Hetz is inventor in patents for the genetic manipulation of the UPR to treat human diseases.

Acknowledgements This work was funded by FONDECYT 1191538, Muscular Dystrophy Association 575897 and ALS Association 19-IIA-456 (DBM), FONDECYT 3180427 (YH), and FONDECYT 1140549, FONDAP program 15150012, Millennium Institute P09-015-F, European Commission R&D MSCA-RISE 734749, FONDEF ID16I10223, FONDEF ID11E1007, CONICYT-Brazil 441921/2016-7, Michael J Fox Foundation for Parkinson's Research—Target

Validation grant 9277, US Office of Naval Research-Global N62909-16-1-2003, US Air Force Office of Scientific Research FA9550-21-1-0096, Department of Defense ALS Research Program Award 81XWH-16-1-0112 and Muscular Dystrophy Association 382453 (CH).

References

Abisambra JF, Jinwal UK, Blair LJ et al (2013) Tau accumulation activates the unfolded protein response by impairing endoplasmic reticulum-associated degradation. J Neurosci 33:9498–9507. https://doi.org/10.1523/JNEUROSCI.5397-12.2013

Acosta-Alvear D, Zhou Y, Blais A et al (2007) XBP1 controls diverse cell type- and condition-specific transcriptional regulatory networks. Mol Cell 27:53–66. https://doi.org/10.1016/j.molcel.2007.06.011

Aguilar M, Bhuket T, Torres S et al (2015) Prevalence of the metabolic syndrome in the United States, 2003-2012. JAMA 313:1973. https://doi.org/10.1001/jama.2015.4260

Ahmed RM, Irish M, Piguet O et al (2016) Amyotrophic lateral sclerosis and frontotemporal dementia: distinct and overlapping changes in eating behaviour and metabolism. Lancet Neurol 15:332–342. https://doi.org/10.1016/S1474-4422(15)00380-4

Ahmed RM, Phan K, Highton-Williamson E et al (2019) Eating peptides: biomarkers of neurodegeneration in amyotrophic lateral sclerosis and frontotemporal dementia. Ann Clin Transl Neurol 6:486–495. https://doi.org/10.1002/acn3.721

Alberti KGMM, Eckel RH, Grundy SM et al (2009) Harmonizing the metabolic syndrome. Circulation 120:1640–1645. https://doi.org/10.1161/CIRCULATIONAHA.109.192644

Allagnat F, Christulia F, Ortis F et al (2010) Sustained production of spliced X-box binding protein 1 (XBP1) induces pancreatic beta cell dysfunction and apoptosis. Diabetologia 53:1120–1130. https://doi.org/10.1007/s00125-010-1699-7

Anttonen A-K, Mahjneh I, Hämäläinen RH et al (2005) The gene disrupted in Marinesco-Sjögren syndrome encodes SIL1, an HSPA5 cochaperone. Nat Genet 37:1309–1311. https://doi.org/10.1038/ng1677

Ash PEA, Bieniek KF, Gendron TF et al (2013) Unconventional translation of C9ORF72 GGGGCC expansion generates insoluble polypeptides specific to c9FTD/ALS. Neuron 77:639–646. https://doi.org/10.1016/j.neuron.2013.02.004

Athauda D, Foltynie T (2016) Insulin resistance and Parkinson's disease: a new target for disease modification? Prog Neurobiol 145–146:98–120. https://doi.org/10.1016/j.pneurobio.2016.10.001

Atkin JD, M a F, Walker AK et al (2008) Endoplasmic reticulum stress and induction of the unfolded protein response in human sporadic amyotrophic lateral sclerosis. Neurobiol Dis 30:400–407. https://doi.org/10.1016/j.nbd.2008.02.009

Aviles-Olmos I, Limousin P, Lees A, Foltynie T (2013) Parkinson's disease, insulin resistance and novel agents of neuroprotection. Brain 136:374–384. https://doi.org/10.1093/brain/aws009

Baek J-H, Mamula D, Tingstam B et al (2019) GRP78 level is altered in the brain, but not in plasma or cerebrospinal fluid in Parkinson's disease patients. Front Neurosci 13:697. https://doi.org/10.3389/fnins.2019.00697

Baleriola J, Walker CA, Jean YY et al (2014) Axonally synthesized ATF4 transmits a neurodegenerative signal across brain regions. Cell 158:1159–1172. https://doi.org/10.1016/j.cell.2014.07.001

Banks WA (2004) The source of cerebral insulin. Eur J Pharmacol 490:5–12. https://doi.org/10.1016/j.ejphar.2004.02.040

Bellucci A, Navarria L, Zaltieri M et al (2011) Induction of the unfolded protein response by α-synuclein in experimental models of Parkinson's disease. J Neurochem 116:588–605. https://doi.org/10.1111/j.1471-4159.2010.07143.x

Bernard-Marissal N, Sunyach C, Marissal T et al (2015) Calreticulin levels determine onset of early muscle denervation by fast motoneurons of ALS model mice. Neurobiol Dis 73:130–136. https://doi.org/10.1016/j.nbd.2014.09.009

Biessels GJ, Reagan LP (2015) Hippocampal insulin resistance and cognitive dysfunction. Nat Rev Neurosci 16:660–671. https://doi.org/10.1038/nrn4019

Biessels GJ, Staekenborg S, Brunner E et al (2006) Risk of dementia in diabetes mellitus: a systematic review. Lancet Neurol 5:64–74. https://doi.org/10.1016/S1474-4422(05)70284-2

Bikman BT, Summers SA (2011) Ceramides as modulators of cellular and whole-body metabolism. J Clin Invest 121:4222–4230. https://doi.org/10.1172/JCI57144

Blackwood EA, Azizi K, Thuerauf DJ et al (2019) Pharmacologic ATF6 activation confers global protection in widespread disease models by reprograming cellular proteostasis. Nat Commun 10:187. https://doi.org/10.1038/s41467-018-08129-2

Brandt C, Nolte H, Henschke S et al (2018) Food perception primes hepatic ER homeostasis via Melanocortin-dependent control of mTOR activation. Cell 175:1321–1335.e20. https://doi.org/10.1016/j.cell.2018.10.015

Bruch J, Kurz C, Vasiljevic A et al (2015) Early Neurodegeneration in the brain of a child without functional PKR-like endoplasmic reticulum kinase. J Neuropathol Exp Neurol 74:850–857. https://doi.org/10.1097/NEN.0000000000000224

Bruch J, Xu H, Rösler TW et al (2017) PERK activation mitigates tau pathology in vitro and in vivo. EMBO Mol Med 9:371–384. https://doi.org/10.15252/emmm.201606664

Burns CM, Chen K, Kaszniak AW et al (2013) Higher serum glucose levels are associated with cerebral hypometabolism in Alzheimer regions. Neurology 80:1557–1564. https://doi.org/10.1212/WNL.0b013e31828f17de

Cabral-Miranda F et al (2020) Control of mammalian brain aging by the unfolded protein response (UPR). bioRxiv. doi:https://doi.org/10.1101/2020.04.13.039172

Cakir I, Nillni EA (2019) Endoplasmic reticulum stress, the hypothalamus, and energy balance. Trends Endocrinol Metab 30:163–176. https://doi.org/10.1016/j.tem.2019.01.002

Calì T, Vanoni O, Molinari M (2008) The endoplasmic reticulum crossroads for newly synthesized polypeptide chains. Prog Mol Biol Transl Sci 83:135–179. https://doi.org/10.1016/S0079-6603(08)00604-1

Carnemolla A, Fossale E, Agostoni E et al (2009) Rrs1 is involved in endoplasmic reticulum stress response in Huntington disease. J Biol Chem 284:18167–18173. https://doi.org/10.1074/jbc.M109.018325

Casas C (2017) GRP78 at the Centre of the Stage in cancer and Neuroprotection. Front Neurosci 11. https://doi.org/10.3389/fnins.2017.00177

Castillo V, Oñate M, Woehlbier U et al (2015) Functional role of the disulfide isomerase ERp57 in axonal regeneration. PLoS One. in press

Chawla A, Nguyen KD, Goh YPS (2011) Macrophage-mediated inflammation in metabolic disease. Nat Rev Immunol 11:738–749. https://doi.org/10.1038/nri3071

Chen Z, Zhong C (2013) Decoding Alzheimer's disease from perturbed cerebral glucose metabolism: implications for diagnostic and therapeutic strategies. Prog Neurobiol 108:21–43. https://doi.org/10.1016/j.pneurobio.2013.06.004

Cheng W, Wang S, Mestre AA et al (2018) C9ORF72 GGGGCC repeat-associated non-AUG translation is upregulated by stress through eIF2α phosphorylation. Nat Commun 9:51. https://doi.org/10.1038/s41467-017-02495-z

Chow H-M, Shi M, Cheng A et al (2019) Age-related hyperinsulinemia leads to insulin resistance in neurons and cell-cycle-induced senescence. Nat Neurosci 22:1806–1819. https://doi.org/10.1038/s41593 019 0505 1

Chung CY, Khurana V, Auluck PK et al (2013) Identification and rescue of–Synuclein toxicity in Parkinson patient-derived neurons. Science 342:983–987. https://doi.org/10.1126/science.1245296

Cissé M, Duplan E, Lorivel T et al (2017) The transcription factor XBP1s restores hippocampal synaptic plasticity and memory by control of the Kalirin-7 pathway in Alzheimer model. Mol Psychiatry 22:1562–1575. https://doi.org/10.1038/mp.2016.152

Clarke JR, Lyra E Silva NM, Figueiredo CP et al (2015) Alzheimer-associated Aβ oligomers impact the central nervous system to induce peripheral metabolic deregulation. EMBO Mol Med 7:190–210. https://doi.org/10.15252/emmm.201404183

Colla E, Coune P, Liu Y et al (2012) Endoplasmic reticulum stress is important for the manifestations of α-synucleinopathy in vivo. J Neurosci 32:3306–3320. https://doi.org/10.1523/JNEUROSCI.5367-11.2012

Cook KM, Hogg PJ (2013) Post-translational control of protein function by disulfide bond cleavage. Antioxid Redox Signal 18:1987–2015. https://doi.org/10.1089/ars.2012.4807

Cooper AA (2006) Synuclein blocks ER-Golgi traffic and Rab1 rescues neuron loss in Parkinson's models. Science 313:324–328. https://doi.org/10.1126/science.1129462

Costa-Mattioli M, Sossin WS, Klann E, Sonenberg N (2009) Translational control of long-lasting synaptic plasticity and memory. Neuron 61:10–26. https://doi.org/10.1016/j.neuron.2008.10.055

Costa-Mattioli M, Walter P (2020) The integrated stress response: from mechanism to disease. Science 368. https://doi.org/10.1126/science.aat5314

Credle JJ, Forcelli PA, Delannoy M et al (2015) α-Synuclein-mediated inhibition of ATF6 processing into COPII vesicles disrupts UPR signaling in Parkinson's disease. Neurobiol Dis 76:112–125. https://doi.org/10.1016/j.nbd.2015.02.005

Cretenet G, Le Clech M, Gachon F (2010) Circadian clock-coordinated 12 Hr period rhythmic activation of the IRE1α pathway controls lipid metabolism in mouse liver. Cell Metab 11:47–57. https://doi.org/10.1016/j.cmet.2009.11.002

Cruchaga C, Kauwe JSK, Harari O et al (2013) GWAS of cerebrospinal fluid tau levels identifies risk variants for Alzheimer's disease. Neuron 78:256–268. https://doi.org/10.1016/j.neuron.2013.02.026

Daniele JR, Higuchi-Sanabria R, Durieux J et al (2020) UPR ER promotes lipophagy independent of chaperones to extend life span. Sci Adv 6:eaaz1441. https://doi.org/10.1126/sciadv.aaz1441

Das I, Krzyzosiak A, Schneider K et al (2015) Preventing proteostasis diseases by selective inhibition of a phosphatase regulatory subunit. Science 348:239–242. https://doi.org/10.1126/science.aaa4484

Dasgupta D, Nakao Y, Mauer AS et al (2020) IRE1A stimulates hepatocyte-derived extracellular vesicles that promote inflammation in mice with Steatohepatitis. Gastroenterology. https://doi.org/10.1053/j.gastro.2020.06.031

de la Monte SM, Tong M (2014) Brain metabolic dysfunction at the core of Alzheimer's disease. Biochem Pharmacol 88:548–559. https://doi.org/10.1016/j.bcp.2013.12.012

Deng Y, Wang ZV, Gordillo R et al (2017a) An adipo-biliary-uridine axis that regulates energy homeostasis. Science 355:eaaf5375. https://doi.org/10.1126/science.aaf5375

Deng Y, Wang ZV, Gordillo R et al (2018) Adipocyte Xbp1s overexpression drives uridine production and reduces obesity. Mol Metab 11:1–17. https://doi.org/10.1016/j.molmet.2018.02.013

Deng Y, Wang ZV, Tao C et al (2013) The Xbp1s/GalE axis links ER stress to postprandial hepatic metabolism. J Clin Invest 123:455–468. https://doi.org/10.1172/JCI62819

Deng J, Yuan F, Guo Y et al (2017b) Deletion of ATF4 in AgRP neurons promotes fat loss mainly via increasing energy expenditure. Diabetes 66:640–650. https://doi.org/10.2337/db16-0954

Duennwald ML, Lindquist S (2008) Impaired ERAD and ER stress are early and specific events in polyglutamine toxicity. Genes Dev 22:3308–3319. https://doi.org/10.1101/gad.1673408

Dukart J, Kherif F, Mueller K et al (2013) Generative FDG-PET and MRI model of aging and disease progression in Alzheimer's disease. PLoS Comput Biol 9:e1002987. https://doi.org/10.1371/journal.pcbi.1002987

Duran-Aniotz C, Cornejo VH, Espinoza S et al (2017) IRE1 signaling exacerbates Alzheimer's disease pathogenesis. Acta Neuropathol. https://doi.org/10.1007/s00401-017-1694-x

Dzhashiashvili Y, Monckton CP, Shah HS et al (2019) The UPR-PERK pathway is not a promising therapeutic target for mutant SOD1-induced ALS. Neurobiol Dis 127:527–544. https://doi.org/10.1016/j.nbd.2019.03.024

Egawa N, Yamamoto K, Inoue H et al (2011) The endoplasmic reticulum stress sensor, ATF6α, protects against neurotoxin-induced dopaminergic neuronal death. J Biol Chem 286:7947–7957. https://doi.org/10.1074/jbc.M110.156430

Ersoy BA, Maner-Smith KM, Li Y et al (2017) Thioesterase-mediated control of cellular calcium homeostasis enables hepatic ER stress. J Clin Invest 128:141–156. https://doi.org/10.1172/JCI93123

Fernandez-Fernandez MR, Ferrer I, Lucas JJ (2011) Impaired ATF6α processing, decreased Rheb and neuronal cell cycle re-entry in Huntington's disease. Neurobiol Dis 41:23–32. https://doi.org/10.1016/j.nbd.2010.08.014

Feyertag F, Alvarez-Ponce D (2017) Disulfide bonds enable accelerated protein evolution. Mol Biol Evol 34:1833–1837. https://doi.org/10.1093/molbev/msx135

Filézac de L'Etang A, Maharjan N, Cordeiro Braña M et al (2015) Marinesco-Sjögren syndrome protein SIL1 regulates motor neuron subtype-selective ER stress in ALS. Nat Neurosci 18:227–238. https://doi.org/10.1038/nn.3903

Fiory F, Perruolo G, Cimmino I et al (2019) The relevance of insulin action in the dopaminergic system. Front Neurosci 13:868. https://doi.org/10.3389/fnins.2019.00868

Fontana L, Partridge L (2015) Promoting health and longevity through diet: from model organisms to humans. Cell 161:106–118. https://doi.org/10.1016/j.cell.2015.02.020

Fregno I, Molinari M (2019) Proteasomal and lysosomal clearance of faulty secretory proteins: ER-associated degradation (ERAD) and ER-to-lysosome-associated degradation (ERLAD) pathways. Crit Rev Biochem Mol Biol 54:153–163. https://doi.org/10.1080/10409238.2019.1610351

Fu H, Hardy J, Duff KE (2018) Selective vulnerability in neurodegenerative diseases. Nat Neurosci 21:1350–1358. https://doi.org/10.1038/s41593-018-0221-2

Fu S, Yang L, Li P et al (2011) Aberrant lipid metabolism disrupts calcium homeostasis causing liver endoplasmic reticulum stress in obesity. Nature 473:528–531. https://doi.org/10.1038/nature09968

Fujisawa T, Homma K, Yamaguchi N et al (2012) A novel monoclonal antibody reveals a conformational alteration shared by amyotrophic lateral sclerosis-linked SOD1 mutants. Ann Neurol 72:739–749. https://doi.org/10.1002/ana.23668

Gasparini L, Gouras GK, Wang R et al (2001) Stimulation of beta-amyloid precursor protein trafficking by insulin reduces intraneuronal beta-amyloid and requires mitogen-activated protein kinase signaling. J Neurosci 21:2561–2570

Gendron TF, Bieniek KF, Zhang Y-J et al (2013) Antisense transcripts of the expanded C9ORF72 hexanucleotide repeat form nuclear RNA foci and undergo repeat-associated non-ATG translation in c9FTD/ALS. Acta Neuropathol 126:829–844. https://doi.org/10.1007/s00401-013-1192-8

Ghaemmaghami S, Huh W-K, Bower K et al (2003) Global analysis of protein expression in yeast. Nature 425:737–741. https://doi.org/10.1038/nature02046

Glass CK, Saijo K, Winner B et al (2010) Mechanisms underlying inflammation in neurodegeneration. Cell 140:918–934. https://doi.org/10.1016/j.cell.2010.02.016

Gonzalez-Perez P, Woehlbier U, Chian R-J et al (2015) Identification of rare protein disulfide isomerase gene variants in amyotrophic lateral sclerosis patients. Gene 566:158–165. https://doi.org/10.1016/j.gene.2015.04.035

Gonzalez-Teuber V, Albert-Gasco H, Auyeung VC et al (2019) Small molecules to improve ER Proteostasis in disease. Trends Pharmacol Sci 40:684–695. https://doi.org/10.1016/j.tips.2019.07.003

Gorbatyuk MS, Shabashvili A, Chen W et al (2012) Glucose regulated protein 78 diminishes α-Synuclein neurotoxicity in a rat model of Parkinson disease. Mol Ther 20:1327–1337. https://doi.org/10.1038/mt.2012.28

Grandjean JMD, Madhavan A, Cech L et al (2020) Pharmacologic IRE1/XBP1s activation confers targeted ER proteostasis reprogramming. Nat Chem Biol. https://doi.org/10.1038/s41589-020-0584-z

Green KM, Glineburg MR, Kearse MG et al (2017) RAN translation at C9orf72-associated repeat expansions is selectively enhanced by the integrated stress response. Nat Commun 8:2005. https://doi.org/10.1038/s41467-017-02200-0

Haan MN (2006) Therapy insight: type 2 diabetes mellitus and the risk of late-onset Alzheimer's disease. Nat Clin Pract Neurol 2:159–166. https://doi.org/10.1038/ncpneuro0124

Halliday M, Radford H, Sekine Y et al (2015) Partial restoration of protein synthesis rates by the small molecule ISRIB prevents neurodegeneration without pancreatic toxicity. Cell Death Dis 6: e1672. https://doi.org/10.1038/cddis.2015.49

Halliday M, Radford H, Zents KAM et al (2017) Repurposed drugs targeting eIF2α-P-mediated translational repression prevent neurodegeneration in mice. Brain 140:1768–1783. https://doi.org/10.1093/brain/awx074

Hamos JE, Oblas B, Pulaski-Salo D et al (1991) Expression of heat shock proteins in Alzheimer's disease. Neurology 41:345–345. https://doi.org/10.1212/WNL.41.3.345

Harada Y, Nakajima K, Masahara-Negishi Y et al (2013) Metabolically programmed quality control system for dolichol-linked oligosaccharides. Proc Natl Acad Sci U S A 110:19366–19371. https://doi.org/10.1073/pnas.1312187110

Hashida K, Kitao Y, Sudo H et al (2012) ATF6alpha promotes astroglial activation and neuronal survival in a chronic mouse model of Parkinson's disease. PLoS One 7:e47950. https://doi.org/10.1371/journal.pone.0047950

Hatahet F, Ruddock LW (2009) Protein disulfide isomerase: a critical evaluation of its function in disulfide bond formation. Antioxid Redox Signal 11:2807–2850. https://doi.org/10.1089/ars.2009.2466

Henderson MX, Trojanowski JQ, Lee VM-Y (2019) α-Synuclein pathology in Parkinson's disease and related α-synucleinopathies. Neurosci Lett 709:134316. https://doi.org/10.1016/j.neulet.2019.134316

Heni M, Kullmann S, Preissl H et al (2015) Impaired insulin action in the human brain: causes and metabolic consequences. Nat Rev Endocrinol 11:701–711. https://doi.org/10.1038/nrendo.2015.173

Henkel A (2018) Unfolded protein response sensors in hepatic lipid metabolism and nonalcoholic fatty liver disease. Semin Liver Dis 38:320–332. https://doi.org/10.1055/s-0038-1670677

Henry FE, Sugino K, Tozer A et al (2015) Cell type-specific transcriptomics of hypothalamic energy-sensing neuron responses to weight-loss. elife 4. https://doi.org/10.7554/eLife.09800

Hetz C, Axten JM, Patterson JB (2019) Pharmacological targeting of the unfolded protein response for disease intervention. Nat Chem Biol 15:764–775. https://doi.org/10.1038/s41589-019-0326-2

Hetz C, Lee A-H, Gonzalez-Romero D et al (2008) Unfolded protein response transcription factor XBP-1 does not influence prion replication or pathogenesis. Proc Natl Acad Sci U S A 105:757–762. https://doi.org/10.1073/pnas.0711094105

Hetz C, Russelakis-Carneiro M, Maundrell K et al (2003) Caspase-12 and endoplasmic reticulum stress mediate neurotoxicity of pathological prion protein. EMBO J 22:5435–5445. https://doi.org/10.1093/emboj/cdg537

Hetz C, Russelakis-Carneiro M, Wälchli S et al (2005) The disulfide isomerase Grp58 is a protective factor against prion neurotoxicity. J Neurosci 25:2793–2802. https://doi.org/10.1523/JNEUROSCI.4090-04.2005

Hetz C, Saxena S (2017) ER stress and the unfolded protein response in neurodegeneration. Nat Rev Neurol 13:477–491. https://doi.org/10.1038/nrneurol.2017.99

Hetz C, Thielen P, Matus S et al (2009) XBP-1 deficiency in the nervous system protects against amyotrophic lateral sclerosis by increasing autophagy. Genes Dev 23:2294–2306. https://doi.org/10.1101/gad.1830709

Hetz C, Zhang K, Kaufman RJ (2020) Mechanisms, regulation and functions of the unfolded protein response. Nat Rev Mol Cell Biol. https://doi.org/10.1038/s41580-020-0250-z

Hipp MS, Park S-H, Hartl FU (2014) Proteostasis impairment in protein-misfolding and -aggregation diseases. Trends Cell Biol 24:506–514. https://doi.org/10.1016/j.tcb.2014.05.003

Ho L, Qin W, Pompl PN et al (2004) Diet-induced insulin resistance promotes amyloidosis in a transgenic mouse model of Alzheimer's disease. FASEB J 18:902–904. https://doi.org/10.1096/fj.03-0978fje

Hollien J, Weissman JS (2006) Decay of endoplasmic reticulum-localized mRNAs during the unfolded protein response. Science 313:104–107. https://doi.org/10.1126/science.1129631

Hoozemans JJM, van Haastert ES, Eikelenboom P et al (2007) Activation of the unfolded protein response in Parkinson's disease. Biochem Biophys Res Commun 354:707–711. https://doi.org/10.1016/j.bbrc.2007.01.043

Hoozemans JJM, van Haastert ES, Nijholt DAT et al (2012) Activation of the unfolded protein response is an early event in Alzheimer's and Parkinson's disease. Neurodegener Dis 10:212–215. https://doi.org/10.1159/000334536

Hoozemans JJM, Veerhuis R, Van Haastert ES et al (2005) The unfolded protein response is activated in Alzheimer's disease. Acta Neuropathol 110:165–172. https://doi.org/10.1007/s00401-005-1038-0

Hotamisligil GS (2010) Endoplasmic reticulum stress and the inflammatory basis of metabolic disease. Cell 140:900–917. https://doi.org/10.1016/j.cell.2010.02.034

Hotamisligil GS (2017) Inflammation, metaflammation and immunometabolic disorders. Nature 542:177–185. https://doi.org/10.1038/nature21363

Hotamisligil G, Shargill N, Spiegelman B (1993) Adipose expression of tumor necrosis factor-alpha: direct role in obesity-linked insulin resistance. Science 259:87–91. https://doi.org/10.1126/science.7678183

Hou H-L, Shen Y-X, Zhu H-Y et al (2006) Alterations of hHrd1 expression are related to hyperphosphorylated tau in the hippocampus in Alzheimer's disease. J Neurosci Res 84:1862–1870. https://doi.org/10.1002/jnr.21081

Hu G, Jousilahti P, Bidel S et al (2007) Type 2 diabetes and the risk of Parkinson's disease. Diabetes Care 30:842–847. https://doi.org/10.2337/dc06-2011

Hyrskyluoto A, Bruelle C, Lundh SH et al (2014) Ubiquitin-specific protease-14 reduces cellular aggregates and protects against mutant huntingtin-induced cell degeneration: involvement of the proteasome and ER stress-activated kinase IRE1. Hum Mol Genet 23:5928–5939. https://doi.org/10.1093/hmg/ddu317

Ilieva EV, Ayala V, Jové M et al (2007) Oxidative and endoplasmic reticulum stress interplay in sporadic amyotrophic lateral sclerosis. Brain 130:3111–3123. https://doi.org/10.1093/brain/awm190

Imanikia S, Sheng M, Castro C et al (2019) XBP-1 remodels lipid metabolism to extend longevity. Cell Rep 28:581–589.e4. https://doi.org/10.1016/j.celrep.2019.06.057

Ito Y, Yamada M, Tanaka H et al (2009) Involvement of CHOP, an ER-stress apoptotic mediator, in both human sporadic ALS and ALS model mice. Neurobiol Dis 36:470–476. https://doi.org/10.1016/j.nbd.2009.08.013

Jawaid A, Khan R, Polymenidou M, Schulz PE (2018) Disease-modifying effects of metabolic perturbations in ALS/FTLD. Mol Neurodegener 13:63. https://doi.org/10.1186/s13024-018-0294-0

Jeon GS, Nakamura T, Lee J-S et al (2014) Potential effect of S-nitrosylated protein disulfide isomerase on mutant SOD1 aggregation and neuronal cell death in amyotrophic lateral sclerosis. Mol Neurobiol 49:796–807. https://doi.org/10.1007/s12035-013-8562-z

Jiang H-Q, Ren M, Jiang H-Z et al (2014) Guanabenz delays the onset of disease symptoms, extends lifespan, improves motor performance and attenuates motor neuron loss in the SOD1 G93A mouse model of amyotrophic lateral sclerosis. Neuroscience 277:132–138. https://doi.org/10.1016/j.neuroscience.2014.03.047

Jin H, Mimura N, Kashio M et al (2014) Late-onset of spinal neurodegeneration in knock-in mice expressing a mutant BiP. PLoS One 9:e112837. https://doi.org/10.1371/journal.pone.0112837
Karagöz GE, Acosta-Alvear D, Walter P (2019) The unfolded protein response: detecting and responding to fluctuations in the protein-folding capacity of the endoplasmic reticulum. Cold Spring Harb Perspect Biol 11:a033886. https://doi.org/10.1101/cshperspect.a033886
Karam JH, Grodsky GM, Forsham PH, McWilliams NB (1963) Excessive insulin response to glucose in obese subjects as measured by immunochemical assay. Diabetes 12:197–204. https://doi.org/10.2337/diab.12.3.197
Kennedy BK, Berger SL, Brunet A et al (2014) Geroscience: linking aging to chronic disease. Cell 159:709–713. https://doi.org/10.1016/j.cell.2014.10.039
Kikuchi H, Almer G, Yamashita S et al (2006) Spinal cord endoplasmic reticulum stress associated with a microsomal accumulation of mutant superoxide dismutase-1 in an ALS model. Proc Natl Acad Sci U S A 103:6025–6030. https://doi.org/10.1073/pnas.0509227103
Kim H-J, Raphael AR, LaDow ES et al (2014) Therapeutic modulation of eIF2α phosphorylation rescues TDP-43 toxicity in amyotrophic lateral sclerosis disease models. Nat Genet 46:152–160. https://doi.org/10.1038/ng.2853
Kimura T, Ishiguro K, Hisanaga S-I (2014) Physiological and pathological phosphorylation of tau by Cdk5. Front Mol Neurosci 7:65. https://doi.org/10.3389/fnmol.2014.00065
Kiskinis E, Sandoe J, Williams LA et al (2014) Pathways disrupted in human ALS motor neurons identified through genetic correction of mutant SOD1. Cell Stem Cell 14:781–795. https://doi.org/10.1016/j.stem.2014.03.004
Koenig AM, Mechanic-Hamilton D, Xie SX et al (2017) Effects of the insulin sensitizer metformin in Alzheimer disease: pilot data from a randomized placebo-controlled crossover study. Alzheimer Dis Assoc Disord 31:107–113. https://doi.org/10.1097/WAD.0000000000000202
Kurz J, Parnham MJ, Geisslinger G, Schiffmann S (2019) Ceramides as novel disease biomarkers. Trends Mol Med 25:20–32. https://doi.org/10.1016/j.molmed.2018.10.009
Kwok CT, Morris AG, Frampton J et al (2013) Association studies indicate that protein disulfide isomerase is a risk factor in amyotrophic lateral sclerosis. Free Radic Biol Med 58:81–86. https://doi.org/10.1016/j.freeradbiomed.2013.01.001
La Joie R, Perrotin A, Barré L et al (2012) Region-specific hierarchy between atrophy, hypometabolism, and β-amyloid (Aβ) load in Alzheimer's disease dementia. J Neurosci 32:16265–16273. https://doi.org/10.1523/JNEUROSCI.2170-12.2012
Labbadia J, Morimoto RI (2015) The biology of proteostasis in aging and disease. Annu Rev Biochem 84:435–464. https://doi.org/10.1146/annurev-biochem-060614-033955
Langston JW, Schüle B, Rees L et al (2015) Multisystem Lewy body disease and the other parkinsonian disorders. Nat Genet 47:1378–1384. https://doi.org/10.1038/ng.3454
Lebeaupin C, Vallée D, Hazari Y et al (2018) Endoplasmic reticulum stress signalling and the pathogenesis of non-alcoholic fatty liver disease. J Hepatol 69:927–947. https://doi.org/10.1016/j.jhep.2018.06.008
Lee A-H, Heidtman K, Hotamisligil GS, Glimcher LH (2011a) Dual and opposing roles of the unfolded protein response regulated by IRE1 and XBP1 in proinsulin processing and insulin secretion. Proc Natl Acad Sci 108:8885–8890. https://doi.org/10.1073/pnas.1105564108
Lee H, Noh J-Y, Oh Y et al (2012) IRE1 plays an essential role in ER stress-mediated aggregation of mutant huntingtin via the inhibition of autophagy flux. Hum Mol Genet 21:101–114. https://doi.org/10.1093/hmg/ddr445
Lee A-H, Scapa EF, Cohen DE, Glimcher LH (2008) Regulation of hepatic Lipogenesis by the transcription factor XBP1. Science 320:1492–1496. https://doi.org/10.1126/science.1158042
Lee S, Shang Y, Redmond SA et al (2016) Activation of HIPK2 promotes ER stress-mediated Neurodegeneration in amyotrophic lateral sclerosis. Neuron 91:41–55. https://doi.org/10.1016/j.neuron.2016.05.021
Lee J, Sun C, Zhou Y et al (2011b) p38 MAPK–mediated regulation of Xbp1s is crucial for glucose homeostasis. Nat Med 17:1251–1260. https://doi.org/10.1038/nm.2449

Liu Z-C, Chu J, Lin L et al (2016a) SIL1 rescued Bip elevation-related tau hyperphosphorylation in ER stress. Mol Neurobiol 53:983–994. https://doi.org/10.1007/s12035-014-9039-4

Liu J, Ibi D, Taniguchi K et al (2016b) Inflammation improves glucose homeostasis through IKKβ-XBP1s interaction. Cell 167:1052–1066.e18. https://doi.org/10.1016/j.cell.2016.10.015

Liu S-Y, Wang W, Cai Z-Y et al (2013) Polymorphism -116C/G of human X-box-binding protein 1 promoter is associated with risk of Alzheimer's disease. CNS Neurosci Ther 19:229–234. https://doi.org/10.1111/cns.12064

López-Otín C, Blasco MA, Partridge L et al (2013) The hallmarks of aging. Cell 153:1194–1217. https://doi.org/10.1016/j.cell.2013.05.039

Lourenco MV, Clarke JR, Frozza RL et al (2013) TNF-α mediates PKR-dependent memory impairment and brain IRS-1 inhibition induced by Alzheimer's β-amyloid oligomers in mice and monkeys. Cell Metab 18:831–843. https://doi.org/10.1016/j.cmet.2013.11.002

Ma T, M a T, Wexler AJ et al (2013) Suppression of eIF2α kinases alleviates Alzheimer's disease-related plasticity and memory deficits. Nat Neurosci 16:1299–1305. https://doi.org/10.1038/nn.3486

Makioka K, Yamazaki T, Fujita Y et al (2010) Involvement of endoplasmic reticulum stress defined by activated unfolded protein response in multiple system atrophy. J Neurol Sci 297:60–65. https://doi.org/10.1016/j.jns.2010.06.019

Maly DJ, Papa FR (2014) Druggable sensors of the unfolded protein response. Nat Chem Biol 10:892–901. https://doi.org/10.1038/nchembio.1664

Martínez G, Vidal RL, Mardones P et al (2016) Regulation of memory formation by the transcription factor XBP1. Cell Rep 14:1382–1394. https://doi.org/10.1016/j.celrep.2016.01.028

Masters CL, Bateman R, Blennow K et al (2015) Alzheimer's disease. Nat Rev Dis Prim 1:15056. https://doi.org/10.1038/nrdp.2015.56

Mattson MP (2012) Energy intake and exercise as determinants of brain health and vulnerability to injury and disease. Cell Metab 16:706–722. https://doi.org/10.1016/j.cmet.2012.08.012

Mattson MP, Arumugam TV (2018) Hallmarks of brain aging: adaptive and pathological modification by metabolic states. Cell Metab 27:1176–1199. https://doi.org/10.1016/j.cmet.2018.05.011

Matus S, Lopez E, Valenzuela V et al (2013) Functional contribution of the transcription factor ATF4 to the pathogenesis of amyotrophic lateral sclerosis. PLoS One 8:e66672. https://doi.org/10.1371/journal.pone.0066672

Medinas DB, Rozas P, Martínez Traub F et al (2018) Endoplasmic reticulum stress leads to accumulation of wild-type SOD1 aggregates associated with sporadic amyotrophic lateral sclerosis. Proc Natl Acad Sci U S A 115:8209–8214. https://doi.org/10.1073/pnas.1801109115

Mercado G, Castillo V, Soto P et al (2018) Targeting PERK signaling with the small molecule GSK2606414 prevents neurodegeneration in a model of Parkinson's disease. Neurobiol Dis 112:136–148. https://doi.org/10.1016/j.nbd.2018.01.004

Montibeller L, de Belleroche J (2018) Amyotrophic lateral sclerosis (ALS) and Alzheimer's disease (AD) are characterised by differential activation of ER stress pathways: focus on UPR target genes. Cell Stress Chaperones 23:897–912. https://doi.org/10.1007/s12192-018-0897-y

Montibeller L, Tan LY, Kim JK et al (2020) Tissue-selective regulation of protein homeostasis and unfolded protein response signalling in sporadic ALS. J Cell Mol Med jcmm:15170. https://doi.org/10.1111/jcmm.15170

Moreno JA, Halliday M, Molloy C et al (2013) Oral treatment targeting the unfolded protein response prevents neurodegeneration and clinical disease in prion-infected mice. Sci Transl Med 5:206ra138. https://doi.org/10.1126/scitranslmed.3006767

Moreno JA, Radford H, Peretti D et al (2012) Sustained translational repression by eIF2α-P mediates prion neurodegeneration. Nature 485:507–511. https://doi.org/10.1038/nature11058

Morita S, Villalta SA, Feldman HC et al (2017) Targeting ABL-IRE1α signaling spares ER-stressed pancreatic β cells to reverse autoimmune diabetes. Cell Metab 25:883–897.e8. https://doi.org/10.1016/j.cmet.2017.03.018

Naidoo N, Zhu J, Galante RJ et al (2018) Reduction of the molecular chaperone binding immunoglobulin protein (BiP) accentuates the effect of aging on sleep-wake behavior. Neurobiol Aging 69:10–25. https://doi.org/10.1016/j.neurobiolaging.2018.04.011

Naranjo JR, Zhang H, Villar D et al (2016) Activating transcription factor 6 derepression mediates neuroprotection in Huntington disease. J Clin Invest 126:627–638. https://doi.org/10.1172/JCI82670

Nardo G, Pozzi S, Pignataro M et al (2011) Amyotrophic lateral sclerosis multiprotein biomarkers in peripheral blood mononuclear cells. PLoS One 6:e25545. https://doi.org/10.1371/journal.pone.0025545

Niccoli T, Cabecinha M, Tillmann A et al (2016) Increased glucose transport into neurons rescues Aβ toxicity in drosophila. Curr Biol 26:2291–2300. https://doi.org/10.1016/j.cub.2016.07.017

Nishitoh H, Kadowaki H, Nagai A et al (2008) ALS-linked mutant SOD1 induces ER stress- and ASK1-dependent motor neuron death by targeting Derlin-1. Genes Dev 22:1451–1464. https://doi.org/10.1101/gad.1640108

Ohtake Y, Matsuhisa K, Kaneko M et al (2018) Axonal activation of the unfolded protein response promotes axonal regeneration following peripheral nerve injury. Neuroscience 375:34–48. https://doi.org/10.1016/j.neuroscience.2018.02.003

Oñate M, Catenaccio A, Martínez G et al (2016) Activation of the unfolded protein response promotes axonal regeneration after peripheral nerve injury. Sci Rep 6:21709. https://doi.org/10.1038/srep21709

Ozcan U (2004) Endoplasmic reticulum stress links obesity, insulin action, and type 2 diabetes. Science 306:457–461. https://doi.org/10.1126/science.1103160

Ozcan L, Ergin AS, Lu A et al (2009) Endoplasmic reticulum stress plays a central role in development of Leptin resistance. Cell Metab 9:35–51. https://doi.org/10.1016/j.cmet.2008.12.004

Parakh S, Jagaraj CJ, Vidal M et al (2018) ERp57 is protective against mutant SOD1-induced cellular pathology in amyotrophic lateral sclerosis. Hum Mol Genet. https://doi.org/10.1093/hmg/ddy041

Parakh S, Shadfar S, Perri ER et al (2020) The redox activity of protein disulfide Isomerase inhibits ALS phenotypes in cellular and Zebrafish models. iScience 23:101097. https://doi.org/10.1016/j.isci.2020.101097

Park K-W, Eun Kim G, Morales R et al (2017) The endoplasmic reticulum chaperone GRP78/BiP modulates prion propagation in vitro and in vivo. Sci Rep 7:44723. https://doi.org/10.1038/srep44723

Park SW, Zhou Y, Lee J et al (2010) The regulatory subunits of PI3K, p85α and p85β, interact with XBP-1 and increase its nuclear translocation. Nat Med 16:429–437. https://doi.org/10.1038/nm.2099

Perurena OH, Festoff BW (1987) Reduction in insulin receptors in amyotrophic lateral sclerosis correlates with reduced insulin sensitivity. Neurology 37:1375–1379. https://doi.org/10.1212/wnl.37.8.1375

Pluvinage JV, Wyss-Coray T (2020) Systemic factors as mediators of brain homeostasis, ageing and neurodegeneration. Nat Rev Neurosci 21:93–102. https://doi.org/10.1038/s41583-019-0255-9

Preissler S, Ron D (2019) Early events in the endoplasmic reticulum unfolded protein response. Cold Spring Harb Perspect Biol 11:a033894. https://doi.org/10.1101/cshperspect.a033894

Radford H, Moreno JA, Verity N et al (2015) PERK inhibition prevents tau-mediated neurodegeneration in a mouse model of frontotemporal dementia. Acta Neuropathol 130:633–642. https://doi.org/10.1007/s00401-015-1487-z

Robblee MM, Kim CC, Abate JP et al (2016) Saturated fatty acids engage an IRE1α-dependent pathway to activate the NLRP3 Inflammasome in myeloid cells. Cell Rep 14:2611–2623. https://doi.org/10.1016/j.celrep.2016.02.053

Rozas P, Pinto C, Martínez Traub F et al (2021) Protein disulfide isomerase ERp57 protects early muscle denervation in experimental ALS. Acta Neuropathol Commun 9:21. https://doi.org/10.1186/s40478-020-01116-z

Sado M, Yamasaki Y, Iwanaga T et al (2009) Protective effect against Parkinson's disease-related insults through the activation of XBP1. Brain Res 1257:16–24. https://doi.org/10.1016/j.brainres.2008.11.104

Sasaki S (2010) Endoplasmic reticulum stress in motor neurons of the spinal cord in sporadic amyotrophic lateral sclerosis. J Neuropathol Exp Neurol 69:346–355. https://doi.org/10.1097/NEN.0b013e3181d44992

Saxena S, Cabuy E, Caroni P (2009) A role for motoneuron subtype-selective ER stress in disease manifestations of FALS mice. Nat Neurosci 12:627–636. https://doi.org/10.1038/nn.2297

Schaum N, Lehallier B, Hahn O et al (2020) Ageing hallmarks exhibit organ-specific temporal signatures. Nature 583:596–602. https://doi.org/10.1038/s41586-020-2499-y

Scheper W, Hoozemans JJM (2015) The unfolded protein response in neurodegenerative diseases: a neuropathological perspective. Acta Neuropathol 130:315–331. https://doi.org/10.1007/s00401-015-1462-8

Schernhammer E, Hansen J, Rugbjerg K et al (2011) Diabetes and the risk of developing Parkinson's disease in Denmark. Diabetes Care 34:1102–1108. https://doi.org/10.2337/dc10-1333

Scheuner D, Kaufman RJ (2008) The unfolded protein response: a pathway that links insulin demand with β-cell failure and diabetes. Endocr Rev 29:317–333. https://doi.org/10.1210/er.2007-0039

Sekar S, Taghibiglou C (2018) Elevated nuclear phosphatase and tensin homolog (PTEN) and altered insulin signaling in substantia nigral region of patients with Parkinson's disease. Neurosci Lett 666:139–143. https://doi.org/10.1016/j.neulet.2017.12.049

Selkoe DJ, Hardy J (2016) The amyloid hypothesis of Alzheimer's disease at 25 years. EMBO Mol Med 8:595–608. https://doi.org/10.15252/emmm.201606210

Senderek J, Krieger M, Stendel C et al (2005) Mutations in SIL1 cause Marinesco-Sjögren syndrome, a cerebellar ataxia with cataract and myopathy. Nat Genet 37:1312–1314. https://doi.org/10.1038/ng1678

Senée V, Chelala C, Duchatelet S et al (2006) Mutations in GLIS3 are responsible for a rare syndrome with neonatal diabetes mellitus and congenital hypothyroidism. Nat Genet 38:682–687. https://doi.org/10.1038/ng1802

Sha H, Yang L, Liu M et al (2014) Adipocyte spliced form of X-box-binding protein 1 promotes Adiponectin Multimerization and systemic glucose homeostasis. Diabetes 63:867–879. https://doi.org/10.2337/db13-1067

Shan B, Wang X, Wu Y et al (2017) The metabolic ER stress sensor IRE1α suppresses alternative activation of macrophages and impairs energy expenditure in obesity. Nat Immunol 18:519–529. https://doi.org/10.1038/ni.3709

Silva RM, Ries V, Oo TF et al (2005) CHOP/GADD153 is a mediator of apoptotic death in substantia nigra dopamine neurons in an in vivo neurotoxin model of parkinsonism. J Neurochem 95:974–986. https://doi.org/10.1111/j.1471-4159.2005.03428.x

Soejima N, Ohyagi Y, Nakamura N et al (2013) Intracellular accumulation of toxic turn amyloid-β is associated with endoplasmic reticulum stress in Alzheimer's disease. Curr Alzheimer Res 10:11–20. https://doi.org/10.2174/1567205011310010003

Soo KY, Halloran M, Sundaramoorthy V et al (2015) Rab1-dependent ER-Golgi transport dysfunction is a common pathogenic mechanism in SOD1, TDP-43 and FUS-associated ALS. Acta Neuropathol 130:679–697. https://doi.org/10.1007/s00401-015-1468-2

Soto C, Pritzkow S (2018) Protein misfolding, aggregation, and conformational strains in neurodegenerative diseases. Nat Neurosci 21:1332–1340. https://doi.org/10.1038/s41593-018-0235-9

Spillantini MG, Schmidt ML, Lee VM et al (1997) Alpha-synuclein in Lewy bodies. Nature 388:839–840. https://doi.org/10.1038/42166

Talbot K, Wang H-Y, Kazi H et al (2012) Demonstrated brain insulin resistance in Alzheimer's disease patients is associated with IGF-1 resistance, IRS-1 dysregulation, and cognitive decline. J Clin Invest 122:1316–1338. https://doi.org/10.1172/JCI59903

Tannous A, Pisoni GB, Hebert DN, Molinari M (2015) N-linked sugar-regulated protein folding and quality control in the ER. Semin Cell Dev Biol 41:79–89. https://doi.org/10.1016/j.semcdb.2014.12.001

Taylor JP, Brown RH, Cleveland DW (2016) Decoding ALS: from genes to mechanism. Nature 539:197–206. https://doi.org/10.1038/nature20413

Taylor RC, Dillin A (2013) XBP-1 is a cell-nonautonomous regulator of stress resistance and longevity. Cell 153:1435–1447. https://doi.org/10.1016/j.cell.2013.05.042

Thapa S, Abdulrahman B, Abdelaziz DH et al (2018) Overexpression of quality control proteins reduces prion conversion in prion-infected cells. J Biol Chem 293:16069–16082. https://doi.org/10.1074/jbc.RA118.002754

Torres M, Medinas DB, Matamala JM et al (2015) The protein disulfide Isomerase ERp57 regulates the steady-state levels of the prion protein. J Biol Chem. https://doi.org/10.1074/jbc.M114.635565

Tsuchiya Y, Saito M, Kadokura H et al (2018) IRE1–XBP1 pathway regulates oxidative proinsulin folding in pancreatic β cells. J Cell Biol 217:1287–1301. https://doi.org/10.1083/jcb.201707143

Tu BP, Weissman JS (2002) The FAD- and O(2)-dependent reaction cycle of Ero1-mediated oxidative protein folding in the endoplasmic reticulum. Mol Cell 10:983–994. https://doi.org/10.1016/s1097-2765(02)00696-2

Turer AT, Scherer PE (2012) Adiponectin: mechanistic insights and clinical implications. Diabetologia 55:2319–2326. https://doi.org/10.1007/s00125-012-2598-x

Turner MR, Hardiman O, Benatar M et al (2013) Controversies and priorities in amyotrophic lateral sclerosis. Lancet Neurol 12:310–322. https://doi.org/10.1016/S1474-4422(13)70036-X

Uehara T, Nakamura T, Yao D et al (2006) S-nitrosylated protein-disulphide isomerase links protein misfolding to neurodegeneration. Nature 441:513–517. https://doi.org/10.1038/nature04782

Unterberger U, Höftberger R, Gelpi E et al (2006) Endoplasmic reticulum stress features are prominent in Alzheimer disease but not in prion diseases in vivo. J Neuropathol Exp Neurol 65:348–357. https://doi.org/10.1097/01.jnen.0000218445.30535.6f

Upton J-P, Wang L, Han D et al (2012) IRE1 cleaves select microRNAs during ER stress to Derepress translation of Proapoptotic Caspase-2. Science

Vaccaro A, Patten SA, Aggad D et al (2013) Pharmacological reduction of ER stress protects against TDP-43 neuronal toxicity in vivo. Neurobiol Dis 55:64–75. https://doi.org/10.1016/j.nbd.2013.03.015

Valdés P, Mercado G, Vidal RL et al (2014) Control of dopaminergic neuron survival by the unfolded protein response transcription factor XBP1. Proc Natl Acad Sci U S A 111:6804–6809. https://doi.org/10.1073/pnas.1321845111

Valenzuela V, Collyer E, Armentano D et al (2012) Activation of the unfolded protein response enhances motor recovery after spinal cord injury. Cell Death Dis 3:272

van der Harg JM, Nölle A, Zwart R et al (2014) The unfolded protein response mediates reversible tau phosphorylation induced by metabolic stress. Cell Death Dis 5:e1393–e1393. https://doi.org/10.1038/cddis.2014.354

Vander Mierde D, Scheuner D, Quintens R et al (2007) Glucose activates a protein Phosphatase-1-mediated signaling pathway to enhance overall translation in pancreatic β-cells. Endocrinology 148:609–617. https://doi.org/10.1210/en.2006-1012

Varela L, Horvath TL (2012) Leptin and insulin pathways in POMC and AgRP neurons that modulate energy balance and glucose homeostasis. EMBO Rep 13:1079–1086. https://doi.org/10.1038/embor.2012.174

Vidal RL, Figueroa A, Court FA et al (2012) Targeting the UPR transcription factor XBP1 protects against Huntington's disease through the regulation of FoxO1 and autophagy. Hum Mol Genet 21:2245–2262. https://doi.org/10.1093/hmg/dds040

Vieira FG, Ping Q, Moreno AJ et al (2015) Guanabenz treatment accelerates disease in a mutant SOD1 mouse model of ALS. PLoS One 10:e0135570. https://doi.org/10.1371/journal.pone.0135570

Wahl D, Solon-Biet SM, Cogger VC et al (2019) Aging, lifestyle and dementia. Neurobiol Dis 130:104481. https://doi.org/10.1016/j.nbd.2019.104481

Wainger BJ, Kiskinis E, Mellin C et al (2014) Intrinsic membrane hyperexcitability of amyotrophic lateral sclerosis patient-derived motor neurons. Cell Rep 7:1–11. https://doi.org/10.1016/j.celrep.2014.03.019

Walker AK, Farg MA, Bye CR et al (2010) Protein disulphide isomerase protects against protein aggregation and is S-nitrosylated in amyotrophic lateral sclerosis. Brain 133:105–116. https://doi.org/10.1093/brain/awp267

Walter P, Ron D (2011) The unfolded protein response: from stress pathway to homeostatic regulation. Science 334:1081–1086. https://doi.org/10.1126/science.1209038

Wang L, Popko B, Roos RP (2011) The unfolded protein response in familial amyotrophic lateral sclerosis. Hum Mol Genet 20:1008–1015. https://doi.org/10.1093/hmg/ddq546

Wang L, Popko B, Roos RP (2014c) An enhanced integrated stress response ameliorates mutant SOD1-induced ALS. Hum Mol Genet 23:2629–2638. https://doi.org/10.1093/hmg/ddt658

Wang L, Popko B, Tixier E, Roos RP (2014a) Guanabenz, which enhances the unfolded protein response, ameliorates mutant SOD1-induced amyotrophic lateral sclerosis. Neurobiol Dis 71:317–324. https://doi.org/10.1016/j.nbd.2014.08.010

Wang J-M, Qiu Y, Yang Z et al (2018) IRE1α prevents hepatic steatosis by processing and promoting the degradation of select microRNAs. Sci Signal 11:eaao4617. https://doi.org/10.1126/scisignal.aao4617

Wang Y, Vera L, Fischer WH, Montminy M (2009) The CREB coactivator CRTC2 links hepatic ER stress and fasting gluconeogenesis. Nature 460:534–537. https://doi.org/10.1038/nature08111

Wang L, Zhai Y-Q, Xu L-L et al (2014b) Metabolic inflammation exacerbates dopaminergic neuronal degeneration in response to acute MPTP challenge in type 2 diabetes mice. Exp Neurol 251:22–29. https://doi.org/10.1016/j.expneurol.2013.11.001

Westergard T, McAvoy K, Russell K et al (2019) Repeat-associated non-AUG translation in C9orf72-ALS/FTD is driven by neuronal excitation and stress. EMBO Mol Med 11. https://doi.org/10.15252/emmm.201809423

Wiersma VI, van Hecke W, Scheper W et al (2016) Activation of the unfolded protein response and granulovacuolar degeneration are not common features of human prion pathology. Acta Neuropathol Commun 4:113. https://doi.org/10.1186/s40478-016-0383-7

Williams KW, Liu T, Kong X et al (2014) Xbp1s in Pomc neurons connects ER stress with energy balance and glucose homeostasis. Cell Metab 20:471–482. https://doi.org/10.1016/j.cmet.2014.06.002

Woehlbier U, Colombo A, Saaranen MJ et al (2016) ALS-linked protein disulfide isomerase variants cause motor dysfunction. EMBO J 35:845–865. https://doi.org/10.15252/embj.201592224

Xiao Y, Deng Y, Yuan F et al (2017) ATF4/ATG5 signaling in hypothalamic Proopiomelanocortin neurons regulates fat mass via affecting energy expenditure. Diabetes 66:1146–1158. https://doi.org/10.2337/db16-1546

Xiao Y, Xia T, Yu J et al (2016) Knockout of inositol-requiring enzyme 1α in pro-opiomelanocortin neurons decreases fat mass via increasing energy expenditure. Open Biol 6:160131. https://doi.org/10.1098/rsob.160131

Yan C, Liu J, Gao J et al (2019) IRE1 promotes neurodegeneration through autophagy-dependent neuron death in the drosophila model of Parkinson's disease. Cell Death Dis 10:800. https://doi.org/10.1038/s41419-019-2039-6

Yang L, Calay ES, Fan J et al (2015) S-Nitrosylation links obesity-associated inflammation to endoplasmic reticulum dysfunction. Science 349:500–506. https://doi.org/10.1126/science.aaa0079

Yang Q, Guo Z (2015) Polymorphisms in protein disulfide isomerase are associated with sporadic amyotrophic lateral sclerosis in the Chinese Han population. Int J Neurosci:1–19. https://doi.org/10.3109/00207454.2015.1050098

Yao T, Deng Z, Gao Y et al (2017) Ire1α in Pomc neurons is required for thermogenesis and Glycemia. Diabetes 66:663–673. https://doi.org/10.2337/db16-0533

Yoon SO, Park DJ, Ryu JC et al (2012) JNK3 perpetuates metabolic stress induced by Aβ peptides. Neuron 75:824–837. https://doi.org/10.1016/j.neuron.2012.06.024

Zhang Y-J, Jansen-West K, Xu Y-F et al (2014) Aggregation-prone c9FTD/ALS poly (GA) RAN-translated proteins cause neurotoxicity by inducing ER stress. Acta Neuropathol. https://doi.org/10.1007/s00401-014-1336-5

Zhang Q, Yu J, Liu B et al (2013) Central activating transcription factor 4 (ATF4) regulates hepatic insulin resistance in mice via S6K1 signaling and the Vagus nerve. Diabetes 62:2230–2239. https://doi.org/10.2337/db12-1050

Zhang X, Zhang G, Zhang H et al (2008) Hypothalamic IKKβ/NF-κB and ER stress link Overnutrition to energy imbalance and obesity. Cell 135:61–73. https://doi.org/10.1016/j.cell.2008.07.043

Zhao W-Q, De Felice FG, Fernandez S et al (2008) Amyloid beta oligomers induce impairment of neuronal insulin receptors. FASEB J 22:246–260. https://doi.org/10.1096/fj.06-7703com

Zhou Y, Lee J, Reno CM et al (2011) Regulation of glucose homeostasis through a XBP-1–FoxO1 interaction. Nat Med 17:356–365. https://doi.org/10.1038/nm.2293

Zuleta A, Vidal RL, Armentano D et al (2012) AAV-mediated delivery of the transcription factor XBP1s into the striatum reduces mutant Huntingtin aggregation in a mouse model of Huntington's disease. Biochem Biophys Res Commun. https://doi.org/10.1016/j.bbrc.2012.03.033

Chapter 13
Endoplasmic Reticulum Homeostasis and Stress Responses in *Caenorhabditis elegans*

Sun-Kyung Lee

Abstract The unfolded protein response (UPR) is an evolutionarily conserved adaptive regulatory pathway that alleviates protein-folding defects in the endoplasmic reticulum (ER). Physiological demands, environmental perturbations and pathological conditions can cause accumulation of unfolded proteins in the ER and the stress signal is transmitted to the nucleus to turn on a series of genes to respond the challenge. In metazoan, the UPR pathways consisted of IRE1/XBP1, PEK-1 and ATF6, which function in parallel and downstream transcriptional activation triggers the proteostasis networks consisting of molecular chaperones, protein degradation machinery and other stress response pathways ((Labbadia J, Morimoto RI, F1000Prime Rep 6:7, 2014); (Shen X, Ellis RE, Lee K, Annu Rev Biochem 28:893-903, 2014)). The integrated responses act on to resolve the ER stress by increasing protein folding capacity, attenuating ER-loading translation, activating ER-associated proteasomal degradation (ERAD), and regulating IRE1-dependent decay of mRNA (RIDD). Therefore, the effective UPR to internal and external causes is linked to the multiple pathophysiological conditions such as aging, immunity, and neurodegenerative diseases. Recent development in the research of the UPR includes cell-nonautonomous features of the UPR, interplay between the UPR and other stress response pathways, unconventional UPR inducers, and noncanonical UPR independent of the three major branches, originated from multiple cellular and molecular machineries in addition to ER. *Caenorhabditis elegans* model system has critically contributed to these unprecedented aspects of the ER UPR and broadens the possible therapeutic targets to treat the ER-stress associated human disorders and time-dependent physiological deterioration of aging.

Keywords Unfolded protein response · Aging · Proteostasis · ER homeostasis · *Caenorhabditis elegans*

S.-K. Lee (✉)
Department of Life Science, Research Institute for Natural Sciences, College of Natural Sciences, Hanyang University, Seoul, Republic of Korea
e-mail: sunkyungl@hanyang.ac.kr

L. B. Agellon, M. Michalak (eds.), *Cellular Biology of the Endoplasmic Reticulum*, Progress in Molecular and Subcellular Biology 59,
https://doi.org/10.1007/978-3-030-67696-4_13

Abbreviations

4E-BP	4E-binding protein
AIRAP	arsenic-inducible proteasomal 19S regulatory particle-associated protein
COPII	coatomer protein complex II
DIC	differential interference contrast
DR	dietary restriction
ECM	extracellular matrix
ERAD	ER-associated proteasomal degradation
ERES	ER-exit sites
ETA	ethanolamine
gf	gain-of-function
GlcNAc	N-acetylglucosamine
HA	hyaluronan
HAase	hyaluronidase
HIF-1	hypoxia-inducible factor-1
HP1	heterochromatin protein 1
HPL-2	heterochromatin protein like-2
KO	knockout
LBS	lipid bilayer stress
lf	loss-of-function
MAPK	mitogen-activated protein kinase
NAC	nascent polypeptide-associated complex
NMD	nonsense-mediated mRNA decay
NMHCIIB	nonmuscle myosin heavy chain IIB
OQC	oxidative quality control
PC	phosphatidylcholine
PD	Parkinson disease
PDI6	disulfide isomerase 6
PE	phosphatidylethanolamine
RIDD	IRE-1-dependent decay of mRNA
RNC	ribosome-nascent chain complexes
ROS	reactive oxygen species
S6K	S6 kinase
sams	S-adenosyl methionine synthetase
SCD	stearoyl-CoA-desaturases
SCV	small clear vesicle
SRP	signal recognition particle
SUMO	small ubiquitin-like modifier
TMEM2	Transmembrane Protein 2
TOR	target of rapamycin
UPR	unfolded protein response
VIT	vitellogenin
α-syn	α-synuclein

13.1 Introduction

Endoplasmic reticulum (ER) is the place where proteins, lipids and oligosaccharides synthesize in eukaryotic cells. ER is constructed with extensive reticular membranous structure. ER membrane is connected with nuclear envelop and some of them is at the close proximity to mitochondria. ER provides lipid bilayer membrane for autophagy vesicles that surround to engulf damaged intracellular organelles and macromolecules. Therefore, ER homeostasis is essential for proper cellular structure, function, and survival in all eukaryotes.

Proteins synthesized in the ER fold and assemble before being shipped to vesicular trafficking. Finely tuned quality control apparatus does the task to ensure proper protein folding. In case unfolded proteins are accumulated in ER, the stress signal is generated and triggers the unfolded protein response (UPR), a highly conserved adaptive regulatory pathway that turns transcription of various stress responsive genes (Ron and Walter 2007; Walter et al. 2011). Not only internal physiological status such as imbalanced metabolism, but also external stimulations such as temperature and pathogen exposure can cause ER stress. In metazoan, IRE1/XBP1, PERK and ATF6, are major components in the UPR pathways and downstream transcriptional regulators activate the gene network of proteostasis including protein folding chaperones, proteasome subunits, as well as other stress response proteins (Labbadia and Morimoto 2014; Shen et al. 2014). The integrated responses increase protein folding capacity, decrease ER-loading protein synthesis, activate ER-associated proteasomal degradation (ERAD), and regulate IRE1-dependent decay of mRNA (RIDD) (Maurel et al. 2014; Labbadia and Morimoto 2015). Therefore, defective UPR can lead to the multiple pathophysiological conditions such as aging, failure in immunity, tumor progression, and neurodegeneration.

The ER is composed of membranous sheets and tubules, forming extensive networks of lipid bilayer. The ER is partitioned into specialized subdomains with different functions. A lipid bilayer sheet of ER membrane forms the outer membrane of nuclear envelop, which shares ER structural elements including the contents of ER lumen. Phospholipids are synthesized in the ER, especially at membrane interfaces with other organelles such as mitochondria. Phospholipid synthesis is critical in ER functions in vesicular trafficking, cholesterol homeostasis, and lipid storage. Therefore, phospholipid metabolism in the ER is regulated by lipid storage, secretion, and stress response, contributing to pathology of obesity, diabetes, atherosclerosis, and neurological disorders.

C. elegans, a free-living nematode, is an ideal system to study the role of ER homeostasis and stress responses in the organismal integration of various physiological processes (Prahlad and Morimoto 2009). Its small body size, large brood size, short life cycle and simple lab-cultivation enable amenable genetic manipulation (Brenner 1974). Full genome sequencing of *C. elegans* reveals extensive sequence homology with mammalian genes and comparative analysis of subcellular localization of proteins indicates conserved residency of the ER apparatus proteins (The C.elegans sequencing consortium 1998; Hazkani-Covo et al. 2004). A fluorescent reporter of intrinsically and ectopically expressed proteins provides a convenient tool to determine cell autonomy of regulation by investigating tissue-specific difference in the stress response, by directly visualizing and assaying protein misfolding and aggregation, lipid droplets, and ER stress marker within various tissues of the live animals under different physiological or environmental conditions. Above all, numerous phenotypes in *C. elegans* growth and developments such as lifespan, behavior, body size, and metabolism can be analyzed as physiological indicators. Many different regimens of stress and drug treatment in combination of genetic approaches on *C. elegans* have provided exceptional platforms to understand how tissue-specific ER response leads to physiological consequences such as longevity, innate immunity, and oxidative stress at whole organismal level (Bar-Ziv et al. 2020). This chapter will overview how *C. elegans* has been utilized to study functions of the ER and what roles of the ER in physiological response and etiology of related diseases we have learned from the simple metazoan of approximately 1000 of somatic cells.

13.2 Brief Biology of *C. elegans* in the Study of ER Homeostasis

C. elegans is tiny with an adult body size of about 1 millimeter long and easily grown on agar plates covered with a lawn of bacterial culture, usually *Escherichia coli* (Corsi A.K., Wightman B. 2015). The nematode is a hermaphrodite that self-fertilizes to populate a culture plate and a single hermaphrodite can produce progenies as many as about 300, and this large brood size makes various genetic analysis feasible. *C. elegans* life cycle from eggs to adults takes approximately three days at 20 °C under the standard laboratory culture condition (Brenner 1974). Fertilized eggs are laid at about 24-cell stage, and embryos inside continue to develop and hatch to become a first stage larva. The worms start to feed on bacterial food and develop through four larval stages (L1 ~ L4), and each stage starts with the molting of the old cuticle that is replaced with a new one. After the last molting of L4, adult worms start to lay eggs and produce progenies for about 3 days until they deplete all the sperm produced by themselves. The sperm-depleted hermaphrodites can produce more progenies when outcrossed to males, which occur at an approximate rate of one to 300 hermaphrodites due to random *X*-chromosome disjunction in meiosis. After finishing reproduction, hermaphrodite worms live several more weeks and finally die from senescence. In a condition of crowding of worms and consequent food depletion, L2 larvae with the activated signal develop into alternative L3 larval stage called the dauer larva, which can survive for several months without food, with enhanced resistance to harmful substances thereby greatly protected from harsh environments. The dauer larvae resume their development into L4 larvae when placed in food back and continue to become reproductive adults.

The worm's moving, eating, mating, egg-laying and developing are easily observed by a simple dissecting microscope, and even subcellular details in individual cells in the worm's transparent body can be visualized using DIC (differential interference contrast, Nomarski) optics. A compound and confocal microscope with high resolution power makes it possible to directly observe the subcellular compartments in individual cells. Tagging various molecular and cellular targets with fluorescent proteins in specific organs and tissues of live worms at different developmental stages is extremely useful to study developmental processes, analyze cellular activity and metabolic status, and characterize protein interactions. Altogether, these features of *C. elegans* model system permit researchers to investigate the structure and function of ER at not only single cell level but also at organismal level.

13.3 Unfolded Protein Response (UPR) in *C. elegans*

In *C. elegans*, all three known UPR pathways of IRE1/XBP1, PERK and ATF6 are conserved, thus the nematode has been a versatile genetic model system to dissect UPR signaling in a multicellular organism (Kirstein-Miles and Morimoto 2010; Nussbaum-Krammer and Morimoto 2014). IRE1 was isolated in a genetic screen using yeast to identify a gene that is required for the UPR (Cox et al. 1993). IRE1 splices *HAC1* mRNA leading to synthesis of HAC1, the bZIP transcription factor activating genes coping with the UPR (Sidrauski and Walter 1997; Shamu 1998). Forward genetic approach using a mutagenized transgenic worm expressing a GFP reporter of *hsp-4*, the ER chaperone BiP homolog in *C. elegans*, identified XBP1, the highly conserved bZIP, as a HAC1-like target of IRE1 (Shen et al. 2001; Calfon et al. 2002). Yeast-one-hybid screening of human cDNA using consensus *cis*-element in a promoter of several mammalian UPR responding genes identified ATF6, which is a type II ER transmembrane protein that contains a bZIP transcription factor domain in its cytoplasmic region cleaved out to be located into nucleus upon activation of the UPR (Yoshida et al. 1998; Haze et al. 1999). To release the UPR, protein synthesis is largely suppressed by phosphorylation of eIFα that reduces initiation rates of translation (Prostko et al. 1993; Brostrom and Brostrom 1997). In an effort to identify a kinase responsible for phosphorylation of eIFα induced by the ER stress, a search of the nucleotide databases pointed out *C. elegans* PERK, *pek-1,* that encodes a type I transmembrane protein residing in ER with a cytoplasmic portion containing a protein-kinase domain and a luminal domain sharing blocks of identity with IRE1 (Harding et al. 1999). Notably, in contrast to mammalian models, all of the *C. elegans* null mutants of *ire-1, xbp-1, pek-1 and atf-6* are viable. Also, many UPR and stress response genes exist as a single homolog in *C.elegans*. Altogether, this systemic simplicity makes it possible to study the UPR at organismal levels, such as counter-interactions between the UPR pathways in different tissues.

In *C. elegans*, *pek-1* and *atf-6* synergistically acts each other to complement the developmental requirement for *ire-1/xbp-1*. All animals lacking either one of *ire-1, xbp-1, pek-1*, and *atf-6* function are viable, but the double mutants or knock-down of *ire-1/xbp-1* and *pek-1* pathways, or those of *ire-1/xbp-1* and *atf-6* are severely larval lethal (Shen et al. 2001, 2005). In contrast, the double mutants of *pek-1;atf-6* shows normal phenotypes. While the *ire-1/xbp-1* pathway regulates the vast majority of UPR transcriptional control, the UPR genes regulated by *pek*-1 are not overlapped with those by *atf-6* (Shen et al. 2005). However, some of the genes regulated by either *pek-1* or *atf-6* are overlapped with those by *ire-1/xbp-1*.The complementary effect is also observed in neuronal development, and exit of glutamate receptors from ER to distal neurites requires ER stress signaling by *ire-1/xbp-1*, but not by *pek-1* or *atf-6* (Shim et al. 2004). Tunicamycin treatment significantly delays organismal development of both *ire-1* and *pek-1* null mutants, while *ire-1* single mutants are more sensitive to tunicamycin than *pek-1* mutants. Despite the moderate resistance to tunicamycin of *pek-1*-null mutants, ER oxidation induced by ER stress significantly increased the level of reactive oxygen species, shortening life span (Harding et al. 2003). Worms lacking a nonmuscle myosin heavy chain IIB (NMHCIIB) homolog are hypersensitive to ER stress (He et al. 2012). Mammalian NMHCIIB directly interacts with IRE1α that oligomerizes to be activated in mammalian cells. Loss of NMHCIIB compromises XBP1 activity and UPR gene expression, but exerts no effect on the PERK pathway. Therefore, specific cellular basis including concerted coordination between the ER and cytoskeleton is likely to contribute to the complementary activity of the three canonical branches of UPR pathways in *C.elegans*.

13.4 UPR and Aging

ER stress caused by the burden of misfolded protein contributes to aging. A line of research using *C. elegans* provides tremendous understanding how adaptation to protein misfolding through UPR is linked to resistance against the time-dependent deterioration of living organisms.

Yun et al. reported that worms lacking *aip-1* were not only sensitive to the insult of arsenite but also reduced their lifespan, and hypersensitive to the toxicity of misfolded polyQ proteins (Yun et al. 2008). Interestingly, overexpression of a truncated AIP-1 without CAAX motif resembling the arsenite-responding human homolog of mammalian AIRAP (arsenic-inducible proteasomal 19S regulatory particle-associated protein), did not rescue the short lifespan or sensitivity to misfolded-protein toxicity, but only a full-length AIP-1 resembling a constitutively expressed ER-associated AIRAP-like protein did, while both versions of AIP-1 proteins were effective in the rescue of arsenite-sensitivity. A single worm homolog AIP-1 possesses both features of stress-inducible and proteasome-adaptive functions in the maintenance of normal lifespan and in neutralizing proteotoxicity that are not related with arsenic exposure. Therefore, a proteasomal adaptation responding to environmental stress contributes to the resistance to proteotoxicity and the maintenance of normal lifespan.

Insulin/IGF-1 signaling is well known to suppress longevity in *C. elegans*. In worm mutants lacking *daf-2* activity, an insulin/IGF-1 receptor homolog, profoundly extends life span, due to activation of DAF-16, FOXO-transcription factor that stabilizes longevity transcriptome (Li et al. 2019). IRE1/XBP1 pathway, but not *pek-1* or *atf-6*, contributes to the long lifespan of *daf-2* mutants (Henis-Korenblit et al. 2010). IRE1/XBP1 pathway collaborates with DAF-16 to enhance ER stress resistance and activates a new set of genes that promotes ER stress resistance and longevity. Interestingly, DAF-16 restores secretory protein metabolism when the UPR is defective, by activating alternative ER-associated degradation systems independent from *ire-1* and ERAD E3 ubiquitin ligase complex *sel-1* and *sel-11* (Safra et al. 2014). Instead, the DAF-16-activated alternative ERAD is mediated by autophagic degradation and by increasing the level of *skr-5*, another component of ERAD E3 ubiquitin ligase complex. The alternative degradation systems act together with the canonical UPR to reinforce ER homeostasis and ER stress resistance. Therefore, it may be possible to improve ER homeostasis by interventions activating DAF-16, a major life span extender.

13.5 Cell-Nonautonomous UPR

In *C. elegans* germline tumor model, the ER stress on apoptosis resistant tumorous cells suppress the progression of germline tumor by activating IRE-1, while inducing germ cell transdifferentiation into ectopic somatic cells (Levi-Ferber et al. 2015). The transdifferentiation restores execution of apoptosis, thus those cells are subsequently removed. Therefore, manipulation of cell identity in tumor cells by the UPR activation may establish a novel therapeutic strategy of tumor suppression.

ER stress response is clearly a cell-intrinsic, adaptive and pro-survival pathway, but also exerts cell-extrinsic effects, as shown in tumor microenvironment where the cancer cell UPR mediates paracrine effects on immune cells (Zanetti et al. 2016). Several studies in *C. elegans* clearly show that the UPR pathway may act in a non-cell-autonomous manner to promote organismal adaptation to environmental stress (Fig. 13.1).

13.5.1 Longevity

Proteostasis is mitigated as organisms age. In *C. elegans*, expression of constitutively active form of XBP-1s, the product of IRE-1-splicing, reverse the age-dependent onset loss of ER proteostasis (Taylor and Dillin 2013). Interestingly, expression of XBP-1s in neurons is sufficient to regain stress resistance, enhance longevity, and activate the UPR in non-neuronal tissues in a cell-nonautonomous

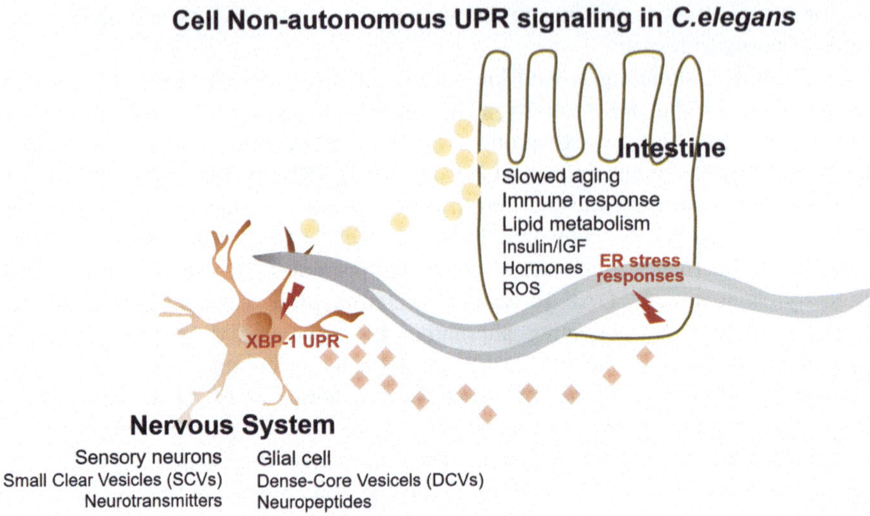

Fig. 13.1 Cell-nonautonomous UPR in *C. elegans*. IRE-1-dependent UPR in the nervous system triggers releases of small molecule neurotransmitters from sensory neurons or neuropeptides from glial cells. Distal tissues such as the intestine receive the signals and activate the UPR to extend lifespan and fight pathogens by mobilizing lipid metabolism and secreting signal molecules such as peptide hormones and ROS, which feedbacks the nervous system

manner, and small clear vesicle (SCV) release is required for the cell-nonautonomous signal from the nervous system expressing XBP-1s. Specifically, neuronal XBP-1 activates intestinal lysosomal pathway, improving cellular proteostasis (Imanikia et al. 2019a). Also, worms expressing XBP-1s either in neurons or in the intestine greatly upregulate lysosomal lipases and fatty acid desaturases in the intestine remodeling lipid metabolism (Imanikia et al. 2019b).

On the other hand, expression of XBP-1s in four astrocyte-like glia extends the life span in *C. elegans* (Frakes et al. 2020). Glial XBP-1s initiates a robust cell nonautonomous activation of the UPR in the intestine which renders worms more resistant to protein aggregation and chronic ER stress. Cell nonautonomous glial UPR requires neuropeptide signaling.

C. elegans larvae in unfavorable environmental conditions enter an alternative larval stage known as dauer, and evolutionarily conserved signaling pathways of neuroendocrine physiology determines the developmental decision. Dominant mutation of neuronal insulin-like peptide *daf-28* causes constitutive entry into dauer diapause and expression of the mutant DAF-28 results in ER stress in the ASI dauer-inducing pheromone sensory neurons. The neuronal ER stress activates PEK-1, which in turn phosphorylates eIF2α, specifically in the ASI neuron pair, to promote entry into dauer diapause. The cell-nonautonomous UPR in the ASI neurons are also responsible for *ire-1*-dependent, but *xbp-1*-independent apoptosis in pluripotent and mitotic germline, which requires both CEP-1/p53 and canonical DNA damage-induced apoptotic genes (Wang et al. 2014). Therefore, the UPR in specific chemosensory neurons promotes stress adaptation during early stage of development and germline apoptosis in cell-nonautnomous manner.

The basic role of UPR pathways in metabolism of secreted proteins under normal unstressed conditions may contribute to the cell-nonautonomous activity of the UPR in the nervous system. Loss of *ire-1/xbp-1* branch compromises ERAD and phagocytosis, hindering clearance of misfolded proteins in ER and letting those proteins secreted and collected into the pseudocoleom (Safra et al. 2013). The basal activity of the UPR is paradoxically harmful in the pathology caused by toxic Aβ protein in a *C. elegans* model of Alzheimer's disease, but may play a role in the UPR response of the nervous system protecting other tissues in the whole organism.

13.5.2 Chromatin Remodeling

Chromatin remodeling plays a critical role in transcriptional response to stress. A member of Rho subfamily, CRP-1, is essential for ER stress response in *C. elegans*. CRP-1 directly interacts with CDC-48, a p96 AAA(+) ATPase homolog, which is implicated in chromatin remodeling and transcriptional control, and promotes both ERAD and transcription of the UPR genes (Caruso et al. 2008). Genome-wide RNAi screen identifies RUVB-2, a Reptin AAA(+) ATPase homolog, as a novel repressor of a subset of UPR (ER) genes, which is degraded by CDC-48 in an XBP-1-dependnet mechanism. Degradation of RUVB-2 by CDC-48 enhances expression of ER stress response genes through an XBP1-dependent mechanism. The functional interaction between CDC-48 and RUVB-2 in controlling transcription of the UPR genes appears evolutionarily conserved, because shRNA knock-down of Reptin in a human liver cell line Huh7 upregulate ER stress response genes. It has been also shown that mutations in HCF-1, a transcriptional regulator that associates with histone modifying enzymes, and SIN-3, a subunit of a histone deacetylase complex, suppress the larval lethality of *xbp-1* mutant worms caused by immune activation in response to infection by pathogenic bacteria (Tillman et al. 2018).

Another study indicates that tissue-specific DNA remodeling regulating the UPR pathway is required for ER stress adaptation in a cell-nonautonomous manner (Kozlowski et al. 2014). HPL-2 (heterochromatin protein like-2), a homolog of heterochromatin protein 1 (HP1), down-regulates the UPR in the intestine. The UPR including activation of ER chaperone expression as well as autophagy are enhanced in *hpl-2(lf)* mutants, and those worms show resistance to ER stress, dependent on *ire-1*, *xbp-1* and *pek-1*, but not *atf-6*. The *hpl-2(lf)* mutant worms expressing *hpl-2* in the intestine respond to ER stress at a comparable level to wild type animals, whereas the mutants expressing *hpl-2* in neurons increase ER stress-induced larval arrest, indicating that *hpl-2*–expressing neurons generate a signal that aggravates ER stress in *hpl-2*–deficient receiving tissues. Retinoblastoma protein homolog *lin-35* and the *lin-13* zinc finger protein act in the same pathway as *hpl-2* suppressing the UPR. The study suggests that tissue-specific chromatin remodeling is responsible for cell-nonautonomous ER stress adaptation by plausible change in tissue-specific transcriptome necessary to maintain ER homeostasis.

These findings in *C. elegans* and the UPR transmission in tumor microenvironment share a strong analogy with quorum sensing, a cell-to-cell communication process by which bacteria monitor their cell population. Therefore, a cell-nonautonomous UPR-based regulation of heterologous cells may be phylogenetically conserved (Zanetti et al. 2016).

13.6 Interplay Between Redox and the UPR

The ER lumen maintains an oxidative environment by the ER oxidoreductase systems that form disulfide bonds and generate reactive oxygen species (ROS), and these oxidizing environments are favorable for oxidative folding for protein (Eletto et al. 2014a). ER stress increases ROS production which eventually leads to cellular oxidative stress that activates oxidative stress defense gene expression. NRF, a highly conserved bZIP-family transcription factor, is a main player in metazoan oxidative and xenobiotic stress responses and the downstream target genes are involved in ROS scavenging and detoxification processes, thereby assuring longevity (Blackwell et al. 2015). Also, a protein disulfide isomerase 6 (PDIA6) attenuates the UPR signaling that is required to be tuned in an appropriate range to adequately cope with ER stress (Groenendyk et al. 2014; Eletto et al. 2014a, b).

In *C. elegans*, SKN-1, a Nrf homolog, is associated the ER, playing a critical modulatory role in the canonical UPR, while functioning as a protective transcription factor to oxidative and xenobiotic stress (Glover-Cutter et al. 2013). Under ER stress, SKN-1 activates the canonical UPR pathway by directly binding to the downstream target genes of XBP-1 and ATF-6, which in turn induce transcription of *skn-1*. The response mediated by SKN-1 is independent from the well-established protective response to oxidative stress. In the absence of *skn-1*, activation of the canonical UPR is largely impaired with significantly reduced XBP-1 splicing, thus SKN-1 is essential for resistance to ER stress. In addition, signaling from the ER is required for SKN-1 to distinctly respond to oxidative stress. IRE-1 is sulfenylated on a cysteine in the active site by ROS and the inhibited IRE-1 suppresses the UPR, in turn initiates the antioxidant response by SKN-1 (Hourihan et al. 2016). Therefore, IRE-1 plays a role of a watchman to alarm SKN-1, which integrates regulation of oxidative stress response and the UPR, coordinating ER and cytoplasmic homeostasis. Lactate and pyruvate promote the UPR and SKN-1 activation by triggering a mild transient increase of ROS, indicating the hormetic ROS signaling in cellular defense mechanism (Tauffenberger et al. 2019). It is also notable that proteasome dysfunction results in proteolytic cleavage or amino acid editing by N-glycanase of ER-associated SKN-1, which in turn induces activation of a series of genes that participate in proteasome homeostasis critical for cytoplasmic UPR, which relieves protein aggregate stress (Lehrbach et al. 2019; Waldherr et al. 2019).

The duration and strength of the UPR signal should be properly regulated to exert the appropriate resolution of ER stress. In a conserved mechanism, the activity of IRE-1 is attenuated by a specific ER resident protein disulfide isomerase PDIA6 (Groenendyk et al. 2014; Eletto et al. 2014b). PDIA6 binds the lumenal domain of activated IRE-1 and oxidizes cysteine 148 residue to turn off the enzymatic activity of IRE-1. PDIA6 deficiency in both mammalian cells and *C. elegans* causes hypersensitivity to ER stress and constitutive UPR. This mechanism allows redox balance in ER controls appropriate level of UPR that is required not only for ER stress but also various physiological status.

Proteostasis network integrates the protein folding and degradation machineries in multiple subcellular compartments. The ER protein quality control is tightly associated with the ER redox homeostasis required for formation of disulfide bond. *C. elegans* expressing two different genetically encoded sensors of redox status (roGFP) in the ER and of hydrogen peroxide (HyPer) in cytoplasm provides valuable information of the ER redox through worm lifetime (Kirstein et al. 2015). The redox state of the ER in *C. elegans* dynamically changes during worm lifetime. In young animals, the ER is oxidizing and this shifts towards reducing conditions during ageing, whereas in the cytosol the redox state becomes more oxidizing with age. The redox state in the cytosol and the ER likewise change in an opposing manner in response to proteotoxic challenges in worms and in HeLa cells expressing the aggregation-prone and disease-associated proteins. Therefore, the intercompartmental redox homeostasis responding to cellular proteostasis is evolutionarily conserved. In addition, proteostasis imbalances induced by expression of the aggregate-toxic proteins in one tissue are communicated to a distal tissue in a bilateral manner between neurons and muscle. Therefore, organelle redox homeostasis regulated across tissues within *C. elegans* provides a new measure of organismal fitness.

Reductive stress which causes breakage of disulfide bonds compromises protein folding in the ER and induces the UPR, which in turn regains proteostasis (Maity et al. 2016). Peroxide accumulation resulted from reductive stress attenuates the UPR by altering translation. A genetic screen in *Saccharomyces cerevisiae* identifies oxidative quality control (OQC) genes, the modulators of the UPR induced by chronic reductive stress but not acute one. Failure of OQC during reductive ER stress shuts off translation selectively of highly expressed proteins, and the process decoupling transcriptional and translational response in both yeasts and worms. PEK-1, the canonical UPR attenuating global translation, helps bypass the non-canonical bipartite translation attenuation dependent on ROS and OQC, by in part decreasing the ROS level that are broadly generated from general translation.

13.7 UPR and Metabolism

C. elegans can be used to uncover biological mechanisms that determine the beneficial effects of dietary restriction (DR) and the role of glycan precursor and phospholipids in the UPR.

13.7.1 Dietary Restriction (DR) and Food Deprivation

The UPR pathway in *C. elegans* plays a role as a key nutritional sensor that regulates organismal energy homeostasis (Jo et al. 2009). Expression of fasting-induced lipases *fil-1* and *fil-2* are impaired in *ire-1* and *hsp-4* mutants which cannot breakdown fat granules during starvation and significantly reduce their motility. Therefore, lipolysis dependent on the UPR pathway is critical for energy supply from stored fat hydrolysis.

DR, the environmental manipulation that robustly extends life span in various species, is dependent on the UPR pathway in *C.elegans* (Chen et al. 2009). Target of rapamycin (TOR) pathway is required for DR-mediated lifespan extension. Sensing nutrients, TOR stimulates protein synthesis by activating S6 kinase (S6K) and inhibiting eukaryotic translation initiation factor 4E-binding protein (4E-BP). Regardless of hypoxia, TOR/S6K leads to translational activation of the hypoxia-inducible factor-1 (HIF-1) (Agani and Jiang 2013). In *C. elegans*, decreased protein levels of HIF-1 extend life span even under rich nutrient conditions, but increased levels of HIF-1 does not allow full extension of life span by DR. To regulate the life span, *hif-1* functions in specific neurons such as serotonergic and oxygen-sensing neurons as well as muscle. The life span extension induced by both DR and *hif-1* modulations is associated with low ER stress and dependent on *ire-1*. The UPR pathway plays a key role in longevity induced by DR in which TOR/S6K signals HIF-1–mediated changes in protein homeostasis in *C. elegans*.

A transient pharmacological ER stress in early development of *C. elegans* enhances proteostasis and prevents induced UPR decline with age and increases adult life span. This ER hormesis is dependent on IRE-1/XBP-1, requires DR-specific FOXA transcription factor PHA-4 and upregulates ERAD pathway Panowski et al. (2007). Interestingly, DR employs this mechanism to promote degradation of ER proteins. Therefore, ER-targeted pharmacological intervention inducing ER hormesis may exert the prolongevity effect of DR.

Newly hatched L1 larvae enters a long developmental arrest of quiescence upon food deprivation, slowing their metabolism and increasing stress resistance, and their resuming of development to become normal adults is dependent on the canonical UPR (Roux et al. 2016). The so-called 'arrested L1 larvae" can live in quiescence for weeks, as twice as long as the never-arrested worms, but showing various aging signs including increased ROS production, and decreased proteotoxic-stress resistance. When fed, the L1 larvae in quiescence resume development and become normal adults, which live normal lifespan thereafter. The reversible aging is dependent on IRE-1/XBP-1, but worms lacking *kgb-1*, a protein kinase, KGB-1, do recover from the quiescence, but fail to reverse aging-associated mitochondrial phenotypes. Therefore, the UPR is critical in the reverse-aging process in animals that have once undergone aging process.

13.7.2 Glycan Metabolism

N-linked glycan oligosaccharides are attached to ER proteins and critical for their correct folding. *C. elegans* with increased synthesis of N-glycan precursors in the hexosamine pathway improves ER protein homeostasis and extends lifespan. Tunicamycin resistance screening identifies *gfat-1* encoding a putative glutamine fructose-6-phosphate aminotransferase, which is required for resistance to ER stress. GFAT-1 is highly conserved enzyme which catalyzes a rate-limiting step in the hexosamine pathway, which generates UDP-N-acetylglucosamine (UDP-GlcNAc), a precursor for N- and O-glycans. GlcNAc supplementation slows aging in wild-type animals by improving ER protein quality control, but independently of *daf-2*/IIS pathway. The precursor supplementation also alleviates neurotoxicity in the model nematode of Huntington's and Parkinson's disease. Autophagy and ERAD are enhanced in *gfat-1(gf)* mutant animals, but the necessary UPR pathway is not further activated. Therefore, hexosamine pathway activation accumulating GlcNAc induces distinct mechanisms of protein quality control employing ERAD and autophagy. The findings suggest nutritional supplementation of N-glycan precursor may be utilized as therapeutic intervention against proteotoxic diseases associated with aging.

13.7.3 Phospholids Metabolism

In *C. elegans*, it is demonstrated that lipid dis-equilibrium activates the UPR, independently from proteostasis. *mdt-15* is the Mediator subunit homolog, involved in fatty acid metabolism (Hou et al. 2014). *mdt-15(lf)* mutant animals compromised in fatty acid desaturation with reduced phosphatidylcholine production activate the UPR without proteostasis imbalance. This finding suggests that membrane composition is a direct input that can activate UPR signaling even when proteostasis in the ER is largely intact. Also, knock-down of the fatty acid metabolism enzymes stearoyl-CoA-desaturases (*SCD*) or S-adenosyl methionine synthetase (*sams-1*) activates the UPR without aggregation of misfolded proteins. These worms with despaired fatty acid metabolism, *mdt-15(lf)* and worms knocked-down of *SCD* or *sams-1*, are neither hypersensitive to tunicamycin nor synthetic lethal with the canonical UPR pathway. Thus, their constitutively activated UPR is not due to proteotoxicity, but likely to changes in ER membrane lipid composition and fluidity.

This is also supported by the study showing that low level of phosphatidylethanolamine (PE) induces ER stress and accelerates degeneration of nematode dopaminergic neurons expressing α-synuclein (α-syn), Parkinson disease (PD)-associated protein (Wang et al. 2014). Supplementation of ethanolamine (ETA), which can be converted to PE via the Kennedy pathway in the ER, prevents the neurodegeneration, having no effect on respiration, but eliminating ER stress. Therefore, the UPR is activated by ER-originated inputs independent from either protein or lipid biosynthesis, and the UPR activated upon ER membrane disequilibrium monitors altered ER homeostasis. Also, the disruption of phosphatidylcholine (PC) synthesis induces lipid bilayer stress (LBS), which activates IRE-1-dependent UPR that selectively regulates a distinct subset of genes independent of proteotoxic stress, and some of them modulates autophagy. Transcriptome and chromatin immunoprecipitation analyses revealed that the UPR as the sum of two independent stress responses to LBS and proteotoxic stress, which deploys divergent transcriptional UPR programs (Ho et al. 2020).

13.7.4 Protein Metabolism

Co-translational delivery to and translocation across the ER membrane of transmembrane and secretory proteins are mediated by the signal recognition particle (SRP). SRP binds to ribosome-nascent chain complexes (RNCs) containing the proteins with hydrophobic signal sequence and targets them to the Sec61 translocon by interacting with the ER membrane-located SRP receptor. A study using *C. elegans* clarifies the role of highly controversial function of the nascent polypeptide-associated complex (NAC) in the process of ER translocation (Gamerdinger et al. 2015). NAC enhances the fidelity of ER translocation by inhibiting additional unspecific contacts between ribosomes and translocons after the delivery of RNCs by SRP. NAC depletion shortens the life span of *C. elegans*, caused global mistargeting of translating ribosomes to the ER, causing impaired protein homeostasis.

Sumoylation covalently links small ubiquitin-like modifier (SUMO) proteins with target proteins to regulate a variety of biological processes. In *C. elegans*, sumoylation plays a regulatory role in the UPR (Lim et al. 2014). Abolishing sumoylation by depleting *ubc-9*, a SUMO-conjugating enzyme, enhances *crt-1*/calreticulin expression in an XBP-1-dependent manner and the increased *crt-1* expression helps relieve ER stress. Also, sumoylation is repressed in worms in ER stress. Knock-down of *ubc-9* mainly affects the expression of genes associated with ER functions, such as lipid and organic acid metabolism. Therefore, sumoylation plays a regulatory role in ER function by controlling the expression of genes required for ER homeostasis.

13.8 UPR and RNA World

ire-1-null worms are viable, but larval lethal in the absence of either *pek-1* or *atf-6* gene. RNAi screen to find genes that are synthetic lethal with *ire-1* mutation, but not with *pek-1* or *atf-6* mutation, identify genes encoding subunits of the exosome complex, which is involved in RNA surveillance (Sakaki et al. 2012). The finding led to the discovery that a series of genes required for nonsense-mediated mRNA decay (NMD) is ER stress responding. Defects in NMD genes cause ER stress and sensitize to tunicamycin in both worms and mammalian cells. ER stress induces gene expression of NMD genes including SMG6, and constitutive localization of NMD regulators to the ER, but does not activate assembly of mRNA surveillance complex. Therefore, NMD pathway controlling mRNA quality participates in UPR process of protein quality control likely in a wide range of metazoan species.

In *C. elegans*, miRNA-233 targets SCA-1, the sarco/endoplasmic reticulum Ca^{2+}-ATPase, to regulate the UPR. Infection of *Pseudomonas aeruginosa* PA14 upregulates miRNA-233 in a PMK-1, p38 MAPK (mitogen-activated protein kinase), –dependent manner, and represses SCA-1 expression at protein level, which in turn activates the UPR. Whereas mir-233 null mutant animals are hypersensitive to PA14 infection, RNAi depletion of *sca-1* enhances resistance to the infection. Therefore, microRNA-dependent pathway is a critical player in innate immunity by regulating the UPR.

13.9 UPR and Innate Immunity

During *C. elegans* development, IRE1/XBP1 pathway plays a protective role for the host in activation of innate immunity (Richardson et al. 2010). PMK-1 p38 mitogen-activated protein kinase (MAPK) pathway is conserved from nematodes to humans, and functions to confer *C. elegans* resistance to pathogenic microbes (Kim et al. 2002). Activation of the PMK-1 pathway induces the XBP-1-dependent UPR. When infected with *P. aeruginosa*, human opportunistic bacteria, normally viable *xbp-1(lf)* mutants show larval lethality accompanied by disruption of ER, but not *pek-1(lf)* or *atf-6(lf)* mutants. The larval lethal phenotype in *xbp-1(lf)* is rescued by the loss of *pmk-1*. Furthermore, *xbp-1(lf)* worms with hyperactive *pmk-1* are larval lethal even without pathogenic PA infection. Therefore, IRE-1/XBP-1 pathway has an ancient and conserved protective role for the host organism to resolve ER stress induced by a mounting innate immune response to microbes during *C. elegans* development.

In *C. elegans*, XBP-1 deficiency constitutively activates ER stress, exhibiting increased basal levels of IRE-1 and PEK-1 activities (Richardson et al. 2011). The enhanced IRE-1 and PEK-1 activities are exacerbated by induction of innate immunity by pathogens, due to increased ER stress level. The double mutants of *xbp-1; pek-1* normally develop similar to wild type at the low range of physiological temperature at 16 to 20 °C, but show larval lethality in a temperature-dependent mode at 20 to 25°C. The larval development of either *xbp-1* or *pek-1* mutant, which is severely compromised on *P. aeruginosa* at 25 °C, but elimination of *pmk-1* activity, the innate immune mediator, rescues the temperature-dependent larval lethal phenotypes. *Xbp-1;pek-1* double mutants also die at young larval stages on non-pathogenic food at higher temperature of 23 to 25 °C, and loss of *pmk-1* rescues the lethal phenotype. Basal immune activity under non-pathogenic conditions presents a low level of ER stress that is severely exacerbated by impaired UPR, leading to larval lethality of the *xbp-1*; *pek-1* mutant. Therefore, the basal immune activation and temperature are distinct contributors to ER stress that function in parallel for organismal survival. The study describes that the UPR plays a more general role in animal physiology and development, as well as the conventional view of UPR function as a stress responsor in protecting against the extreme accumulation of misfolded proteins.

Lipoproteins vitellogenins (VITs), homologs of human apolipoprotein B-100, are required for the UPR to support immunity in *C. elegans* (Singh and Aballay 2017). Mutations in vitellogenins result in abnormal accumulation of the proteins in the intestine, which causes the ER stress, reducing lifespan and immunity to pathogens. In those *vit* mutants, expression of immune proteins secreted by the intestinal cells are largely suppressed in an IRE-1-independent manner. The study provides some answers how abnormal protein aggregates and ER stress negatively affects organismal immunity and also shows a clear example of IRE-1-independent suppression of gene expression of secretory proteins.

13.10 Noncanonical UPR Response in *C. elegans*

In *C. elegans*, several noncanonical UPRs have been reported, independent from three major canonical UPR pathways. *C. elegans* noncanonical UPR pathways function in supporting immune defense and longevity.

13.10.1 PQN/ABU Genes in Parallel with IRE-1/XBP-1

In the absence of the canonical UPR pathway of *ire-1/xbp-1*, a set of genes are upregulated by tunicamycin treatment and function to relieve ER stress (Urano et al. 2002). The upregulation is associated with survival of animals under UPR stress, because RNAi to the genes results in death of viable *ire-1* or *xbp-1* null mutants and downregulates genes required for ERAD. The noncanonical UPR pathway includes *pqn/abu* genes, which encode ER-resident type I transmembrane proteins whose luminal region containing prion-like glutamine/asparagine (Q/N)-rich domains. PQN/ABU are also similar to mammalian scavenger receptors that bind damaged extracellular proteins and target them to lysosomal degradation, thus suggested to interact with misfolded ER luminal proteins. Therefore, these noncannonical UPR pathways of PQN/ABU function either in a pathway parallel to the canonical UPR or in the ERAD.

SIR-2.1, a member of the Sir-2 family of NAD(+)-dependent protein deacetylases, regulates aging via the insulin/IGF pathway transcription factor *daf-16*, the well-established longevity control pathway in *C. elegans* (Kenyon 2005). Resveratrol, a polyphenolic small molecule showing various therapeutic benefits, extends nematodes life span, fully dependent on *sir-2.1* but independent from *daf-16. sir-2.1* suppresses gene expression of PQN/ABU, which are upregulated in worms treated with resveratrol (Viswanathan et al. 2005). The knock-down of some PQN/ABU genes compromises life extension whereas their overexpression further extends life span. Therefore, the noncannonical UPR genes, negatively regulated by *sir-2.1,* mediates longevity in *C. elegans.*

C. elegans has evolved defense mechanisms to recognize and respond against various pathogens, and apoptotic signaling is also involved in the worm's response to pathogen infection. The infection of human pathogenic bacteria, *Samonella enterica*, activates the well-established *ced-3* caspase apoptotic pathway that triggers programmed cell death of germline cells (Aballay and Ausubel 2001). The dead cells are recognized and subsequently engulfed by phagocytic cells through the signaling mediated by the apoptotic transmembrane receptor CED-1 that recognizes cell corpses with extracellular EGF repeats (Zhou et al. 2001). The extracellular domain of CED-1 also shares homologous sequence structure with PQN/ABU proteins (Urano et al. 2002). Loss-of-function (lf) *ced-1* mutant worms are immunocompromised and severely ill to die when infected with live *Salmonella enterica. Pqn/abu* genes are downregulated in *ced-1(lf)* mutants (Haskins et al. 2008). While abrogation of the gene expression of *pqn/abu* makes wildtype animals more vulnerable to *Salmonella* exposure, overexpression of *pqn/abu* genes confers protection against pathogen-mediated killing. Therefore, the noncanonical UPR genes are regulated in a CED-1-dependent manner, and required for the *C. elegans* immune response to pathogenic bacteria, while the question whether *pqn/abu* genes directly affect maturation and function of *ced-1* awaits to be answered.

Interestingly, OCTR-1, an octopamine G protein-coupled catecholamine receptor (GPCR, G protein–coupled receptor), functions in specific sensory neurons, suppressing the innate immune response to pathogenic bacteria via p38 MAPK pathway (Sun et al. 2011). When stimulated by octopamine release from RIC neurons that are activated by pathogens, *octr-1* activity in ASH and ASI neurons, which detect soluble chemicals and volatile odorants, down-regulates the non-canonical UPR genes *pqn/abu. octr-1* activity is supported by *arr-1*/arrestin-1, the only GPCR adaptor in *C. elegans* (Singh and Aballay 2012). These genes are highly expressed in non-neuronal tissues such as the pharynx and the intestine, the primary interfaces between host cells involved in immune responses and bacterial pathogens. Also, importantly, OCTR-1 regulates the canonical UPR pathway of IRE-1/XBP-1, not during development, only at the adult stage (Sun et al. 2012). Therefore, the nervous system can sense inflammatory responses and reacts by temporally controlling stress-response pathways during development at the organismal level. Neuronal control of immune pathway is present in diverse species including humans and the studies in *C. elegans* provides clues for underlying mechanisms in vertebrates (Hoffman and Aballay 2019).

13.10.2 Autophagy Induction Independent of IRE-1

Cells undergo a finite number of cell divisions before irreversibly exiting mitosis. This mitotic senescence is regulated by intricate pathways that in concert limit the number of cell divisions and the onset of complex phenotypes associated with mitotic exit. Regulatory failure in the mitotic lifespan can lead to cell immortalization resulting in tumor growth. A large-scale screen in yeast *Saccharomyces cerevisiae* for mutants that significantly increase the number of cell division before the exit of mitosis reveals a cross talk between ER homeostasis and mitotic longevity (Ghavidel et al. 2015). *RER1* is a receptor that maintains ER compartmentalization by Golgi-to-ER retrieving components of the vesicles in ER-to-Golgi transport. In both yeast and *C. elegans,* inactivation of RER1 induces the ER stress, which in turn activates autophagy that remove bulky cytoplasmic protein aggregates in an IRE-1-independent manner. The IRE-1-independent ER stress response enhances mitotic longevity in yeast and reproductive lifespan in worms. ER stress induced by inhibition of Golgi-to-ER transport of ER components activates catabolism of protein aggregates which is a mitotic longevity driver.

On the other hand, blocking ER-to-Golgi transport is well-known to activate the canonical UPR. ULK1/ATG1, *unc-51*-like autophagy activating kinase 1, phosphorylates SEC16A that facilitates the recruitment of soluble subunits of the coatomer protein complex II (COPII) at the ER-exit sites (ERES) where ER-to-Golgi trafficking cargos bud off (Joo et al. 2016). Deficiency of ULK activity results in abnormal development of neurons that often degenerate in both worms and mice, activating the canonical UPR, but not accumulating abnormal membranous structures, a hallmark of defective autophagy. Therefore, the ER-to-Golgi trafficking regulation by ULK1, which is not dependent on autophagy, is essential for cellular homeostasis.

These evolutionarily conserved ER stress responses associated the trafficking between ER and Golgi suggest that the release of cargos from ER relieve the ER stress through the canonical UPR, but the ER stress resulted from depletion of ER components is resolved via autophagy pathway likely removing wrecked ER undergoing disintegration.

13.10.3 ER Stress Response in Differentiation

Protein folding chaperones are selectively upregulated in specific cells to match the cell-type–specific folding needs during development. Differentiation of secretory cells dramatically increases in protein synthesis, challenging ER proteostasis, and the canonical UPR prepares those cells for the onset of secretory function. In *C. elegans*, HSP-4, a BiP homolog that is ubiquitously expressed in adult worms, is selectively induced in alae-secreting daughter cells, but repressed in hypodermal daughter cells during differentiation of seam cells (Zha et al. 2019). Interestingly, the lineage-specific induction of *hsp-4* does not require the canonical UPR signaling but is controlled by a specific developmental program. A worm homolog of B-Lymphocyte-Induced Maturation Protein 1 (BLMP-1/BLIMP1), a regulator of mammalian secretory cell differentiation, is required in the repression of *hsp-4* in hypodermal daughter cells. Thus, a chaperone for a broad range of substrates that is normally the canonical UPR-dependent can be selectively regulated during differentiation that generates cells demanding a large capacity of ER proteostasis.

13.10.4 The UPR from outside

A whole-genome CRISPR-based knockout (KO) screen in *C. elegans* identifies the cell-surface hyaluronidase (HAase), Transmembrane Protein 2 (TMEM2), as a potent modulator of ER stress resistance (Schinzel et al. 2019). TMEM2 breakdown of the glycosaminoglycan, hyaluronan (HA), within extracellular matrix (ECM) changes ER stress resistance, independent of canonical UPR pathways but dependent upon the cell-surface receptor, CD44, a putative HA receptor, and the MAPK cell-signaling components, ERK and p38. Surprisingly, ectopic expression of human TMEM2 in *C. elegans* protects them from ER stress and increased both longevity and pathogen resistance independent of canonical UPR but dependent on the ERK ortholog *mpk-1* and the p38 ortholog *pmk-1*.

13.11 Conclusions

C. elegans, a simple but versatile animal model, has provided a powerful platform for studying ER homeostasis and associated stress responses. The studies have clearly revealed that ER stress is caused by multiple different molecular and cellular conditions in addition to accumulation of protein overload in ER. At organismal level, the evolutionarily highly conserved ER UPR pathway of IRE-1/XBP-1, PEK-1, and ATF6 functions not only cell-autonomously but also cell-nonautonomously to regulate longevity and innate immunity against pathogens. In addition, noncanonical UPRs independent of the major three branches of UPR play a critical role in dealing with certain types of ER stress in certain contexts, in which canonical UPR does not particularly function. Comprehensive investigation of the ER homeostasis and stress response using *C. elegans* will continue to provide full of fascinating insights into answering the question how internal and external ER effectors coordinate balanced responses to maintain survival and reproduction to ensure ultimate wellness of living organisms.

13.12 Summary Points

- *C. elegans*, a simple nematode model utilizing a variety of genetic tools and phenotypic analysis, is an ideal system to study the ER homeostasis and related stress responses at organismal level to investigate integration of various physiological and environmental inputs.
- The nematode has highly conserved canonical UPR branches, IRE-1/XBP-1, PERK, and ATF6. In a synergistic mode, PERK or ATF6 pathways are complementary to IRE-1/XBP-1, which regulates most of the transcriptional response. IRE-1/XBP-1 pathway in worms is critical for macromolecules metabolism, development, lifespan extension and innate immunity.
- *C. elegans* UPR functions in a cell-nonautonomous manner. Via regular neurotransmission, IRE-1/XBP-1 in a small subset of either neurons or glia in the nervous system affects the UPR in distal tissues such as the intestine by activating a lysosomal degradation pathway.
- The UPR interplays with oxidative stress, and SKN-1, a Nrf homolog that is physically associated with ER, orchestrates regulation of the UPR and oxidative stress and signal from stressed ER is required for the integrating function of SKN-1.
- Noncanonical UPR pathways in worms involve expression of distinct set of ER membrane proteins, autophagy, vesicular trafficking between ER and Golgi, and extracellular matrix integrity, and play a critical role especially in case canonical UPRs are not activated.

13.13 Future Issues

- It is crucial to unravel how different types of cell-nonautonomous signaling (eg. SCV or DCV) from different tissues (eg, sensory neurons or glia) are integrated to exert similar outcomes on ER homeostasis at organismal level.
- It is also critical to understand the physiological consequences of the cell-nonautonomous UPR network on regulation of various stress responses in local tissues.
- It will be necessary to establish the circuitry of the nervous system downstream of sensory neurons and glia and molecular networks sensing the signals to clarify the signaling path from the nervous system to the target at the distal tissues.
- Characterization of the biological effects of the tissue-to-tissue UPR network throughout development, adulthood, and aging will greatly help identify critical molecules that detect and treat altered UPR often associated with a variety of human disorders.
- Full investigation of non-canonical UPR signaling will provide rich information to explore alternative ways to treat altered UPR caused in specific conditions that are not controlled by therapeutic interventions targeting canonical UPR.

Conflict-of-Interest Disclosure Statement
The author declares no conflict-of-interest.

Acknowledgments This research was supported by the National Research Foundation of Korea (NRF) funded by the Ministry of Education, Science and Technology (2018R1A2A3074987). I sincerely apologize our eminent colleagues in case unwittingly omitting their valuable works.

References

Aballay A, Ausubel FM (2001) Programmed cell death mediated by ced-3 and ced-4 protects Caenorhabditis elegans from salmonella typhimurium-mediated killing. Proc Natl Acad Sci U S A 98:2735–2739. https://doi.org/10.1073/pnas.041613098

Agani F, Jiang B-H (2013) Oxygen-independent regulation of HIF-1: novel involvement of PI3K/AKT/mTOR pathway in cancer. Curr Cancer Drug Targets 13:245–251. https://doi.org/10.2174/1568009611313030003

Bar-Ziv R, Frakes AE, Higuchi-Sanabria R et al (2020) Measurements of physiological stress responses in C. Elegans J Vis Exp. https://doi.org/10.3791/61001

Blackwell TK, Steinbaugh MJ, Hourihan JM et al (2015) SKN-1/Nrf, stress responses, and aging in Caenorhabditis elegans. Free Radic Biol Med 88:290–301. https://doi.org/10.1016/j.freeradbiomed.2015.06.008

Brenner S (1974) The genetics of Caenorhabditis elegans. Genetics 77:71–94

Brostrom CO, Brostrom MA (1997) Regulation of translational initiation during cellular responses to stress. In: Progress in nucleic acid research and molecular biology. Academic, pp 79–125

Calfon M, Zeng H, Urano F et al (2002) IRE1 couples endoplasmic reticulum load to secretory capacity by processing the XBP-1 mRNA. Nature 415:92–96. https://doi.org/10.1038/415092a

Caruso M-E, Jenna S, Bouchecareilh M et al (2008) GTPase-mediated regulation of the unfolded protein response in Caenorhabditis elegans is dependent on the AAA+ ATPase CDC-48. Mol Cell Biol 28:4261–4274. https://doi.org/10.1128/MCB.02252-07

Chen D, Thomas EL, Kapahi P (2009) HIF-1 modulates dietary restriction-mediated lifespan extension via IRE-1 in Caenorhabditis elegans. PLoS Genet 5:e1000486. https://doi.org/10.1371/journal.pgen.1000486

Corsi A.K., Wightman B. and CMA (2015) A Transparent window into biology: A primer on Caenorhabditis elegans. In: Community TC elegans R (ed) WormBook, ed., Ed. WormBook

Cox JS, Shamu CE, Walter P (1993) Transcriptional induction of genes encoding endoplasmic reticulum resident proteins requires a transmembrane protein kinase. Cell 73:1197–1206. https://doi.org/10.1016/0092-8674(93)90648-A

Dai LL, Gao JX, Zou CG, Ma YC, Zhang KQ (2015) mir-233 modulates the unfolded protein response in C. elegans during Pseudomonas aeruginosa infection. PLoS Pathog 11(1): e1004606. https://doi.org/10.1371/journal.ppat.1004606. PMID: 25569229; PMCID: PMC4287614

Eletto D, Chevet E, Argon Y, Appenzeller-Herzog C (2014a) Redox controls UPR to control redox. J Cell Sci 127:3649–3658. https://doi.org/10.1242/jcs.153643

Eletto D, Eletto D, Dersh D et al (2014b) Protein disulfide isomerase A6 controls the decay of IRE1α signaling via disulfide-dependent association. Mol Cell 53:562–576. https://doi.org/10.1016/j.molcel.2014.01.004

Frakes AE, Metcalf MG, Tronnes SU et al (2020) Four glial cells regulate ER stress resistance and longevity via neuropeptide signaling in C. elegans. Science 367:436–440. https://doi.org/10.1126/science.aaz6896

Gamerdinger M, Hanebuth MA, Frickey T, Deuerling E (2015) The principle of antagonism ensures protein targeting specificity at the endoplasmic reticulum. Science 348:201–207. https://doi.org/10.1126/science.aaa5335

Ghavidel A, Baxi K, Ignatchenko V et al (2015) A genome scale screen for mutants with delayed exit from mitosis: ire 1-independent induction of autophagy integrates ER homeostasis into mitotic lifespan. PLoS Genet 11:e1005429. https://doi.org/10.1371/journal.pgen.1005429

Glover-Cutter KM, Lin S, Blackwell TK (2013) Integration of the unfolded protein and oxidative stress responses through SKN-1/Nrf. PLoS Genet 9:e1003701. https://doi.org/10.1371/journal.pgen.1003701

Groenendyk J, Peng Z, Dudek E et al (2014) Interplay between the oxidoreductase PDIA6 and microRNA-322 controls the response to disrupted endoplasmic reticulum calcium homeostasis. Sci Signal 7:ra54. https://doi.org/10.1126/scisignal.2004983

Harding HP, Zhang Y, Ron D (1999) Protein translation and folding are coupled by an endoplasmic-reticulum-resident kinase. Nature 397:271–274. https://doi.org/10.1038/16729

Harding HP, Zhang Y, Zeng H et al (2003) An integrated stress response regulates amino acid metabolism and resistance to oxidative stress. Mol Cell 11:619–633. https://doi.org/10.1016/S1097-2765(03)00105-9

Haskins KA, Russell JF, Gaddis N et al (2008) Unfolded protein response genes regulated by CED-1 are required for Caenorhabditis elegans innate immunity. Dev Cell 15:87–97. https://doi.org/10.1016/j.devcel.2008.05.006

Haze K, Yoshida H, Yanagi H et al (1999) Mammalian transcription factor ATF6 is synthesized as a transmembrane protein and activated by proteolysis in response to endoplasmic reticulum stress. Mol Biol Cell 10:3787–3799. https://doi.org/10.1091/mbc.10.11.3787

Hazkani-Covo E, Levanon EY, Rotman G et al (2004) Evolution of multicellularity in Metazoa: comparative analysis of the subcellular localization of proteins in saccharomyces, drosophila and Caenorhabditis. Cell Biol Int 28:171–178. https://doi.org/10.1016/j.cellbi.2003.11.016

He Y, Beatty A, Han X et al (2012) Nonmuscle myosin IIB links cytoskeleton to IRE1α signaling during ER stress. Dev Cell 23:1141–1152. https://doi.org/10.1016/j.devcel.2012.11.006

Henis-Korenblit S, Zhang P, Hansen M et al (2010) Insulin/IGF-1 signaling mutants reprogram ER stress response regulators to promote longevity. Proc Natl Acad Sci U S A 107:9730–9735. https://doi.org/10.1073/pnas.1002575107

Ho N, Yap WS, Xu J et al (2020) Stress sensor Ire1 deploys a divergent transcriptional program in response to lipid bilayer stress. J Cell Biol 219. https://doi.org/10.1083/jcb.201909165

Hoffman C, Aballay A (2019) Role of neurons in the control of immune defense. Curr Opin Immunol 60:30–36. https://doi.org/10.1016/j.coi.2019.04.005

Hou NS, Gutschmidt A, Choi DY et al (2014) Activation of the endoplasmic reticulum unfolded protein response by lipid disequilibrium without disturbed proteostasis in vivo. Proc Natl Acad Sci U S A 111:E2271–E2280. https://doi.org/10.1073/pnas.1318262111

Hourihan JM, Moronetti Mazzeo LE, Fernández-Cárdenas LP, Blackwell TK (2016) Cysteine Sulfenylation directs IRE-1 to activate the SKN-1/Nrf2 antioxidant response. Mol Cell 63:553–566. https://doi.org/10.1016/j.molcel.2016.07.019

Imanikia S, Özbey NP, Krueger C et al (2019a) Neuronal XBP-1 activates intestinal lysosomes to improve Proteostasis in C. elegans. Curr Biol 29:2322–2338.e7. https://doi.org/10.1016/j.cub.2019.06.031

Imanikia S, Sheng M, Castro C et al (2019b) XBP-1 remodels lipid metabolism to extend longevity. Cell Rep 28:581–589.e4. https://doi.org/10.1016/j.celrep.2019.06.057

Jo H, Shim J, Lee JH et al (2009) IRE-1 and HSP-4 contribute to energy homeostasis via fasting-induced lipases in C. elegans. Cell Metab 9:440–448. https://doi.org/10.1016/j.cmet.2009.04.004

Joo JH, Wang B, Frankel E et al (2016) The noncanonical role of ULK/ATG1 in ER-to-Golgi trafficking is essential for cellular homeostasis. Mol Cell 62:491–506. https://doi.org/10.1016/j.molcel.2016.04.020

Kenyon C (2005) The plasticity of aging: insights from long-lived mutants. Cell 120:449–460. https://doi.org/10.1016/j.cell.2005.02.002

Kim DH, Feinbaum R, Alloing G et al (2002) A conserved p38 MAP kinase pathway in Caenorhabditis elegans innate immunity. Science 297:623–626. https://doi.org/10.1126/science.1073759

Kirstein-Miles J, Morimoto RI (2010) Caenorhabditis elegans as a model system to study intercompartmental proteostasis: interrelation of mitochondrial function, longevity, and neurodegenerative diseases. Dev Dyn an Off Publ Am Assoc Anat 239:1529–1538. https://doi.org/10.1002/dvdy.22292

Kirstein J, Morito D, Kakihana T et al (2015) Proteotoxic stress and ageing triggers the loss of redox homeostasis across cellular compartments. EMBO J 34:2334–2349. https://doi.org/10.15252/embj.201591711

Kozlowski L, Garvis S, Bedet C, Palladino F (2014) The Caenorhabditis elegans HP1 family protein HPL-2 maintains ER homeostasis through the UPR and hormesis. Proc Natl Acad Sci U S A 111:5956–5961. https://doi.org/10.1073/pnas.1321698111

Labbadia J, Morimoto RI (2014) Proteostasis and longevity: when does aging really begin? F1000Prime Rep 6:7. https://doi.org/10.12703/P6-7

Labbadia J, Morimoto RI (2015) The biology of proteostasis in aging and disease. Annu Rev Biochem 84:435–464. https://doi.org/10.1146/annurev-biochem-060614-033955

Lehrbach NJ, Breen PC, Ruvkun G (2019) Protein sequence editing of SKN-1A/Nrf1 by peptide: N-Glycanase controls proteasome gene expression. Cell 177:737–750.e15. https://doi.org/10.1016/j.cell.2019.03.035

Levi-Ferber M, Gian H, Dudkevich R, Henis-Korenblit S (2015) Transdifferentiation mediated tumor suppression by the endoplasmic reticulum stress sensor IRE-1 in C. elegans. elife 4. https://doi.org/10.7554/eLife.08005

Li S-T, Zhao H-Q, Zhang P et al (2019) DAF-16 stabilizes the aging transcriptome and is activated in mid-aged Caenorhabditis elegans to cope with internal stress. Aging Cell 18:e12896. https://doi.org/10.1111/acel.12896

Lim Y, Lee D, Kalichamy K et al (2014) Sumoylation regulates ER stress response by modulating calreticulin gene expression in XBP-1-dependent mode in Caenorhabditis elegans. Int J Biochem Cell Biol 53:399–408. https://doi.org/10.1016/j.biocel.2014.06.005

Maity S, Rajkumar A, Matai L et al (2016) Oxidative homeostasis regulates the response to reductive endoplasmic reticulum stress through translation control. Cell Rep 16:851–865. https://doi.org/10.1016/j.celrep.2016.06.025

Maurel M, Chevet E, Tavernier J, Gerlo S (2014) Getting RIDD of RNA: IRE1 in cell fate regulation. Trends Biochem Sci 39:245–254. https://doi.org/10.1016/j.tibs.2014.02.008

Nussbaum-Krammer CI, Morimoto RI (2014) Caenorhabditis elegans as a model system for studying non-cell-autonomous mechanisms in protein-misfolding diseases. Dis Model Mech 7:31–39. https://doi.org/10.1242/dmm.013011

Panowski S, Wolff S, Aguilaniu H et al (2007) PHA-4/Foxa mediates diet-restriction-induced longevity of C. elegans. Nature 447:550–555. https://doi.org/10.1038/nature05837

Prahlad V, Morimoto RI (2009) Integrating the stress response: lessons for neurodegenerative diseases from C. elegans. Trends Cell Biol 19:52–61. https://doi.org/10.1016/j.tcb.2008.11.002

Prostko CR, Brostrom MA, Brostrom CO (1993) Reversible phosphorylation of eukaryotic initiation factor 2α in response to endoplasmic reticular signaling. Mol Cell Biochem 127–128:255–265. https://doi.org/10.1007/BF01076776

Richardson CE, Kinkel S, Kim DH (2011) Physiological IRE-1-XBP-1 and PEK-1 signaling in Caenorhabditis elegans larval development and immunity. PLoS Genet 7:e1002391. https://doi.org/10.1371/journal.pgen.1002391

Richardson CE, Kooistra T, Kim DH (2010) An essential role for XBP-1 in host protection against immune activation in C. elegans. Nature 463:1092–1095. https://doi.org/10.1038/nature08762

Ron D, Walter P (2007) Signal integration in the endoplasmic reticulum unfolded protein response. Nat Rev Mol Cell Biol 8:519–529. https://doi.org/10.1038/nrm2199

Roux AE, Langhans K, Huynh W, Kenyon C (2016) Reversible age-related phenotypes induced during larval quiescence in C. elegans. Cell Metab 23:1113–1126. https://doi.org/10.1016/j.cmet.2016.05.024

Safra M, Ben-Hamo S, Kenyon C, Henis-Korenblit S (2013) The ire-1 ER stress-response pathway is required for normal secretory-protein metabolism in C. elegans. J Cell Sci 126:4136–4146. https://doi.org/10.1242/jcs.123000

Safra M, Fickentscher R, Levi-Ferber M et al (2014) The FOXO transcription factor DAF-16 bypasses ire-1 requirement to promote endoplasmic reticulum homeostasis. Cell Metab 20:870–881. https://doi.org/10.1016/j.cmet.2014.09.006

Sakaki K, Yoshina S, Shen X et al (2012) RNA surveillance is required for endoplasmic reticulum homeostasis. Proc Natl Acad Sci U S A 109:8079–8084. https://doi.org/10.1073/pnas.1110589109

Schinzel RT, Higuchi-Sanabria R, Shalem O et al (2019) The Hyaluronidase, TMEM2, promotes ER homeostasis and longevity independent of the UPR(ER). Cell 179:1306–1318.e18. https://doi.org/10.1016/j.cell.2019.10.018

Shamu CE (1998) Splicing: HACking into the unfolded-protein response. Curr Biol 8:R121–R123. https://doi.org/10.1016/s0960-9822(98)70986-8

Shen X, Ellis RE, Lee K et al (2014) The biology of proteostasis in aging and disease. Annu Rev Biochem 28:893–903. https://doi.org/10.1016/j.molcel.2015.06.027

Shen X, Ellis RE, Lee K et al (2001) Complementary signaling pathways regulate the unfolded protein response and are required for C. elegans development. Cell 107:893–903. https://doi.org/10.1016/S0092-8674(01)00612-2

Shen X, Ellis RE, Sakaki K, Kaufman RJ (2005) Genetic interactions due to constitutive and inducible gene regulation mediated by the unfolded protein response in C. elegans. PLoS Genet 1:e37

Shim J, Umemura T, Nothstein E, Rongo C (2004) The unfolded protein response regulates glutamate receptor export from the endoplasmic reticulum. Mol Biol Cell 15:4818–4828. https://doi.org/10.1091/mbc.e04-02-0108

Sidrauski C, Walter P (1997) The transmembrane kinase Ire1p is a site-specific endonuclease that initiates mRNA splicing in the unfolded protein response. Cell 90:1031–1039. https://doi.org/10.1016/S0092-8674(00)80369-4

Singh J, Aballay A (2017) Endoplasmic reticulum stress caused by lipoprotein accumulation suppresses immunity against bacterial pathogens and contributes to Immunosenescence. MBio 8:e00778–e00717. https://doi.org/10.1128/mBio.00778-17

Singh V, Aballay A (2012) Endoplasmic reticulum stress pathway required for immune homeostasis is neurally controlled by arrestin-1. J Biol Chem 287:33191–33197. https://doi.org/10.1074/jbc.M112.398362

Sun J, Liu Y, Aballay A (2012) Organismal regulation of XBP-1-mediated unfolded protein response during development and immune activation. EMBO Rep 13:855–860. https://doi.org/10.1038/embor.2012.100

Sun J, Singh V, Kajino-Sakamoto R, Aballay A (2011) Neuronal GPCR controls innate immunity by regulating noncanonical unfolded protein response genes. Science 332:729–732. https://doi.org/10.1126/science.1203411

Tauffenberger A, Fiumelli H, Almustafa S, Magistretti PJ (2019) Lactate and pyruvate promote oxidative stress resistance through hormetic ROS signaling. Cell Death Dis 10:653. https://doi.org/10.1038/s41419-019-1877-6

Taylor RC, Dillin A (2013) XBP-1 is a cell-nonautonomous regulator of stress resistance and longevity. Cell 153:1435–1447. https://doi.org/10.1016/j.cell.2013.05.042

The C.elegans sequencing consortium (1998) Genome sequence of the Nematode C. elegans: a platform for investigating biology. Science 282:2012–2018. https://doi.org/10.1126/science.282.5396.2012

Tillman EJ, Richardson CE, Cattie DJ et al (2018) Endoplasmic reticulum homeostasis is modulated by the Forkhead transcription factor FKH-9 during infection of Caenorhabditis elegans. Genetics 210:1329–1337. https://doi.org/10.1534/genetics.118.301450

Urano F, Calfon M, Yoneda T et al (2002) A survival pathway for Caenorhabditis elegans with a blocked unfolded protein response. J Cell Biol 158:639–646. https://doi.org/10.1083/jcb.200203086

Viswanathan M, Kim SK, Berdichevsky A, Guarente L (2005) A role for SIR-2.1 regulation of ER stress response genes in determining C. elegans life span. Dev Cell 9:605–615. https://doi.org/10.1016/j.devcel.2005.09.017

Waldherr SM, Strovas TJ, Vadset TA et al (2019) Constitutive XBP-1s-mediated activation of the endoplasmic reticulum unfolded protein response protects against pathological tau. Nat Commun 10:4443. https://doi.org/10.1038/s41467-019-12070-3

Walter P, Ron D, Kanno T et al (2011) The unfolded protein response: from stress pathway to homeostatic regulation. Science 221:1081–1086. https://doi.org/10.1126/science.1209038

Wang S, Zhang S, Liou L-C et al (2014) Phosphatidylethanolamine deficiency disrupts α-synuclein homeostasis in yeast and worm models of Parkinson disease. Proc Natl Acad Sci U S A 111: E3976–E3985. https://doi.org/10.1073/pnas.1411694111

Yoshida H, Haze K, Yanagi H et al (1998) Identification of the cis-acting endoplasmic reticulum stress response element responsible for transcriptional induction of mammalian glucose-regulated proteins. Involvement of basic leucine zipper transcription factors. J Biol Chem 273:33741–33749. https://doi.org/10.1074/jbc.273.50.33741

Yun C, Stanhill A, Yang Y et al (2008) Proteasomal adaptation to environmental stress links resistance to proteotoxicity with longevity in Caenorhabditis elegans. Proc Natl Acad Sci U S A 105:7094–7099. https://doi.org/10.1073/pnas.0707025105

Zanetti M, Rodvold JJ, Mahadevan NR (2016) The evolving paradigm of cell-nonautonomous UPR-based regulation of immunity by cancer cells. Oncogene 35:269–278. https://doi.org/10.1038/onc.2015.108

Zha J, Ying M, Alexander-Floyd J, Gidalevitz T (2019) HSP-4/BiP expression in secretory cells is regulated by a developmental program and not by the unfolded protein response. PLoS Biol 17: e3000196. https://doi.org/10.1371/journal.pbio.3000196

Zhou Z, Hartwieg E, Horvitz HR (2001) CED-1 is a transmembrane receptor that mediates cell corpse engulfment in C. elegans. Cell 104:43–56. https://doi.org/10.1016/s0092-8674(01)00190-8

Chapter 14
Tardigrada: An Emerging Animal Model to Study the Endoplasmic Reticulum Stress Response to Environmental Extremes

Łukasz Kaczmarek

Abstract Tardigrada (also known as "water bears") are hydrophilous micro-invertebrates with a bilaterally symmetrical body and four pairs of legs usually terminating with claws. Water bears are quite complex animals and range from 50 to 1200 μm in length. Their body is divided into a head segment and four trunk segments, each bearing a pair of legs. They inhabit almost all terrestrial and aquatic environments, from the ocean depths to highest mountains ranges. However, one of their best known and unusual features is their capability for cryptobiosis. In this state tardigrades are able to survive extremely low and high temperatures and atmospheric pressures, complete lack of water, high doses of radiation, high concentrations of toxins and even a cosmic vacuum. The cellular mechanisms enabling cryptobiosis are poorly understood, although it appears the synthesis of certain types of molecules (sugars and proteins) enable the prevention of cellular damage at different levels. The endoplasmic reticulum (ER) is a morphologically and functionally diverse organelle able to integrate multiple extracellular and internal signals and generate adaptive cellular responses. However, the ER morphology and activity in the case of tardigrades has been studied rarely and in the context of oogenesis, functioning of the digestive system, and in the role and function of storage cells. Thus, there are no direct studies on the contribution of the ER in the ability of this organism to cope with environmental stress during cryptobiosis. Nevertheless, it is highly probable that the ER has a crucial role in this uncommon process. Since water bears are easy to handle laboratory animals, they may represent an ideal model organism to uncover the important role of the ER in the cell response to extreme environmental stress conditions.

Keywords Anhydrobiosis · Cryptobiosis · Cytoprotective strategies · Endoplasmic reticulum · Environmental stress · Extremophiles · Water bears

---

Ł. Kaczmarek (✉)
Department of Animal Taxonomy and Ecology, Faculty of Biology, Adam Mickiewicz University, Poznań, Poland
e-mail: kaczmar@amu.edu.pl

L. B. Agellon, M. Michalak (eds.), *Cellular Biology of the Endoplasmic Reticulum*, Progress in Molecular and Subcellular Biology 59,
https://doi.org/10.1007/978-3-030-67696-4_14

14.1 Introduction: Tardigrada

Tardigrades are a great example of animals responding in a very unique way to different kinds of environmental stresses. Although we know quite a lot about the stress response at the level of entire tardigrade organism, we know very little how it works at the cellular and molecular level. These small invertebrates have typical intracellular organelles, including endoplasmic reticulum (ER), but there are no studies to date on ER stress in tardigrades. In this paper, I want to suggest tardigrades as an excellent emerging animal model to study their ER stress response at the molecular level for a better understanding of the unique protection mechanisms to extreme environmental and nutritional stresses.

Tardigrada are commonly called water bears due to their bear-like appearance. They were first observed in 1773 and officially described in 1776. Tardigrades are most closely related to Arthropoda and Onychophora and these three phyla are included in the monophyletic taxon Panarthropoda within the Ecdysozoa group of moulting animals (Nelson et al. 2015). Today more than 1300 species of tardigrades subdivided into two classes have been described, which include Eutardigrada (Apochela and Parachela) (Figs. 14.1a, 14.2a) and Heterotardigrada (Fig. 14.1b) (Degma et al. 2020). The lifespan of active tardigrades range from one month to 2–3 years, although their lifespan may be greatly extended by periods of cryptobiosis (Jönsson and Bertolani 2001; Rebecchi et al. 2007; Jørgensen et al. 2007; Tsujimoto et al. 2016). Water bears inhabit marine, freshwater, and terrestrial habitats and can be found in the deepest oceans and on the highest mountain ranges, from polar regions to the tropics. Terrestrial tardigrades inhabit mosses, liverworts, cushion-shaped flowering plants, lichens, soil and leaf litter in xerothermic, moist and very wet habitats. Freshwater and marine species can be found in sand, sediments, cryoconite holes on glaciers, on aquatic plants, and occasionally in the water column. Tardigrade densities vary widely, from a few to several hundred specimens per 1 cm^2 (Jönsson 2003). Moreover, densities of aquatic species fluctuate seasonally with the highest densities in autumn to spring and the lowest in the summer

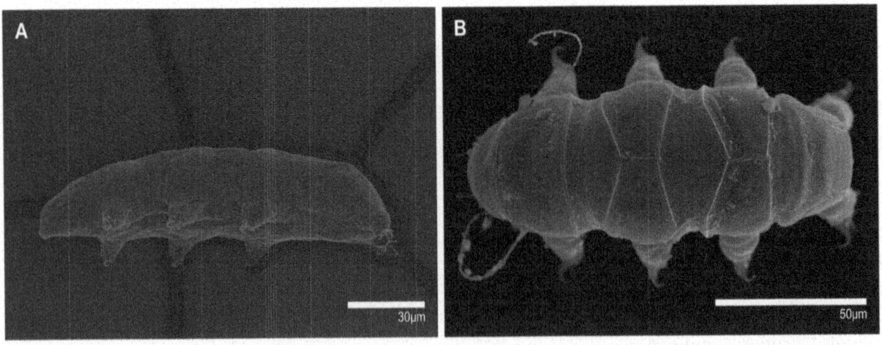

Fig. 14.1 Adult tardigrades: (**a**) *Diphascon* sp. (Eutardigrada; Parachela); (**b**) *Claxtonia wendti* (Richters 1903) (Heterotardigrada). SEM

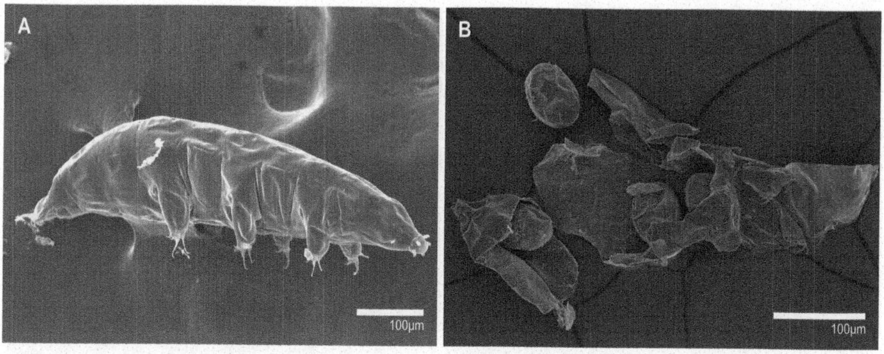

Fig. 14.2 (**a**) *Milnesium* sp. (Eutardigrada; Apochela); (**b**) smooth eggs in exuviae (*Milnesium* sp.) (Eutardigrada; Apochela). SEM

(e.g. Kathman and Nelson 1987; Jakubowska-Krępska et al. 2018). Tardigrada in general have a worldwide distribution. It may be anticipated that with water bears being such small organisms they should not have a true biogeography (e.g. the everything small is everywhere hypothesis) (Fontaneto 2011). Some species are considered to be cosmopolitan (e.g. Kaczmarek et al. 2020a), whereas others have a surprisingly narrow environmental tolerance and are rare or endemic. Many terrestrial and freshwater species have very restricted distributions and show significant biogeographical patterns (e.g. Pilato and Binda 2001; Kaczmarek et al. 2014, 2015a, b, 2016; McInnes et al. 2017; Gąsiorek et al. 2019a).

Tardigrade taxonomy is based mainly on the morphology of their buccal–pharyngeal apparatuses, claws, cuticle structure and egg shell. The buccal–pharyngeal apparatus is important from taxonomic point of view only in the class Eutardigrada. Most important is the structure of the buccal tube, oral cavity armature, structure and number of peribuccal structures, shape of stylet sheets and number of placoids. In the case of claws the most important are their number, structure, size, structure of lunules and presence or absence of cuticular bars close to the claws. Tardigrade cuticle may be smooth, sculptured (Eutardigrada), or thickened into cuticular plates with sculpture (Heterotardigrada). Moreover, in Heterotardigrada, the cuticle can be equipped with a different number of spines or filaments on the dorsal and/or later side of the body. In Eutardigrada the cuticle can be covered by pores, granulation, reticulation, gibbosities or spines. In many eutardigrades the structure of the egg chorion is one of the most taxonomically important character. Eggs can be covered by different types of egg processes, sculpture and areolation (Figs. 14.3a-d) (e.g. Kristensen 1987; Pilato and Binda 2010; Nelson et al. 2015; Gąsiorek et al. 2019b).

Tardigrades have bilaterally symmetrical body and four pairs of legs usually terminating with claws. The body of water bears is divided into a head segment and four trunk segments each bearing a pair of legs. Tardigrades range from 50–1200 µm in length. Internal organs are quite complex with well-developed digestive, nervous and reproductive systems. Because of miniaturization, respiratory and circulatory

Fig. 14.3 Ornamented tardigrade eggs: (**a**) *Mesobiotus* cf. *harmsworthi* (Eutardigrada; Parachela); (**b**) *Macrobiotus* cf. *hufelandi* (Eutardigrada; Parachela); (**c**) *Paramacrobiotus* sp. (Eutardigrada; Parachela); (**d**) *Mesobiotus* sp. (Eutardigrada; Parachela). SEM

systems are absent, and respiration occurs through the cuticle while circulation is accomplished by the movement of fluid and storage cells in the body cavity when the animal moves (Nelson et al. 2015).

Tardigrades can be carnivorous, herbivorous, bacteriovorous or omnivorous (e.g. Schill et al. 2011; Bryndová et al. 2020). Their digestive system starts with the buccal–pharyngeal apparatus with pharynx and buccal glands (foregut), later is divided into a midgut and a hindgut and finishes with an anus or cloaca. Excretion and osmoregulation is strictly connected to the cuticle, and/or Malpighian organs (Avdonina et al. 2007; Biserova and Mustafina 2015; Nelson et al. 2015).

The nervous system consists of a dorsal brain, circumesophageal connective around the buccal tube, a subesophageal ganglion, circumpharyngeal connective around the pharynx, and four ventral ganglia joined by paired nerve tracts. Sensory organs are represented by simple photoreceptors (eyespots), cephalic sensory organs or oral papillae (e.g. Greven 2007; Persson et al. 2012; Mayer et al. 2013; Schulze and Schmidt-Rhaesa 2013).

The tardigrade reproductive system is composed of a single dorsal-caudal gonad (in female), one or two gonoducts (in male) and one gonopore or cloaca. Additionally, many seminal receptacles and seminal vesicles are also present especially in marine species (Nelson et al. 2015; Poprawa and Janelt 2019). Sexual dimorphism is well visible in Heterotardigrada and Apochela, but poorly marked in Parachela (e.g. Ciobanu et al. 2015; Hansen et al. 2016; Gąsiorek et al. 2020). Heterotardigrada as well as many Eutardigrada produce smooth oval eggs with a thin egg capsule and deposit them in the moulted cuticle (Fig. 14.2b). The other Eutardigrada e.g. from the genera *Macrobiotus*, *Paramacrobiotus*, *Minibiotus* and *Mesobiotus* produce ornamented eggs which are laid directly to the environment (singly or in small groups) (Figs. 14.3a-d). Parental care has been observed in only a few freshwater species. Eggs are found in mosses, lichens, leaf litter, sediments and soil during all seasons of the year. The embryonic development varies in different species and according to the temperature, and takes from 5 to more than 100 days. Postembryonic development takes place by moulting and growth of the individual cells. This means that the number of cells in juveniles and adults is the same. Larvae, juveniles and adult eutardigrades are similar in morphology, but heterotardigrades and some eutardigrades exhibit some morphological differences especially in the claws and filaments (e.g. Hansen et al. 2016; Surmacz et al. 2020). Moulting usually occurs every 5–30 days throughout the life of tardigrades, and is dependent on food availability. Marine tardigrades are mainly gonochoric (bisexual) or very rarely hermaphroditic, parthenogenetic species are absent. Terrestrial and freshwater species can be gonochoric, unisexual (parthenogenetic) and sporadically hermaphroditic. In some species, males are present only during some seasons of the year and for the other periods females reproduce parthenogenetically (e.g. Bertolani 2001; Nelson et al. 2015; Poprawa and Janelt 2019; Sugiura et al. 2019; Kaczmarek et al. 2020b).

A unique process, referred to as cyclomorphosis, is observed in the marine eutardigrade *Halobiotus crispae* Kristensen 1982. Cyclomorphosis is an annual cycle of morphological and physiological changes which probably occurs in response to changing environmental conditions like temperature or oxygen level (Halberg et al. 2009).

14.2 Cryptobiosis

One of the best known adaptation of tardigrades to fast changing environmental conditions is cryptobiosis. At present, four types of cryptobiosis have been described, i.e. anhydrobiosis, anoxybiosis, cryobiosis and osmobiosis (e.g. Guidetti et al. 2011b).

In general, all tardigrades are aquatic, which means that they need a film of water surrounding their body to be active. The lack of water in the environment induces anhydrobiosis which is the best known form of cryptobiosis (for review see e.g. Rebecchi et al. 2007; Hengherr et al. 2008a; Wełnicz et al. 2011). Most, if not all, terrestrial tardigrades are capable of undergoing anhydrobiosis in all life stages, and can survive in this state for up to 20 years (Jönsson and Bertolani 2001; Jørgensen et al. 2007). In contrast, many marine and true freshwater tardigrades lack this ability, but experimental proof is scarce. In the anhydrobiotic state tardigrades lose up to 95% of water, strongly reduce or suspend their metabolism, contract their body, retract the head and legs, and form an immobile barrel-shaped tun. Survival in this state is correlated with the synthesis of different types of cell protectants (see below). In this state, tardigrades are highly resistant to extreme environmental conditions (see below). When environmental conditions return to normal and liquid water is again available tardigrades can return to full activity within a few minutes. The lifespan of anhydrobiotic tardigrades can be extended since during anhydrobiosis they do not age, in which support of the "Sleeping Beauty" hypothesis (for review see Kaczmarek et al. 2019).

Anoxybiosis is induced by the absence of oxygen or low oxygen levels in the water surrounding the tardigrade body (Clegg 2001). In this state tardigrades become immobile, rigid and transparent due to water absorption resulting from loss of osmotic control. In most cases, in the anoxybiotic state, tardigrades are able to survive from a few hours to a few days. This strategy is used e.g. by Antarctic freshwater tardigrades to survive winters in deep anoxic water (Nelson et al. 2015). The metabolic rate in anoxybiosis is unknown.

Cryobiosis is induced by very low temperature (below 0 °C), and this form of cryptobiosis enables tardigrades to survive freezing and thawing (Wright 2001). During cryobiosis tardigrades also form a barrel-shaped tun and synthetize different types of cell protectants (see below). They are able to survive in a frozen state up to 30 years (Tsujimoto et al. 2016).

Osmobiosis is the tardigrade response to changes in the osmotic pressure. Some tardigrades in this state tolerate variations in salinity. Terrestrial and freshwater tardigrades placed in salt solutions form tuns and can survive in this state up to 24 h. In contrast, some marine heterotardigrades are able to survive in an osmobiotic state while in freshwater for up to three days. When the appropriate osmotic pressure returns tardigrades became active again (Nelson et al. 2015; Heidemann et al. 2016).

14.3 Resistance to Extreme Environmental Conditions

In cryptobiosis, some species of tardigrades are very resistant to extreme temperatures. Currently, they are able to survive from −272.8 °C (Becquerel 1950) to ca. 60 °C (Guidetti et al. 2008; Ramløv and Westh 2001), and even up to 15 min at 150 °C (Rahm 1923, 1924, 1926). Cryobiotic and anhydrobiotic tardigrade specimens are able to withstand immersion in liquid air (ca. -190 °C), liquid nitrogen (ca. -253 °C) and liquid helium (ca. -272 °C) (Rahm 1923, 1924, 1926). In more recent studies, arctic tardigrades inhabiting soil were able to survive up to six years at −80 °C (Newsham et al. 2006). Interestingly, the longest duration of cryoprotective dehydration after which tardigrades were able to return to active metabolic functioning, was recently extended to more than 30 years. Tsujimoto et al. (2016) examined a frozen moss sample from the Antarctica stored at −20 °C for more than 30 years, and found active specimens of *Acutuncus antarcticus* (Richters 1904). Water bears also exhibit a significant resistance to a very wide range of atmospheric pressures (from

200–280 hPa to 7500 MPa [which equals 74,019 atmospheres]), which suggests that they can survive a meteorite impact (Jönsson et al. 2008; Ono et al. 2008). In the case of high doses of ionising radiation and X-rays, tardigrades are able to withstand between 5000–8000 Gy (May et al. 1964; Horikawa et al. 2006), or even between 1000–9000 Gy of gamma radiation (Jönsson et al. 2005). Tardigrada species are also resistant to high doses of UV radiation, and are able to survive doses up to 88 kJ/m^2 (Altiero et al. 2011). Moreover, some tardigrades are also resistant to chemical toxins such as hydrogen sulphide, carbon dioxide, ethanol (for ca. 10 min), 1-hexanol, or biocide methyl bromide gas (Baumann 1922; Jönsson and Guidetti 2001; Ramløv and Westh 2001). Even more fascinating, tardigrades are able to survive in outer space (space vacuum) at low earth orbit (for review see Guidetti et al. 2012; Erdmann and Kaczmarek 2016). Such extreme resistance of tardigrades to environmental stresses, while in the anhydrobiotic state, can be connected with their low rate of metabolism, and an almost lack of internal water (Wright 2001). However, tardigrades are also very resistant in their active state unlike other multicellular extremophiles. Active tardigrades can survive in temperatures from −196 °C to 38 °C (Ramløv and Westh 2001; Li and Wang 2005; Rebecchi et al. 2009a). They are also significantly resistant to high atmospheric pressures (up to 100 MPa) (Seki and Toyoshima 1998), and their resistance to radiation while in the active state is similar to their anhydrobiotic state (Jönsson et al. 2005; Horikawa et al. 2006, 2009; Altiero et al. 2011). This list includes only the extremes in conditions that tardigrades are known to withstand, but the true list of extreme stressors is probably much longer.

14.4 Protective Mechanisms

Tardigrades have morphological, behavioural and cytoprotective strategies to survive extreme environmental conditions. Morphological and behavioural strategies includes small body size, lack of aging when in anhydrobiosis, and different thicknesses of the cuticle that help anterior-posterior body contraction, reduction of the cuticle permeability and tun formation (see e.g. Jönsson 2001; Rebecchi et al. 2007).

The cytoprotective strategies include synthesis of different types of molecules to prevent damage to the membrane and macromolecules by stabilizing their structures or scavenging free radicals. For example, trehalose, a non-reducing disaccharide, is known to counteract the loss of water and protect cell organelles (e.g. Hengherr et al. 2008b). Trehalose replaces water and protects biomolecules and the integrity of membranes during dehydration (Crowe et al. 1984). Moreover, trehalose is also responsible for formation of a glassy matrix that reduces the rates of chemical reactions and inhibits free radical release. However, it should be remembered that there are a large number of both known and still unknown substances capable of replacing water in anhydrobiotic tardigrades (e.g. Halberg et al. 2013; Wharton 2015).

Other molecules involved in cytoprotection include unique heat-soluble proteins, responsible for cell protection and are defined as intrinsically disordered proteins. They are represented by heat shock proteins (HSP), cytoplasmic abundant heat soluble (CAHS) proteins, mitochondrial abundant heat soluble (MAHS) proteins, secretory abundant heat soluble (SAHS) proteins and hydrophilic late embryogenesis abundant (LEA) proteins. The synthesis of different classes of HSP was reported under stress conditions for many invertebrates including tardigrades (e.g. Boothby et al. 2017). Heat shock proteins are highly conservative and work as molecular chaperones supporting the repair of misfolded proteins. The CAHS, MAHS, and SAHS proteins appear to be tardigrade-specific proteins as they have only been observed in tardigrades (Tanaka et al. 2015). The MAHS proteins are suggested to be mitochondrial-specific protectants, whereas CAHS and SAHS function as molecular shields under dry conditions, inside and outside cells, respectively. Recently, it was shown that silencing of CAHS and SAHS genes reduces tardigrade tolerance to desiccation. Conversely, the heterologous expression of CAHS in prokaryotic and eukaryotic cells bestow tolerance to desiccation (Boothby et al. 2017). Moreover, purified CAHS proteins are able to protect desiccation-sensitive proteins *in vitro*. It has been proposed that under desiccation these proteins form a glassy matrix, the integrity of which is correlated with their protective abilities (Boothby et al. 2017). It was recently shown that the distribution of CAHS, MAHS and SAHS and encoding genes is scattered among tardigrades, suggesting species-specific adaptations (e.g. Yoshida et al. 2017). The LEA proteins are hydrophilic and probably have a similar protective function as non-reducing disaccharides in sequestering ions and acting as a hydration buffer (e.g. Tunnacliffe and Wise 2007). However, LEA proteins are not common in all tardigrades (Wang et al. 2014b; Tanaka et al. 2015). Additionally, Hashimoto et al. (2016) showed that the novel unique eutardigrade (*Ramazzottius varieornatus* Bertolani and Kinchin 1993) protein damage suppressor (Dsup) increases radiotolerance in tardigrades. Chavez et al. (2019) found that Dsup is a nucleosome-binding protein that protects the chromatin from oxidative damage and thus promotes survival of tardigrades under various conditions. The Dsup has also been shown to confer radiotolerance to cultured human cells, decreasing X-ray-induced DNA damage by up to ca. 40%. Tanaka et al. (2015), demonstrated that tardigrade RvLEAM and MAHS proteins increased hyperosmotic tolerance of cultured human cells. Such a finding indicates that it is likely many other unique tardigrade proteins might be able to protect human cells against desiccation, radiation, and other stress factors (Tanaka et al. 2015). Kamilari et al. (2019) showed that cytoprotectants like CAHS, SAHS, MAHS and RvLEAM are missing in heterotardigrades. For example, strong cryptobiont marine species *Echiniscoides* cf. *sigismundi* is depleted of the mentioned protein encoding genes which means that cryptobiosis in general cannot be attributed solely to these proteins. Accordingly, they found in this species a unique and highly expressed cold shock protein domain (CSP) and hypothesized that the CSP acts as RNA-chaperone involved in regulation of translation following cryobiosis (Kamilari et al. 2019).

The other cytoprotective strategy proposed for anhydrobiotic tardigrade species is removal of damaged or dysfunctional molecules. This removal can be realized e.g. via ubiquitin-mediated protein degradation. It should be mentioned that protein degradation is one of the functions of the ER (see below). Expression of the ubiquitin-encoding gene in tardigrades is up-regulated during rehydration (Wang et al. 2014b).

An efficient cytoprotective strategy may also include an increased tolerance to oxidative stress caused by reactive oxygen species (ROS), which seems to be very important during anhydrobiosis (e.g. Rebecchi 2013). The ROS cause numerous oxidative modifications and damage of cellular molecules such as nucleic acids, proteins and lipids (Hansen et al. 2006). Protection against this damage can be supported by antioxidant mechanisms classified as primary and secondary elements of the defence systems (e.g. Wełnicz et al. 2011; Rebecchi 2013). The primary system precludes oxidative damage directly by intercepting ROS before they damage cellular molecules and is mediated by a few well-known extracellular and intracellular enzymes, including superoxide dismutases and glutathione-related enzymes, as well as various catalases, peroxidases, transferases, reductases and glutaredoxins. The secondary system includes enzymes which are able to repair ROS-modified molecules (e.g. Radak et al. 2011). In tardigrades, it has been reported that expression of DNA-repair proteins is induced by desiccation, and these proteins are able to repair DNA through translesion synthesis, base excision repair, mismatch repair, nucleotide excision repair and photoactivation (Wang et al. 2014b). These two systems are supplemented by low-molecular weight compounds, which include ascorbic acid, bilirubin, glutathione, coenzyme Q and vitamin precursors, such as carotenoids and tocopherols (e.g. Rebecchi 2013; Niki 2014).

We know much about the tardigrade's ability to undergo anhydrobiosis, but we are far from understanding the processes involved at the molecular and cellular levels. The mechanisms involved in anhydrobiosis are probably also linked to the processes of aging, as increased longevity may be a result of resistance against external environmental stressors (e.g. Clegg 2001; Wełnicz et al. 2011; Kaczmarek et al. 2019).

A better understanding of protective strategies which are used by anhydrobiotic animals may provide important answers in many branches of applied sciences, like space biology, medicine or biotechnology. The knowledge may be can applied, for example, to the preservation of biological materials for transplantation, food production, enzymes working in a small amount of water, mechanisms of DNA protection and repair, or even dry vaccines (e.g. Bradbury 2001; Schill et al. 2009; Jönsson 2019).

14.5 Endoplasmic Reticulum, Stress Response and Tardigrades

The ER is a cellular organelle present in two main forms, the smooth endoplasmic reticulum (SER) and rough endoplasmic reticulum (RER). The ER is found in almost all eukaryotic cells and forms an interconnected network of flattened, membrane-enclosed sacs termed cisternae (RER) and/or tubular structures (SER) (Goodman 2008). A specific form of SER is sarcoplasmic reticulum (SR) which is present in muscle cells.

The ER synthesis function includes, protein synthesis, folding and transport in vesicles to the Golgi apparatus (Goodman 2008). The correct folding of proteins is possible due to ER chaperone proteins including e.g. protein disulfide isomerase (PDI), the Hsp70 family member BiP/Grp78, calnexin, calreticulin, and the peptidylpropyl isomerase family. However, the ER is also responsible for degradation of unfolded or mutated proteins in the process of endoplasmic-reticulum-associated protein degradation (ERAD). The entire process is divided into three stages (i) recognition of misfolded or mutated proteins in the endoplasmic reticulum, (ii) retro-translocation into the cytosol and (iii) ubiquitin-dependent degradation by the proteasome. Correctly folded proteins are transported from the RER to the Golgi apparatus, whereas, unfolded or mis-folded proteins cause a stress response in the ER [so called the unfolded protein response (UPR)]. Transmembrane receptor proteins like inositol-requiring enzyme 1 (IRE1), protein kinase R (PKR)-like endoplasmic reticulum kinase (PERK) and activating transcription factor 6 (ATF6) are very important proteins in the signaling pathways of the UPR. The PERK is a type I membrane protein which is induced by ER stress that is caused by mis-folded proteins (e.g. Harding et al. 1999) and plays very important role in the ability of cells to survive during these stressful conditions (Harding et al. 2000). The IRE1 activates the UPR and, in mammalian, cells promotes cell survival. It can also initiate apoptosis due to decay of antiapoptotic miRNAs (e.g. Chen and Brandizzi 2013). During accumulation of misfolded proteins ATF6 is proteolytically cleaved. The cytosolic portion of ATF6 is then moved to the nucleus and acts as a transcription factor which causes the transcription of ER chaperones (McKimpson and Kitsis 2017). Also, serious disturbances in calcium regulation, redox regulation, glucose deprivation, viral infection, or the over-expression of proteins can induce the ER stress response (Xu et al. 2005; Kober et al. 2012). This, however, slows down folding of proteins and increases the amounts of unfolded proteins. Such a stress may contribute to damage in hypoxia/ischemia, insulin resistance, and other disorders (Ozcan et al. 2004). As a result, the UPR plays a crucial role in the pathophysiology of the heart, but modulation may also be promising in anti-cancer therapy (Zhao and Ackerman 2006; Groenendyk et al. 2010; Wang et al. 2014a; Dicks et al. 2015).

The SER also participates in lipid and steroids synthesis, transformation of carbohydrates and inactivation of toxins and drugs (Alberts et al. 2002; Goodman 2008; Nowakowska-Gołacka et al. 2019). It was also shown that some cytoprotectant LEA proteins (ArLEA1A and ArLEA1B), in bdelloid rotifers, are localised in the ER and provide partial protection against desiccation induced damage (Tripathi et al. 2012). Similarly, in the anhydrobiotic midge *Polypedilum vanderplanki* Hinton 1951, PvLEA3 proteins (chimeras combining LEA-like parts and transmembrane domains, shared with PvLIL proteins) are located in the ER and contribute to desiccation tolerance (Voronina et al. 2020). However, the exact role of LEA proteins targeted to the ER of anhydrobiotic animals still awaits its explanation (Menze and Hand 2009). Furthermore, an important role of the ER in desiccation tolerance was also discussed for the anhydrobiotic nematode *Panagrolaimus superbus* (Fuchs 1930). It was suggested that accumulation of unfolded proteins in the ER, arising from physiological or abiotic stress, leads to the expression of several protein folding chaperones, including HSP90 and HSP70 families and their co-chaperones (Tyson et al. 2012). One of the important pathway in the ER that is responsible for mis-folded protein degradation is ERAD. Cellular stress coping response strategies responding to internal stressors like accumulation of unfolded or mutated proteins, may also be important in responding to extreme external environmental stress including the lack of water, high doses of radiation, or extremes in temperatures.

To our knowledge studies focused strictly on ER (Figs. 14.4a-b) function in tardigrades has never been conducted. However, ER morphology and function has occasionally been investigated in research focused on oogenesis, functioning of the digestive system, and the role and function of storage cells. It has been shown for a few tardigrade species that the ER in the cells of the gonad wall is involved in synthesis of the precursors of egg shell (chorion) (Węglarska 1982; Poprawa 2005a, 2011; Poprawa et al. 2015b, c; Janelt et al. 2019). It also participates in the synthesis of precursors of yolk in oocytes and trophocytes. The ER morphology, especially RER, in germline cells undergo change in the course of oogenesis. During previtellogenesis single short cisterns of ER are present in the cell cytoplasm while during vitellogenesis the ER takes the form of a branched network of cisternae. Sometimes the cisternae form concentric systems (Węglarska 1979, 1987; Poprawa 2005b; Poprawa et al. 2015a, b, c; Janelt et al. 2019). Beside its role in the synthesis of storage material, the ER also participates in the synthesis of digestive enzymes and yolk precursors in digestive cells of the midgut. The ER is usually located in the basal and perinuclear parts of the cell but not in the apical part (Rost-Roszkowska et al. 2011, 2013, 2018, 2019, 2020). In storage cells, which are very important during tardigrade cryptobiosis, the ER is responsible for the synthesis of yolk precursors and reserve material being a source of energy (Węglarska 1975; Poprawa 2006; Hyra et al. 2016; Czerneková et al. 2018). The presence of active ER was also reported in tardigrade cells during anhydrobiosis and encystation, however, their strict function was not studied (Rost-Roszkowska and Poprawa 2008; Czerneková et al. 2017, 2018). In the tardigrade *Hys exemplaris* (Gąsiorek et al. 2018), a very dense RER network was observed in secretory cells, suggesting the existence of numerous secreted cytoprotectant proteins in the tuns (anhydrobiotic specimens) in comparison to hydrated specimens where RER was not as well developed (Richaud et al. 2020). Moreover, a raw searching of a database shared by Yoshida et al. (2017) concerning proteins differently expressed in active tardigrade specimens and

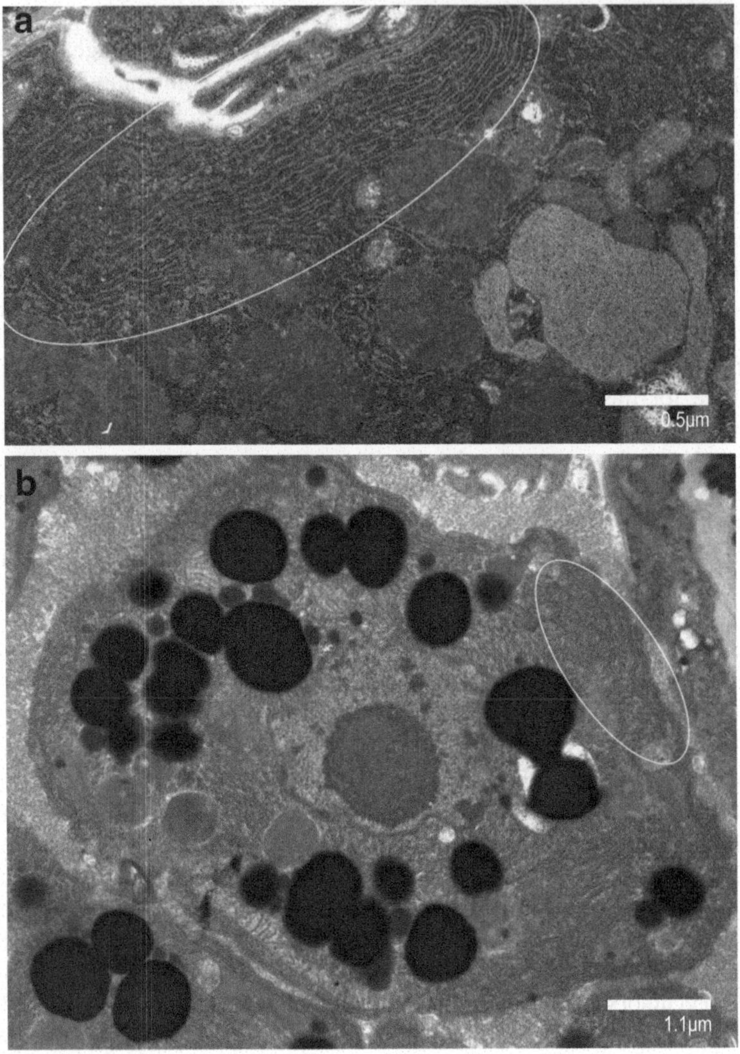

Fig. 14.4 (**a-b**) Endoplasmic reticulum in cells of *Hypsibius exemplaris* Gąsiorek et al. (2018). TEM

anhydrobiotic specimens (tuns) indicated the presence of ER proteins (e.g. reticulon) and proteins quantitatively modulated by ER-stress (e.g. lipocalin) as the proteins most highly expressed in tuns (Hanna Kmita pers. comm.). At present, little is known about the role and function of the ER in tardigrades, especially in the case of the environmental stress response. However, taking into consideration the role and function of the ER in eukaryotic cells, it is highly likely that the ER plays a crucial role in the tardigrade response to stress induced by extreme environmental conditions.

14.6 Tardigrades as an Animal Model to Study the ER Stress Response

Many species of tardigrades have been used in different types of experiments focused mainly on their ability to survive extreme conditions. A few years ago they were proposed as excellent model organisms for the study of space biology or astrobiology (e.g. Horikawa et al. 2008; Guidetti et al. 2012; Erdmann and Kaczmarek 2016). Until now, a few different tardigrade taxa, like *Hys. exemplaris* (former *Hys. dujardini* Doyère 1840), species of the *Macrobiotus hufelandi* group, *Milnesium inceptum* Morek et al. 2019 (former *Mil. tardigradum* Doyère 1840), species of the *Paramacrobiotus richtersi* group and *Pam. areolatus* group, *Pam. experimentalis* Kaczmarek et al. 2020b, *Ram. oberhaeuseri* (Doyère 1840), and *Ram. varieornatus*, *Richtersius coronifer* (Richters 1903) were used in research focused mainly on their extremophilic abilities (e.g. Jönsson et al. 2001; Hengherr et al. 2008a; Horikawa et al. 2008; Rebecchi et al. 2009b; Reuner et al. 2010; Altiero et al. 2011; Guidetti et al. 2011a; Persson et al. 2011; Horikawa et al. 2012; Czerneková et al. 2017).

In general, tardigrades have many features that make them ideal model organisms for studies focused on cytoprotective mechanisms, synthesis of unique proteins and other molecules, and to understand DNA protection and repair mechanisms. All tardigrade species are rather simple animals and small in size, but despite this, they have quite complex internal structures and their cellular architecture is the same as in other animals, including humans. Moreover, many water bear species are very easy to culture in the laboratory (e.g. Altiero and Rebecchi 2001; Suzuki 2003; Horikawa et al. 2008; Stec et al. 2015; Tsujimoto et al. 2016; Erdmann et al. 2017; McNuff 2018), because they need a little space and have minimal environmental and nutritional requirements. Nevertheless, different tardigrade species have different dietary requirements. They can be carnivorous, herbivorous, bacteriovorous or omnivorous. Some tardigrades are dioecious whereas others are unisexual, reproducing by parthenogenesis. This provides the opportunity to study mechanisms in males and females and look for potential differences. For parthenogenetic species it is very easy to establish isogenic strains that can produce large number of clonal specimens. Tardigrades are also characterized by a very diverse lifespan, from few weeks (e.g. *Hys. exemplaris* or *Acu. Antarcticus*) to even two-three years in some species of *Paramacrobiotus* or *Dactylobiotus*. Moreover, their total lifespan can be expanded by anhydrobiosis and this can be very useful for different types of short or long term studies. What is more, most tardigrade species develop very fast, reach adulthood very quickly and produce large number of eggs which hatch within a few to several days. Eutardigrades are very often transparent making it possible to observe internal processes *in vivo*. The entire and partial genomes of several Tardigrada species have been determined, which makes comparative studies possible at the molecular level. Currently, a large number of unique tardigrade proteins and other molecules which are considered cytoprotectants have been identified and studied. These proteins may be useful for biotechnological or medical research (e.g. Bradbury 2001; Schill et al. 2009; Jönsson 2019). Additionally, some tardigrade species are commercially available (e.g. Sciento (Manchester, UK)). Moreover, the ability for cryptobiosis in tardigrades provides the opportunity to study specimens in at least a few metabolic stages, i.e. cryotobiotic, anhydrobiotic, anoxybiotic or osmobiotic as well as in full active stage and all intermediate stages. It can be speculated that UPR signaling pathways in tardigrades are similar to those in other organisms, and general ER functions are likely the same. Notwithstanding, it is highly possible that new players, like e.g. the LEA proteins, may introduce a new, unique tardigrade regulation of evolutionally conserved signaling pathways (UPR) supporting extreme resistance to environmental or nutritional stresses in tardigrades or possibly also other anhydrobiotic animals like nematodes, rotifers or some arthropods. Additionally, Tardigrades tolerate high oxidative stress caused by ROS, but the mechanism of this tolerance is still not well understand. However, some research suggest an important role for ER localized PERK or/and ATF6, in the antioxidant response of cells (e.g. Verfaillie et al. 2012; McKimpson and Kitsis 2017). It is also possible that Dsup, which increases radiotolerance in tardigrades, by reduction of DNA damage, may help reduce stress-induced damage to the genome (Dicks et al. 2015).

What is more, it was shown, the same proteins or molecules synthetized by tardigrades work in human cells (Hashimoto et al. 2016), suggesting that understanding the ER role and function of the ER in tardigrades will provide answers concerning human ER involvement in environmental stress, and provide information enabling stress reduction or elimination.

14.7 Summary Points

- There is not much known about the role and functioning of the ER in tardigrades.
- A well-functioning ER plays a major role in appropriate cellular metabolism and understanding ER function in tardigrades.
- Tardigrades show high tolerance to non-physiological and abiotic stress conditions.
- Some tardigrade proteins can function within human cells and protect them against radiation and DNA damage.
- The use of tardigrades as model organisms may be useful in understanding human diseases caused by stress.

14.8 Future Issues

- Understanding the exact ER morphology in different types of tardigrade cells.
- Understanding the role of the ER in tardigrades, especially in response to environmental stress.

- Understanding proteins and other molecules synthesized by tardigrades and their role in the ER response to environmental stress.
- Understanding the specific pathways involved in assisting tardigrades cope with physiological and environmental stress.
- To determine whether similar mechanisms function in the cells of other animals, including humans.
- To determine whether the results obtained in research on tardigrades can be used to identify new medical treatments for diseases caused by inappropriate ER functioning.

Acknowledgements I want to thank to Professor Izabela Poprawa (Silesian University, Katowice, Poland) and Professor Hanna Kmita (Adam Mickiewicz University in Poznań, Poznań, Poland).

I also want thank to Professor Izabela Poprawa, Msc Milena Roszkowska (Adam Mickiewicz University in Poznań, Poznań, Poland), Dr. Magdalena Gawlak (Institute of Plant Protection – National Research Institute, Poznań, Poland), Professor Łukasz Michalczyk (Jagiellonian University, Kraków, Poland) and MSc student Tomasz Bartylak (Adam Mickiewicz University in Poznań, Poznań, Poland) for the preparation of microphotographs and figures for this paper.

The studies were conducted in the framework of activities of BARg (Biodiversity and Astrobiology Research group).

References

Alberts B, Johnson A, Lewis J et al (2002) Molecular biology of the cell. In: The endoplasmic reticulum, 4th edn. Garland Science, New York. Available from: https://www.ncbi.nlm.nih.gov/books/NBK26841/

Altiero T, Guidetti R, Caselli V, Cesari M, Rebecchi L (2011) Ultraviolet radiation tolerance in hydrated and desiccated eutardigrades. J Zool Syst Evol Res 49:104–110. https://doi.org/10.1111/j.1439-0469.2010.00607.x

Altiero T, Rebecchi L (2001) Rearing tardigrades: results and problems. Zool Anz 240:217–221. https://doi.org/10.1078/0044-5231-00028

Avdonina AM, Biserova NM, Bertolani R, Rebecchi L (2007) Ultrastructure of the digestive system of *Ramazzottius tribulosus* and *Macrobiotus richtersi* (Eutardigrada) in relationship with diet. J Limnol 66:5–11. https://doi.org/10.4081/jlimnol.2007.s1.5

Baumann H (1922) Die Anabiose der Tardigraden. Zool Jahrb 45:501–556

Becquerel P (1950) La suspension de la vie au desessous de 1/20 K absolu par demagnetisation adiabatique de l'alun de fer dans le vide les plus elève. C R Acad Séan Paris 231:261–263

Bertolani R (2001) Evolution of the reproductive mechanisms in tardigrades – a review. Zool Anz 240(3–4):247–252. https://doi.org/10.1078/0044-5231-00032

Bertolani R, Kinchin IM (1993) A new species of *Ramazzottius* (Tardigrada, Hypsibiidae) in a rain gutter sediment from England. Zool J Linnean Soc 109:327–333. https://doi.org/10.1111/j.1096-3642.1993.tb02538.x

Biserova NM, Mustafina AR (2015) Comparative midgut ultrastructure in three species of Tardigrada. Arthropoda Sel 24(4):373–385. https://doi.org/10.15298/arthsel.24.4.01

Boothby TC, Tapia H, Brozena AH, Piszkiewicz S, Smith AE, Giovannini I, Rebecchi L, Pielak GJ, Koshland D, Goldstein B (2017) Tardigrades use intrinsically disordered proteins to survive desiccation. Mol Cell 65:975–984. https://doi.org/10.1016/j.molcel.2017.02.018

Bradbury J (2001) Of tardigrades, trehalose, and tissue engineering. Lancet 358(9279):392. https://doi.org/10.1016/S0140-6736(01)05595-7

Bryndová M, Stec D, Schill RO, Michalczyk Ł, Devetter M (2020) Tardigrade dietary preferences and diet effects on tardigrade life history traits. Zool J Linnean Soc 188(3):865–877. https://doi.org/10.1093/zoolinnean/zlz146

Chavez C, Cruz-Becerra G, Fei J, Kassavetis GA, Kadonaga JT (2019) The tardigrade damage suppressor protein binds to nucleosomes and protects DNA from hydroxyl radicals. eLife 8: e47682. https://doi.org/10.7554/eLife.47682

Chen Y, Brandizzi F (2013) IRE1: ER stress sensor and cell fate executor. Trends Cell Biol 23 (11):547–555. https://doi.org/10.1016/j.tcb.2013.06.005

Ciobanu DA, Roszkowska M, Kaczmarek Ł (2015) Two new tardigrade species from Romania (Eutardigrada: Milnesiidae, Macrobiotidae), with some remarks on secondary sex characters in *Milnesium dornensis* sp. nov. Zootaxa 3941:542–564. https://doi.org/10.11646/zootaxa.3941.4.4

Clegg JS (2001) Cryptobiosis – a peculiar state of biological organization. Comp Biochem Physiol B128:613–624. https://doi.org/10.1016/S1096-4959(01)00300-1

Crowe JH, Crowe LM, Chapman D (1984) Preservation of membranes in anhydrobiotic organisms: the role of trehalose. Science 223:701–703. https://doi.org/10.1126/science.223.4637.701

Czerneková M, Janelt K, Student S, Jönsson KI, Poprawa I (2018) A comparative ultrastructure study of storage cells in the eutardigrade *Richtersius coronifer* in the hydrated state and after desiccation and heating stress. PLoS One 13(8):e0201430. https://doi.org/10.1371/journal.pone.0201430

Czerneková M, Jönsson KI, Chajec L, Student S, Poprawa I (2017) The structure of the desiccated *Richtersius coronifer* (Richters, 1903). Protoplasma 254:1367–1377. https://doi.org/10.1007/s00709-016-1027-2

Degma P, Bertolani R, Guidetti R (2020) Actual checklist of Tardigrada species. (2009–2020, 37th Edition: 08-07-2020). https://iris.unimore.it/retrieve/270444/Actual%20checklist%20of%20Tardigrada%2037th%20Edition%2008_07_20.pdf

Dicks N, Gutierrez K, Michalak M, Bordignon V, Agellon LB (2015) Endoplasmic reticulum stress, genome damage, and cancer. Front Oncol 5:11. https://doi.org/10.3389/fonc.2015.00011

Doyère PLN (1840) Mémoire sur les Tardigrades. Ann Sci Nat Zool Paris Ser 2(14):269–362

Erdmann W, Idzikowski B, Kowalski W, Szymański B, Kosicki JZ, Kaczmarek Ł (2017) Can the tardigrade *Hypsibius dujardini* survive in the absence of the geomagnetic field? PLoS One 12 (9):e0183380. https://doi.org/10.1371/journal.pone.0183380

Erdmann W, Kaczmarek Ł (2016) Tardigrades in space research – past and future. Orig Life Evol Biospheres 47:545–553. https://doi.org/10.1007/s11084-016-9522-1

Fontaneto D (2011) Biogeography of microscopic organisms: is everything small everywhere? Cambridge University Press, Cambridge, pp 1–365. https://doi.org/10.1111/boj.12137

Fuchs AG (1930) Neue an Borken- und Russelkäfer gebundene Nematoden, halb-parasitische und Wohnungseinmieter. Freilebende Nematoden aus Moos und Walderde in Borken- und Russelkäfergängen. Zool Jahrb Abt Syst Ökol Geogr Tiere 59:505–646

Gąsiorek P, Stec D, Morek W, Michalczyk Ł (2018) An integrative redescription of *Hypsibius dujardini* (Doyère, 1840), the nominal taxon for Hypsibioidea (Tardigrada: Eutardigrada). Zootaxa 4415(1):45–75. https://doi.org/10.11646/zootaxa.4415.1.2

Gąsiorek P, Vončina K, Michalczyk Ł (2019a) *Echiniscus testudo* (Doyère, 1840) in New Zealand: anthropogenic dispersal or evidence for the "everything is everywhere" hypothesis? New Zeal J Zool 46(2):174–181. https://doi.org/10.1080/03014223.2018.1503607

Gąsiorek P, Stec D, Morek W, Michalczyk Ł (2019b) Deceptive conservatism of claws: distinct phyletic lineages concealed within Isohypsibioidea (Eutardigrada) revealed by molecular and morphological evidence. Contrib Zool 88(1):78–132. https://doi.org/10.1163/18759866-20191350

Gąsiorek P, Vončina K, Michalczyk Ł (2020) An overview of the sexual dimorphism in *Echiniscus* (Heterotardigrada, Echiniscoidea), with the description of *Echiniscus masculinus* sp. nov. (the *virginicus* complex) from Borneo. Zoosyst Evol 96(1):103–113. https://doi.org/10.3897/zse.96.49989

Goodman SR (2008) Medical Cell Biology (3rd). Academic. ISBN 9780080919317

Greven H (2007) Comments on the eyes of tardigrades. Arthropod Struct Dev 36(4):401–407. https://doi.org/10.1016/j.asd.2007.06.003

Groenendyk J, Sreenivasaiah PK, Kim DH, Agellon LB, Michalak M (2010) Biology of endoplasmic reticulum stress in the heart. Circ Res 107(10):1185–1197. https://doi.org/10.1161/CIRCRESAHA.110.227033

Guidetti R, Altiero T, Bertolani R, Grazioso P, Rebecchi L (2011a) Survival of freezing by hydrated tardigrades inhabiting terrestrial and freshwater habitats. Zoology 114:123–128. https://doi.org/10.1016/j.zool.2010.11.005

Guidetti R, Altiero T, Rebecchi L (2011b) On dormancy strategies in tardigrades. J Insect Physiol 57:567–576. https://doi.org/10.1016/j.jinsphys.2011.03.003

Guidetti R, Boschini D, Altiero T, Bertolani R, Rebecchi L (2008) Diapause in tardigrades: a study of factors involved in encystment. J Exp Biol 211:2296–2302. https://doi.org/10.1242/jeb.015131

Guidetti R, Rizzo AM, Altiero T, Rebecchi L (2012) What can we learn from the toughest animals of the earth? Water bears (tardigrades) as multicellular model organisms in order to perform scientific preparations for lunar exploration. Planet Space Sci 74:97–102. https://doi.org/10.1016/j.pss.2012.05.021

Halberg KA, Larsen KW, Jørgensen A, Ramløv H, Møbjerg N (2013) Inorganic ion composition in Tardigrada: cryptobionts contain a large fraction of unidentified organic solutes. J Exp Biol 216:1235–1243. https://doi.org/10.1242/jeb.075531

Halberg KA, Persson D, Ramløv H, Westh P, Møbjerg-Kristensen R, Møbjerg N (2009) Cyclomorphosis in Tardigrada: adaption to environmental constrains. J Exp Biol 212:2803–2811. https://doi.org/10.1242/jeb.029413

Hansen JM, Go YM, Jones DP (2006) Nuclear and mitochondrial compartmentation of oxidative stress and redox signalling. Annu Rev Pharmacol Toxicol 46:215–234. https://doi.org/10.1146/annurev.pharmtox.46.120604.141122

Hansen JG, Kristensen RM, Jørgensen A, Accogli G, D'Addabbo R, Gallo M (2016) Postembryonic development, paedomorphosis, secondary sexual dimorphism and population structure of a new *Florarctus* species (Tardigrada, Heterotardigrada). Zool J Linnean Soc 178(4):871–877. https://doi.org/10.1111/zoj.12436

Harding HP, Zhang Y, Bertolotti A, Zeng H, Ron D (2000) Perk is essential for translational regulation and cell survival during the unfolded protein response. Mol Cell 5(5):897–904. https://doi.org/10.1016/s1097-2765(00)80330-5

Harding HP, Zhang Y, Ron D (1999) Protein translation and folding are coupled by an endoplasmic-reticulum-resident kinase. Nature 397(6716):271–274. https://doi.org/10.1038/16729

Hashimoto T, Horikawa DD, Saito Y, Kuwahara H, Kozuka-Hata H, Shin-I T, Minakuchi Y, Ohishi K, Motoyama A, Aizu T, Enomoto A, Kondo K, Tanaka S, Hara Y, Koshikawa S, Sagara H, Miura T, Yokobori S, Miyagawa K, Suzuki Y, Kubo T, Oyama M, Kohara Y, Fujiyama A, Arakawa K, Katayama T, Toyoda A, Kunieda T (2016) Extremotolerant tardigrade genome and improved radiotolerance of human cultured cells by tardigrade-unique protein. Nat Commun 7:12808. https://doi.org/10.1038/ncomms12808

Heidemann NWT, Smith DK, Hygum TL, Stapane L, Clausen LKB, Jørgensen A, Hélix-Nielsen C, Møbjerg N (2016) Osmotic stress tolerance in semi-terrestrial tardigrades. Zool J Linnean Soc 178(4):912–918. https://doi.org/10.1111/zoj.12502

Hengherr S, Brummer F, Schill RO (2008a) Anhydrobiosis in tardigrades and its effects on longevity traits. J Zool 275:216–220. https://doi.org/10.1111/j.1469-7998.2008.00427.x

Hengherr S, Heyer AG, Köhler HR, Schill RO (2008b) Trehalose and anhydrobiosis in tardigrades – evidence for divergence in responses to dehydration. FEBS J 275:281–288. https://doi.org/10.1111/j.1742-4658.2007.06198.x

Hinton HE (1951) A new chironomid from Africa, the larvae of which can be dehydrated without injury. Proc Zool Soc Lond 121:371–380. https://doi.org/10.1111/j.1096-3642.1951.tb00801.x

Horikawa DD, Iwata K, Kawai K, Koseki S, Okuda T, Yamamoto K (2009) High hydrostatic pressure tolerance of four different anhydrobiotic animal species. Zool Sci 26:238–242. https://doi.org/10.2108/zsj.26.238

Horikawa D, Kunieda T, Abe W, Watanabe M, Nakahara Y, Yukuhiro F, Sakashita T, Hamada N, Wada S, Funayama T, Katagiri C, Kobayashi Y, Higashi S, Okuda T (2008) Establishment of a rearing system of the extremotolerant tardigrade *Ramazzottius varieornatus*: a new model animal for astrobiology. Astrobiology 8:549–556. https://doi.org/10.1089/ast.2007.0139

Horikawa DD, Sakashita T, Katagiri C, Watanabe M, Kikawada T, Nakahara Y, Hamada N, Wada S, Funayama T, Higashi S, Kobayashi Y, Okuda T, Kuwabara M (2006) Radiation tolerance in the tardigrade *Milnesium tardigradum*. Int J Radiat Biol 82:843–848. https://doi.org/10.1080/09553000600972956

Horikawa DD, Yamaguchi A, Sakashita T, Tanaka D, Hamada N, Yukuhiro F, Kuwahara H, Kunieda T, Watanabe M, Nakahara Y, Wada S, Funayama T, Katagiri C, Higashi S, Yokobori S-I, Kuwabara M, Rothschild LJ, Okuda T, Hashimoto H, Kobayashi Y (2012) Tolerance of anhydrobiotic eggs of the tardigrade *Ramazzottius varieornatus* to extreme environments. Astrobiology 12(4):283–289. https://doi.org/10.1089/ast.2011.0669

Hyra M, Rost-Roszkowska MM, Student S, Włodarczyk A, Deperas M, Janelt K, Poprawa I (2016) Body cavity cells of Parachela during their active life. Zool J Linnean Soc 178:878–887. https://doi.org/10.1111/zoj.12463

Jakubowska-Krępska N, Gołdyn B, Krzemińska-Wowk P, Kaczmarek Ł (2018) Tardigrades as potential bioindicators in biological wastewater treatment plants. Eur J Ecol 4(2):124–130. https://doi.org/10.2478/eje-2018-0019

Janelt K, Jezierska M, Poprawa I (2019) The female reproductive system and oogenesis in *Thulinius ruffoi* (Tardigrada, Eutardigrada, Isohypsibiidae). Arthropod Struct Dev 50:53–63. https://doi.org/10.1016/j.asd.2019.04.003

Jönsson KI (2001) The nature of selection on anhydrobiotic capacity in tardigrades. Zool Anz 240 (3–4):409–417. https://doi.org/10.1078/0044-5231-00049

Jönsson KI (2003) Population density and species composition of moss-living tardigrades in a boreo-nemoral forest. Ecography 26:356–364. https://doi.org/10.1034/j.1600-0587.2003.03458.x

Jönsson KI (2019) Radiation tolerance in tardigrades: current knowledge and potential applications in medicine. Cancers 11(9):1333. https://doi.org/10.3390/cancers11091333

Jönsson KI, Bertolani R (2001) Facts and fiction about long-term survival in tardigrades. J Zool 255:121–123. https://doi.org/10.1017/S0952836901001169

Jönsson KI, Borsari S, Rebecchi L (2001) Anhydrobiotic survival in populations of the tardigrades *Richtersius coronifer* and *Ramazzottius oberhaeuseri* from Italy and Sweden. Zool Anz 240 (3–4):419–423. https://doi.org/10.1078/0044-5231-00050

Jönsson KI, Guidetti R (2001) Effects of methyl-bromide fumigation on anhydrobiotic micrometazoans. Ecotox Environ Safe 50:72–75. https://doi.org/10.1006/eesa.2001.2090

Jönsson KI, Harms-Ringdhal M, Torudd J (2005) Radiation tolerance in the eutardigrade *Richtersius coronifer*. Int J Radiation Biol 81:649–656. https://doi.org/10.1080/09553000500368453

Jönsson KI, Rabbow E, Schill RO, Harms-Ringdahl M, Rettberg P (2008) Tardigrades survive exposure to space in low earth orbit. Curr Biol 18:R729–R731. https://doi.org/10.1016/j.cub.2008.06.048

Jørgensen A, Møbjerg N, Kristensen RM (2007) A molecular study of the tardigrade *Echiniscus testudo* (Echiniscidae) reveals low DNA sequence diversity over a large geographical area. J Limnol 66:77–85. https://doi.org/10.4081/jlimnol.2007.s1.77

Kaczmarek Ł, Bartels PJ, Roszkowska M, Nelson DR (2015b) The zoogeography of marine tardigrades. Zootaxa 4037:1–189. https://doi.org/10.11646/zootaxa.4037.1.1

Kaczmarek Ł, Michalczyk Ł, McInnes SJ (2014) Annotated zoogeography of non-marine Tardigrada. Part I: Central America. Zootaxa 3763(1):1–62. https://doi.org/10.11646/zootaxa.3763.1.1

Kaczmarek Ł, Michalczyk Ł, McInnes SJ (2015a) Annotated zoogeography of non-marine Tardigrada. Part II: South America. Zootaxa 3923(1):1–107. https://doi.org/10.11646/zootaxa.3923.1.1

Kaczmarek Ł, Michalczyk Ł, McInnes SJ (2016) Annotated zoogeography of non-marine Tardigrada. Part III: North America and Greenland. Zootaxa 4203(1):1–249. https://doi.org/10.11646/zootaxa.4203.1.1

Kaczmarek Ł, Mioduchowska M, Kačarević U, Kubska K, Parnikoza I, Gołdyn B, Roszkowska M (2020a) New records of Antarctic Tardigrada with comments on interpopulation variability of the *Paramacrobiotus fairbanksi* Schill, Förster, Dandekar and wolf, 2010. Diversity 12:108. https://doi.org/10.3390/d12030108

Kaczmarek Ł, Roszkowska M, Fontaneto D, Jezierska M, Pietrzak B, Wieczorek R, Poprawa I, Kosicki JZ, Karachitos A, Kmita H (2019) Staying young and fit? Ontogenetic and phylogenetic consequences of animal anhydrobiosis. J Zool 309(1):1–11. https://doi.org/10.1111/jzo.12677

Kaczmarek Ł, Roszkowska M, Poprawa I, Janelt K, Kmita H, Gawlak M, Fiałkowska E, Mioduchowska M (2020b) Integrative description of bisexual *Paramacrobiotus experimentalis* sp. nov. (Macrobiotidae) from republic of Madagascar (Africa) with microbiome analysis. Mol Phylogenet Evol 145:106730. https://doi.org/10.1016/j.ympev.2019.106730

Kamilari M, Jørgensen A, Schiøtt M, Møbjerg N (2019) Comparative transcriptomics suggest unique molecular adaptations within tardigrade lineages. BMC Genomics 20(1):607. https://doi.org/10.1186/s12864-019-5912-x

Kathman RD, Nelson DR (1987) Population trends in the aquatic tardigrade *Pseudobiotus augusti* (Murray). In: Bertolani R (ed) Biology of Tardigrades. Select Symp and Monogr UZI, vol 1, pp 155–168

Kober L, Zehe C, Bode J (2012) Development of a novel ER stress based selection system for the isolation of highly productive clones. Biotechnol Bioeng 109(10):2599–2611. https://doi.org/10.1002/bit.24527

Kristensen RM (1982) The first record of cyclomorphosis in Tardigrada based on a new genus and species from arctic meiobenthos. Z Zool Syst Evolut-forsch 20:249–270. https://doi.org/10.1111/j.1439-0469.1983.tb00552.x

Kristensen RM (1987) Generic revision of the Echiniscidae (Heterotardigrada), with a discussion of the origin of the family. In: Bertolani R (Ed.) biology of Tardigrades. Select Symp and Monogr UZI 1:261–335

Li X, Wang L (2005) Effect of thermal acclimation on preferred temperature, avoidance temperature and lethal thermal maximum of *Macrobiotus harmsworthi* Murrray (Tardigrada, Macrobiotidae). J Therm Biol 30:443–448. https://doi.org/10.1016/j.jtherbio.2005.05.003

May RM, Maria M, Guimard J (1964) Actions différentielles des rayons x et ultraviolets sur le tardigrade *Macrobiotus areolatus*, a l'état et desséché. Bull Biol France Belgique 98:349–367

Mayer G, Martin C, Rüdiger J, Kauschke S, Stevenson PA, Poprawa I, Hohberg K, Schill RO, Pflüger H-J, Schlegel M (2013) Selective neuronal staining in tardigrades and onychophorans provides insights into the evolution of segmental ganglia in panarthropods. BMC Evol Biol 13:230. https://doi.org/10.1186/1471-2148-13-230

McInnes SJ, Michalczyk Ł, Kaczmarek Ł (2017) Annotated zoogeography of non-marine Tardigrada. Part IV: Africa. Zootaxa 4284(1):1–74. https://doi.org/10.11646/zootaxa.4284.1.1

McKimpson WM, Kitsis RN (2017) A new role for the ER unfolded protein response mediator ATF6: induction of a generalized antioxidant program. Circ Res 120(5):759–761. https://doi.org/10.1161/CIRCRESAHA.117.310577

McNuff R (2018) Laboratory culture of *Hypsibius exemplaris*. Cold Spring Harb Protoc (11):867–870. https://doi.org/10.1101/pdb.prot102319

Menze MA, Hand SC (2009) How do animal mitochondria tolerate water stress? Commun Integr Biol 2:428–430. https://doi.org/10.4161/cib.2.5.9027

Morek W, Suzuki AC, Schill RO, Georgiev D, Yankova M, Marley NJ, Michalczyk Ł (2019) Redescription of *Milnesium alpigenum* Ehrenberg, 1853 (Tardigrada: Apochela) and a description of *Milnesium inceptum* sp. nov., a tardigrade laboratory model. Zootaxa 4586 (1):35–64. https://doi.org/10.11646/zootaxa.4586.1.2

Nelson DR, Guidetti R, Rebecchi L (2015) Phylum Tardigrada. In: Thorp J, Rogers DC (eds) Ecology and general biology. Thorp and Covich's Freshwater Invertebrates. Academic Press, London, pp 347–380. https://doi.org/10.1016/B978-0-12-385026-3.00017-6

Newsham KK, Maslen NR, McInnes SK (2006) Survival of Antarctic soil metazoans at −80° C for six years. Cryo Letters 27(5):269–280

Niki E (2014) Antioxidants: basic principles, emerging concepts, and problems. Biom J 37:106. https://doi.org/10.4103/2319-4170.128727

Nowakowska-Gołacka J, Sominka H, Sowa-Rogozińska N, Słomińska-Wojewódzka M (2019) Toxins utilize the endoplasmic reticulum-associated protein degradation pathway in their intoxication process. Int J Mol Sci 20(6):1307. https://doi.org/10.3390/ijms20061307

Ono F, Saigusa M, Uozumi T, Matsushima Y, Ikeda H, Saini NL, Yamashita M (2008) Effect of high hydrostatic pressure onto life of the tiny animal tardigrade. J Phys Chem Solids 69:2297–2300. https://doi.org/10.1016/j.jpcs.2008.04.019

Ozcan U, Cao Q, Yilmaz E, Lee AH, Iwakoshi NN, Ozdelen E, Tuncman G, Görgün C, Glimcher LH, Hotamisligil GS (2004) Endoplasmic reticulum stress links obesity, insulin action, and type 2 diabetes. Science 306(5695):457–461. https://doi.org/10.1126/science.1103160

Persson DK, Halberg KA, Jørgensen A, Møbjerg N, Kristensen RM (2012) Neuroanatomy of *Halobiotus crispae* (Eutardigrada: Hypsibiidae): tardigrade brain structure supports the clade Panarthropoda. J Morphol 273:1227–1245. https://doi.org/10.1002/jmor.20054

Persson D, Halberg KA, Jorgensen A, Ricci C, Mobjerg N, Kristensen RM (2011) Extreme stress tolerance in tardigrades: surviving space conditions in low earth orbit. J Zool Syst Evol Res 49:90–97. https://doi.org/10.1111/j.1439-0469.2010.00605.x

Pilato G, Binda MG (2001) Biogeography and limno-terrestrial tardigrades: are they truly incompatible binomials? Zool Anz 240:511–516. https://doi.org/10.1078/0044-5231-00061

Pilato G, Binda MG (2010) Definition of families, subfamilies, genera and subgenera of the Eutardigrada, and keys to their identification. Zootaxa 2404:1–52. https://doi.org/10.5281/zenodo.194138

Poprawa I (2005a) The structure and the formation of egg shells in the parthenogenetic species *Dactylobiotus dispar* Murray, 1907 (Tardigrada: Eutardigrada). Fol Biol 53(3–4):173–177. https://doi.org/10.3409/173491605775142828

Poprawa I (2005b) The ovary structure, previtellogenic and vitellogenic stages in parthenogenetic species *Dactylobiotus dispar* (Murray, 1907) (Tardigrada: Eutardigrada). Tiss Cell 37 (5):385–392. https://doi.org/10.1016/j.tice.2005.06.003

Poprawa I (2006) Ultrastructural changes of the storage cells during oogenesis in *Dactylobiotus dispar* (Murray, 1907) (Tardigrada: Eutardigrada). Zool Pol 51(1–4):13–18

Poprawa I (2011) Ultrastructural studies of the formation of the egg capsule in the hermaphroditic species, *Isohypsibius granulifer granulifer* Thulin, 1928 (Eutardigrada: Hypsibiidae). Zool Sci 28(1):37–40. https://doi.org/10.2108/zsj.28.37

Poprawa I, Hyra M, Kszuk-Jendrysik M, Rost-Roszkowska MM (2015a) Ultrastructural changes and programmed cell death of trophocytes in the gonad of *Isohypsibius granulifer granulifer* Thulin, 1928 (Tardigrada, Eutardigrada, Isohypsibiidae). Micron 70:26–33. https://doi.org/10.1016/j.micron.2014.11.008

Poprawa I, Hyra M, Rost-Roszkowska MM (2015c) Germ cell clusters organization and oogenesis in the tardigrade *Dactylobiotus parthenogeneticus* Bertolani, 1982 (Eutardigrada, Murrayidae). Protoplasma 252(4):1019–1029. https://doi.org/10.1007/s00709-014-0737-6

Poprawa I, Janelt K (2019) Reproduction, gonad structure and oogenesis in tardigrades. In: Tworzydło W, Biliński S (eds) Evo-Devo: non-model species in cell and developmental biology, vol 68. Springer, pp 495–513

Poprawa I, Schlechte-Wełnicz W, Hyra M (2015b) Ovary organization and oogenesis in the tardigrade *Macrobiotus polonicus* Pilato, Kaczmarek, Michalczyk & Lisi, 2003 (Eutardigrada,

Macrobiotidae): ultrastructural and histochemical analysis. Protoplasma 252:857–865. https://doi.org/10.1007/s00709-014-0725-x

Radak Z, Zhao Z, Goto S, Koltai E (2011) Age-associated neurodegeneration and oxidative damage to lipids, proteins and DNA. Mol Asp Med 32:305–315. https://doi.org/10.1016/j.mam.2011.10.010

Rahm PG (1923) Biologische und Physiologische Beiträge zur Kenntnis der Moosfauna. Z Alg Physiol 20:1–34

Rahm PG (1924) Weitere physiologische Versuche mit niederen Temperaturen Ein Beitrag zur Lösung des Kalteproblems. Verh dtsch zool Ges 29:106–111

Rahm PG (1926) Die Trochenstarre (Anabiose) de moostierwelt (ihr Verlauf. ihre Bedeutung und ihr Unterschied on der Cystebildung). Biol Zbl 46:452–477

Ramløv H, Westh P (2001) Cryptobiosis in the eutardigrade *Adorybiotus* (*Richtersius*) *coronifer*: tolerance to alcohols, temperature and de novo protein synthesis. Zool Anz 240:517–523. https://doi.org/10.1078/0044-5231-00062

Rebecchi L (2013) Dry up and survive: the role of antioxidant defences in anhydrobiotic organisms. J Limnol 72:62–72. https://doi.org/10.4081/jlimnol.2013.s1.e8

Rebecchi L, Altiero T, Guidetti R (2007) Anhydrobiosis: the extreme limit of desiccation tolerance. ISJ 4:65–81

Rebecchi L, Boschini D, Cesari M, Lencioni V, Bertolani R, Guidetti R (2009a) Stress response of a boreo-alpine species of tardigrade, *Borealibius zetlandicus* (Eutardigrada, Hypsibiidae). J Limnol 68:64–70. https://doi.org/10.4081/jlimnol.2009.64

Rebecchi L, Cesari M, Altiero T, Frigieri A, Guidetti R (2009b) Survival and DNA degradation in anhydrobiotic tardigrades. J Exp Biol 212:4033–4039. https://doi.org/10.1242/jeb.033266

Reuner A, Hengherr S, Brummer F, Schill RO (2010) Comparative studies on storage cells in tardigrades during starvation and anhydrobiosis. Curr Zool 56(2):259–263. https://doi.org/10.1093/czoolo/56.2.259

Richaud M, Le Goff E, Cazevielle C, Ono F, Mori Y, Saini NL, Cuq P, Baghdiguian S, Godefroy N, Galas S (2020) Ultrastructural analysis of the dehydrated tardigrade *Hypsibius exemplaris* unveils an anhydrobiotic-specific architecture. Sci Rep 10:4324. https://doi.org/10.1038/s41598-020-61165-1

Richters F (1903) Nordische Tardigraden. Zool Anz 27:168–172

Richters F (1904) Vorläufiger Bericht über die antarktische Moosfauna. Verh Dtsch Zool Ges 1904:236–239

Rost-Roszkowska MM, Janelt K, Poprawa I (2018) The role of autophagy in the midgut epithelium of Parachela (Tardigrada). Zoomorphology 137(4):501–509. https://doi.org/10.1007/s00435-018-0407-x

Rost-Roszkowska MM, Janelt K, Poprawa I (2019) Fine structure of the midgut epithelium of *Thulinius ruffoi* (Tardigrada, Eutardigrada, Parachela) in relation to oogenesis and simplex stages. Arthropod Struct Dev 49:128–136. https://doi.org/10.1016/j.asd.2018.12.002

Rost-Roszkowska M, Janelt K, Poprawa I (2020) Ultrastructure of the midgut epithelium in three species of Macrobiotidae (Tardigrada: Eutardigrada: Parachela). Zool J Linnean Soc 188(3):788–796. https://doi.org/10.1093/zoolinnean/zlz052

Rost-Roszkowska MM, Poprawa I (2008) Ultrastructure of the midgut epithelium in *Dactylobiotus dispar* (Tardigrada: Eutardigrada) during encystation. Zool Pol 53:19–25. https://doi.org/10.2478/v10049-008-0002-7

Rost-Roszkowska MM, Poprawa I, Hyra M, Marek-Sędzioł M, Kaczmarek Ł (2013) The fine structure of the midgut epithelium in *Xerobiotus pseudohufelandi* (Iharos, 1966) (Tardigrada, Eutardigrada, Macrobiotidae). J Limnol 72:54–61. https://doi.org/10.4081/jlimnol.2013.s1.e7

Rost-Roszkowska MM, Poprawa I, Wojtowicz M, Kaczmarek Ł (2011) Ultrastructural changes of the midgut epithelium in *Isohypsibius granulifer granulifer* Thulin, 1928 (Tardigrada: Eutardigrada) during oogenesis. Protoplasma 248(2):405–414. https://doi.org/10.1007/s00709-010-0186-9

Schill RO, Jonsson KI, Pfannkuchen M, Brummer F (2011) Food of tardigrades: a case study to understand food choice, intake and digestion. J Zool Syst Evol Res 49:66–70. https://doi.org/10.1111/j.1439-0469.2010.00601.x

Schill RO, Mali B, Dandekar T, Schnolzer M, Reuter D, Frohme M (2009) Molecular mechanisms of tolerance in tardigrades: new perspectives for preservation and stabilization of biological material. Biotechnol Adv 27:348–352. https://doi.org/10.1016/j.biotechadv.2009.01.011

Schulze C, Schmidt-Rhaesa A (2013) The architecture of the nervous system of *Echiniscus testudo* (Echiniscoidea, Heterotardigrada). J Limnol 72:44–53. https://doi.org/10.4081/jlimnol.2013.s1.e6

Seki K, Toyoshima M (1998) Preserving tardigrades under pressure. Nature 395:853–854. https://doi.org/10.1038/27576

Stec D, Smolak R, Kaczmarek Ł, Michalczyk Ł (2015) An integrative description of *Macrobiotus paulinae* sp. nov. (Tardigrada: Eutardigrada: Macrobiotidae: *hufelandi* group) from Kenya. Zootaxa 4052(5):501–526. https://doi.org/10.11646/zootaxa.4052.5.1

Sugiura K, Minato H, Suzuki AC, Arakawa K, Kunieda T, Matsumoto M (2019) Comparison of sexual reproductive behaviors in two species of Macrobiotidae (Tardigrada: Eutardigrada). Zool Sci 36(2):120–127. https://doi.org/10.2108/zs180103

Surmacz B, Morek W, Michalczyk Ł (2020) What to do when ontogenetic tracking is unavailable: a morphometric method to classify instars in *Milnesium* (Tardigrada). Zool J Linnean Soc 188 (3):797–808. https://doi.org/10.1093/zoolinnean/zlz099

Suzuki AC (2003) Life history of *Milnesium tardigradum* Doyère (Tardigrada) under rearing environment. Zool Sci 20:49–57. https://doi.org/10.2108/zsj.20.49

Tanaka S, Tanaka J, Miwa Y, Horikawa DD, Katayama T, Arakawa K, Toyoda A, Kubo T, Kunieda T (2015) Novel mitochondria-targeted heat-soluble proteins identified in the anhydrobiotic tardigrade improve osmotic tolerance of human cells. PLoS One 10:e0118272. https://doi.org/10.1371/journal.pone.0118272

Tripathi R, Boschetti C, McGee B, Tunnacliffe A (2012) Trafficking of bdelloid rotifer late embryogenesis abundant proteins. J Exp Biol 215:2786–2794. https://doi.org/10.1242/jeb.071647

Tsujimoto M, Imura S, Kanda H (2016) Recovery and reproduction of an Antarctic tardigrade retrieved from a moss sample frozen for over 30 years. Cryobiology 72(1):78–81. https://doi.org/10.1016/j.cryobiol.2015.12.003

Tunnacliffe A, Wise MJ (2007) The continuing conundrum of the LEA proteins. Naturwissenschaften 94:791–812. https://doi.org/10.1007/s00114-007-0254-y

Tyson T, O'Mahony Zamora G, Wong S, Skelton M, Daly B, Jones JT, Mulvihill ED, Elsworth B, Phillips M, Blaxter M, Burnell AM (2012) A molecular analysis of desiccation tolerance mechanisms in the anhydrobiotic nematode *Panagrolaimus superbus* using expressed sequenced tags. BMC Res Notes 5:68. https://doi.org/10.1186/1756-0500-5-68

Verfaillie T, Rubio N, Garg AD, Bultynck G, Rizzuto R, Decuypere J-P, Piette J, Linehan C, Gupta S, Samali A, Agostinis P (2012) PERK is required at the ER-mitochondrial contact sites to convey apoptosis after ROS-based ER stress. Cell Death Differ 19(11):1880–1891. https://doi.org/10.1038/cdd.2012.74

Voronina TA, Nesmelov AA, Kondratyeva SA, Deviatiiarov RM, Miyata Y, Tokumoto S, Cornette R, Gusev OA, Kikawada T, Shagimardanova EI (2020) New group of transmembrane proteins associated with desiccation tolerance in the anhydrobiotic midge *Polypedilum vanderplanki*. Sci Rep 10:11633. https://doi.org/10.1038/s41598-020-68330-6

Wang WA, Groenendyk J, Michalak M (2014a) Endoplasmic reticulum stress associated responses in cancer. Biochim Biophys Acta 1843(10):2143–2149. https://doi.org/10.1016/j.bbamcr.2014.01.012

Wang C, Grohme MA, Mali B, Schill RO, Frohme M (2014b) Towards decrypting cryptobiosis – analyzing anhydrobiosis in the tardigrade *Milnesium tardigradum* using transcriptome sequencing. PLoS One 9:e92663. https://doi.org/10.1371/journal.pone.0092663

Węglarska B (1975) Studies on the morphology of *Macrobiotus richtersi* Murray, 1911. Mem Ist Ital Idrobiol 32:445–464

Węglarska B (1979) Electron microscope study of previtellogenesis and vitellogenesis in *Macrobiotus richtersi* J. Murr. (Eutardigrada). Zesz Nauk Uniw Jagiell Pr Zool 25:169–189

Węglarska B (1982) Ultrastructural study of the formation of egg envelopes in *Macrobiotus richtersi* (Eutardigrada). In: Nelson DR (ed) Proc third Int Symp Tardigrada. East Tennessee State Univ Press, Johnson City, pp 115–128

Węglarska B (1987) Yolk formation in *Isohypsibius* (Eutardigrada). Zoomorphology 107:287–292

Wełnicz W, Grohme MA, Kaczmarek Ł, Schill RO, Frohme M (2011) Anhydrobiosis in tardigrades – the last decade. J Insect Physiol 57:577–583. https://doi.org/10.1016/j.jinsphys.2011.03.019

Wharton DA (2015) Anhydrobiosis. Curr Biol 25:R1114–R1116. https://doi.org/10.1016/j.cub.2015.09.047

Wright JC (2001) Cryptobiosis 300 years on from van Leuwenhoek: what have we learned about tardigrades? Zool Anz 240:563–582. https://doi.org/10.1078/0044-5231-00068

Xu C, Bailly-Maitre B, Reed JC (2005) Endoplasmic reticulum stress: cell life and death decisions. J Clin Invest 115(10):2656–2664. https://doi.org/10.1172/JCI26373

Yoshida Y, Koutsovoulos G, Laetsch DR, Stevens L, Kumar S, Horikawa DD, Ishino K, Komine S, Kunieda T, Tomita M, Blaxter M, Arakawa K (2017) Comparative genomics of the tardigrades *Hypsibius dujardini* and *Ramazzottius varieornatus*. PLoS Biol 15:e2002266. https://doi.org/10.1371/journal.pbio.2002266

Zhao L, Ackerman SL (2006) Endoplasmic reticulum stress in health and disease. Curr Opin Cell Biol 18(4):444–452. https://doi.org/10.1016/j.ceb.2006.06.005

The manufacturer's authorised representative in the EU is Springer Nature Customer Service Centre GmbH, Europaplatz 3, 69115 Heidelberg, Germany. If you have any concerns regarding our products, please contact ProductSafety@springernature.com

Printed and bound by CPI Group (UK) Ltd, Croydon, CR0 4YY

22/09/2026

02220981-0013